Bayesian Statistical Methods

Bayesian Statistical Methods: With Applications to Machine Learning provides data scientists with the foundational and computational tools needed to carry out a Bayesian analysis. Compared to others, this book is more focused on Bayesian methods applied routinely in practice, including multiple linear regression, mixed effects models and generalized linear models. This second edition includes a new chapter on Bayesian machine learning methods to handle large and complex datasets and several new applications to illustrate the benefits of the Bayesian approach in terms of uncertainty quantification.

Readers familiar with only introductory statistics will find this book accessible, as it includes many worked examples with complete R code, and comparisons are presented with analogous frequentist procedures. The book can be used as a one-semester course for advanced undergraduate and graduate students and can be used in courses comprising undergraduate statistics majors, as well as non-statistics graduate students from other disciplines such as engineering, ecology and psychology. In addition to thorough treatment of the basic concepts of Bayesian inferential methods, the book covers many general topics:

- Advice on selecting prior distributions
- Computational methods including Markov chain Monte Carlo (MCMC) sampling
- Model-comparison and goodness-of-fit measures, including sensitivity to priors.

To illustrate the flexibility of the Bayesian approaches for complex data structures, the latter chapters provide case studies covering advanced topics:

- Handling of missing and censored data
- Priors for high-dimensional regression models
- Machine learning models including Bayesian adaptive regression trees and deep learning
- Computational techniques for large datasets
- Frequentist properties of Bayesian methods.

The advanced topics are presented with sufficient conceptual depth that the reader will be able to carry out such analysis and argue the relative merits of Bayesian and classical methods. A repository of R code, motivating data sets and complete data analyses is made available on the book's website.

Brian J. Reich, Gertrude M. Cox Distinguished Professor of Statistics at North Carolina State University, applies Bayesian statistical methods in a variety of fields including environmental epidemiology, engineering, weather and climate. He is a Fellow of the American Statistical Association, former Editor-in-Chief of the Journal of Agricultural, Biological, and Environmental Statistics and recipient of the LeRoy & Elva Martin Teaching Award at NC State University.

Sujit K. Ghosh, Professor of Statistics at North Carolina State University, has worked in advanced research fields such as Bayesian inference, spatial statistics, survival analysis and shape-constrained inference, addressing complex inferential challenges in biomedical and environmental sciences, econometrics, and engineering. At NC State, he has been honored with the D.D. Mason Faculty Award and the Cavell Brownie Mentoring Award, reflecting his excellence in research, mentoring and teaching. His leadership includes impactful service as Program Director at NSF's Division of Mathematical Sciences, Deputy Director at SAMSI and President of the IISA.

CHAPMAN & HALL/CRC
Texts in Statistical Science Series

Recently Published Titles

Applied Categorical and Count Data Analysis, Second Edition
Wan Tang, Hua He, and Xin M. Tu

Geographic Data Science with Python
Sergio Rey, Dani Arribas-Bel, and Levi John Wolf

Models for Multi-State Survival Data
Rates, Risks, and Pseudo-Values
Per Kragh Andersen and Henrik Ravn

Spatio–Temporal Methods in Environmental Epidemiology with R, Second Edition
Gavin Shaddick, James V. Zidek, and Alex Schmidt

A Course in the Large Sample Theory of Statistical Inference
W. Jackson Hall and David Oakes

Statistical Inference, Second Edition
George Casella and Roger Berger

Nonparametric Statistical Methods Using R, Second Edition
John Kloke and Joesph McKean

Generalized Linear Mixed Models
Modern Concepts, Methods and Applications, Second Edition
Walter W. Stroup, Marina Ptukhina, and Julie Garai

Analysis of Categorical Data with R, Second Edition
Christopher R. Bilder, Thomas M. Loughin

Applied Nonparametric Statistical Methods, Fifth Edition
Nigel C. Smeeton, Neil H. Spencer, and Peter Sprent

Linear Models with R, Third Edition
Julian J. Faraway

Statistics in Survey Sampling
Jae Kwang Kim

Time Series: A Data Analysis Approach Using R
Robert H. Shumway and David S. Stoffer

Bayesian Statistical Methods: With Applications to Machine Learning, Second Edition
Brian J. Reich and Sujit K. Ghosh

For more information about this series, please visit: https://www.routledge.com/Chapman--HallCRC-Texts-in-Statistical-Science/book-series/CHTEXSTASCI

Bayesian Statistical Methods

With Applications to Machine Learning

Second Edition

Brian J. Reich and Sujit K. Ghosh

CRC Press
Taylor & Francis Group
Boca Raton London New York

CRC Press is an imprint of the
Taylor & Francis Group, an **informa** business
A CHAPMAN & HALL BOOK

Designed Cover Image: © Swagata Sarkar

Second edition published 2026
by CRC Press
2385 NW Executive Center Drive, Suite 320, Boca Raton FL 33431

and by CRC Press
4 Park Square, Milton Park, Abingdon, Oxon, OX14 4RN

CRC Press is an imprint of Taylor & Francis Group, LLC

First edition published by CRC Press 2019

ISBN: 978-1-032-48632-1 (hbk)
ISBN: 978-1-032-48658-1 (pbk)
ISBN: 978-1-003-39011-4 (ebk)

DOI: 10.1201/9781003390114

Typeset in LM Roman
by KnowledgeWorks Global Ltd.

Publisher's Note: This book has been prepared from camera-ready copy provided by the author.

Access the Support Material: http://go.ncsu.edu/bsm2

To Michelle, Sophie, Swagata and Sabita

Contents

Preface

Bayesian methods are standard in various fields of science including biology, engineering, finance and genetics, and thus they are an essential addition to an analyst's toolkit. In this book, we cover the material we deem indispensable for a practicing Bayesian data analyst. The book covers the most common statistical methods including the t-test, multiple linear regression, mixed models and generalized linear models from a Bayesian perspective and includes many examples and code to implement the analyses. To illustrate the flexibility of the Bayesian approach, the later chapters explore advanced topics such as nonparametric regression, missing data and hierarchical models. In addition to these important practical matters, we provide sufficient depth so that the reader can defend his/her analysis and argue the relative merits of Bayesian and classical methods.

The book is intended to be used as a one-semester course for advanced undergraduate and graduate students. At North Carolina State University (NCSU) this book is used for a course comprising undergraduate statistics majors, non-statistics graduate students from all over campus (e.g., engineering, ecology, psychology, etc.) and students from the Masters of Science in Statistics Program. Statistics PhD students take a separate course that covers much of the same material but at a more theoretical and technical level. We hope this book and associated computing examples also serve as a useful resource to practicing data analysts. Throughout the book we have included case studies from several fields to illustrate the flexibility and utility of Bayesian methods in practice.

It is assumed that the reader is familiar with undergraduate-level calculus including limits, integrals and partial derivatives and some basic linear algebra. Derivations of some key results are given in the main text when this helps to communicate ideas, but the vast majority of derivations are relegated to the Appendix for interested readers. Knowledge of introductory statistical concepts through multiple regression would also be useful to contextualize the material, but this background is not assumed and thus not required. Fundamental concepts are covered in detail but with references held to a minimum in favor of clarity; advanced concepts are described concisely at a high level with references provided for further reading.

The book begins with a review of probability in the first section of Chapter 1. A solid understanding of this material is essential to proceed through the book, but this section may be skipped for readers with the appropriate background. The remainder of Chapter 1 through Chapter 6 form the core of the book. Chapter 1 introduces the central concepts of and motivation for Bayesian inference. Chapter 2 provides ways to select the prior distribution, which is the genesis of a Bayesian analysis. For all but the most basic methods, advanced computing tools are required, and Chapter 3 covers these methods with the most weight given to Markov chain Monte Carlo, which is used for the remainder of the book. Chapter 4 applies these tools to common statistical models including multiple linear regression, generalized linear models and mixed effects models. After cataloging numerous statistical methods in Chapter 4, Chapters 5 and 6 treat the problems of testing hypotheses and comparing models for a given dataset, and verifying and validating that the model fits the data well. Chapter 7 introduces hierarchical modeling as a general framework to extend Bayesian methods to complex problems, and illustrates this approach using detailed case studies. The final chapter investigates the theoretical properties of Bayesian methods, which is important to justify their use but can be omitted if the course is intended for non-PhD students.

The key addition to the second edition is the inclusion of emerging topics in the rapidly evolving field of Bayesian statistics. The second edition expands the coverage of computational methods in Chapter 3. While MCMC remains the primary computational approach, the second edition now includes advanced tools needed to handle large datasets and/or complex models, including Approximate Bayesian Computing (ABC), Simulation-Based Inference (SBI) and Variational Bayesian (VB) methods. Chapter 5 now includes descriptions and examples of A/B and multiple testing. Most of the additions are in Chapter 8, an entirely new chapter on Bayesian Machine learning. The chapter begins with methods for high-dimensional data including linear regression with many covariates and graphical models. The chapter then turns to advanced regression models including Bayesian Adaptive Regression Trees (BART) and deep learning. These methods are illustrated using an example to improve short-term forecasts of hurricane intensity. After establishing these core machine learning tools, the focus turns to applications of machine learning including causal inference and equation learning. The chapter emphasizes the balance between predictive accuracy and interpretable machine learning. These new concepts are introduced at a high-level through examples, with the goal of showcasing the potential of Bayesian machine learning and pointing readers to the rapidly-evolving literature for details.

The choice of software is crucial for any modern textbook or statistics course. We elected to use `R` as the software platform due to its immense popularity in the statistics community, and access to online tutorials and assistance. Fortunately, there are now many options within `R` to conduct a Bayesian analysis, and we compare several including `JAGS`, `BUGS`, `STAN` and `NIMBLE`. We selected the package `JAGS` as the primary package for no particularly strong reason other than we have found it works well for the courses taught at our university. In our assessment, `JAGS` provides the nice combination of ease of use, speed and flexibility for the size and complexity of problems we consider. A repository of code and datasets used in the book is available at

http://go.ncsu.edu/bsm2.

Throughout the book we use `R/JAGS`, but favor conceptual discussions over computational details, and these concepts transcend software. The course webpage also includes homework solutions and latex/beamer slides.

We wish to thank our NCSU colleagues Kirkwood Cloud, Rosemary Danaher, Qian Guan, Margaret Johnson, Ryan Martin, Krishna Pacifici and Ana-Maria Staicu for providing valuable feedback. We also thank the students in Bayesian courses taught at NCSU for their probing questions that helped shape the material in the book. Finally, we thank our families and friends for their enduring support. We gratefully acknowledge the contribution of Swagata Sarkar, whose distinctive mixed media artwork enriches and enhances the visual appeal of the book's front cover.

1

Basics of Bayesian inference

CONTENTS

1.1 Probability background

Understanding basic probability theory is essential to any study of statistics. Generally speaking, the field of probability assumes a mathematical model for the process of interest and uses the model to compute the probability of events (e.g., what is the probability of flipping five straight heads using a fair coin?). In contrast, the field of statistics uses data to refine the probability model and test hypotheses related to the underlying process that generated the data (e.g., given we observe five straight heads, can we conclude the coin is biased?). Therefore, probability theory is a key ingredient to a statistical analysis, and in this section we review the most relevant concepts of probability for a Bayesian analysis.

Before developing probability mathematically, we briefly discuss probability from a conceptual perspective. The objective is to compute the probability of an event, $\mathcal{A}$, denoted $\text{Prob}(\mathcal{A})$. For example, we may be interested in the probability that the random variable X (random variables are generally represented with capital letters) takes the specific value x (lower-case letter), denoted $\text{Prob}(X = x)$, or the probability that X will fall in the interval $[a, b]$, denoted $\text{Prob}(X \in [a, b])$. There are two leading interpretations of this statement: objective and subjective. An objective interpretation views $\text{Prob}(\mathcal{A})$ as a purely mathematical statement. A frequentist interpretation is that if we repeated the experiment many times and recorded the sample proportion of the times $\mathcal{A}$ occurred, this proportion would eventually converge to the number $\text{Prob}(\mathcal{A}) \in [0, 1]$ as the number of samples increases. A subjective interpretation is that $\text{Prob}(\mathcal{A})$ represents an individual's degree of belief, which is often quantified in terms of the amount the individual would be willing to wager that $\mathcal{A}$ will occur. As we will see, these two conceptual interpretations of probability parallel the two primary statistical frameworks: frequentist and Bayesian. However, a Bayesian analysis makes use of both of these concepts.

1.1.1 Univariate distributions

The random variable X's support $\mathcal{S}$ is the smallest set so that $X \in \mathcal{S}$ with probability one. For example, if X is the number of successes in n trials then $\mathcal{S} = \{0, 1, ..., n\}$. Probability equations differ based on whether $\mathcal{S}$ is a countable set: X is a discrete random variable if $\mathcal{S}$ is countable, and X is continuous otherwise. Discrete random variables can have a finite (rainy days in the year) or an infinite (number of lightning strikes in a year) number of possible outcomes as long as the number is countable, e.g., a random count $X \in \mathcal{S} = \{0, 1, 2, ...\}$ has an infinite but countable number of possible outcomes and is thus discrete. An example of a continuous random variable is the amount of rain on a given day, which can be any real non-negative number and so $\mathcal{S} = [0, \infty)$.

1.1.1.1 Discrete distributions

For a discrete random variable, the probability mass function (PMF) $f(x)$ assigns a probability to each element of X's support, that is,

$$\text{Prob}(X = x) = f(x). \tag{1.1}$$

A PMF is valid if all probabilities are non-negative, $f(x) \geq 0$, and sum to one, $\sum_{x \in \mathcal{S}} f(x) = 1$. The PMF can also be used to compute probabilities of more complex events by summing over the PMF. For example, the probability that X is either x_1 or x_2, i.e., $X \in \{x_1, x_2\}$, is

$$\text{Prob}(X = x_1 \text{ or } X = x_2) = f(x_1) + f(x_2). \tag{1.2}$$

Generally, the probability of the event that X falls in a set $\mathcal{S}' \subset \mathcal{S}$ is the sum over elements in $\mathcal{S}'$,

$$\text{Prob}(X \in \mathcal{S}') = \sum_{x \in \mathcal{S}'} f(x). \tag{1.3}$$

Using this fact defines the cumulative distribution function (CDF)

$$F(x) = \text{Prob}(X \leq x) = \sum_{c \leq x} f(c). \tag{1.4}$$

A PMF is a function from the support of X to the probability of events. It is often useful to summarize the function using a few interpretable quantities such as the mean and variance. The expected value or mean value is

$$\text{E}(X) = \sum_{x \in \mathcal{S}} x f(x) \tag{1.5}$$

and measures the center of the distribution. The variance measures the spread around the mean via the expected squared deviation from the center of the distribution,

$$\text{Var}(X) = \text{E}\{[X - \text{E}(X)]^2\} = \sum_{x \in \mathcal{S}} [x - E(X)]^2 f(x). \tag{1.6}$$

The variance is often converted to the standard deviation $\text{SD}(X) = \sqrt{\text{Var}(X)}$ to express the variability on the same scale as the random variable.

The central concept of statistics is that the PMF and its summaries (such as the mean) describe the population of interest, and a statistical analysis uses a sample from the population to estimate these functions of the population. For example, we might take a sample of size n from the population. Denote the ith sample value as $X_i \sim f$ ("$\sim$" means "is distributed as"), and $X_1, ..., X_n$ as the complete sample. We might then approximate the population mean $\text{E}(X)$ with the sample mean $\bar{X} = \frac{1}{n}\sum_{i=1}^{n} X_i$, the probability of an outcome $f(x)$ with the sample proportion of the n observations that equal x, and the entire PMF $f(x)$ with a sample histogram. However, even for a large sample, $\bar{X}$ will likely not equal $\text{E}(X)$, and if we repeat the sampling procedure again we might get a different $\bar{X}$ while $\text{E}(X)$ does not change. The distribution of a statistic, i.e., a summary of the sample, such as $\bar{X}$ across random samples from the population, is called the statistic's sampling distribution.

A statistical analysis to infer about the population from a sample often proceeds under the assumption that the population belongs to a parametric family of distributions. This is called a parametric statistical analysis. In this type of analysis, the entire PMF is assumed to be known up to a few unknown parameters denoted $\boldsymbol{\theta} = (\theta_1, ..., \theta_p)$ (or simply θ if there is only $p = 1$ parameter). We then denote the PMF as $f(x|\boldsymbol{\theta})$. The vertical bar "$|$" is read as "given," and so $f(x|\boldsymbol{\theta})$ gives the probability that $X = x$ given the parameters $\boldsymbol{\theta}$. For example, a common parametric model for count data with $\mathcal{S} = \{0, 1, 2, ...\}$ is the Poisson family. The Poisson PMF with unknown parameter $\theta \geq 0$ is

$$\text{Prob}(X = x|\theta) = f(x|\theta) = \frac{\exp(-\theta)\theta^x}{x!}. \tag{1.7}$$

Clearly $f(x|\theta) > 0$ for all $x \in \mathcal{S}$, and it can be shown that $\sum_{x=0}^{\infty} f(x|\theta) = 1$, so this is a valid PMF. As shown in Figure 1.1, changing the parameter θ changes the PMF, and so the Poisson is not a single distribution but rather a family of related distributions indexed by θ.

A parametric assumption greatly simplifies the analysis because we only have to estimate a few parameters and we can compute the probability of any x in $\mathcal{S}$. Of course, this assumption is only useful if the assumed distribution provides a reasonable fit to the observed data, and thus a statistician needs a large catalog of distributions to be able to find an appropriate family for a given analysis. Appendix A.1 provides a list of parametric distributions, and we discuss a few discrete distributions below.

Bernoulli: If X is binary, i.e., $\mathcal{S} = \{0, 1\}$, then X follows a Bernoulli(θ) distribution. A binary random variable is often used to model the result of a trial where a success is recorded as a one and a failure as zero. The parameter $\theta \in [0, 1]$ is the success probability $\text{Prob}(X = 1|\theta) = \theta$, and to be a valid PMF we must have the failure probability $\text{Prob}(X = 0|\theta) = 1 - \theta$. These two cases can be written concisely as

$$f(x|\theta) = \theta^x(1 - \theta)^{1-x}. \tag{1.8}$$

This gives mean

$$\text{E}(X|\theta) = \sum_{x=0}^{1} xf(x|\theta) = f(1|\theta) = \theta \tag{1.9}$$

and variance $\text{Var}(X|\theta) = \theta(1 - \theta)$.

Binomial: The binomial distribution is a generalization of the Bernoulli to the case of $n \geq 1$ independent trials. Specifically, if $X_1, ..., X_n$ are the binary results of the n independent trials each with success probability θ (so that $X_i \sim \text{Bernoulli}(\theta)$ for all $i = 1, ..., n$)

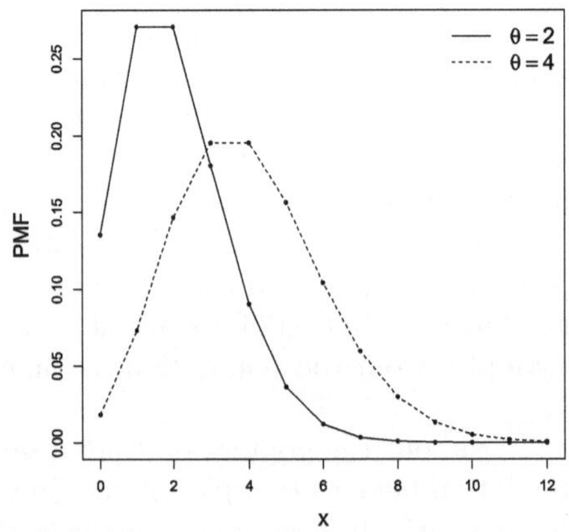

FIGURE 1.1
Poisson probability mass function. Plot of the PMF $f(x|\theta) = \frac{\exp(-\theta)\theta^x}{x!}$ for $\theta = 2$ and $\theta = 4$. The PMF is connected by lines for visualization, but the probabilities are only defined for $x = \{0, 1, 2, ...\}$.

and $X = \sum_{i=1}^{n} X_i$ is the total number of successes, then X's support is $\mathcal{S} = \{0, 1, ..., n\}$ and $X \sim \text{Binomial}(n, \theta)$. The PMF is

$$f(x|\theta) = \binom{n}{x}\theta^x(1-\theta)^{n-x}, \tag{1.10}$$

where $\binom{n}{x}$ is the binomial coefficient. This gives mean and variance $\text{E}(X|\theta) = n\theta$ and $\text{Var}(X|\theta) = n\theta(1-\theta)$. It is certainly reasonable that the expected number of successes in n trials is n times the success probability for each trial, and Figure 1.2 shows that the variance is maximized with $\theta = 0.5$ when the outcome of each trial is the least predictable. Appendix A.1 provides other distributions with support $\mathcal{S} = \{0, 1, ..., n\}$ including the discrete uniform and beta-binomial distributions.

Poisson: When the support is the counting numbers $\mathcal{S} = \{0, 1, 2, ...\}$, a common model is the Poisson PMF defined above (and plotted in Figure 1.1). The Poisson PMF can be motivated as the distribution of the number of events that occur in a time interval of length T if the events are independent and equally likely to occur at any time with rate θ/T events per unit of time. The mean and variance are $\text{E}(X|\theta) = \text{Var}(X|\theta) = \theta$. Assuming that the mean and variance are equal is a strong assumption, and Appendix A.1 provides alternatives with support $\mathcal{S} = \{0, 1, 2, ...\}$ including the geometric and negative binomial distributions.

1.1.1.2 Continuous distributions

The PMF does not apply for continuous distributions with support $\mathcal{S}$ that is a subinterval of the real line. To see this, assume that X is the daily rainfall (inches) and can thus be any non-negative real number, $\mathcal{S} = [0, \infty)$. What is the probability that X is exactly $\pi/2$ inches? Well, within some small range around $\pi/2$, $\mathcal{T} = (\pi/2 - \epsilon, \pi/2 + \epsilon)$ with say $\epsilon = 0.001$, it seems reasonable to assume that all values in $\mathcal{T}$ are equally likely, say $\text{Prob}(X = x) = q$ for all $x \in \mathcal{T}$. But since there are an uncountable number of values in $\mathcal{T}$ when we sum the

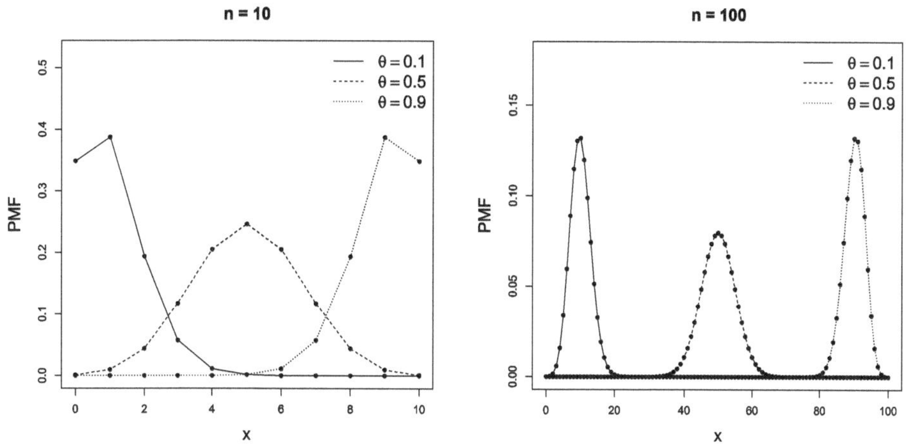

FIGURE 1.2
Binomial probability mass function. Plot of the PMF $f(x|\theta) = \binom{n}{x}\theta^x(1-\theta)^{n-x}$ for combinations of the number of trials n and the success probability θ. The PMF is connected by lines for visualization, but the probabilities are only defined for $x = \{0, 1, ..., n\}$.

probability over the values in $\mathcal{T}$, we get infinity and thus the probabilities are invalid unless $q = 0$. Therefore, for continuous random variables $\text{Prob}(X = x) = 0$ for all x, and we must use a more sophisticated method for assigning probabilities.

Instead of defining the probability of outcomes directly using a PMF, for continuous random variables we define probabilities indirectly through the Cumulative Distribution Function (CDF)

$$F(x) = \text{Prob}(X \le x). \tag{1.11}$$

The CDF can be used to compute probabilities for any interval, e.g., in the rain example $\text{Prob}(X \in \mathcal{S}) = \text{Prob}(X < \pi/2 + \epsilon) - \text{Prob}(X < \pi/2 - \epsilon) = F(\pi/2 + \epsilon) - F(\pi/2 - \epsilon)$, which converges to zero as ϵ shrinks if F is a continuous function. Defining the probability of X falling in an interval resolves the conceptual problems discussed above, because it is easy to imagine the proportion of days with rainfall in an interval converging to a non-zero value as the sample size increases.

The probability on a small interval is

$$\text{Prob}(x - \epsilon < X < x + \epsilon) = F(x + \epsilon) - F(x - \epsilon) \approx 2\epsilon f(x) \tag{1.12}$$

where $f(x)$ is the derivative of $F(x)$ and is called the probability density function (PDF). If $f(x)$ is the PDF of X, then the probability of $X \in [a, b]$ is the area under the density curve between a and b (Figure 1.3),

$$\text{Prob}(a \le X \le b) = F(b) - F(a) = \int_a^b f(x)dx. \tag{1.13}$$

The distributions of random variables are usually defined via the PDF. To ensure that the PDF produces valid probability statements we must have $f(x) \ge 0$ for all x and $\int_{-\infty}^{\infty} f(x)dx = 1$. Because $f(x)$ is not a probability, but rather a function used to compute probabilities via integration, the PDF can be greater than one for some x so long as it integrates to one.

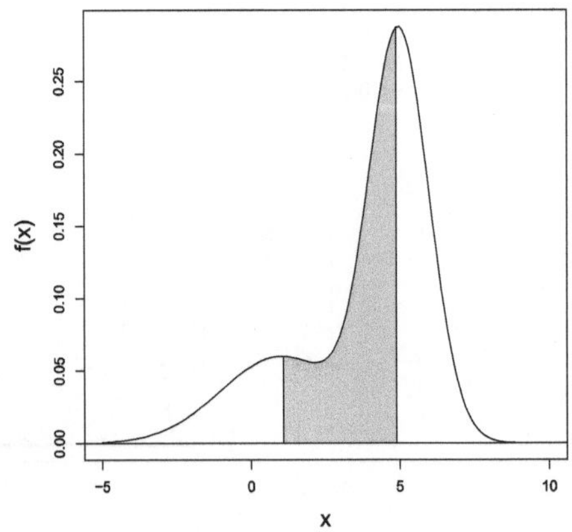

FIGURE 1.3
Computing probabilities using a PDF. The curve is a PDF $f(x)$, and the shaded area is $\text{Prob}(1 < X < 5) = \int_1^5 f(x)dx$.

The formulas for the mean and variance of a continuous random variable resemble those for a discrete random variable except that the sum over the PMF is replaced with an integral over the PDF,

$$\begin{aligned} \text{E}(X) &= \int_{-\infty}^{\infty} x f(x)dx \\ \text{Var}(X) &= \text{E}\{[X - E(X)]^2\} = \int_{-\infty}^{\infty} [x - \text{E}(X)]^2 f(x)dx. \end{aligned}$$

Another summary that is defined for both discrete and continuous random variables (but is much easier to define in the continuous case) is the quantile function $Q(\tau)$. For $\tau \in [0,1]$, $Q(\tau)$ is the solution to

$$\text{Prob}[X \le Q(\tau)] = F[Q(\tau)] = \tau. \tag{1.14}$$

That is, $Q(\tau)$ is the value so that the probability of X being no larger than $Q(\tau)$ is τ. The quantile function is the inverse of the distribution function, $Q(\tau) = F^{-1}(\tau)$, and gives the median $Q(0.5)$ and a $(1-\alpha)\%$ equal-tailed interval $[Q(\alpha/2), Q(1-\alpha/2)]$ so that $\text{Prob}[Q(\alpha/2) \le X \le Q(1-\alpha/2)] = 1-\alpha$.

Gaussian: As with discrete data, parametric models are typically assumed for continuous data and practitioners must be familiar with several parametric families. The most common parametric family with support $\mathcal{S} = (-\infty, \infty)$ is the normal (Gaussian) family. The normal distribution has two parameters, the mean $\text{E}(X|\boldsymbol{\theta}) = \mu$ and variance $\text{Var}(X|\boldsymbol{\theta}) = \sigma^2$, and the familiar bell-shaped PDF

$$f(x|\boldsymbol{\theta}) = \frac{1}{\sqrt{2\pi}\sigma} \exp\left[-\frac{1}{2\sigma^2}(x-\mu)^2\right], \tag{1.15}$$

where $\boldsymbol{\theta} = (\mu, \sigma^2)$. The Gaussian distribution is famous because of the Central Limit Theorem (CLT). The CLT applies to the distribution of the sample mean $\bar{X}_n = \frac{\sum_{i=1}^n X_i}{n}$, where

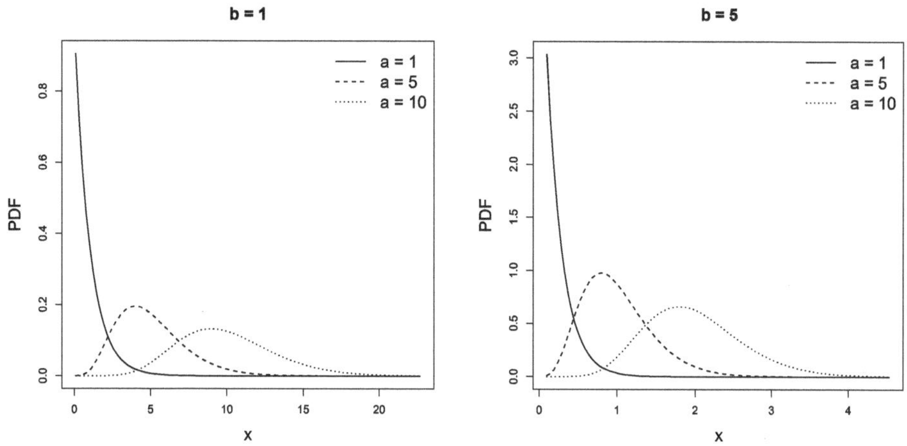

FIGURE 1.4
Plots of the gamma PDF. Plots of the gamma density function $f(x|\boldsymbol{\theta}) = \frac{b^a}{\Gamma(a)}x^a \exp(-bx)$ for several combinations of a and b.

$X_1, ..., X_n$ are independent samples from some distribution $f(x)$. The CLT says that under fairly general conditions, for large n the distribution of $\bar{X}_n$ is approximately normal even if $f(x)$ is not. Therefore, the Gaussian distribution is a natural model for data that are defined as averages, but can be used for other data as well. Appendix A.1 gives other continuous distributions with $\mathcal{S} = (-\infty, \infty)$ including the double exponential and Student's t-distributions.

Gamma: The gamma distribution has $\mathcal{S} = [0, \infty)$. The PDF is

$$f(x|\boldsymbol{\theta}) = \begin{cases} \frac{b^a}{\Gamma(a)}x^{a-1}\exp(-bx) & x \geq 0 \\ 0 & x < 0 \end{cases} \tag{1.16}$$

where Γ is the gamma function and $a > 0$ and $b > 0$ are the two parameters in $\boldsymbol{\theta} = (a, b)$. Beware that the gamma PDF is also written with b in the denominator of the exponential function, but we use the parameterization above. Under the parameterization in (1.16) the mean is a/b and the variance is a/b^2. As shown in Figure 1.4, a is the shape parameter and b is the scale. Setting $a = 1$ gives the exponential distribution with PDF $f(x|\boldsymbol{\theta}) = b\exp(-bx)$ which decays exponentially from the origin and large a gives approximately a normal distribution. Varying b does not change the shape of the PDF but only affects its spread. Appendix A.1 gives other continuous distributions with $\mathcal{S} = [0, \infty)$ including the inverse-gamma distribution.

Beta: The beta distribution has $\mathcal{S} = [0, 1]$ and PDF

$$f(x|\boldsymbol{\theta}) = \begin{cases} \frac{\Gamma(a+b)}{\Gamma(a)\Gamma(b)}x^{a-1}(1-x)^{b-1} & x \in [0, 1] \\ 0 & x < 0 \text{ or } x > 1, \end{cases} \tag{1.17}$$

where $a > 0$ and $b > 0$ are the two parameters in $\boldsymbol{\theta} = (a, b)$. As shown in Figure 1.5, the beta distribution is quite flexible and can be left-skewed, right-skewed or symmetric. The beta distribution also includes the uniform distribution $f(x) = 1$ for $x \in [0, 1]$ by setting $a = b = 1$.

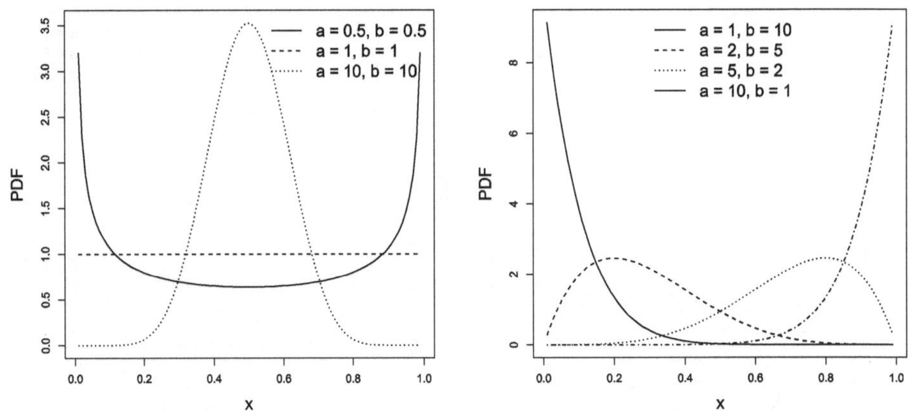

FIGURE 1.5
Plots of the beta PDF. Plots of the beta density function $f(x|\boldsymbol{\theta}) = \frac{\Gamma(a+b)}{\Gamma(a)\Gamma(b)}x^{a-1}(1-x)^{b-1}$ for several a and b.

TABLE 1.1
Hypothetical joint PMF. This PMF $f(x_1, x_2)$ gives the probabilities that a patient has a positive ($X_1 = 1$) or negative ($X_1 = 0$) primary health outcome and the patient having ($X_2 = 1$) or not having ($X_2 = 0$) a negative side effect.

		X_2		
		0	1	$f_1(x_1)$
X_1	0	0.06	0.14	0.20
	1	0.24	0.56	0.80
	$f_2(x_2)$	0.30	0.70	

1.1.2 Multivariate distributions

Most statistical analyses involve multiple variables with the objective of studying relationships between variables. To model relationships between variables we need multivariate extensions of mass and density functions. Let $X_1, ..., X_p$ be p random variables, $\mathcal{S}_j$ be the support of X_j so that $X_j \in \mathcal{S}_j$, and $\mathbf{X} = (X_1, ..., X_p)$ be the random vector. Table 1.1 describes the joint distribution between the $p = 2$ variables: X_1 indicates that the patient has the primary health outcome and X_2 indicates the patient has a side effect. If all variables are discrete, then the joint PMF is

$$\text{Prob}(X_1 = x_1, ..., X_p = x_p) = f(x_1, ..., x_p). \tag{1.18}$$

To be a valid PMF, f must be non-negative $f(x_1, ..., x_p) \geq 0$ and sum to one $\sum_{x_1 \in \mathcal{S}_1}, ..., \sum_{x_p \in \mathcal{S}_1} f(x_1, ..., x_p) = 1$.

As in the univariate case, probabilities for continuous random variables are computed indirectly via the PDF $f(x_1, ..., x_p)$. In the univariate case, probabilities are computed as the area under the density curve. For $p > 1$, probabilities are computed as the volume under the density surface. For example, for $p = 2$ random variables, the probability of $a_1 < X_1 < b_1$ and $a_2 < X_2 < b_2$ is

$$\int_{a_1}^{b_1}\int_{a_2}^{b_2} f(x_1, x_2)dx_1dx_2. \tag{1.19}$$

This gives the probability of the random vector $\mathbf{X} = (X_1, X_2)$ lying in the rectangle defined by the endpoints a_1, b_1, a_2, and b_2. In general, the probability of the random vector $\mathbf{X}$ falling in region $\mathcal{A}$ is the p-dimensional integral $\int_{\mathcal{A}} f(x_1, ..., x_p)dx_1, ..., dx_p$. As an example, consider the bivariate PDF on the unit square with $f(x_1, x_2) = 1$ for $\mathbf{X}$ with $x_1 \in [0, 1]$ and $x_2 \in [0, 1]$ and $f(x_1, x_2) = 0$ otherwise. Then $\text{Prob}(X_1 < .5, X_2 < .1) = \int_0^{0.5}\int_0^{0.1} f(x_1, x_2)dx_1dx_2 = 0.05$.

1.1.3 Marginal and conditional distributions

The marginal and conditional distributions that follow from a multivariate distribution are key to a Bayesian analysis. To introduce these concepts we assume discrete random variables, but extensions to the continuous case are straightforward by replacing sums with integrals. Further, we assume a bivariate PMF with $p = 2$. Again, extensions to high dimensions are conceptually straightforward by replacing sums over one or two dimensions with sums over $p - 1$ or p dimensions.

The marginal distribution of X_j is simply the distribution of X_j if we consider only a univariate analysis of X_j and disregard the other variable. Denote $f_j(x_j) = \text{Prob}(X_j = x_j)$ as the marginal PMF of X_j. The marginal distribution is computed by summing over the other variable in the joint PMF

$$f_1(x_1) = \sum_{x_2} f(x_1, x_2) \quad \text{and} \quad f_2(x_2) = \sum_{x_1} f(x_1, x_2). \tag{1.20}$$

These are referred to as the marginal distributions because in a two-way table such as Table 1.1, the marginal distributions are the row and column totals of the joint PMF that appear along the table's margins.

As with any univariate distribution, the marginal distribution can be summarized with its mean and variance,

$$\begin{aligned}
\mu_j &= \text{E}(X_j) = \sum_{x_j \in \mathcal{S}_j} x_j f_j(x_j) = \sum_{x_1}\sum_{x_2} x_j f(x_1, x_2)\\
\sigma_j^2 &= \text{Var}(X_j) = \sum_{x_j}(x_j - \mu_j)^2 f_j(x_j) = \sum_{x_1}\sum_{x_2}(x_j - \mu_j)^2 f(x_1, x_2).
\end{aligned}$$

The marginal mean and variance measure the center and spread of the marginal distributions, respectively, but do not capture the relationship between the two variables. The most common one-number summary of the joint relationship is covariance, defined as

$$\begin{aligned}
\sigma_{12} &= \text{Cov}(X_1, X_2) = \text{E}[(X_1 - \mu_1)(X_2 - \mu_2)] \qquad (1.21)\\
&= \sum_{x_1}\sum_{x_2}(x_1 - \mu_1)(x_2 - \mu_2)f(x_1, x_2).
\end{aligned}$$

The covariance is often hard to interpret because it depends on the scale of both X_1 and X_2, i.e., if we double X_1 we double the covariance. Correlation is a scale-free summary of the joint relationship, $\rho_{12} = \frac{\sigma_{12}}{\sigma_1\sigma_2}$. In vector notation, the mean of the random vector $\mathbf{X}$ is $\text{E}(\mathbf{X}) = (\mu_1, \mu_2)^T$, the covariance matrix is $\text{Cov}(\mathbf{X}) = \begin{bmatrix}\sigma_1^2 & \sigma_{12}\\ \sigma_{12} & \sigma_2^2\end{bmatrix}$, and the correlation

TABLE 1.2
Table of the Atlantic hurricanes that made landfall between 1990 and 2016. The counts are tabulated by their maximum Saffir–Simpson intensity category and whether they made landfall in the US or elsewhere. The counts are downloaded from `http://www.aoml.noaa.gov/hrd/hurdat/`.

(a) Counts

	Category					
	1	2	3	4	5	Total
US	14	13	10	1	1	39
Not US	46	19	20	17	3	105
Total	60	32	30	18	4	144

(b) Sample proportions

	Category					
	1	2	3	4	5	Total
US	0.0972	0.0903	0.0694	0.0069	0.0069	0.2708
Not US	0.3194	0.1319	0.1389	0.1181	0.0208	0.7292
Total	0.4167	0.2222	0.2083	0.1250	0.0278	1.0000

matrix is $\text{Cor}(\mathbf{X}) = \begin{bmatrix} 1 & \rho_{12} \\ \rho_{12} & 1 \end{bmatrix}$. Generalizing for $p > 2$, the mean vector becomes $\text{E}(\mathbf{X}) = (\mu_1, ..., \mu_p)^T$, and the covariance matrix becomes the symmetric $p \times p$ matrix with diagonal elements $\sigma_1^2, ..., \sigma_p^2$ and (i, j) off-diagonal element σ_{ij}.

While the marginal distributions sum over columns or rows of a two-way table, the conditional distributions focus only on a single column or row. The conditional distribution of X_1 given that the random variable X_2 is fixed at x_2 is denoted $f_{1|2}(x_1|X_2 = x_2)$ or simply $f(x_1|x_2)$. Referring to Table 1.1, the knowledge that $X_2 = x_2$ restricts the domain to a single column of the two-way table. However, the probabilities in a single column do not define a valid PMF because their sum is less than one. We must rescale these probabilities to sum to one by dividing by the column total, which we have previously defined as $f_2(x_2)$. Therefore, the general expression for the conditional distributions of $X_1|X_2 = x_2$, and similarly $X_2|X_1 = x_1$, is

$$f_{1|2}(x_1|X_2 = x_2) = \frac{f(x_1, x_2)}{f_2(x_2)} \quad \text{and} \quad f_{2|1}(x_2|X_1 = x_1) = \frac{f(x_1, x_2)}{f_1(x_1)}. \tag{1.22}$$

Atlantic hurricanes example: Table 1.2 provides the counts of Atlantic hurricanes that made landfall between 1990 and 2016 tabulated by their intensity category (1–5) and whether they hit the US or elsewhere. Of course, these are only sample proportions and not true probabilities, but for this example we treat Table 1.2b as the joint PMF of location, $X_1 \in \{\text{US}, \text{Not US}\}$, and intensity category, $X_2 \in \{1, 2, 3, 4, 5\}$. The marginal distribution of X_1 is given in the final column and is simply the row sums of the joint PMF. The marginal probability of a hurricane making landfall in the US is $\text{Prob}(X_1 = \text{US}) = 0.2708$, which is the proportion calculation as if we had never considered the storms' categories. Similarly, the column sums are the marginal probability of intensity averaging over location, e.g., $\text{Prob}(X_2 = 5) = 0.0278$.

The conditional distributions tell us about the relationship between the two variables. For example, the marginal probability of a hurricane reaching category 5 is 0.0278, but given that the storm hits the US, the conditional distribution is slightly lower, $f_{2|1}(5|\text{US}) = \text{Prob}(X_1 = \text{US}, X_2 = 5)/\text{Prob}(X_1 = \text{US}) = 0.0069/0.2708 = 0.0255$. By definition, the conditional probabilities sum to one,

$$\begin{aligned} f_{2|1}(1|\text{US}) &= \frac{0.0972}{0.2708} = 0.3589,\ f_{2|1}(2|\text{US}) = \frac{0.0903}{0.2708} = 0.3334 \\ f_{2|1}(3|\text{US}) &= \frac{0.0694}{0.2708} = 0.2562,\ f_{2|1}(4|\text{US}) = \frac{0.0069}{0.2708} = 0.0255 \\ f_{2|1}(5|\text{US}) &= \frac{0.0069}{0.2708} = 0.0255. \end{aligned}$$

Given that a storm hits the US, the probability of a category 2 or 3 storm increases, while the probability of a category 4 or 5 storm decreases, and so there is a relationship between landfall location and intensity.

Independence example: Consider the joint PMF of the primary health outcome (X_1) and side effect (X_2) in Table 1.1. In this example, the marginal probability of a positive primary health outcome is $\text{Prob}(X_1 = 1) = 0.80$, as is the conditional probability $f_{1|2}(1|X_2 = 1) = 0.56/0.70 = 0.80$ given the patient has the side effect and the conditional probability $f_{1|2}(1|0) = 0.24/0.30 = 0.80$ given the patient does not have the side effect. In other words, both with and without knowledge of side effect status, the probability of a positive health outcome is 0.80, and thus side effect is not informative about the primary health outcome. This is an example of two random variables that are independent.

Generally, X_1 and X_2 are independent if and only if the joint PMF (or PDF for continuous random variables) factors into the product of the marginal distributions,

$$f(x_1, x_2) = f_1(x_1)f_2(x_2). \tag{1.23}$$

From this expression it is clear that if X_1 and X_2 are independent then $f_{1|2}(x_1|X_2 = x_2) = f(x_1, x_2)/f_2(x_2) = f_1(x_1)$, and thus X_2 is not informative about X_1. A special case of joint independence is if all marginal distributions are the same, $f_j(x) = f(x)$ for all j and x. In this case, we say that X_1 and X_2 are independent and identically distributed ("iid"), which is denoted $X_j \stackrel{iid}{\sim} f$. If variables are not independent then they are said to be dependent.

Multinomial: Parametric families are also useful for multivariate distributions. A common parametric family for discrete data is the multinomial, which, as the name implies, is a generalization of the binomial. Consider the case of n independent trials where each trial results in one of p possible outcomes (e.g., $p = 3$ and each result is either win, lose or draw). Let $X_j \in \{0, 1, ..., n\}$ be the number of trials that resulted in outcome j, and $\mathbf{X} = (X_1, ..., X_p)$ be the vector of counts. If we assume that θ_j is the probability of outcome j for each trial, with $\boldsymbol{\theta} = (\theta_1, ..., \theta_p)$ and $\sum_{j=1}^{p} \theta_j = 1$, then $\mathbf{X}|\boldsymbol{\theta} \sim \text{Multinomial}(n, \boldsymbol{\theta})$ with

$$f(x_1, ..., x_p) = \frac{n!}{x_1! \cdot ... \cdot x_p!}\theta_1^{x_1} \cdot ... \cdot \theta_p^{x_p} \tag{1.24}$$

where $n = \sum_{j=1}^{p} x_j$. If there are only $p = 2$ categories, then this would be a binomial experiment $X_2 \sim \text{Binomial}(n, \theta_2)$.

Multivariate normal: The multivariate normal distribution is a generalization of the normal distribution to $p > 1$ random variables. For $p = 2$, the bivariate normal has five parameters, $\boldsymbol{\theta} = (\mu_1, \mu_2, \sigma_1^2, \sigma_2^2, \rho)$: a mean parameter for each variable $\text{E}(X_j) = \mu_j$, a variance for each variable $\text{Var}(X_j) = \sigma_j^2 > 0$, and the correlation $\text{Cor}(X_1, X_2) = \rho \in (-1, 1)$. The density function is

$$f(x_1, x_2) = \frac{1}{2\pi\sigma_1\sigma_2\sqrt{1-\rho^2}} \exp\left[-\frac{z_1^2 - 2\rho z_1 z_2 + z_2^2}{2(1-\rho^2)}\right], \tag{1.25}$$

where $z_j = (x_j - \mu_j)/\sigma_j$. As shown in Figure 1.6 the density surface is elliptical with center determined by $\boldsymbol{\mu} = (\mu_1, \mu_2)$ and shape determined by the covariance matrix $\boldsymbol{\Sigma} = \begin{bmatrix} \sigma_1^2 & \sigma_1\sigma_2\rho \\ \sigma_1\sigma_2\rho & \sigma_2^2 \end{bmatrix}$.

A convenient feature of the bivariate normal distribution is that the marginal and conditional distributions are also normal. The marginal distribution of X_j is Gaussian with mean μ_j and variance σ_j^2. The conditional distribution, shown in Figure 1.7, is

$$X_2|X_1 = x_1 \sim \text{Normal}\left[\mu_2 + \rho\frac{\sigma_2}{\sigma_1}(x_1 - \mu_1), (1-\rho^2)\sigma_2^2\right]. \tag{1.26}$$

If $\rho = 0$, then the conditional distribution is the marginal distribution, as expected. If $\rho > 0$ ($\rho < 0$), then the conditional mean increases (decreases) with x_1. Also, the conditional variance $(1-\rho^2)\sigma_2^2$ is less than the marginal variance σ_2^2, especially for ρ near -1 or 1, and so conditioning on X_1 reduces uncertainty in X_2 when there is strong correlation.

The multivariate normal PDF for $p > 2$ is most concisely written using matrix notation. The multivariate normal PDF for the random vector $\mathbf{X}$ with mean vector $\boldsymbol{\mu}$ and covariance matrix $\boldsymbol{\Sigma}$ is

$$f(\mathbf{X}) = (2\pi)^{-p/2}|\boldsymbol{\Sigma}|^{-1/2}\exp\left[-\frac{1}{2}(\mathbf{X}-\boldsymbol{\mu})^T\boldsymbol{\Sigma}^{-1}(\mathbf{X}-\boldsymbol{\mu})\right] \tag{1.27}$$

where $|\mathbf{A}|$, $\mathbf{A}^T$ and $\mathbf{A}^{-1}$ are the determinant, transpose and inverse, respectively, of the matrix $\mathbf{A}$. From this expression it is clear that the contours of the log PDF are elliptical. All conditional and marginal distributions are normal, as are all linear combinations $\sum_{j=1}^p w_jX_j$ for any $w_1, ..., w_p$.

1.2 Bayes' rule

As the name implies, Bayes' rule (or Bayes' theorem) is fundamental to Bayesian statistics. However, this rule is a general result from probability and follows naturally from the definition of a conditional distribution. Consider two random variables X_1 and X_2 with joint PMF (or PDF, the result holds for both discrete and continuous data) density function $f(x_1, x_2)$. The definition of a conditional distribution gives $f(x_1|x_2) = f(x_2, x_1)/f(x_2)$ and $f(x_2|x_1) = f(x_1, x_2)/f(x_1)$. Combining these two expressions gives Bayes' rule

$$f(x_2|x_1) = \frac{f(x_1|x_2)f(x_2)}{f(x_1)}. \tag{1.28}$$

This result is useful as a means to reverse conditioning from $X_1|X_2$ to $X_2|X_1$, and also indicates the need to define a joint distribution for this inversion to be valid.

1.2.1 Discrete example of Bayes' rule

You have a scratchy throat, and so you go to the doctor who administers a rapid strep throat test. Let $Y \in \{0, 1\}$ be the binary indicator of a positive test; i.e., $Y = 1$ if the test is positive for strep and $Y = 0$ if the test is negative. The test is not perfect. The false positive rate $p \in [0, 1]$ is the probability of testing positive if you do not have strep, and the false negative rate $q \in [0, 1]$ is the probability of testing negative given that you actually have strep. To express these probabilities mathematically we must define the true disease

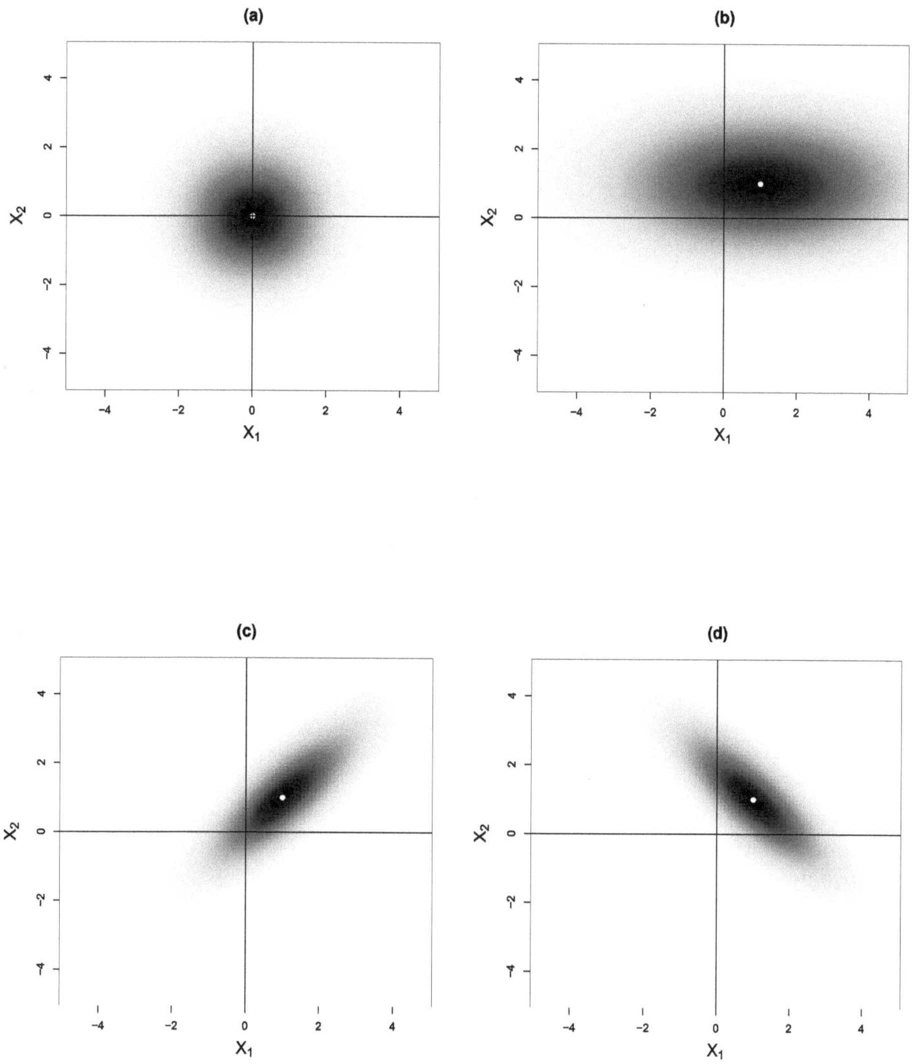

FIGURE 1.6
Plots of the bivariate normal PDF. Panel (a) plots the bivariate normal PDF for $\boldsymbol{\mu} = (0, 0)$, $\sigma_1 = 1$, $\sigma_2 = 1$ and $\rho = 0$. The other panels modify Panel (a) as follows: (b) has $\mu = (1, 1)$ and $\sigma_1 = 2$, (c) has $\mu = (1, 1)$ and $\rho = 0.8$, and (d) has $\mu = (1, 1)$ and $\rho = -0.8$. The plots are shaded according to the PDF with white indicating the PDF near zero and black indicating the areas with highest PDF; the white dot is the mean vector $\boldsymbol{\mu}$.

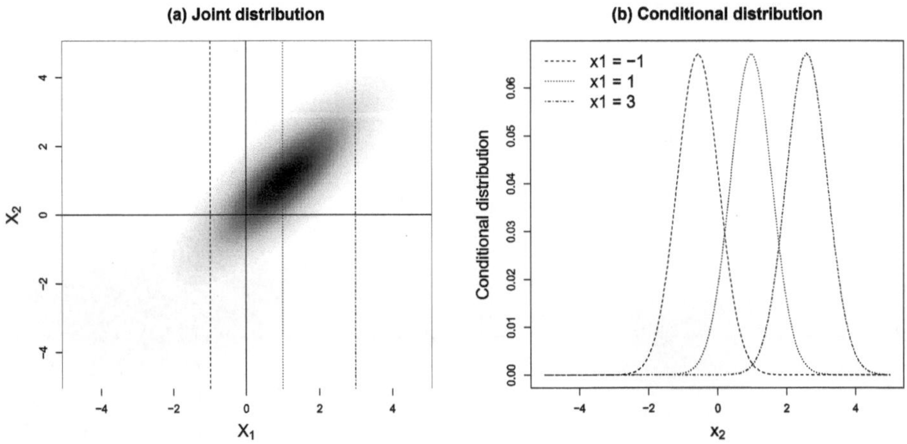

FIGURE 1.7
Plots of the joint and conditional bivariate normal PDF. Panel (a) plots the bivariate normal PDF for $\boldsymbol{\mu} = (1, 1)$, $\sigma_1 = \sigma_2 = 1$ and $\rho = 0.8$. The plot is shaded according to the PDF with white indicating the PDF near zero and black indicating the areas with highest PDF; the vertical lines represent $x_1 = -1$, $x_1 = 1$ and $x_1 = 3$. The conditional distribution of $X_2|X_1 = x_1$ for these three values of x_1 are plotted in Panel (b).

status $\theta \in \{0, 1\}$, where $\theta = 1$ if you are truly infected and $\theta = 0$ otherwise. This unknown variable, which we hope to estimate, is called a parameter. Given these error rates and the definition of the model parameter, the data distribution can be written

$$\text{Prob}(Y = 1|\theta = 0) = p \qquad \text{and} \qquad \text{Prob}(Y = 0|\theta = 1) = q. \tag{1.29}$$

Generally, the PMF (or PDF) of the observed data given the model parameters is called the likelihood function.

To formally analyze this problem we must determine which components should be treated as random variables. Is the test result Y a random variable? Before the exam, Y is clearly random and (1.29) defines its conditional distribution. This is aleatoric uncertainty because the results may differ if we repeat the test. However, after learning of the test results, Y is determined, and you must decide how to proceed given the value of Y at hand. In this sense, Y is known and no longer random at the analysis stage.

Is the true disease status θ a random variable? Certainly θ is not a random variable in the sense that it changes from second-to-second or minute-to-minute, and so it is reasonable to assume that the true disease status is a fixed quantity for the purpose of this analysis. However, because our test is imperfect, we do not know θ. This is epistemic uncertainty because θ is a quantity that we could theoretically know, but at the analysis stage we do not and cannot know θ using only noisy data. Despite our uncertainty about θ, we have to decide what to do next and so it is useful to quantify our uncertainty using the language of probability. If the test is reliable and p and q are both small, then in light of a positive test we might conclude that θ is more likely to be one than zero. But how much more likely? Twice as likely? Three times? **In Bayesian statistics, we quantify uncertainty about fixed but unknown parameters using probability theory by treating them as**

random variables. As (1.28) suggests, inversion of conditional probabilities requires we treat both the data and parameters as random.

The probabilities in (1.29) provide the distribution of the test result given disease status, $Y|\theta$. However, we would like to quantify uncertainty in the disease status given the test results; that is, we require the distribution of $\theta|Y$. Since this is the uncertainty distribution after collecting the data this is referred to as the posterior distribution. As discussed above, Bayes' rule can be applied to reverse the order of conditioning,

$$\text{Prob}(\theta = 1|Y = 1) = \frac{\text{Prob}(Y = 1|\theta = 1)\text{Prob}(\theta = 1)}{\text{Prob}(Y = 1)}, \tag{1.30}$$

where the marginal probability $\text{Prob}(Y = 1)$ is

$$\sum_{\theta=0}^{1} f(1,\theta) = \text{Prob}(Y = 1|\theta = 1)\text{Prob}(\theta = 1) + \text{Prob}(Y = 1|\theta = 0)\text{Prob}(\theta = 0). \tag{1.31}$$

To apply Bayes' rule requires specifying the unconditional probability of having strep throat, $\text{Prob}(\theta = 1) = \pi \in [0, 1]$. Since this is the probability of infection before we conduct the test, we refer to this as the prior probability. We can then compute the posterior using Bayes' rule,

$$\text{Prob}(\theta = 1|Y = 1) = \frac{(1-q)\pi}{(1-q)\pi + p(1-\pi)}. \tag{1.32}$$

To understand this equation consider a few extreme scenarios. Assuming the error rates p and q are not zero or one, if $\pi = 1$ ($\pi = 0$), then the posterior probability of $\theta = 1$ ($\theta = 0$) is one for any value of Y. That is, if we have no prior uncertainty then the imperfect data does not update the prior. Conversely, if the test is perfect and $q = p = 0$, then for any prior $\pi \in (0, 1)$ the posterior probability that θ equals Y is one. That is, with perfect data the prior is irrelevant. Finally, if $p = q = 1/2$, then the test is a random coin flip, and the posterior probability is the prior probability $\text{Prob}(\theta = 1|Y) = \pi$.

For a more realistic scenario we use the data in Table 1.3 taken from [51]. We plug in the sample error rates from these data for $p = 37/(37 + 610) = 0.057$ and $q = 48/(48 + 123) = 0.281$. Of course these data represent only a sample, and the sample proportions are not exactly the true error rates, but for illustration we assume these error rates are correct. Then if we assume prior probability of disease is $\pi = 0.5$, the posterior probabilities are $\text{Prob}(\theta = 1|Y = 0) = 0.230$ and $\text{Prob}(\theta = 1|Y = 1) = 0.927$. Therefore, beginning with a prior probability of 0.5, a negative test moves the probability down to 0.230, and a positive test increases the probability to 0.927.

Of course, in reality the way individuals process test results is complicated and subjective. If you have had strep many times before and you went to the doctor because your current symptoms resemble previous bouts with the disease, then perhaps your prior is $\pi = 0.8$, and the posterior is $\text{Prob}(\theta = 1|Y = 1) = 0.981$. On the other hand, if you went to the doctor only at the urging of your friend and your prior probability is $\pi = 0.2$, then $\text{Prob}(\theta = 1|Y = 1) = 0.759$.

This simple example illustrates a basic Bayesian analysis. The objective is to compute the posterior distribution of the unknown parameters θ. The posterior has two ingredients: the likelihood of the data given the parameters and the prior distribution. Selection of these two distributions is thus largely the focus of the remainder of this book.

1.2.2 Continuous example of Bayes' rule

Let $\theta \in [0, 1]$ be the proportion of the population in a county that has health insurance. It is known that the proportion varies across counties following a $\text{Beta}(a, b)$ distribution,

TABLE 1.3
Strep throat data. Number of patients that are truly positive and tested positive in the rapid strep throat test data taken from Table 1 of [51].

	Truly positive, test positive	Truly positive, test negative	Truly negative, test positive	Truly negative, test negative
Children	80	38	23	349
Adults	43	10	14	261
Total	123	48	37	610

and so the prior is $\theta \sim \text{Beta}(a, b)$. We take a sample of size $n = 20$ from your county and assume that the number of respondents with insurance, $Y \in \{0, 1, ..., n\}$, is distributed as $Y|\theta \sim \text{Binomial}(n, \theta)$. Joint probabilities for θ and Y can be computed from

$$\begin{aligned} f(\theta, y) &= f(y|\theta)f(\theta) \\ &= \left\{\binom{n}{y}\theta^y(1-\theta)^{n-y}\right\}\left\{\frac{\Gamma(a+b)}{\Gamma(a)\Gamma(b)}\theta^{a-1}(1-\theta)^{b-1}\right\} \\ &= c\theta^{y+a-1}(1-\theta)^{n-y+b-1} \end{aligned}$$

where $c = \binom{n}{y}\Gamma(a+b)/[\Gamma(a)\Gamma(b)]$ is a constant that does not depend on θ. Figure 1.8 plots $f(\theta, y)$ and the marginal distributions for θ and Y. Given the way we have defined the problem, the marginal distribution of θ, $f(\theta)$, is a $\text{Beta}(a, b)$ PDF, which could also be derived by summing $f(\theta, y)$ over y. The marginal distribution of Y plotted on the right of Figure 1.8 is $f(y) = \int_0^1 f(\theta, y)d\theta$. In this case the marginal distribution of Y follows a beta-binomial distribution, but as we will see, this is not needed in the Bayesian analysis.

In this problem we are given the unconditional distribution of the disease rate (prior) and the distribution of the sample given the true proportion (likelihood). Bayes' rule gives the (posterior) distribution of the true proportion given the sample. Say we observe $Y = 12$ positive cases. The horizontal line in Figure 1.8 traces over the conditional distribution $f(\theta|Y = 12)$. The conditional distribution is centered around the sample proportion $Y/n = 0.60$ but has non-trivial mass from 0.4 to 0.8. More formally, the posterior is

$$\begin{aligned} f(\theta|y) &= \frac{f(y|\theta)f(\theta)}{f(y)} \\ &= \left[\frac{c}{f(y)}\right]\theta^{y+a-1}(1-\theta)^{n-y+b-1} \\ &= C\theta^{A-1}(1-\theta)^{B-1} \end{aligned} \tag{1.33}$$

where $C = c/f(y)$, $A = y + a$, and $B = n - y + b$.

We note the resemblance between $f(\theta|y)$ and the PDF of a $\text{Beta}(A, B)$ density. Both include $\theta^{A-1}(1-\theta)^{B-1}$ but differ in the normalizing constant, C, for $f(\theta|y)$ compared to $\Gamma(A+B)/[\Gamma(A)\Gamma(B)]$ for the $\text{Beta}(A, B)$ PDF. Since both $f(\theta|y)$ and the $\text{Beta}(A, B)$ PDF are proper, they both integrate to one, and thus

$$\int_0^1 C\theta^{A-1}(1-\theta)^{B-1}d\theta = \int_0^1 \frac{\Gamma(A+B)}{\Gamma(A)\Gamma(B)}\theta^{A-1}(1-\theta)^{B-1}d\theta = 1 \tag{1.34}$$

and so

$$C\int_0^1 \theta^{A-1}(1-\theta)^{B-1}d\theta = \frac{\Gamma(A+B)}{\Gamma(A)\Gamma(B)}\int_0^1 \theta^{A-1}(1-\theta)^{B-1}d\theta \tag{1.35}$$

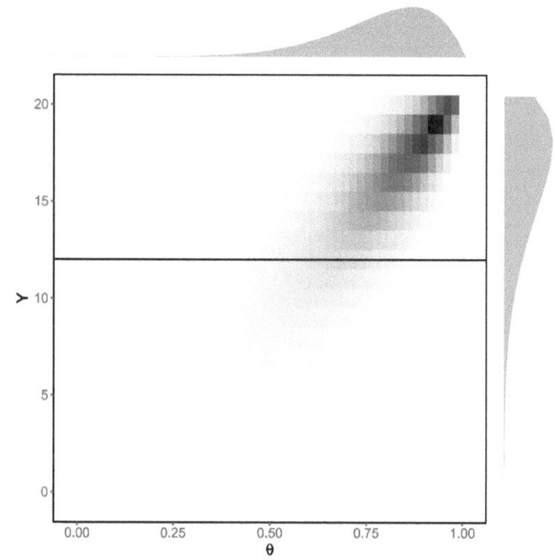

FIGURE 1.8
Joint distribution for the beta-binomial example. Plot of $f(\theta, y)$ for the example with $\theta \sim \text{Beta}(8, 2)$ and $Y|\theta \sim \text{Binomial}(20, \theta)$. The marginal distributions $f(\theta)$ (top) and $f(y)$ (right) are plotted in the margins. The horizontal line is the $Y = 12$ line.

and thus $C = \Gamma(A + B)/[\Gamma(A)\Gamma(B)]$. Therefore, $f(\theta|y)$ is in fact the Beta(A, B) PDF and $\theta|Y = y \sim \text{Beta}(y + a, n - y + b)$.

Dealing with the normalizing constant makes posterior calculations quite tedious. Fortunately this can often be avoided by discarding terms that do not involve the parameter of interest and comparing the remaining terms with known distributions. The derivation above can be simplified to (using "$\propto$" to mean "proportional to")

$$f(\theta|y) \propto f(y|\theta)f(\theta) \propto \theta^{(y+a)-1}(1 - \theta)^{(n-y+b)-1},$$

immediately concluding that $\theta|Y = y \sim \text{Beta}(y + a, n - y + b)$.

Figure 1.9 plots the posterior distribution for two priors and $Y \in \{0, 5, 10, 15, 20\}$. The plots illustrate how the posterior combines information from the prior and the likelihood. In both plots, the peak of the posterior distribution increases with the observation Y. Comparing the plots shows that the prior also contributes to the posterior. When we observe $Y = 0$ successes, the posterior under the Beta(8,2) prior (left) is pulled from zero to the right by the prior (thick line). Under the Beta(1,1), i.e., the uniform prior, when $Y = 0$ the posterior is concentrated around $\theta = 0$.

1.3 Introduction to Bayesian inference

A parametric statistical analysis models the random process that produced the data, $\mathbf{Y} = (Y_1, ..., Y_n)$, in terms of fixed but unknown parameters $\boldsymbol{\theta} = (\theta_1, ..., \theta_p)$. The PDF (or PMF) of the data given the parameters, $f(\mathbf{Y}|\boldsymbol{\theta})$, is called the likelihood function and links the

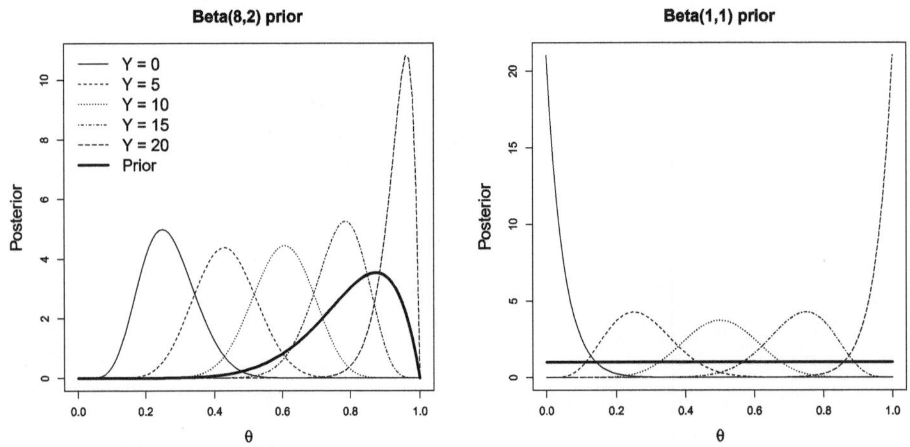

FIGURE 1.9
Posterior distribution for the beta-binomial example. The thick lines are the beta prior for success probability θ, and the thin lines are the posterior assuming $Y|\theta \sim \text{Binomial}(20, \theta)$ for various values of Y.

observed data with the unknown parameters. Statistical inference is concerned with the inverse problem of using the likelihood function to estimate $\boldsymbol{\theta}$. Of course, if the data are noisy, then we cannot perfectly estimate $\boldsymbol{\theta}$, and a Bayesian quantifies uncertainty about the unknown parameters by treating them as random variables. Treating $\boldsymbol{\theta}$ as a random variable requires specifying the prior distribution, $\pi(\boldsymbol{\theta})$, which represents our uncertainty about the parameters before we observe the data.

If we view $\boldsymbol{\theta}$ as a random variable, we can apply Bayes' rule to obtain the posterior distribution

$$p(\boldsymbol{\theta}|\mathbf{Y}) = \frac{f(\mathbf{Y}|\boldsymbol{\theta})\pi(\boldsymbol{\theta})}{\int f(\mathbf{Y}|\boldsymbol{\theta})\pi(\boldsymbol{\theta})d\theta} \propto f(\mathbf{Y}|\boldsymbol{\theta})\pi(\boldsymbol{\theta}). \tag{1.36}$$

The posterior is proportional to the likelihood times the prior, and quantifies the uncertainty about the parameters that remain after accounting for prior knowledge and the new information in the observed data.

Table 1.4 establishes the notation we use throughout for the prior, likelihood and posterior. We will not adhere to the custom (e.g., Section 1.1) that random variables are capitalized because in a Bayesian analysis more often than not it is the parameters that are the random variables, and capital Greek letters, e.g., $\text{Prob}(\Theta = \theta)$, are unfamiliar to most readers. We will however follow the custom to use bold font to represent vectors and matrices. Also, assume independence unless otherwise noted. For example, if we say "the priors are $\theta_1 \sim \text{Uniform}(0, 1)$ and $\theta_2 \sim \text{Gamma}(1, 1)$," you should assume that θ_1 and θ_2 have independent priors.

The Bayesian framework provides a logically consistent framework to use all available information to quantify uncertainty about model parameters. However, to apply Bayes' rule requires specifying the prior distribution and the likelihood function. *How do we pick the prior?* In many cases prior knowledge from experience, expert opinion or similar studies is available and can be used to specify an informative prior. It would be a waste to discard this information. In other cases where prior information is unavailable, then the prior should

TABLE 1.4
Notation used throughout the book for distributions involving the parameter vector $\boldsymbol{\theta} = (\theta_1, ..., \theta_p)$ and data vector $\mathbf{Y} = (Y_1, ..., Y_n)$.

Prior density of $\boldsymbol{\theta}$	$\pi(\boldsymbol{\theta})$
Likelihood function of $\mathbf{Y}$ given $\boldsymbol{\theta}$	$f(\mathbf{Y}\vert\boldsymbol{\theta})$
Marginal density of $\mathbf{Y}$	$m(\mathbf{Y}) = \int f(\mathbf{Y}\vert\boldsymbol{\theta})\pi(\boldsymbol{\theta})d\boldsymbol{\theta}$
Posterior density of $\boldsymbol{\theta}$ given $\mathbf{Y}$	$p(\boldsymbol{\theta}\vert\mathbf{Y}) = f(\mathbf{Y}\vert\boldsymbol{\theta})\pi(\boldsymbol{\theta})/m(\mathbf{Y})$

be uninformative to reflect this uncertainty. For instance, in the beta-binomial example in Section 1.2 we might use a uniform prior that puts equal mass on all possible parameter values. The choice of prior distribution is subjective, i.e., driven by the analyst's past experience and personal preferences. If a reader does not agree with your prior, then they are unlikely to be persuaded by your analysis. Therefore, the prior, especially an informative prior, should be carefully justified, and a sensitivity analysis comparing the posteriors under different priors should be presented.

How do we pick the likelihood? The likelihood function is the same as in a classical analysis, e.g., a maximum likelihood analysis. The likelihood function for multiple linear regression is the product of Gaussian PDFs defined by the model

$$Y_i|\boldsymbol{\theta} \overset{indep}{\sim} \text{Normal}\left(\beta_0 + \sum_{j=1}^{p} X_{ij}\beta_j, \sigma^2\right) \tag{1.37}$$

where X_{ij} is the value of the jth covariate for the ith observation and $\boldsymbol{\theta} = (\beta_0, ..., \beta_p, \sigma^2)$ are the unknown parameters. A thoughtful application of multiple linear regression must consider many questions, including

- Which covariates should we include?
- Are the errors Gaussian? Independent? Do they have equal variance?
- Should we include quadratic or interaction effects?
- Should we consider a transformation of the response (e.g., model $\log(Y_i)$)?
- Which observations are outliers? Should we remove them?
- How should we handle the missing observations?
- What p-value threshold should be used to define statistical significance?

As with specifying the prior, these concerns are arguably best resolved using subjective subject-matter knowledge. For example, while there are statistical methods to select covariates (Chapter 6), a more reliable strategy is to ask a subject-matter expert which covariates are the most important to include, at least as an initial list to be refined in the statistical analysis. As another example, it is hard to determine (without a natural ordering as in times series data) whether the observations are independent without consulting someone familiar with the data collection and the study population. Other decisions are made based on visual inspections of the data (such as scatter plots and histograms of the residuals) or ad hoc rules of thumb (threshold on outliers' z-scores or p-values for statistical significance).

Therefore, in a typical statistical analysis there are many subjective choices to be made, and the choice of prior is far from the most important.

Bayesian statistical methods are often criticized as being subjective. Perhaps an objective analysis that is free from personal preferences or beliefs is an ideal we should strive for (and this is the aim of objective Bayesian methods, see Section 2.4), but it is hard to make the case that non-Bayesian methods are objective, and it can be argued that almost any scientific knowledge and theories are subjective in nature. In an interesting article by [124], the authors cite many scientific theories (mainly from physics) where subjectivity played a major role and they concluded, "*Subjectivity occurs, and should occur, in the work of scientists; it is not just a factor that plays a minor role that we need to ignore as a flaw...*" and they further added that, "*Total objectivity in science is a myth. Good science inevitably involves a mixture of subjective and objective parts.*" The Bayesian inferential framework provides a logical foundation to accommodate objective and subjective parts involved in data analysis. Hence, a good scientific practice would be to state upfront all assumptions and then make an effort to validate such assumptions using the current data or preferable future test cases. There is nothing wrong to have a subjective but reasonably flexible model as long as we can exhibit some form of sensitivity analysis when the assumptions of the model are mildly violated.

In addition to explicitly acknowledging subjectivity, another important difference between Bayesian and frequentist (classical) methods is their notion of uncertainty. While a Bayesian considers only the data at hand, a frequentist views uncertainty as arising from repeating the process that generated the data many times. That is, a Bayesian might give a posterior probability that the population mean μ (a parameter) is positive given the data we have observed, whereas a frequentist would give a probability that the sample mean $\bar{Y}$ (a statistic) exceeds a threshold given a specific value of the parameters if we repeated the experiment many times (as is done when computing a p-value).

The frequentist view of uncertainty is well-suited for developing procedures that have desirable error rates when applied broadly. This is reasonable in many settings. For instance, a regulatory agency might want to advocate statistical procedures that ensure only a small proportion of the medications made available to the public have adverse side effects. In some cases, however, it is hard to see why repeating the sampling is a useful thought experiment. For example, [22] study the relationship between a region's climate and the type of religion that emerged from that region. Assuming the data set consists of the complete list of known cultures, it is hard to imagine repeating the process that led to these data as it would require replaying thousands of years of human history.

Bayesians can and do study frequentist properties. This is critical to build trust in the methods. If a Bayesian weather forecaster gives the posterior predictive 95% interval every day for a year, but at the end of the year these intervals included the observed temperature only 25% of the time, then the forecaster would lose all credibility. It turns out that Bayesian methods often have desirable frequentist properties, and Chapter 9 examines these properties.

Developing Bayesian methods with good frequentist properties is often called calibrated Bayes (e.g., [98]). According to Rubin [98, 133]: "*The applied statistician should be Bayesian in principle and calibrated to the real world in practice – appropriate frequency calculations help to define such a tie... frequency calculations are useful for making Bayesian statements scientific, scientific in the sense of capable of being shown wrong by empirical test; here the technique is the calibration of Bayesian probabilities to the frequencies of actual events.*"

Bayesian inference is fundamentally about combining prior knowledge with observed data to update our understanding of the underlying phenomenon or parameter values. This process reduces uncertainty in a way that can be quantified using concepts from information theory. The fundamental question is, *Are we learning more by combining the information brought by the data* $\mathbf{Y}$ *through* $f(\mathbf{Y}|\boldsymbol{\theta})$ *and the prior information brought by* $\pi(\boldsymbol{\theta})$*?*

In order to quantify the information gain/loss, we need a measure of information by quantifying the uncertainty. Below we use the variance because it is a classic measure of uncertainty, but the concepts can be extended to other metrics such as Fisher information. For simplicity, we assume univariate θ with $p = 1$. It can be shown that

$$\mathrm{E}\{\mathrm{Var}(\theta|\mathbf{Y})\} < \mathrm{Var}(\theta)$$

by using the well-known conditional variance identity $\mathrm{Var}(\theta) = \mathrm{E}\{\mathrm{Var}(\theta|\mathbf{Y})\} + \mathrm{Var}\{\mathrm{E}(\theta|\mathbf{Y})\}$, as the second term is necessarily positive unless the marginal distribution of $\mathbf{Y}$ is degenerate or uninformative. Thus, if reduction in variance is a key goal of Bayesian updating, indicating increased precision in parameter estimation, then prior uncertainty is reduced in the light of data on average. However, it is interesting to note that it is possible to construct examples for which $\mathrm{Var}(\theta|\mathbf{Y}) > \mathrm{Var}(\theta)$ for a pathological observed dataset.

This result extends to a sequential analysis where data arrives in batches, $\mathbf{Y}_1, ..., \mathbf{Y}_K$. Defining $V_0 = \mathrm{Var}(\theta)$ as the prior variance and $V_k = \mathrm{E}\{\mathrm{Var}(\theta|\mathbf{Y}_1, ..., \mathbf{Y}_k)\}$ as the expected (over the data) posterior variance at step k, then

$$V_0 > V_1 > ... > V_K. \tag{1.38}$$

Thus, as we gather data, on average the posterior variance of θ decreases, as one would expect, irrespective of the initial prior distribution used.

Through variance reduction, Bayesian inference systematically reduces uncertainty. The prior contributes initial knowledge, the likelihood integrates new data, and the posterior represents an updated understanding, which is more precise and information-rich than the prior alone. This dynamic process aligns with the core goal of statistical inference: learning and decision-making under uncertainty. Thus, instead of arguing about Bayesian versus frequentist inference, it is probably more prudent to consider the integrative view of Bayesian and frequentist inference by taking advantage of information gain brought by both the data and prior information.

1.4 Summarizing the posterior

The final output of a Bayesian analysis is the posterior distribution of the model parameters. The posterior contains all the relevant information from the data and the prior, and thus all statistical inference should be based on the posterior distribution. However, when there are many parameters, the posterior distribution is a high-dimensional function that is difficult to display graphically, and for complicated statistical models the mathematical form of the posterior may be challenging to work with. In this section, we discuss some methods to summarize a high-dimensional posterior with low-dimensional summaries.

1.4.1 Point estimation

One approach to summarizing the posterior is to use a point estimate, i.e., a single value that represents the best estimate of the parameters given the data and (for a Bayesian analysis) the prior. The posterior mean, median and mode are all sensible choices. Thinking of the Bayesian analysis as a procedure that can be applied to any dataset, the point estimator is an example of an *estimator*, i.e., a function that takes the data as input and returns an estimate of the parameter of interest. Bayesian estimators such as the posterior mean can then be seen as competitors to other estimators such as the sample mean estimator for a population mean or a sample variance for a population variance, or more generally as a competitor to the maximum likelihood estimator. We study the properties of these estimators in Chapter 9.

A common point estimator is the maximum a posteriori (MAP) estimator, defined as the value that maximizes the posterior (i.e., the posterior mode),

$$\hat{\boldsymbol{\theta}}_{MAP} = \underset{\boldsymbol{\theta}}{\arg\max} \log[p(\boldsymbol{\theta}|\mathbf{Y})] = \underset{\boldsymbol{\theta}}{\arg\max} \log[f(\mathbf{Y}|\boldsymbol{\theta})] + \log[\pi(\boldsymbol{\theta})]. \tag{1.39}$$

The second equality holds because the normalizing constant $m(\mathbf{Y})$ does not depend on the parameters and thus does not affect the optimization.

If the prior is uninformative, i.e., mostly flat as a function of the parameters, then the MAP estimator should be similar to the maximum likelihood estimator (MLE)

$$\hat{\boldsymbol{\theta}}_{MLE} = \underset{\boldsymbol{\theta}}{\arg\max} \log[f(\mathbf{Y}|\boldsymbol{\theta})]. \tag{1.40}$$

In fact, this relationship is often used to intuitively justify maximum likelihood estimation. The addition of the log prior, $\log[\pi(\boldsymbol{\theta})]$, in (1.39) can be viewed as a regularization or penalty term to add stability or prior knowledge.

Point estimators are often useful as fast methods to estimate the parameters for purpose of making predictions. However, a point estimate alone does not quantify uncertainty about the parameters. Sections 1.4.2 and 1.4.3 provide more thorough summaries of the posterior for univariate and multivariate problems, respectively.

1.4.2 Univariate posteriors

A univariate posterior (i.e., from a model with $p = 1$ parameter) is best summarized with a plot because this retains all information about the parameter. Figure 1.10 shows a hypothetical univariate posterior with PDF centered at 0.8 and most of its mass on $\theta > 0.4$.

Point estimators such as the posterior mean or median summarize the center of the posterior, and should be accompanied by a posterior variance or standard deviation to convey uncertainty. The posterior standard deviation resembles a frequentist standard error in that if the posterior is approximately Gaussian then the posterior probability that the parameter is within two posterior standard deviation units of the posterior mean is roughly 0.95. However, the standard error is the standard deviation of the estimator (e.g., the sample mean) if we repeatedly sample different data sets and compute the estimator for each data set. In contrast, the posterior standard deviation quantifies uncertainty about the parameter given only the single data set under consideration.

The interval $\text{E}(\theta|\mathbf{Y}) \pm 2\text{SD}(\theta|Y)$ is an example of a credible interval. A $(1-\alpha)\%$ credible interval is any interval (l, u) so that $\text{Prob}(l < \theta < u|\mathbf{Y}) = 1 - \alpha$. There are infinitely many intervals with this coverage, but the easiest to compute is the equal-tailed interval with l and u set to the $\alpha/2$ and $1 - \alpha/2$ posterior quantiles. An alternative is the highest posterior density interval which searches over all l and u to minimize the interval width $u - l$ while maintaining the appropriate posterior coverage. The HPD thus has the highest average posterior density of all intervals of the form (l, u) that have the nominal posterior probability. As opposed to equal-tailed intervals, the HPD requires an additional optimization step, but this can be computed using the `R` package `HDInterval`.

Interpreting a posterior credible interval is fairly straightforward. If (l, u) is a posterior 95% interval, this means "given my prior and the observed data, I am 95% sure that θ

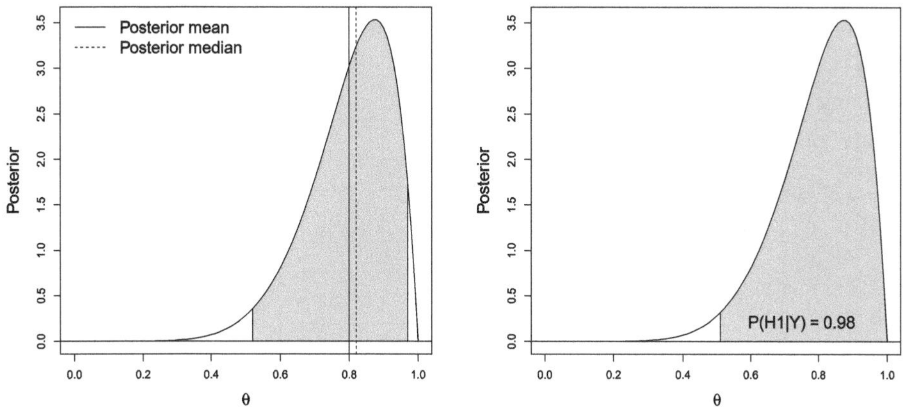

FIGURE 1.10
Summaries of a univariate posterior. The plot on the left gives the posterior mean (solid vertical line), median (dashed vertical line) and 95% equal-tailed interval (shaded area). The plot on the right shades the posterior probability of the hypothesis that $\theta > 0.5$.

is between l and u." In a Bayesian analysis we express our subjective uncertainty about unknown parameters by treating them as random variables, and in this subjective sense it is reasonable to assign probabilities to θ. This is in contrast with frequentist confidence intervals which have a more nuanced interpretation. A confidence interval is a procedure that defines an interval for a given data set in a way that ensures the procedure's intervals will include the true value 95% of the time when applied to random datasets.

The posterior distribution can also be used for hypothesis testing. Since the hypotheses are functions of the parameters, we can assign posterior probabilities to each hypothesis. Figure 1.10 (right) plots the posterior probability of the null hypothesis that $\theta < 0.5$ (white) and the posterior probability of the alternative hypothesis that $\theta > 0.5$ (shaded). These probabilities summarize the weight of evidence in support of each hypothesis, and can be used to guide future decisions. Hypothesis testing, and more generally model selection, is discussed in greater detail in Chapter 6.

Summarizing a univariate posterior using `R`: We have seen that if $Y|\theta \sim \text{Binomial}(n,\theta)$ and $\theta \sim \text{Beta}(a,b)$, then $\theta|Y \sim \text{Beta}(A,B)$, where $A = Y + a$ and $B = n - Y + b$. Listing 1.1 specifies a data set with $Y = 40$ and $n = 100$ and summarizes the posterior using `R`. Since the posterior is the beta density, the functions `dbeta`, `pbeta` and `qbeta` can be used to summarize the posterior. The posterior median and 95% credible set are 0.401 and (0.309, 0.498).

Monte Carlo sampling: Although univariate posterior distributions are best summarized by a plot, higher dimensional posterior distributions call for other methods such as Monte Carlo (MC) sampling and so we introduce this approach here. MC sampling draws S samples from the posterior, $\theta^{(1)},...,\theta^{(S)}$, and uses these samples to approximate the posterior. For example, the posterior mean is approximated using the mean of the S samples, $\text{E}(\theta|\mathbf{Y}) \approx \sum_{s=1}^{S} \theta^{(s)}/S$, the posterior 95% credible set is approximated using the 0.025 and 0.975 quantiles of the S samples, etc. Listing 1.1 provides an example of using MC sampling to approximate the posterior mean and 95% credible set.

Listing 1.1
Summarizing a univariate posterior in R.

```
# Load the data
 > n <- 100
 > Y <- 40
 > a <- 1
 > b <- 1
 > A <- Y + a
 > B <- n - Y + b

# Define a grid of points for plotting
 > theta <- seq(0,1,.001)

# Evaluate the density at these points
 > pdf <- dbeta(theta,A,B)

# Plot the posterior density
 > plot(theta,pdf,type="l",ylab="Posterior",xlab=expression(theta))

# Posterior mean
 > A/(A + B)
 [1] 0.4019608

# Posterior median (0.5 quantile)
 > qbeta(0.5,A,B)
 [1] 0.4013176

# Posterior probability P(theta<0.5|Y)
 > pbeta(0.5,A,B)
 [1] 0.976978

# Equal-tailed 95% credible interval
 > qbeta(c(0.025,0.975),A,B)
 [1] 0.3093085 0.4982559

# Monte Carlo approximation
 > S       <- 100000
 > samples <- rbeta(S,A,B)
 > mean(samples)
 [1] 0.402181
 >   quantile(samples,c(0.025,0.975))
     2.5%     97.5%
 0.3092051 0.4973871
```

Sensitivity analysis: Selecting a prior distribution (Chapter 2) is a key step in a Bayesian analysis. As with any subjective decision, including the choice of the likelihood, removal of outliers, variable transformations, etc., it is important to understand the impact of prior choice on the results. A prior sensitivity analysis varies the prior distribution to assess its impact on the posterior. This is an important step both if the prior is informative or uninformative. If the prior is informative, then a sensitivity analysis can illustrate the degree of reliance on the prior distribution, which is useful if there are questions about the informative prior. If the prior is intended to be uninformative, a sensitivity analysis is important because there is no unique uninformative prior, and ideally the choice of prior has little impact.

Of course, drastic changes to the prior will always affect the posterior. Typically, a sensitivity analysis studies the effect of small to moderate changes to the prior distribution on the main findings of the analysis. For the binomial example above with $Y = 40$ and $n = 100$, say the objective is to determine if $\theta < 0.5$. Under the Beta(a, b) prior with $a = b = 1$, the posterior probability is Prob$(\theta < 0.5|Y) = 0.977$. Repeating this with (a, b) equal (5,1), (1,5) and (5,5) gives Prob$(\theta < 0.5|Y)$ equal to 0.941, 0.991 and 0.973, respectively. In all cases, the conclusion is that θ is less than 0.5 with high probability, and therefore the main findings are not sensitive to the prior.

1.4.3 Multivariate posteriors

Unlike the univariate case, a simple plot of the posterior will not suffice, especially for large p, because plotting in high dimensions is challenging. The typical remedy for this is to marginalize out other parameters and summarize univariate marginal distributions with plots, point estimates and credible sets, and perhaps plots of a few bivariate marginal distributions (i.e., integrating over the other $p - 2$ parameters) of interest.

Consider the model $Y_i|\mu, \sigma \overset{iid}{\sim}$ Normal(μ, σ^2) with independent priors $\mu \sim$ Normal$(0, 100^2)$ and $\sigma \sim$ Uniform$(0, 10)$ (other priors are discussed in Section 4.1.1). The likelihood

$$f(\mathbf{Y}|\mu, \sigma) \propto \prod_{i=1}^{n} \frac{1}{\sigma} \exp\left[-\frac{(Y_i - \mu)^2}{2\sigma^2}\right] \propto \sigma^{-n} \exp\left[-\frac{\sum_{i=1}^{n}(Y_i - \mu)^2}{2\sigma^2}\right] \tag{1.41}$$

factors as the product of n terms because the observations are assumed to be independent. The prior is $f(\mu, \sigma) = f(\mu)f(\sigma)$ because μ and σ have independent priors and since $f(\sigma) = 1/10$ for all $\sigma \in [0, 10]$, the prior becomes

$$\pi(\mu, \sigma) \propto \exp\left(-\frac{\mu^2}{2 \cdot 100^2}\right) \tag{1.42}$$

for $\sigma \in [0, 10]$ and $f(\mu, \sigma) = 0$ otherwise. The posterior is proportional to the likelihood times the prior, and thus

$$p(\mu, \sigma|\mathbf{Y}) \propto \sigma^{-n} \exp\left[-\frac{\sum_{i=1}^{n}(Y_i - \mu)^2}{2\sigma^2}\right] \exp\left(-\frac{\mu^2}{2 \cdot 100^2}\right) \tag{1.43}$$

for $\sigma \in [0, 10]$. Figure 1.11 plots this bivariate posterior assuming there are $n = 5$ observations: $Y_1 = 2.68$, $Y_2 = 1.18$, $Y_3 = -0.97$, $Y_4 = -0.98$, $Y_5 = -1.03$.

The two parameters in Figure 1.11 depend on each other. If $\sigma = 1.5$ (i.e., the conditional distribution traced by the horizontal line at $\sigma = 1.5$ in Figure 1.11) then the posterior of μ concentrates between -1 and 1, whereas if $\sigma = 3$, the posterior of μ spreads from -3 to

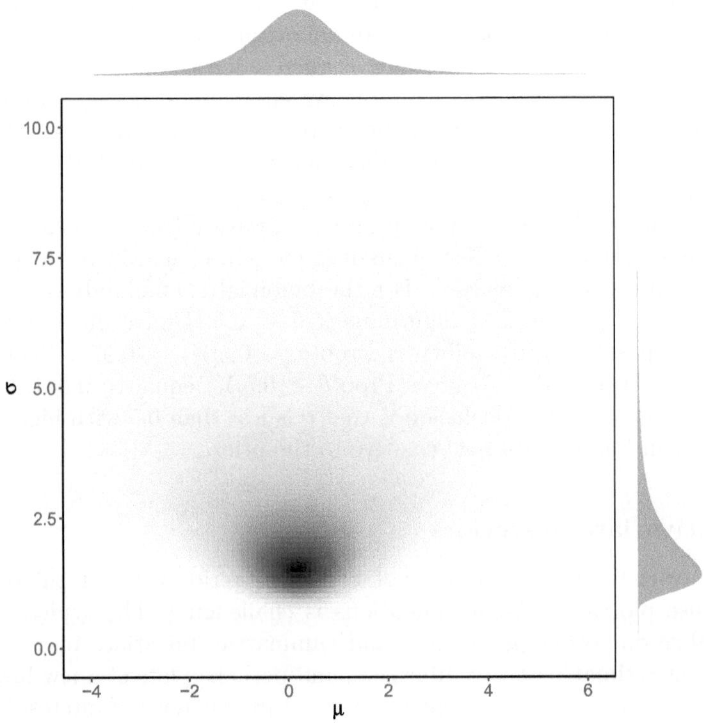

FIGURE 1.11
Bivariate posterior distribution. The bivariate posterior (center) and univariate marginal posteriors (top for μ and right for σ) for the model with $Y_i \stackrel{iid}{\sim} \text{Normal}(\mu, \sigma^2)$, priors $\mu \sim \text{Normal}(0, 100^2)$ and $\sigma \sim \text{Uniform}(0, 10)$, and $n = 5$ observations $Y_1 = 2.68$, $Y_2 = 1.18$, $Y_3 = -0.97$, $Y_4 = -0.98$, $Y_5 = -1.03$.

3. It is difficult to describe this complex bivariate relationship, so we often summarize the univariate marginal distributions instead. The marginal distributions

$$p(\mu|\mathbf{Y}) = \int_0^{10} p(\mu, \sigma|\mathbf{Y})d\sigma \quad \text{and} \quad p(\sigma|\mathbf{Y}) = \int_{-\infty}^{\infty} p(\mu, \sigma|\mathbf{Y})d\mu. \tag{1.44}$$

are plotted on the top (for μ) and right (for σ) of Figure 1.11; they are the row and column sums of the joint posterior. By integrating over the other parameters, the marginal distribution of a parameter accounts for posterior uncertainty in the remaining parameters. The marginal distributions are usually summarized with point and interval estimates as in Table 1.5.

The marginal distributions and their summaries above were computed by evaluating the joint posterior (1.43) for values of (μ, σ) that form a grid (i.e., pixels in Figure 1.11) and then simply summing over columns or rows of the grid. This is a reasonable approximation for $p = 2$ variables but quickly becomes unfeasible as p increases. Thus, it was only with the advent of more efficient computing algorithms in the 1990s that Bayesian statistics became

TABLE 1.5
Bivariate posterior distribution. Summaries of the marginal posterior distributions for the model with $Y_i \overset{iid}{\sim} \text{Normal}(\mu, \sigma^2)$, priors $\mu \sim \text{Normal}(0, 100^2)$ and $\sigma \sim \text{Uniform}(0, 10)$, and $n = 5$ observations $Y_1 = 2.68$, $Y_2 = 1.18$, $Y_3 = -0.97$, $Y_4 = -0.98$, $Y_5 = -1.03$.

Parameter	Posterior mean	Posterior SD	95% credible set
μ	0.17	1.31	(−2.49, 2.83)
σ	2.57	1.37	(1.10, 6.54)

feasible for even medium-sized applications. These exciting computational developments are the subject of Chapter 3.

1.5 The posterior predictive distribution

Often the objective of a statistical analysis is to build a stochastic model that can be used to make predictions of future events or impute missing values. Let Y^* be the future observation we would like to predict. Assuming that the observations are independent given the parameters and that Y^* follows the same model as the observed data, then given $\boldsymbol{\theta}$ we have $Y^* \sim f(y|\boldsymbol{\theta})$, and prediction is straightforward. Unfortunately, we do not know $\boldsymbol{\theta}$ exactly, even after observing $\mathbf{Y}$. A remedy for this is to plug in a value of $\boldsymbol{\theta}$, say, the posterior mean $\hat{\boldsymbol{\theta}} = \text{E}(\boldsymbol{\theta}|\mathbf{Y})$, and then sample $Y^* \sim f(Y|\hat{\boldsymbol{\theta}})$. However, this ignores uncertainty about the unknown parameters. If the posterior variance of $\boldsymbol{\theta}$ is small, then its uncertainty is negligible; otherwise a better approach is needed.

For the sake of prediction, the parameters are not of interest themselves, but rather they serve as vehicles to transfer information from the data to the predictive model. We would rather bypass the parameters altogether and simply use the Posterior Predictive Distribution (PPD)

$$Y^*|\mathbf{Y} \sim f^*(Y^*|\mathbf{Y}). \tag{1.45}$$

The PPD is the distribution of a new outcome given the observed data.

In a parametric model, the PPD naturally accounts for uncertainty in the model parameters; this is an advantage of the Bayesian framework. The PPD accounts for parametric uncertainty because it can be written

$$f^*(Y^*|\mathbf{Y}) = \int f^*(Y^*, \boldsymbol{\theta}|\mathbf{Y})d\boldsymbol{\theta} = \int f(Y^*|\boldsymbol{\theta})p(\boldsymbol{\theta}|\mathbf{Y})d\boldsymbol{\theta}, \tag{1.46}$$

where f is the likelihood density (here we assume that the observations are independent given the parameters, and so $f(Y^*|\boldsymbol{\theta}) = f(Y^*|\boldsymbol{\theta}, \mathbf{Y})$, and p is the posterior density.

To further illustrate how the PPD accounts for parameteric uncertainty, we consider how to make a sample from the PPD. If we first draw posterior sample $\boldsymbol{\theta}^* \sim p(\boldsymbol{\theta}|\mathbf{Y})$ and then a prediction from the likelihood, $Y^*|\boldsymbol{\theta}^* \sim f(Y|\boldsymbol{\theta}^*)$, then Y^* follows the PPD. A Monte Carlo approximation (Section 3.2.1) repeats these steps many times to approximate the PPD. Unlike the plug-in predictor, each predictive uses a different value of the parameters and thus accurately reflects parametric uncertainty. It can be shown that $\text{Var}(Y^*|\mathbf{Y}) \geq \text{Var}(Y^*|\mathbf{Y}, \boldsymbol{\theta})$ with equality holding only if there is no posterior uncertainty in the mean of $Y^*|\mathbf{Y}, \boldsymbol{\theta}$.

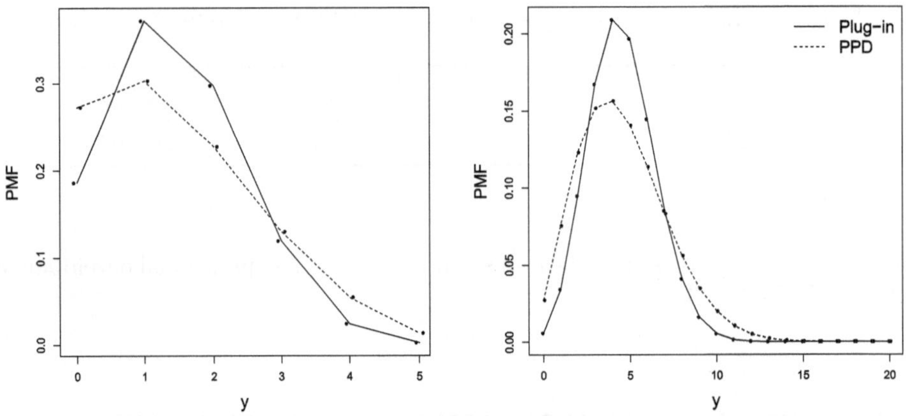

FIGURE 1.12
Posterior predictive distribution for a beta-binomial example. Plots of the posterior predictive distribution (PPD) from the model $Y|\theta \sim \text{Binomial}(n, \theta)$ and $\theta \sim \text{Beta}(1, 1)$. The "plug-in" PMF is the binomial density evaluated at the posterior mean $\hat{\theta}$, $f(y|\hat{\theta})$. This is compared with the full PPD for $Y = 1$ success in $n = 5$ trials (left) and $Y = 4$ successes in $n = 20$ trials (right). The PMFs are connected by lines for visualization, but the probabilities are only defined for $y = \{0, 1, ..., n\}$.

As an example, consider the model $Y|\theta \sim \text{Binomial}(n, \theta)$ and $\theta \sim \text{Beta}(1, 1)$. Given the data we have observed (Y and n), we would like to predict the outcome if we repeat the experiment, $Y^* \in \{0, 1, ..., n\}$. The posterior of θ is $\theta|Y \sim \text{Beta}(Y + 1, n - Y + 1)$, and the posterior mean is $\hat{\theta} = (Y + 1)/(n + 2)$. The solid lines in Figure 1.12 show the plug-in prediction $Y^* \sim \text{Binomial}(n, \hat{\theta})$ versus the full PPD (Listing 1.2) that accounts for uncertainty in θ (which is a Beta-Binomial$(n, Y+1, n+1)$ distribution). For both $n = 5$ and $n = 20$, the PPD is considerably wider than the plug-in predictive distribution, as expected.

If the observation is $Y|\theta \sim \text{Binomial}(n, \theta)$ and the prior distribution is $\theta \sim \text{Beta}(a, b)$, then the posterior is $\theta|Y \sim \text{Beta}(A, B)$ for $A = Y + a$ and $B = n - Y + b$. Now we wish to predict a new observation from the model $Y^*|\theta \sim \text{Binomial}(n^*, \theta)$ with $\theta \sim \text{Beta}(A, B)$. Averaging over the posterior uncertainty for θ gives, by definition (see the beta-binomial distribution in Section 9.4), $Y^*|Y \sim \text{BetaBinomial}(n^*, A, B)$. The ratio of the variance of Y^* from the PPD and the plug-in distribution $Y^*|Y \sim \text{Binomial}(n^*, \hat{\theta})$ for $\hat{\theta} = A/(A + B)$ is

$$\frac{n^* + n + a + b}{1 + n + a + b},$$

which is large if both n (small dataset) and $a + b$ (uninformative prior) are small.

As a second example, say there are n observations $Y_1, ..., Y_n|\mu \stackrel{iid}{\sim} \text{Normal}(\mu, \sigma^2)$ with variance σ^2 that is known and fixed. The unknown mean parameter has prior distribution $\mu \sim \text{Normal}(0, \sigma^2/m)$ so that if m is taken to be small, then the prior variance is large and vice versa. It can be shown (Section 2.1) that the posterior distribution given the data $\mathbf{Y} = (Y_1, ..., Y_n)$ is

$$\mu|\mathbf{Y} \sim \text{Normal}\left(\hat{\mu}, \frac{\sigma^2}{n + m}\right)$$

Listing 1.2
Summarizing a posterior predictive distribution (PPD) in R.

```
> # Load the data
> n <- 5
> Y <- 1
> a <- 1
> b <- 1
> A <- Y + a
> B <- n - Y + b

> # Plug-in estimator
> theta_hat   <- A/(A+B)
> y           <- 0:5
> PPD         <- dbinom(y,n,theta_hat)
> names(PPD) <- y
> round(PPD,2)
   0    1    2    3    4    5
0.19 0.37 0.30 0.12 0.02 0.00

> # Draws from the PPD, Y_star[i]~Binomial(n,theta_star[i])
> S            <- 100000
> theta_star  <- rbeta(S,A,B)
> Y_star       <- rbinom(S,n,theta_star)
> PPD          <- table(Y_star)/S
> round(PPD,2)
   0    1    2    3    4    5
0.27 0.30 0.23 0.13 0.05 0.01
```

for sample mean $\bar{Y} = \sum_{i=1}^{n} Y_i/n$ and posterior mean estimator $\hat{\mu} = n\bar{Y}/(n+m)$. Therefore, the plug-in prediction would take and $Y^*|Y \sim \text{Normal}(\hat{\mu}, \sigma^2)$. In contrast, the PPD that accounts for both variability in the response given μ and uncertainty in μ can be shown to be

$$Y^*|\mathbf{Y} \sim \text{Normal}\left(\hat{\mu}, \sigma^2 + \frac{\sigma^2}{n+m}\right).$$

This can be derived by writing $Y^* = \mu + \varepsilon$ where $\varepsilon \sim \text{Normal}(0, \sigma^2)$, and applying the property of the normal distribution that the sum of two independent normal random variables is also normal with the mean and variance as the sum of the components' means and variances. Therefore, accounting for uncertainty in μ results in the addition of $\sigma^2/(n+m)$ to the PPD variance, which could be substantial if both the data are limited (small n) and the prior variance is large (small m).

1.6 Exercises

1. If X has support $X \in \mathcal{S} = [1, \infty]$, find the constant c (as a function of θ) that makes $f(x) = c\exp(-x/\theta)$ a valid PDF.
2. Argue that for continuous random variable X (e.g., normal), $\text{Prob}(X = x)$ must be zero for any single value of x.
3. Assume that $X \sim \text{Uniform}(a, b)$ so the support is $\mathcal{S} = [a, b]$, and the PDF is $f(x) = 1/(b - a)$ for any $x \in \mathcal{S}$.
   (a) Prove that this is a valid PDF.
   (b) Derive the mean and variance of X.
4. Select an appropriate family of distributions for the following quantities. Justify your choice.
   (a) Time between entering Starbucks and having your order.
   (b) Heart rate of a randomly selected patient.
   (c) Number of hurricanes that make landfall in a given year.
   (d) Number of the 100 patients in a clinical trial that experience an adverse event
   (e) The data plotted in each of these histograms

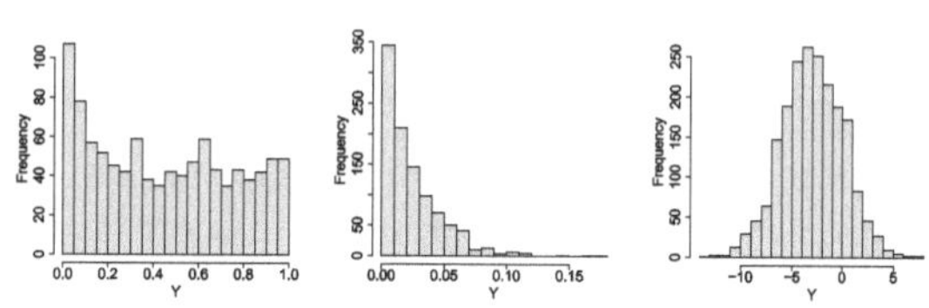

   (f) The data in this table

Failure	Success
799	202

5. Expert knowledge dictates that a parameter must be positive and that its prior distribution should have the mean 5 and variance 3. Find a prior distribution that satisfies these constraints.
6. X_1 and X_2 have joint PMF

x_1	x_2	$\text{Prob}(X_1 = x_1, X_2 = x_2)$
0	0	0.15
1	0	0.15
2	0	0.15
0	1	0.15
1	1	0.20
2	1	0.20

   (a) Compute the marginal distribution of X_1.
   (b) Compute the marginal distribution of X_2.
   (c) Compute the conditional distribution of $X_1|X_2$.
   (d) Compute the conditional distribution of $X_2|X_1$.

(e) Are X_1 and X_2 independent? Justify your answer.

7. If (X_1, X_2) is bivariate normal with $\text{E}(X_1) = \text{E}(X_2) = 0$, $\text{Var}(X_1) = \text{Var}(X_2) = 1$, and $\text{Cor}(X_1, X_2) = \rho$:

   (a) Derive the marginal distribution of X_1.

   (b) Derive the conditional distribution of $X_1|X_2$.

8. Assume (X_1, X_2) have support $X_1, X_2 \in \mathcal{R}$ and bivariate PDF

$$f(x_1, x_2) = \frac{1}{2\pi}\left(1 + x_1^2 + x_2^2\right)^{-3/2}.$$

   (a) Plot the conditional distribution of $X_1|X_2 = x_2$ for $x_2 \in \{-3, -2, -1, 0, 1, 2, 3\}$ (preferably on the same plot).

   (b) Do X_1 and X_2 appear to be correlated? Justify your answer.

   (c) Do X_1 and X_2 appear to be independent? Justify your answer.

9. According to `insurance.com`, the 2017 auto theft rate was 135 per 10,000 residents in Raleigh, NC compared to 214 per 10,000 residents in Durham/Chapel Hill, NC. Assuming Raleigh's population is twice as large as Durham/Chapel Hill and a car has been stolen somewhere in the triangle (i.e., one of these two areas), what the probability it was stolen in Raleigh?

10. Say 60% of the emails that arrive in your inbox are spam. Your account correctly flags spam emails as spam with probability 95% and incorrectly flags other emails as spam with probability 2%. Given that an email is flagged as spam, what is the probability it is actually a spam email?

11. Your daily commute is distributed uniformly between 15 and 20 minutes if there is no convention downtown. However, conventions are scheduled for roughly 1 in 4 days, and your commute time is distributed uniformly from 15 to 30 minutes if there is a convention. Let Y be your commute time this morning.

    (a) What is the probability that there was a convention downtown given $Y = 18$?

    (b) What is the probability that there was a convention downtown given $Y = 28$?

12. Consider the problem of short-term weather forecasting. Describe (i) a potential source of aleatoric uncertainty and (ii) a potential source of epistemic uncertainty.

13. For this problem pretend we are dealing with a language with a six-word dictionary

$$\{\text{fun, sun, sit, sat, fan, for}\}.$$

    An extensive study of literature written in this language reveals that all words are equally likely except that "for" is α times as likely as the other words. Further study reveals that:

    i. Each keystroke is an error with probability θ.

    ii. All letters are equally likely to produce errors.

    iii. Given that a letter is typed incorrectly it is equally likely to be any other letter.

    iv. Errors are independent across letters.

For example, the probability of correctly typing "fun" (or any other word) is $(1-\theta)^3$, the probability of typing "pun" or "fon" when intending to type is "fun" is $\theta(1-\theta)^2$, and the probability of typing "foo" or "nnn" when intending to type "fun" is $\theta^2(1-\theta)$. Use Bayes' rule to develop a simple spell checker for this language. For each of the typed words "sun", "the", "foo", give the probability that each word in the dictionary was the intended word. Perform this for the parameters below:

(a) $\alpha = 2$ and $\theta = 0.1$.
(b) $\alpha = 50$ and $\theta = 0.1$.
(c) $\alpha = 2$ and $\theta = 0.95$.

Comment on the changes you observe in these three cases.

14. If Y follows the Zero-Inflated Poisson (ZIP) distribution, it can be written in terms of two variables $Y = UV$, where $U \sim \text{Bernoulli}(p)$ independent of $V \sim \text{Poisson}(\lambda)$.

    (a) Generate and plot a sample of size $n = 10,000$ with $p = 0.2$ and $\lambda = 4$.
    (b) What is a real-life example of a variable that might follow this distribution? In your example, what are the interpretations of U and V?
    (c) What is the probability that $Y = 0$?
    (d) What is the probability that $Y = 4$?
    (e) What is the probability that $U = 0$ given that $Y = 0$?
    (f) What is the probability that $U = 0$ given that $Y = 4$?

15. Let $X_1 \sim \text{Bernoulli}(\theta)$ be the indicator that a tree species occupies a forest and $\theta \in [0,1]$ denote the prior occupancy probability. The researcher gathers a sample of n trees from the forest, and X_2 belong to the species of interest. The model for the data is $X_2|X_1 \sim \text{Binomial}(n, \lambda X_1)$ where $\lambda \in [0,1]$ is the probability of detecting the species given it is present. Give expressions in terms of n, θ and λ for the following joint, marginal and conditional probabilities:

    (a) $\text{Prob}(X_1 = X_2 = 0)$.
    (b) $\text{Prob}(X_1 = 0)$.
    (c) $\text{Prob}(X_2 = 0)$.
    (d) $\text{Prob}(X_1 = 0|X_2 = 0)$.
    (e) $\text{Prob}(X_2 = 0|X_1 = 0)$.
    (f) $\text{Prob}(X_1 = 0|X_2 = 1)$.
    (g) $\text{Prob}(X_2 = 0|X_1 = 1)$.
    (h) Provide intuition for how (d)-(g) change with n, θ and λ.
    (i) Assuming $\theta = 0.5$, $\lambda = 0.1$, and $X_2 = 0$, how large must n be before we can conclude with 95% confidence that the species does not occupy the forest?

16. In a study that uses Bayesian methods to forecast the number of species that will be discovered in future years, [47] report that the number of marine bivalve species discovered each year from 2010-2015 was 64, 13, 33, 18, 30 and 20. Denoting Y_t as the number of species discovered in year t and assuming $Y_t|\lambda \stackrel{iid}{\sim} \text{Poisson}(\lambda)$ and $\lambda \sim \text{Uniform}(0,100)$, plot the posterior distribution of λ.

17. Assume that (X,Y) follow the bivariate normal distribution and that both X and Y have marginal mean zero and marginal variance one. We observe six independent and identically distributed data points: $(-3.3,-2.6)$, $(0.1,-0.2)$, $(-1.1,-1.5)$, $(2.7,1.5)$, $(2.0,1.9)$ and $(-0.4,-0.3)$. Make a scatter plot of the data, and, assuming the correlation parameter ρ has a Uniform$(-1,1)$ prior, plot the posterior distribution of ρ.

18. The normalized difference vegetation index (NDVI) is commonly used to classify land cover using remote sensing data. Hypothetically, say that NDVI follows a Beta(25, 10) distribution for pixels in a rain forest, and a Beta(10, 15) distribution for pixels in a deforested area now used for agriculture. Assuming about 10% of the rain forest has been deforested, your objective is to build a rule to classify individual pixels as deforested based on their NDVI.

    (a) Plot the PDF of NDVI for forested and deforested pixels, and the marginal distribution of NDVI averaging over categories.

    (b) Give an expression for the probability that a pixel is deforested given its NDVI value, and plot this probability by NDVI.

    (c) You will classify a pixel as deforested if you are at least 90% sure it is deforested. Following this rule, give the range of NDVI that will lead to a pixel being classified as deforested.

19. Let n be the unknown number of customers who visit a store on the day of a sale. The number of customers who make a purchase is $Y|n \sim \text{Binomial}(n, \theta)$ where θ is the known probability of making a purchase given the customer visited the store. The prior is $n \sim \text{Poisson}(5)$. Assuming θ is known and n is the unknown parameter, plot the posterior distribution of n for all combinations of $Y \in \{0, 5, 10\}$ and $\theta \in \{0.2, 0.5\}$, and comment on the effect of Y and θ on the posterior.

20. Last spring your lab planted ten seedlings, and two survived the winter. Let θ be the probability that a seedling survives the winter.

    (a) Assuming a uniform prior distribution for θ, compute its posterior mean and standard deviation.

    (b) Assuming the same prior as in (a), compute and compare the equal-tailed and highest density 95% posterior credible intervals.

    (c) If you plant another 10 seedlings next year, what is the posterior predictive probability that at least one will survive the winter?

21. X_1 and X_2 are binary indicators of failure for two parts of a machine. Independent tests have shown that $X_1 \sim \text{Bernoulli}(1/2)$ and $X_2 \sim \text{Bernoulli}(1/3)$. Y_1 and Y_2 are binary indicators of two system failures. We know that $Y_1 = 1$ if both $X_1 = 1$ and $X_2 = 1$ and $Y_1 = 0$ otherwise, and $Y_2 = 0$ if both $X_1 = 0$ and $X_2 = 0$ and $Y_2 = 1$ otherwise. Compute the following probabilities:

    (a) The probability that $X_1 = 1$ and $X_2 = 1$ given $Y_1 = 1$.

    (b) The probability that $X_1 = 1$ and $X_2 = 1$ given $Y_2 = 1$.

    (c) The probability that $X_1 = 1$ given $Y_1 = 1$.

    (d) The probability that $X_1 = 1$ given $Y_2 = 1$.

22. The table below has the overall free throw proportion and results of free throws taken in pressure situations, defined as "clutch" (`https://stats.nba.com/`), for ten National Basketball Association players (those who received the most votes for the Most Valuable Player Award) for the 2016–2017 season. Since the overall proportion is computed using a large sample size, assume it is fixed, and analyze the clutch data for each player separately using Bayesian methods. Assume a uniform prior throughout this problem.

Player	Overall proportion	Clutch makes	Clutch attempts
Russell Westbrook	0.845	64	75
James Harden	0.847	72	95
Kawhi Leonard	0.880	55	63
LeBron James	0.674	27	39
Isaiah Thomas	0.909	75	83
Stephen Curry	0.898	24	26
Giannis Antetokounmpo	0.770	28	41
John Wall	0.801	66	82
Anthony Davis	0.802	40	54
Kevin Durant	0.875	13	16

(a) Describe your model for studying the clutch success probability including the likelihood and prior.

(b) Plot the posteriors of the clutch success probabilities.

(c) Summarize the posteriors in a table.

(d) Do you find evidence that any of the players have a different clutch percentage than overall percentage?

(e) Are the results sensitive to your prior? That is, do small changes in the prior lead to substantial changes in the posterior?

23. In the early twentieth century, it was generally agreed that Hamilton and Madison (ignore Jay for now) wrote 51 and 14 Federalist Papers, respectively. There was dispute over how to attribute 12 other papers between these two authors. In the 51 papers attributed to Hamilton, the word "upon" was used 3.24 times per 1,000 words, compared to 0.23 times per 1,000 words in the 14 papers attributed to Madison (for historical perspective on this problem, see [110]).

    (a) If the word "upon" is used three times in a disputed text of length $1,000$ words and we assume the prior probability 0.5, what is the posterior probability the paper was written by Hamilton?

    (b) Give one assumption you are making in (a) that is likely unreasonable. Justify your answer.

    (c) In (a), if we changed the number of instances of "upon" to one, do you expect the posterior probability to increase, decrease or stay the same? Why?

    (d) In (a), if we changed the text length to $10,000$ words and number of instances of "upon" to 30, do you expect the posterior probability to increase, decrease or stay the same? Why?

    (e) Let Y be the number of observed number of instances of "upon" in 1,000 words. Compute the posterior probability the paper was written by Hamilton for each $Y \in \{0, 1, ..., 20\}$, plot these posterior probabilities versus Y and give a rule for the number of instances of "upon" needed before the paper should be attributed to Hamilton.

24. Say $n = 1,000$ high school students are randomly selected to enter a tutorial program. It is known that 70% of the population from which they are drawn graduate from high school. After the program, it is found that $Y = 725$ of the n students graduate high school. Defining the true graduation rate for students in the program as $\theta \in (0, 1)$, we wish to test the null and alternative hypotheses

$$\mathcal{H}_o : \theta \leq 0.7 \text{ versus } \mathcal{H}_a : \theta > 0.7.$$

(a) Conduct a classical/frequentist test, and explain the results as if you are presenting them to a non-statistician.
(b) List a few subjective decisions in the analysis in (a).
(c) Conduct a Bayesian test, and explain the results as if you are presenting them to a non-statistician.
(d) Define a p-value and posterior probability of the null hypothesis, and describe how they are different.

25. The initial Moderna COVID-19 vaccine trial [9] produced the data in the table below

Placebo		Vaccine	
Infected	Participants	Infected	Participants
185	14073	11	14134

Let θ_0 be the probability of getting infected under placebo and θ_1 be the probability under vaccine.

(a) Can we say $\theta_0 = 185/14073 = 0.01315$ and $\theta_1 = 11/14134 = 0.00078$? Why or why not?
(b) What prior distributions would you select for θ_0 and θ_1?
(c) Conduct a separate Bayesian analysis of the data from each treatment group. Give the likelihood, priors and posterior, and summarize the results in a table and a figure. Does it appear that the vaccine is effective?
(d) What are some key assumptions you are making, and how might you justify them?
(e) Conduct a prior sensitivity analysis by comparing the results for 2-3 different prior distributions. Are the results sensitive to the prior?

2

From prior information to posterior inference

CONTENTS

One of the most controversial and yet crucial aspects of Bayesian analysis is the construction of a prior distribution. Often a user is faced with the questions like, "*Where does the prior distribution come from?*" or "*What is the true or correct prior distribution?*" and so on. There is no concept of the "true, correct or best" prior distribution, but rather the prior distribution can be viewed as an initialization of a statistical (in this case a Bayesian) inferential procedure that gets updated as the data accrue. The choice of a prior distribution is necessary (as you would need to initiate the inferential machine) but there is no notion of the 'optimal' prior distribution. Choosing a prior distribution is similar in principle to initializing any other sequential procedure (e.g., iterative optimization methods like Newton–Raphson, EM, etc.). The choice of such initialization can be good or bad in the sense of the rate of convergence of the procedure to its final value, but as long as the procedure is guaranteed to converge, the choice of prior does not have a permanent impact. As discussed in Section 9.2, the posterior is guaranteed to converge to the true value under very general conditions on the prior distribution.

In this chapter we discuss several general approaches for selecting prior distributions. We begin with conjugate priors. Conjugate priors lead to simple expressions for the posterior distribution and thus illustrate how prior information affects the Bayesian analysis. Conjugate priors are useful when prior information is available, but can also be used when it is not. We conclude with default (objective) Bayesian priors that attempt to remove the subjectivity in prior selection through conventions to be adopted when prior information is not available.

2.1 Conjugate priors

Conjugate priors are the most convenient choice. A prior and likelihood pair are conjugate if the resulting posterior is a member of the same family of distributions as the prior. Therefore, conjugacy refers to a mathematical relationship between two distributions and not to a deeper theme of the appropriate way to express prior beliefs. For example, if we select a beta prior, then Figure 1.5 shows that by changing the hyperparameters (i.e., the parameters that define the prior distribution, in this case a and b), the prior could be concentrated around a single value or spread equally across the unit interval.

Because both the prior and posterior are members of the same family, the update from the prior to the posterior affects only the parameters that index the family. This provides an opportunity to build intuition about Bayesian learning through simple examples. Conjugate priors are not unique. For example, the beta prior is conjugate for both the binomial and negative binomial likelihood and both a gamma prior and a Bernoulli prior (trivially) are conjugate for a Poisson likelihood. Also, conjugate priors are somewhat limited because not all likelihood functions have a known conjugate prior, and most conjugacy pairs are for small examples with only a few parameters. These limitations are abated through hierarchical modeling (Chapter 7) and Gibbs sampling (Chapter 3), which provide a framework to build rich statistical models by layering simple conjugacy pairs.

Below we discuss several conjugate priors and mathematical tools needed to derive the corresponding posteriors. Detailed derivation of many of the posterior distributions is deferred to Appendix A.3, and Appendix A.2 has an abridged list of conjugacy results.

2.1.1 Beta-binomial model for a proportion

Returning to the beta-binomial example in Section 1.2, the data $Y \in \{0, 1, ..., n\}$ is the number of successes in n independent trials, each with success probability θ. The likelihood is then $Y|\theta \sim \text{Binomial}(n, \theta)$. Since we are only interested in terms in the likelihood that involve θ, we focus only on its kernel, i.e., the terms that involve θ. The kernel of the binomial PMF is

$$f(Y|\theta) \propto \theta^Y (1-\theta)^{n-Y}. \tag{2.1}$$

If we view the likelihood as a function of θ, it resembles a beta distribution, and so we might suspect that a beta distribution is the conjugate prior for the binomial likelihood. If we select the prior $\theta \sim \text{Beta}(a, b)$, then (as shown in Section 1.2) this combination of likelihood and prior leads to the posterior distribution

$$\theta|Y \sim \text{Beta}(A, B), \tag{2.2}$$

where the updated parameters are $A = Y + a$ and $B = n - Y + b$. Since both the prior and posterior belong to the beta family of distributions, this is an example of a conjugate prior.

The $\text{Beta}(A, B)$ distribution has mean $A/(A + B)$, and so the prior and posterior means are

$$\text{E}(\theta) = \hat{\theta}_0 = \frac{a}{a+b} \quad \text{and} \quad \text{E}(\theta|Y) = \hat{\theta}_1 = \frac{Y+a}{n+a+b}. \tag{2.3}$$

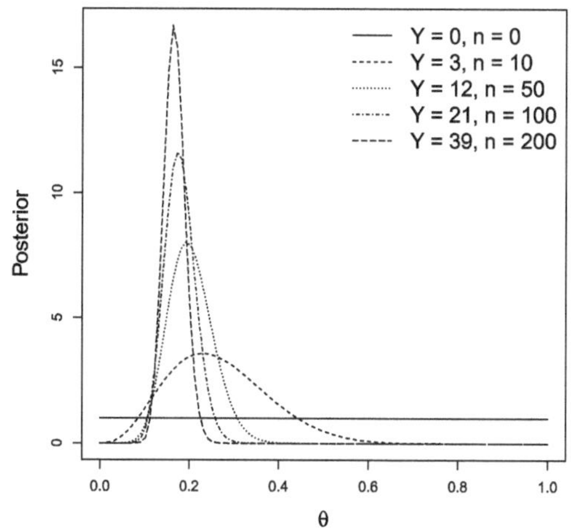

FIGURE 2.1
Posterior distributions from the beta-binomial model. Plot of the posterior of θ from the model $Y|\theta \sim \text{Binomial}(n, \theta)$ and $\theta \sim \text{Beta}(1, 1)$ for various n and Y.

The prior and posterior means are both estimators of the population proportion, and thus we denote them as $\hat{\theta}_0$ and $\hat{\theta}_1$, respectively. The prior mean $\hat{\theta}_0$ is an estimator of θ before observing the data, and this is updated to $\hat{\theta}_1$ by the observed data. A natural estimator of the population proportion θ is the sample proportion $\hat{\theta} = Y/n$, which is the number of successes divided by the number of trials. Comparing the sample proportion to the posterior mean, the posterior mean adds a to the number of successes in the numerator and $a + b$ to the number of trials in the denominator. Therefore, we can think of a as the prior number of successes, $a + b$ as the prior number of trials, and thus b as the prior number of failures (see Section 2.1.5 for a tie to natural conjugate priors).

Viewing the hyperparameters as the prior number of successes and failures provides a means of balancing the information in the prior with the information in the data. For example, the $\text{Beta}(0.5, 0.5)$ prior in Figure 1.5 has one prior observation, and the uniform $\text{Beta}(1, 1)$ prior contributes one prior success and one prior failure. If the prior is meant to reflect prior ignorance, then we should select small a and b, and if there is strong prior information that θ is approximately θ_0, then we should select a and b so that $\theta_0 = a/(a+b)$ and $a + b$ is large.

The posterior mean can also be written as

$$\hat{\theta}_1 = (1 - w_n)\hat{\theta}_0 + w_n\hat{\theta} \tag{2.4}$$

where $w_n = n/(n + a + b)$ is the weight given to the sample proportion and $1 - w_n$ is the weight given to the prior mean. This confirms the intuition that for any prior (i.e., for any a and b), if the sample size n is small, the posterior mean is approximately the prior mean, and as the sample size increases, the posterior mean becomes closer to the sample proportion. Also, as $a + b \to 0$, $w_n \to 1$, so as we make the prior vague with large variance, the posterior mean coincides with the sample proportion for any sample size n.

Figure 2.1 plots the posterior for various n and Y. This plot is meant to illustrate a sequential analysis. In most cases the entire data set is analyzed in a single analysis, but in some cases data are analyzed as they arrive, and a Bayesian analysis provides a framework to analyze data sequentially. In Figure 2.1, before any data are collected, θ has a uniform prior. After 10 observations, there are three successes, and the posterior concentrates below 0.5. After an additional 40 samples, the posterior centers on the sample proportion 12/50. As data accrues, the posterior converges to $\theta \approx 0.2$, and the posterior variance decreases.

Generalization to multiple categories: In the beta/binomial model, the outcome of each experiment has two possibilities, 0 (failure) and 1 (success). This can be generalized to an p outcomes, say $p = 3$ for win, lose or draw. The outcomes are gathered across the n independent trials as $\mathbf{Y} = (Y_1, ..., Y_p)$ where Y_j is the number of the n trials that resulted in outcome j. The distribution of $\mathbf{Y}$ is determined by p probabilities $\boldsymbol{\theta} = (\theta_1, ..., \theta_p)$ with θ_j equal to the probability that a given trial has outcome j. The conjugate prior for the probabilities $\boldsymbol{\theta}$ is the Dirichlet$(a_1, ..., a_p)$ prior with PDF

$$\pi(\theta_1, ..., \theta_p) \propto \prod_{j=1}^{p} \theta_j^{a_j - 1}.$$

The Dirichlet distribution generalizes the beta distribution in the sense that if $p = 2$, then $\theta_1 \sim \text{Beta}(a_1, a_2)$ and $\theta_2 \sim \text{Beta}(a_2, a_1)$. It can be shown that the posterior is

$$\boldsymbol{\theta}|\mathbf{Y} \sim \text{Dirichlet}(Y_1 + a_1, ..., Y_p + a_p).$$

Note that if $p = 2$ and success and failure are coded as 1 and 2, then this is identical to the beta/binomial result because $\theta_1|\mathbf{Y} \sim \text{Beta}(Y_1 + a_1, Y_2 + a_2)$ and $Y_2 = n - Y_1$.

2.1.2 Poisson-gamma model for a rate

The Poisson-gamma conjugate pair is useful for estimating the rate of events per unit of observation effort, denoted θ. For example, an ecologist may survey N acres of a forest and observe $Y \in \{0, 1, ...\}$ individuals of the species of interest, or a company may observe Y employee injuries in N person-hours on the job. For these data, we might assume the model $Y|\theta \sim \text{Poisson}(N\theta)$ where N is the known sampling effort and $\theta > 0$ is the unknown event rate. The likelihood is

$$f(Y|\theta) = \frac{\exp(-N\theta)(N\theta)^Y}{Y!} \propto \exp(-N\theta)\theta^Y. \tag{2.5}$$

The kernel of the likelihood resembles a Gamma(a, b) distribution

$$\pi(\theta) = \frac{b^a}{\Gamma(a)} \exp(-b\theta)\theta^{a-1} \propto \exp(-b\theta)\theta^{a-1} \tag{2.6}$$

in the sense that θ appears in the PDF raised to a power and in the exponent. Combining the likelihood and prior gives posterior

$$p(\theta|Y) \propto [\exp(-N\theta)\theta^Y] \cdot [\exp(-b\theta)\theta^{a-1}] \propto \exp(-B\theta)\theta^{A-1}, \tag{2.7}$$

where $A = Y + a$ and $B = N + b$. The posterior of θ is thus $\theta|Y \sim \text{Gamma}(A, B)$, and the gamma prior is conjugate for the Poisson likelihood.

A simple estimate of the expected number of events per unit of effort is the sample event rate $\hat{\theta} = Y/N$. The mean and variance of a Gamma(A, B) distribution are A/B and A/B^2, respectively, and so the posterior mean under the Poisson-gamma model is

$$\text{E}(\theta|Y) = \frac{Y + a}{N + b}. \tag{2.8}$$

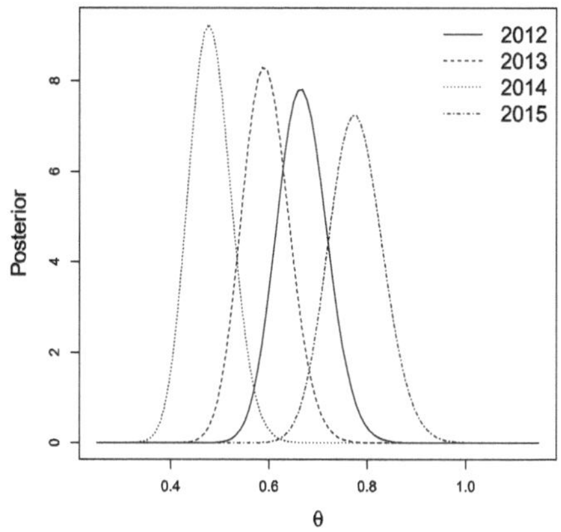

FIGURE 2.2
Posterior distributions for the NFL concussion example. Plot of the posterior of θ from the model $Y|\theta \sim \text{Poisson}(N\theta)$ and $\theta \sim \text{Gamma}(a, b)$, where $N = 256$ is the number of games played in an NFL season, Y is the number of concussions in a year, and $a = b = 0.1$ are the hyperparameters. The plot gives the posterior for years 2012–2015, which had 171, 152, 123, and 199 concussions, respectively.

Therefore, compared to the sample event rate we add a events to the numerator and b units of effort to the denominator. As in the beta-binomial example, the posterior mean can be written as a weighted average of the sample rate and the prior mean,

$$E(\theta|Y) = (1 - w_N)\frac{a}{b} + w_N\frac{Y}{N}, \tag{2.9}$$

where $w_N = N/(N + b)$ is the weight given to the sample rate Y/N and $1 - w_N$ is the weight given to the prior mean a/b. As $b \to 0$, $w_n \to 1$, and so for a vague prior with large variance the posterior mean coincides with the sample rate. A common setting for the hyperparameters is $a = b = \epsilon$ for some small value ϵ, which gives prior mean 1 and large prior variance $1/\epsilon$.

NFL concussions example: Concussions are an increasingly serious concern in the National Football League (NFL). The NFL has 32 teams and each team plays 16 regular-season games per year, for a total of $N = 32 \cdot 16/2 = 256$ games. According to Frontline/PBS (`http://apps.frontline.org/concussion-watch`) there were $Y_1 = 171$ concussions in 2012, $Y_2 = 152$ concussions in 2013, $Y_3 = 123$ concussions in 2014, and $Y_4 = 199$ concussions in 2015. Figure 2.2 plots the posterior of the concussion rate for each year assuming $a = b = 0.1$. Comparing only 2014 with 2015 there does appear to be some evidence of an increase in the concussion rate per game; for 2014 the posterior mean and 95% credible interval are 0.48 (A/B) and (0.40, 0.57), respectively, (computed using the Gamma(A, B) quantile function), compared to 0.78 and (0.67, 0.89) for 2015. However, 2015 does not appear to be significantly different than 2012 or 2013.

2.1.3 Normal-normal model for a mean

Gaussian responses play a key role in applied and theoretical statistics. The t-test, ANalysis Of VAriance (ANOVA), and linear regression all assume Gaussian responses. We develop these methods in Chapter 4, but here we discuss conjugate priors for the simpler model $Y_i|\mu,\sigma^2 \stackrel{iid}{\sim} \text{Normal}(\mu,\sigma^2)$ for $i = 1,...,n$. This model has two parameters, the mean μ and the variance σ^2. In this subsection we assume that σ^2 is fixed and focus on estimating μ; in the following subsection we analyze σ^2 given μ, and Chapter 4 derives the joint posterior of both parameters.

Assuming σ^2 to be fixed, a conjugate prior for the unknown mean μ is $\mu \sim \text{Normal}(\mu_0, \sigma^2/m)$, where the prior variance is proportional to σ^2 to express prior uncertainty on the scale of the data. The hyperparameter $m > 0$ controls the strength of the prior, with small m giving large prior variance and vice versa. Appendix A.3 shows that the posterior is

$$\mu|\mathbf{Y} \sim \text{N}\left(w\bar{Y} + (1-w)\mu_0, \frac{\sigma^2}{n+m}\right) \tag{2.10}$$

where $w = n/(n+m) \in [0,1]$ is the weight given to the sample mean $\bar{Y} = \sum_{i=1}^n Y_i/n$. Letting $m \to 0$ gives $w_n \to 1$ so the posterior mean coincides with the sample mean as the prior variance increases.

The posterior standard deviation is $\sigma/\sqrt{n+m}$, which is less than the standard error of the sample mean, $\sigma/\sqrt{n}$. The prior with hyperparameter m reduces the posterior standard deviation by the same amount as adding an additional m observations, and this stabilizes a Bayesian analysis. For Gaussian data analyses such as ANOVA or regression the prior mean μ_0 is often set to zero. In the normal-normal model with $\mu_0 = 0$ the posterior mean estimator is $\text{E}(\mu|\mathbf{Y}) = w\bar{Y}$. More generally, $\text{E}(\mu|\mathbf{Y}) = w(\bar{Y} - \mu_0) + \mu_0$. This is an example of a *shrinkage* estimator because the sample mean is shrunk towards the prior mean by the shrinkage factor $w \in [0,1]$. As will be discussed in Chapter 4, shrinkage estimators have advantages, particularly in hard problems such as regression with many predictors and/or small sample size.

Blood alcohol concentration (BAC) example: The BAC level is the percent of your blood that is concentrated with alcohol. The legal limit for operating a vehicle is BAC ≤ 0.08 in most US states. Let Y be the measured BAC and μ be your true BAC. Of course, the BAC test has error, and the error standard deviation for a sample near the legal limit has been established in (hypothetical) laboratory tests to be $\sigma = 0.01$, so that the likelihood of the data is $Y|\mu \sim \text{Normal}(\mu, 0.01^2)$. Your BAC is measured to be $Y = 0.082$, which is above the legal limit.

Your defense is that you had two drinks, and that the BAC of someone your size after two drinks has been shown to follow a $\text{Normal}(0.05, 0.02^2)$ distribution depending on the person's metabolism and the timing of the drinks. Figure 2.3 plots the prior $\mu \sim \text{Normal}(0.05, 0.02^2)$, i.e., with $m = 0.25$. The prior probability that your BAC exceeds 0.08 is 0.067. The posterior distribution with $n = 1$, $Y = 0.082$, $\sigma = 0.01$ and $m = 0.25$ is

$$\mu|Y \sim \text{Normal}(0.0756, 0.0089^2). \tag{2.11}$$

Therefore, $\text{Prob}(\mu > 0.08|Y) = 0.311$, and so there is considerable uncertainty about whether your BAC exceeds the legal limit, and in fact the posterior odds that your BAC is below the legal limit,

$$\frac{\text{Prob}(\mu \leq 0.08|Y)}{\text{Prob}(\mu > 0.08|Y)},$$

are greater than two.

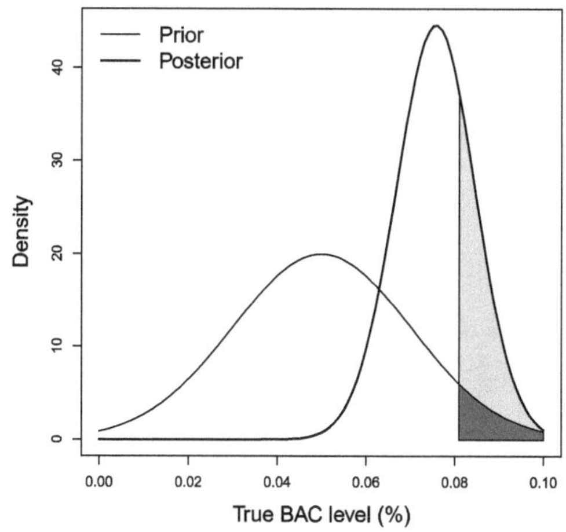

FIGURE 2.3
Posterior distributions for the blood alcohol content example. Plot of the prior and posterior PDF, with the prior (0.067) and posterior (0.311) probabilities of exceeding the legal limit of 0.08 shaded.

2.1.4 Normal-inverse gamma model for a variance

Next we turn to estimating a Gaussian variance assuming the mean is fixed. As before, the sampling density is $Y_i|\sigma^2 \overset{iid}{\sim} \text{Normal}(\mu, \sigma^2)$ for $i = 1, ..., n$. With the mean fixed, the likelihood is

$$f(\mathbf{Y}|\sigma^2) \propto \prod_{i=1}^{n} \frac{1}{\sigma} \exp\left[-\frac{(Y_i - \mu)^2}{2\sigma^2}\right] \propto (\sigma^2)^{-n/2} \exp\left(-\frac{SSE}{2\sigma^2}\right) \tag{2.12}$$

where $SSE = \sum_{i=1}^{n}(Y_i - \mu)^2$. The likelihood has σ^2 raised to a negative power and σ^2 in the denominator of the exponent. Of the distributions in Appendix A.1 with support $[0, \infty)$, only the inverse gamma PDF has these properties. Taking the prior $\sigma^2 \sim \text{InvGamma}(a, b)$ gives

$$\pi(\sigma^2) \propto (\sigma^2)^{-(a+1)} \exp\left(-\frac{b}{\sigma^2}\right). \tag{2.13}$$

Combining the likelihood and prior gives the posterior

$$p(\sigma^2|\mathbf{Y}) \propto f(\mathbf{Y}|\sigma^2)\pi(\sigma^2) \propto (\sigma^2)^{-(A+1)} \exp\left(-\frac{B}{\sigma^2}\right), \tag{2.14}$$

where $A = n/2 + a$ and $B = SSE/2 + b$ are the updated parameters, and therefore

$$\sigma^2|\mathbf{Y} \sim \text{InvGamma}(A, B). \tag{2.15}$$

The posterior mean (if $n/2 + a > 1$) is

$$E(\sigma^2|\mathbf{Y}) = \frac{SSE/2 + b}{n/2 + a - 1} = \frac{SSE + 2b}{n - 1 + 2a - 1}. \tag{2.16}$$

Therefore, if we take the hyperparameters to be $a = 1/2 + m/2$ and $b = \epsilon/2$ for small m and ϵ, then the posterior-mean estimator is $\text{E}(\sigma^2|\mathbf{Y}) = (SSE + \epsilon)/(n - 1 + m)$, and compared to the usual sample variance $SSE/(n-1)$ the small values ϵ and m are added to the numerator and denominator, respectively. In this sense, the prior adds an additional m degrees of freedom for estimating the variance, which can stabilize the estimator if n is small.

Conjugate prior for a precision: We have introduced the Gaussian distribution as having two parameters, the mean and the variance. However, it can also be parameterized in terms of its mean and precision (inverse variance), $\tau = 1/\sigma^2$. In particular, the `JAGS` package used in this book employs this parameterization. In this parameterization, $Y|\mu, \tau \sim \text{Normal}(\mu, \tau)$ has PDF

$$f(Y|\mu, \tau) = \frac{\tau^{1/2}}{\sqrt{2\pi}} \exp\left[-\frac{\tau}{2}(Y - \mu)^2\right]. \tag{2.17}$$

This parameterization makes derivations and computations slightly easier, especially for the multivariate normal distribution where using a precision matrix (inverse covariance matrix) avoids some matrix inversions.

Not surprisingly, the conjugate prior for the precision is the gamma family. If $Y_i \stackrel{iid}{\sim} \text{Normal}(\mu, \tau)$ then the likelihood is

$$f(\mathbf{Y}|\tau) \propto \prod_{i=1}^{n} \tau^{1/2} \exp\left[-\frac{\tau}{2}(Y_i - \mu)^2\right] \propto \tau^{n/2} \exp\left(-\frac{\tau}{2}SSE\right), \tag{2.18}$$

and the Gamma(a, b) prior is $\pi(\tau) \propto \tau^{a-1}\exp(-\tau b)$. Combining the likelihood and prior gives

$$p(\tau|\mathbf{Y}) \propto f(\mathbf{Y}|\tau)\pi(\tau) \propto \tau^{A-1}\exp(-\tau B), \tag{2.19}$$

where $A = n/2 + a$ and $B = SSE/2 + b$ are the updated parameters. Therefore,

$$\tau|\mathbf{Y} \sim \text{Gamma}(A, B). \tag{2.20}$$

The InvGamma(a, b) prior for the variance and the Gamma(a, b) prior for the precision give the exact same posterior distribution. That is, if we use the InvGamma(a, b) prior for the variance and then convert this to obtain the posterior of $1/\sigma^2$, the results are identical as if we have conducted the analysis with a Gamma(a, b) prior for the precision. Throughout the book we use the mean-variance parameterization except for cases involving `JAGS` code, when we adopt their mean-precision parameterization.

2.1.5 Natural conjugate priors

We have thus far catalogued a series of popular sampling distributions and corresponding choices of the conjugate family of priors. Is there a natural way of constructing a class of conjugate priors given a specific sampling density $f(y|\theta)$? It turns out that for many of these familiar choices of sampling densities (e.g., exponential family) we can construct a class of conjugate priors, and priors constructed in this manner are called natural conjugate priors.

Let $Y_i \stackrel{iid}{\sim} f(y|\theta)$ for $i = 1, \ldots, n$, and consider a class of priors defined by

$$\pi(\theta|y_1^0, \ldots, y_m^0, m) \propto \prod_{j=1}^{m} f(y_j^0|\theta) \tag{2.21}$$

where y_j^0 are some arbitrary fixed values in the support of the sampling distribution for $j = 1, \ldots, m$ and $m \geq 1$ is a fixed integer. The pseudo-observations y_j^0 can be seen as the hyperparameters of the prior distribution, and such a prior is a proper distribution if there exists m such that $\int \prod_{j=1}^{m} f(y_j^0|\theta)\, d\theta < \infty$. To see that the prior defined in (2.21) is indeed conjugate notice that

$$p(\theta|\mathbf{Y}) \propto \pi(\theta|y_1^0, \ldots, y_m^0, m) \prod_{i=1}^{n} f(Y_i|\theta) \propto \prod_{j=1}^{M} f(y_j^*|\theta) \tag{2.22}$$

where $M = m + n$ and $y_j^* = y_j^0$ for $j = 1, \ldots, m$ and $y_j^* = Y_{j-m}$ for $j = m+1, \ldots, m+n$. Thus, the posterior distribution belongs to the same class of distributions as in (2.21).

Below we revisit some previous examples using this method of creating natural conjugate priors.

Bernoulli trials (Section 2.1.1): When $Y|\theta \sim \text{Bernoulli}(\theta)$, we have $f(y|\theta) \propto \theta^y(1-\theta)^{1-y}$ and so the natural conjugate prior with the first s_0 pseudo observations set to $y_j^0 = 1$ and the remaining $m - s_0$ pseudo-observations set to $y_j^0 = 0$ gives the prior $\pi(\theta|y_1^0, \ldots, y_m^0, m) \propto \theta^{s_0}(1-\theta)^{m-s_0}$, that is, $\theta \sim \text{Beta}(s_0 + 1, m - s_0 + 1)$. This beta prior is restricted to have integer-valued hyperparameters, but once we see the form, we can relax the assumption that m and s_0 are integer valued.

Poisson counts (Section 2.1.2): When $Y|\theta \sim \text{Poisson}(\theta)$, we have $f(y|\theta) \propto \theta^y e^{-\theta}$, and so the natural conjugate prior by using equation (2.21) is given by $\pi(\theta|y_1^0, \ldots, y_m^0, m) \propto \theta^{s_0} e^{-m\theta}$, where $s_0 = \sum_{j=1}^{m} y_j^0$, and this naturally leads to a gamma prior distribution $\theta \sim \text{Gamma}(s_0+1, m)$. Therefore, we can view the prior as consisting of m pseudo observations with sample rate s_0/m. Again, the restriction that m is an integer can be relaxed once it is revealed that the prior is from the gamma family.

Normal distribution with fixed variance (Section 2.1.3): Assuming $Y|\theta \sim \text{Normal}(\mu, 1)$, we have $f(y|\mu) \propto \exp\{-(y-\mu)^2/(2\sigma^2)\}$, and so the natural conjugate prior is $\pi(\mu|y_1^0, \ldots, y_m^0, m) \propto \exp\{-\sum_{j=1}^{m}(y_j^0 - \mu)^2/(2\sigma^2)\} \propto \exp\{-m(\mu - \bar{y}^0)^2/(2\sigma^2)\}$, where $\bar{y}^0 = \sum_{j=1}^{m} y_j^0/m$ and this naturally leads to the prior $\theta \sim \text{Normal}(\bar{y}^0, \sigma^2/m)$. Therefore, the prior can be viewed as consisting of m pseudo-observations with mean $\bar{y}^0$.

This systematic way of obtaining a conjugate prior takes away the mystery of first guessing the form of the conjugate prior and then verifying its conjugacy. The procedure works well when faced with problems that do not have a familiar likelihood and works even when we have vector-valued parameters.

2.1.6 Normal-normal model for a mean vector

In Bayesian linear regression the parameter of interest is the vector of regression coefficients. This is discussed extensively in Chapter 4, but here we provide the conjugacy relationship that underlies the regression analysis. Although still a bit cumbersome, linear regression notation is far more concise using matrices. Say the n-vector $\mathbf{Y}$ is multivariate normal

$$\mathbf{Y}|\boldsymbol{\beta} \sim \text{Normal}(\mathbf{X}\boldsymbol{\beta}, \boldsymbol{\Sigma}). \tag{2.23}$$

The mean of $\mathbf{Y}$ is decomposed as the known $n \times p$ matrix $\mathbf{X}$ and unknown p-vector $\boldsymbol{\beta}$, and $\boldsymbol{\Sigma}$ is the $n \times n$ covariance matrix. The prior for $\boldsymbol{\beta}$ is multivariate normal with mean $\boldsymbol{\mu}$ and $p \times p$ covariance matrix $\boldsymbol{\Omega}$,

$$\boldsymbol{\beta} \sim \text{Normal}(\boldsymbol{\mu}, \boldsymbol{\Omega}). \tag{2.24}$$

As shown in Appendix A.3, the posterior of $\boldsymbol{\beta}$ is multivariate normal

$$\boldsymbol{\beta}|\mathbf{Y} \sim \text{Normal}\left[\boldsymbol{\Sigma}_\beta(\mathbf{X}^T\boldsymbol{\Sigma}^{-1}\mathbf{Y} + \boldsymbol{\Omega}^{-1}\boldsymbol{\mu}), \boldsymbol{\Sigma}_\beta\right], \tag{2.25}$$

where $\mathbf{\Sigma}_\beta = (\mathbf{X}'\mathbf{\Sigma}^{-1}\mathbf{X} + \mathbf{\Omega}^{-1})^{-1}$. In standard linear regression the errors are assumed to be independent and identically distributed, and thus the covariance is proportional to the identity matrix, $\mathbf{\Sigma} = \sigma^2\mathbf{I}_n$. In this case, if the prior is uninformative with $\mathbf{\Omega}^{-1} \approx 0$, then the posterior mean reduces to the familiar least squares estimator $(\mathbf{X}^T\mathbf{X})^{-1}\mathbf{X}^T\mathbf{Y}$ and the posterior covariance reduces to the covariance of the sampling distribution of the least squares estimator, $\sigma^2(\mathbf{X}^T\mathbf{X})^{-1}$.

2.1.7 Normal-inverse Wishart model for a covariance matrix

Say $\mathbf{Y}_1, ..., \mathbf{Y}_n$ are vectors of length p that are independently distributed as multivariate normal with known mean vectors $\boldsymbol{\mu}_i$ and $p \times p$ unknown covariance matrix $\mathbf{\Sigma}$. The conjugate prior for the covariance matrix is the inverse Wishart prior. The inverse Wishart family's support is symmetric positive definite matrices (i.e., covariance matrices), and reduces to the inverse gamma family if $p = 1$. The inverse Wishart prior with degrees of freedom $\nu > p - 1$ and $p \times p$ symmetric and positive definite scale matrix $\mathbf{R}$ has PDF

$$\pi(\mathbf{\Sigma}) \propto |\mathbf{\Sigma}|^{-(\nu+p+1)/2} \exp\left[-\frac{1}{2}\text{Trace}(\mathbf{\Sigma}^{-1}\mathbf{R})\right] \tag{2.26}$$

and mean $\text{E}(\mathbf{\Sigma}) = \mathbf{R}/(\nu - p - 1)$ assuming $\nu > p + 1$. The prior concentration around $\mathbf{R}/(\nu - p - 1)$ increases with ν, and therefore small ν, say $\nu = p - 1 + \epsilon$ for small $\epsilon > 0$, gives the least informative prior. An interesting special case is when $\nu = p + 1$ and $\mathbf{R}$ is a diagonal matrix. This induces a uniform prior on each off-diagonal element of the correlation matrix corresponding to covariance matrix $\mathbf{\Sigma}$.

As shown in Appendix A.3, the posterior is

$$\mathbf{\Sigma}|\mathbf{Y} \sim \text{InvWishart}_p\left(n + \nu, \sum_{i=1}^{n}(\mathbf{Y}_i - \boldsymbol{\mu}_i)(\mathbf{Y}_i - \boldsymbol{\mu}_i)^T + \mathbf{R}\right). \tag{2.27}$$

The posterior mean is $[\sum_{i=1}^n(\mathbf{Y}_i - \boldsymbol{\mu}_i)(\mathbf{Y}_i - \boldsymbol{\mu}_i)^T + \mathbf{R}]/[n + \nu - p - 1]$. For $\mathbf{R} \approx 0$ and $\nu \approx p + 1$, the posterior mean is approximately $[\sum_{i=1}^n(\mathbf{Y}_i - \boldsymbol{\mu}_i)(\mathbf{Y}_i - \boldsymbol{\mu}_i)^T]/n$, which is the sample covariance matrix assuming the means are known and not replaced by the sample means.

Marathon example: Figure 2.4 plots the data for several of the top female runners in the 2016 Boston Marathon. Let Y_{ij} be the speed (minutes/mile) for runner $i = 1, ..., n$ and mile $j = 1, ..., p = 26$. For this analysis, we have discarded all runners with missing data (for a missing data analysis see Section 7.1.4), leaving $n = 59$ observations $\mathbf{Y}_i = (Y_{i1}, ..., Y_{ip})^T$. We analyze the covariance of the runners' data to uncover patterns and possibly strategy.

For simplicity we conduct the analysis conditioning on $\mu_{ij} = \bar{Y}_j = \sum_{i=1}^n Y_{ij}/n$, i.e., the sample mean for mile j. For the prior, we take $\nu = p+1$ and $\mathbf{R} = \mathbf{I}_p/\nu$ so that elements of the correlation matrix have Uniform$(-1, 1)$ priors. The code in Listing 2.1 generates S samples from the posterior of $\mathbf{\Sigma}$ and uses the samples to approximate the posterior mean. To avoid storage problems, rather than storing all S samples, the code simply retains the running mean of the samples. However, Monte Carlo sampling produces the full joint distribution of all elements of $\mathbf{\Sigma}$. In particular, for each draw from the posterior, Listing 2.1 converts the simulated covariance matrix to the corresponding correlation matrix, and computes the posterior mean of the correlation matrix.

The estimated (posterior mean) variance (the diagonal elements of Figure 2.4b) increases with the mile as the top runners separate themselves from the pack. The correlation between speeds at different miles (Figure 2.4c) is high for all pairs of miles in the first half (miles 1–13) of the race as the runners maintain a fairly steady pace. The correlation is low between

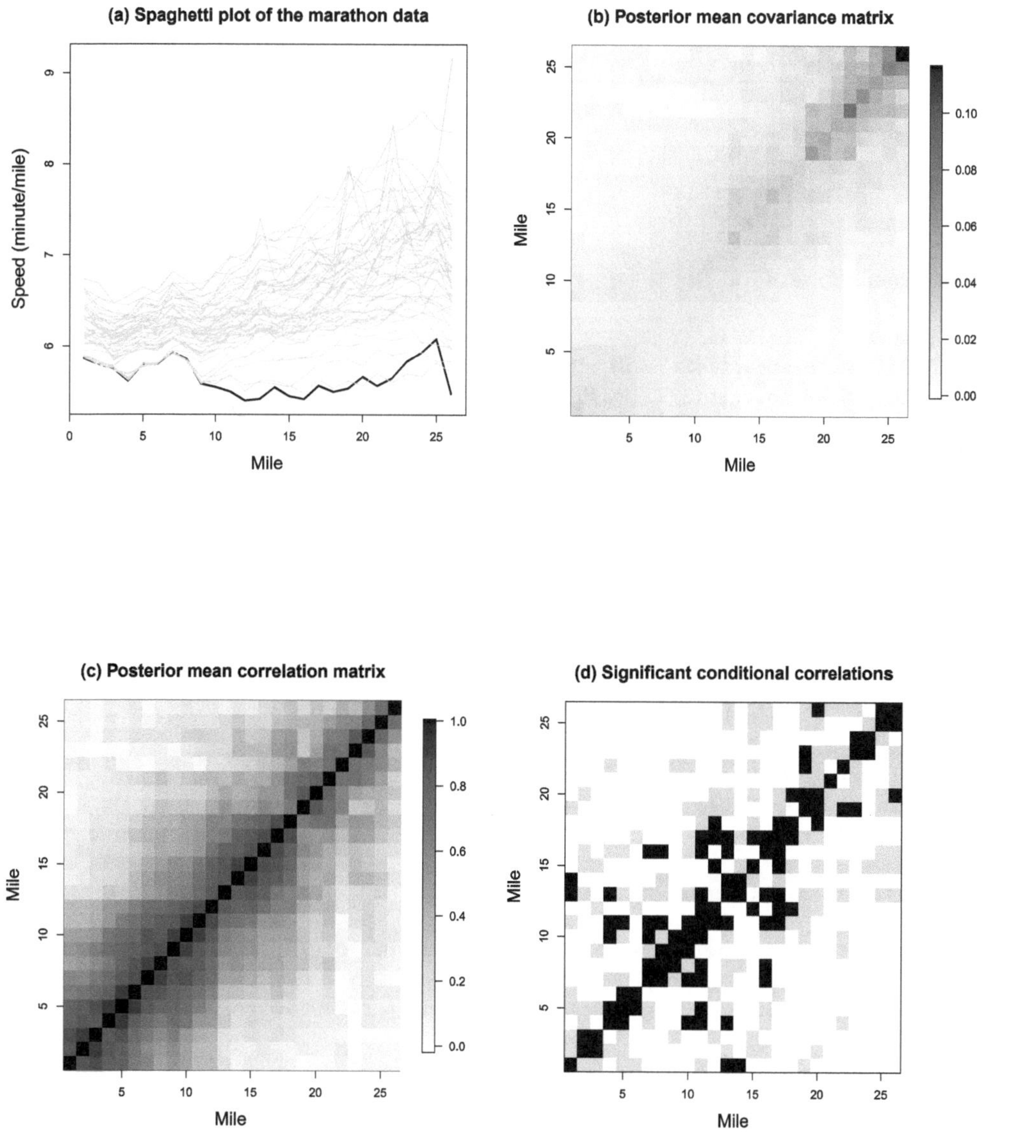

FIGURE 2.4

Covariance analysis of the 2016 Boston Marathon data. Panel (a) plots the minute/mile for each runner (the winner is in black), Panels (b) and (c) show the posterior mean of the covariance and correlation matrices, respectively, and Panel (d) is shaded gray (black) if the 95% (99%) credible interval for the elements of the precision matrix excludes zero.

Listing 2.1
Monte Carlo analysis of the Boston Marathon data.

```

# Hyperpriors

 nu       <- p+1
 R        <- diag(p)/(p+1)

# Process the data

 Ybar     <- colMeans(Y)
 SSE      <- sweep(Y,2,Ybar,"-")
 SSE      <- solve(t(SSE)%*%SSE)

# Monte Carlo settings from the posterior

 S        <- 10000
 Sigma_mn <- Cor_mn <- Omega_pos <- 0

# Monte Carlo sampling

 for(s in 1:S){
   Omega     <- rwish(n+nu,SSE+R)
   Sigma     <- solve(Omega)
   Sigma_mn  <- Sigma_mn + Sigma/S
   Cor_mn    <- Cor_mn + cov2cor(Sigma)/S
   Omega_pos <- Omega_pos + (Omega>0)/S
 }

# Evaluate significance of the precision matrix

 Omega_sig <- ifelse(Omega_pos<0.025 | Omega_pos>0.975,
                     "gray","white")
 Omega_sig <- ifelse(Omega_pos<0.005 | Omega_pos>0.995,
                     "black",Omega_sig)

# Plot some of the results

 library(fields)
 image.plot(1:p,1:p,Sigma_mn,
            xlab="Mile",ylab="Mile",
            main="Posterior mean covariance matrix")
```

a runner's first- and second-half mile times, and the strongest correlations in the second half of the race are between subsequent miles.

Normal-Wishart model for a precision matrix: Correlation is clearly a central concept in statistics, but it should not be confused with causation. For example, consider the three variables that follow the distribution $Z_1 \sim \text{Normal}(0,1)$, $Z_2|Z_1 \sim \text{Normal}(Z_1,1)$ and $Z_3|Z_1,Z_2 \sim \text{Normal}(Z_2,1)$. In this toy example, the first variable has a causal effect on the second, and the second has a causal effect on the third. The shared relationship with Z_2 results in a correlation of 0.57 between Z_1 and Z_3. However, if we condition on Z_2, then Z_1 and Z_3 are independent since the conditional distribution Z_3 depends on Z_2 but not Z_1.

Statistical inference about the precision matrix (inverse covariance matrix) $\mathbf{\Omega} = \mathbf{\Sigma}^{-1}$ is a step closer to uncovering causality than inference about the correlation/covariance matrix. For a normal random variable $\mathbf{Y} = (Y_1, ..., Y_p)^T \sim \text{Normal}(\boldsymbol{\mu}, \mathbf{\Omega}^{-1})$, the (j,k) element of the precision matrix $\mathbf{\Omega}$ measures the strength of the correlation between Y_j and Y_k *after accounting for the effects of the other* $p-2$ *elements of* $\mathbf{Y}$. We say that variables j and k are conditionally correlated if and only if the (j,k) element of $\mathbf{\Omega}$ is non-zero. Therefore, association tests based on $\mathbf{\Omega}$ rather than $\mathbf{\Sigma}$ eliminate spurious correlation (e.g., between Z_1 and Z_3) induced by lingering variables (e.g., Z_2). Assuming all variables relevant to the problem are included in the p variables under consideration, then these conditional correlations have causal interpretations.

The Wishart distribution is the conjugate prior for a normal precision matrix. If $\mathbf{\Sigma} \sim \text{InvWishart}(\nu, \mathbf{R}^{-1})$, then $\mathbf{\Omega} = \mathbf{\Sigma}^{-1} \sim \text{Wishart}(\nu, \mathbf{R})$ and has prior density

$$\pi(\mathbf{\Omega}) \propto |\mathbf{\Omega}|^{(p-\nu-1)/2} \exp\left[-\text{Trace}\left(\mathbf{\Omega}\mathbf{R}^{-1}\right)/2\right]. \tag{2.28}$$

Given a sample $\mathbf{Y}_1, ..., \mathbf{Y}_n \sim \text{Normal}(\boldsymbol{\mu}, \mathbf{\Omega}^{-1})$ and conditioning on the mean vectors $\boldsymbol{\mu}_i$, the posterior is

$$\mathbf{\Omega}|\mathbf{Y} \sim \text{Wishart}\left(n+\nu, \left[\sum_{i=1}^{n}(\mathbf{Y}_i - \boldsymbol{\mu}_i)(\mathbf{Y}_i - \boldsymbol{\mu}_i)^T + \mathbf{R}^{-1}\right]^{-1}\right). \tag{2.29}$$

Marathon example: Listing 2.1 computes and Figure 2.4d plots the conditional correlations with credible sets that exclude zero for the Boston Marathon example. Many of the non-zero correlations in Figure 2.4c do not have significant conditional correlations. For example, if the paces of the first 25 miles are known, then only miles 20 and 25 have strong conditional correlation with the final mile time. Therefore, despite the many strong correlations between the final mile and the other 25 miles, for the purpose of predicting the final mile time given all other mile times, using only miles 20 and 25 may be sufficient.

2.1.8 Mixtures of conjugate priors

A limitation of a conjugate prior is that restricting the prior to a parametric family limits how accurately prior uncertainty can be expressed. For example, the normal distribution is a conjugate prior for a normal mean, but a normal prior must be symmetric and unimodal, which may not describe the available prior information. A mixture of conjugate priors provides a much richer class of priors. A mixture prior is

$$\pi(\boldsymbol{\theta}) = \sum_{j=1}^{J} q_j \pi_j(\boldsymbol{\theta}) \tag{2.30}$$

where $q_1, ..., q_J$ are the mixture probabilities with $q_j \geq 0$ and $\sum_{j=1}^{J} q_j = 1$, and π_j are valid PDFs (or PMFs). Each term in the mixture can be thought of as the opinion of a different expert. That is, with probability q_j you select the prior from expert j, denoted π_j.

The posterior distribution under a mixture prior is

$$p(\boldsymbol{\theta}|\mathbf{Y}) \propto f(\mathbf{Y}|\boldsymbol{\theta})\pi(\boldsymbol{\theta}) \propto \sum_{j=1}^{p} q_j f(\mathbf{Y}|\boldsymbol{\theta})\pi_j(\boldsymbol{\theta}) \propto \sum_{j=1}^{p} Q_j p_j(\boldsymbol{\theta}|\mathbf{Y}), \tag{2.31}$$

where $p_j(\boldsymbol{\theta}|\mathbf{Y}) \propto f(\mathbf{Y}|\boldsymbol{\theta})\pi_j(\boldsymbol{\theta})$ is the posterior under prior π_j and the updated mixture weights Q_j are positive and sum to one. The posterior mixture weights Q_j are not the same as the prior mixture weights q_j and unfortunately have a fairly complicated form. However, since each mixture component is conjugate, the p_j belong to the same family as the π_j. Therefore, the posterior is also a mixture distribution over the same parametric family as the prior, and thus the prior is conjugate.

ESP example: As discussed in [106], [14] presents the results of several experiments on extrasensory perception (ESP). In one experiment (Experiment 2 in [14]), subjects were asked to say which of two pictures they preferred before being shown the pictures. Unknown to the subject, one of the pictures was a subliminally exposed negative picture, and the subjects avoided this picture in 2,790 of the 5,400 trials (51.7%). Even more dramatically, participants that scored high in stimulus seeking avoided the picture in $Y = 963$ of the $n = 1,800$ trials (53.5%).

The data are then $Y \sim \text{Binomial}(n, \theta)$, and the objective is to test whether $\theta = 0.5$, and thus the subjects do not have ESP. A skeptical prior that presumes the subjects do not have ESP should be concentrated around $\theta = 0.5$, say π_1 is the Beta(2000, 2000) PDF. At the other extreme, if we approach the analysis with an open mind, then we may select an uninformative prior, say π_2 is the uniform Beta(1, 1) PDF. The mixture prior combines these two extremes,

$$\pi(\theta) = q\pi_1(\theta) + (1 - q)\pi_2(\theta), \tag{2.32}$$

as shown in Figure 2.5 (right) for $q = 0.5$. The mixture prior is simply the average of the two beta prior PDFs in the left panel. The prior has a peak at 0.5 corresponding to π_1, but does not go completely to zero as θ departs from 0.5 because of the addition of the uniform prior π_2. The posterior distribution under the mixture prior (Figure 2.5, right) is also a mixture of two beta distributions. The posterior is a weighted average of the two posterior densities in the left panel. The posterior is bimodal with one mode near 0.51 corresponding to the skeptical prior and a second mode near 0.53 corresponding to the uniform prior. This shape of the density is not possible within the standard beta family.

Randomized clinical trials example: [159] analyze data from over 20,000 randomized clinical trials (RCT) to formulate a prior distribution to be used in future trials. For RCT $i = 1, ..., n$, they observed a treatment effect b_i and standard error s_i: e.g., b_i might be the difference in sample means between treatment groups and s_i the associated standard error from a two-sample t-test. Let β_i be the true effect size for trial i and $\delta_i = \beta_i/s_i$ be the signal-to-noise ratio. [159] estimated the distribution of the δ_i across trials to be the mixture distribution (Table 1 of [159])

$$\begin{aligned}\pi(\delta) &= 0.32\phi(\delta; 0.00, 1.17^2) + 0.31\phi(\delta; 0.00, 1.74^2) \\ &+ 0.30\phi(\delta; 0.00, 2.38^2) + 0.07\phi(\delta; 0.00, 5.73^2),\end{aligned} \tag{2.33}$$

where $\phi(d; m, v)$ is the Gaussian PDF with mean m and variance v. This prior is plotted in Figure 2.6 and is symmetric around $\delta = 0$ and decreasing in δ.

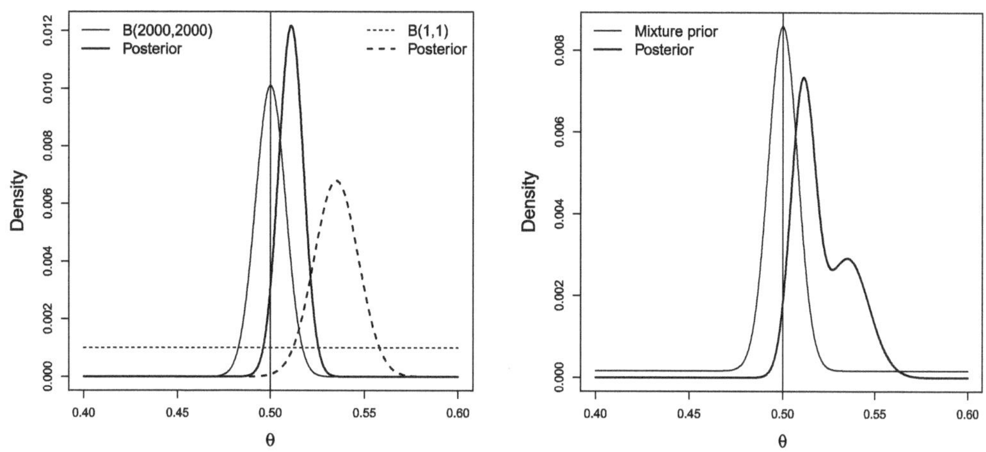

FIGURE 2.5
Prior and posterior distributions for the ESP example. The data are binomial with $Y = 963$ successes in $n = 1800$ trials. The left panel plots the Beta(2000,2000) and Beta(1,1) priors and resulting posteriors, and the right panel plots the mixture prior that equally weights the two beta priors and the resulting posterior.

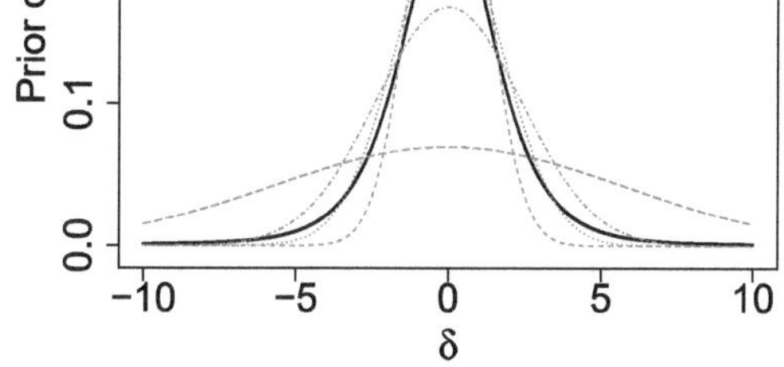

FIGURE 2.6
Mixture prior for a clinical trial. Four component ("Comp") mixture prior (2.33) for the signal-to-noise ratio of a randomized clinical trial (see [159]).

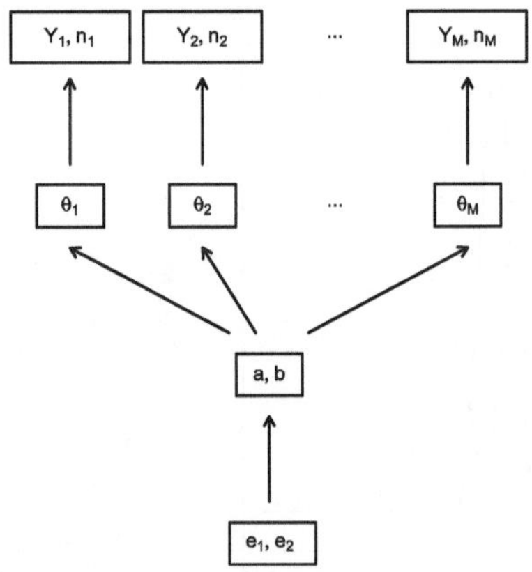

FIGURE 2.7
Hierarchical priors. The data (Y_i and n_i) depends on parameters (θ_i) whose priors depend on two layers of hyperparameters (a and b and then e_1 and e_2).

2.2 Hierarchical priors

Most prior distributions for model parameters depend on other parameters, called hyperparameters. For example, if we select prior $\theta \sim \text{Gamma}(a, b)$, then a and b are hyperparameters that need to be specified. Typically they are fixed based either on prior knowledge or to give an uninformative prior with large variance. However, there are no "right" values for the hyperparameters, and this leads to a difficult and subjective decision.

There are several approaches to selecting hyperparameters. The most straightforward is a sensitivity analysis where several hyperparameters are applied and the posterior is compared across hyperparameters; if the posterior does not change substantially with the hyperparameters, then we can be confident that the subjective prior is not affecting the results. Another approach is to avoid the subjective decision of selecting the prior altogether using default (objective) priors described in Section 2.4. Below we describe a middle path using hierarchical prior distributions.

Say $Y|\theta \sim \text{Binomial}(n, \theta)$ and $\theta \sim \text{Beta}(a, b)$, but we do not know the hyperparameters a and b. Following the central tenet of Bayesian analysis, if a and b are unknown, then we should select prior distributions for them to reflect this uncertainty. A hierarchical prior is $\theta|a, b \sim \text{Beta}(a, b)$ and $a, b \overset{iid}{\sim} \text{Gamma}(e_1, e_2)$ (of course, other families of distributions could be used). This is called a hierarchical prior because it is defined in layers/stages with the distribution of parameters in one stage depending on parameters from the previous stage (see Figure 2.7 for an illustration under a more complex model). Because the prior is built in stages, the joint distribution of θ, a and b is guaranteed to be a valid probability distribution.

A hierarchical model (the subject of Chapter 7) includes both the data model and hierarchical prior, e.g.,

$$\begin{aligned} Y|\theta, a, b &\sim \text{Binomial}(n, \theta) \\ \theta|a, b &\sim \text{Beta}(a, b) \\ a, b &\sim \text{Gamma}(e_1, e_2). \end{aligned}$$

In this hierarchical prior, we replace the task of selecting values for a and b with selecting prior distributions for a and b to quantify our certainty. However, this only leads to the obvious question of how to select the hyperparameters e_1 and e_2, which is arguably more difficult because parameters in this layer are harder to interpret than those in layers closer to the data.

While a hierarchical prior does not resolve the prior selection issue in the simple beta-binomial model above, hierarchical priors are very useful for more complicated problems. To illustrate this, say data are collected for M basketball players, similar to the NBA clutch free throw data in Section 1.6. Player i makes n_i attempts and is successful $Y_i \in \{0, 1, ..., n_i\}$ times. If we let $\theta_i \in (0, 1)$ be the true success probability for player i, a hierarchical model (Figure 2.7) is

$$\begin{aligned} Y_i|\theta_i, a, b &\sim \text{Binomial}(n_i, \theta_i) \\ \theta_i|a, b &\sim \text{Beta}(a, b) \\ a, b &\sim \text{Gamma}(e_1, e_2). \end{aligned}$$

The hyperparameters a and b determine the distribution of success probabilities across players. The idea of a hierarchical prior is to leverage these data to estimate a and b by giving them priors and studying their posteriors.

Is it possible to estimate a and b from the data? For that matter, is it possible to estimate e_1 and e_2? The best case is a large number of attempts n_i for all players so that θ_i can be estimated with high precision. Further assuming many players M, then a and b can be estimated to match the distribution of $\theta_1, ..., \theta_M$ across players. However, even the ideal case that we know a and b exactly provides insufficient information to estimate e_1 and e_2 as there are many combinations of these hyperparameters that can fit the distribution of the two points a and b. Therefore, these hyperparameters cannot be estimated and must be fixed. As in the simple beta-binomial model, a sensitivity analysis or default priors could be used.

Defining and interpreting hierarchical models requires concepts beyond the scope of this section and will be revisited in the discussion of random effects (Section 4.4) and hierarchical modeling (Chapter 7). Also, estimating parameters such as a and b requires advanced computational tools that will be introduced in Chapter 3.

2.3 Improper priors

Section 1.1 introduced the concept of a probability density function (PDF). A *proper* PDF is non-negative and integrates to one. If the prior is selected from a common family of distributions as in Section 2.1 then the prior is ensured to be proper. In this chapter we explore the possibility of using an improper prior, i.e., a function $\pi(\theta)$ that does not integrate to one over the parameter's support.

As a concrete example, assume that $Y_1, ..., Y_n|\mu, \sigma^2 \stackrel{iid}{\sim} \text{Normal}(\mu, \sigma^2)$ with σ taken to be fixed and μ is to be estimated. If we truly have no prior information, then it is tempting to fit the model with prior $\pi(\mu) = c > 0$ for all possible $\mu \in (-\infty, \infty)$. This can be viewed as approximating either a uniform prior with bounds tending to infinity or a Gaussian prior with variance tending to infinity. However, for any positive c, $\int_{-\infty}^{\infty} \pi(\mu)d\mu = \infty$, and so the prior is improper.

Despite the fact that the prior is not a proper PDF, it is still possible to use the density function

$$p(\mu|\mathbf{Y}) \propto f(\mathbf{Y}|\mu)\pi(\mu) \tag{2.34}$$

to summarize uncertainty about μ. That is, we can treat this product of the likelihood and improper prior as a posterior distribution. Applying this rule for the normal mean example gives $\mu|\mathbf{Y} \sim \text{Normal}(\bar{Y}, \sigma^2/n)$, which is a proper PDF as long as $n \geq 1$.

As this example shows, using improper priors can lead to a proper posterior distribution. However, *improper priors must be used with caution.* It must be verified that the resulting posterior distribution is proper; i.e., it must be mathematically proven that the posterior integrates to one over the parameter's support. This step is not needed if the prior is proper because any proper prior leads to a proper posterior. Also, from a conceptual perspective, it is difficult to apply the subjective Bayesian argument that the prior and posterior distributions represent the analyst's uncertainty about the parameters when the prior uncertainty cannot be summarized by a valid PDF. One way to justify Bayesian inference with an improper prior is to view the posterior under an improper prior as the limit of the posterior for proper priors (with, say, the prior variance increasing to infinity) [146].

An obvious remedy to these concerns is to replace the improper prior with a uniform prior with bounds that almost surely include the true value of a parameter. In fact, a uniform prior was the original prior used by Laplace in his formulation of Bayesian statistics. While this alleviates mathematical problems of ensuring a proper posterior and conceptual problems with Bayesian interpretation, it raises a concern about invariance to parameterization. For example, let $\theta \in (0, 1)$ be the probability of an event and $r = \theta/(1 - \theta) > 0$ be the corresponding odds of the event. A Uniform(0,100) prior for the odds r appears to be uninformative, but in fact the induced prior mean of $\theta = r/(1 + r)$ is 0.95, which is clearly not uninformative for θ.

There are some cases when improper priors will necessarily lead to improper posteriors. One such situation is the mixture distributions where Y_i is Gaussian with mean 0 and variance 1 with probability 1/2 and Gaussian with mean θ and variance 1 with probability 1/2. The PDF is

$$f(y|\theta) = \frac{1}{2}\phi(y; 0, 1) + \frac{1}{2}\phi(y; \theta, 1),$$

where $\phi(y; \mu, \sigma^2)$ is the Gaussian PDF with mean μ and variance σ^2. It can be shown that if $Y_1, ..., Y_n$ are independent and all follow this PDF, then an improper prior for θ always leads to an improper posterior for any n. Again, we emphasize that improper priors must be used with caution.

2.4 Default priors

In an idealized subjective Bayesian analysis, the system (physical, biological, economic, etc.) is understood except for the values of a few parameters, and studies are conducted sequentially to refine estimates of the unknown parameters. In this case, the ability to incorporate the current state of knowledge as prior information in the analysis of the current study is a strength of the Bayesian paradigm and should be embraced. However, in many cases, there is no prior information, leaving the analyst to select, say, conjugate uninformative priors in an ad hoc manner.

In the absence of prior information, selecting the prior can be viewed as a nuisance to be avoided if possible. Default priors, also called objective priors, offer a solution by providing systematic approaches to formulate the prior. Berger [15] argues that objectivity is essential in many analyses, such as obtaining approval from regulatory agencies, where any apparent personal bias can be a huge red flag. Default methods also provide useful baselines for comparison for subjective methods and make application of Bayesian methods more approachable for non-statisticians [15]. Of course, eliminating the subjectivity of prior selection is only part of the story; there remains subjectivity in selecting the data to be analyzed, the inference to be conducted, the likelihood to be used, and indeed, the form of the default prior. For this reason, we refer to this style of analysis as a "Bayesian analysis using default prior" rather than an "objective Bayesian analysis".

In the remainder of this section we discuss several default approaches to setting priors. Jeffreys prior (Section 2.4.1) is arguably the most common and so we describe this method in the most detail. We provide only brief conceptual discussions for the other approaches because deriving the exact expressions of the priors even in moderately complex cases often requires tedious calculus.

2.4.1 Jeffreys prior

A Jeffreys prior (JP) is a method to construct a prior distribution that is invariant to reparameterization. This is a necessary condition for a method to be objective. For example, if one statistician places a prior on the standard deviation (σ), and another places a prior on the variance (σ^2), and the two get different results, then the method is subjective because it depends on the choice of parameterization.

The univariate JP for θ is

$$\pi(\theta) \propto \sqrt{I(\theta)} \tag{2.35}$$

where $I(\theta)$ is the expected Fisher information defined as

$$I(\theta) = -\text{E}\left(\frac{d^2 \log f(\mathbf{Y}|\theta)}{d\theta^2}\right) \tag{2.36}$$

and the expectation is with respect to $\mathbf{Y}|\theta$. The key feature of this construction is that it is invariant to transformation. That is, say statistician one uses a JP on θ, denoted $\pi_1(\theta)$, and statistician two places a JP on $\gamma = g(\theta)$, denoted $\pi_2(\gamma)$. Using the change of variables formula, the prior for γ induced by π_1 is

$$\pi_3(\gamma) = \pi_1[h(\gamma)]\left|\frac{dh(\gamma)}{d\gamma}\right| \tag{2.37}$$

where $h(\gamma) = \theta$ is the inverse of g (assuming g is invertable). It can be shown that $\pi_2 = \pi_3$, that is, placing a JP on γ is equivalent to placing a JP on θ and then reparameterizing to γ; this establishes invariance to reparameterization.

Once the likelihood function f has been determined, then the JP is determined; in this sense the prior is objective because any two analysts working with the same likelihood will get the same prior and thus the same posterior. An important point is that the prior does not depend on the data, $\mathbf{Y}$, because the expectation in $I(\theta)$ removes the dependence on the data. Finally, the JP is often improper, and therefore before using JP in practice it must be verified that the posterior is a valid PDF that integrates to one (Section 2.3).

JP can also be applied in multivariate problems. The multivariate JP for $\boldsymbol{\theta} = (\theta_1, ..., \theta_p)$ is

$$\pi(\boldsymbol{\theta}) \propto \sqrt{|\mathbf{I}(\boldsymbol{\theta})|} \tag{2.38}$$

where $\mathbf{I}(\boldsymbol{\theta})$ is the $p \times p$ expected Fisher information matrix with (i, j) element

$$-\mathrm{E}\left(\frac{\partial^2 \log f(\mathbf{Y}|\boldsymbol{\theta})}{\partial\theta_i \partial\theta_j}\right). \tag{2.39}$$

As with the univariate JP, this prior is invariant to transformation and can be improper.

Binomial proportion: As a univariate example, consider the binomial model $Y \sim$ Binomial(n, θ). The second derivative of the binomial log likelihood is

$$\begin{aligned}
\frac{d^2 \log f(\mathbf{Y}|\theta)}{d\theta^2} &= \frac{d^2}{d\theta^2} \log \binom{n}{Y} + Y \log(\theta) + (n - Y) \log(1 - \theta) \\
&= \frac{d}{d\theta} \frac{Y}{\theta} - \frac{n - Y}{1 - \theta} \\
&= -\frac{Y}{\theta^2} - \frac{n - Y}{(1 - \theta)^2}.
\end{aligned}$$

The information is the expected value of the negative second derivative. Under the binomial model, the expected value of Y is $n\theta$, and since the second derivative is linear in Y, the expectation passes into each term as

$$I(\theta) = \frac{\mathrm{E}(Y)}{\theta^2} + \frac{n - \mathrm{E}(Y)}{(1 - \theta)^2} = \frac{n}{\theta} + \frac{n}{1 - \theta} = \frac{n}{\theta(1 - \theta)}. \tag{2.40}$$

The JP is then

$$\pi(\theta) \propto \sqrt{\frac{n}{\theta(1 - \theta)}} \propto \theta^{1/2-1}(1 - \theta)^{1/2-1}, \tag{2.41}$$

which is the kernel of the Beta(1/2,1/2) PDF. Therefore, the JP for the binomial proportion is $\theta \sim$ Beta$(1/2, 1/2)$. In Section 2.1 we saw that a beta prior was conjugate, and the JP provides an default way to set the hyperparameters in the beta prior.

Normal model: As a multivariate example, consider the Gaussian model $Y_i|\mu, \sigma^2 \stackrel{iid}{\sim}$ Normal(μ, σ^2). As shown in Appendix A.3, the bivariate JP is

$$\pi(\mu, \sigma^2) \propto \frac{1}{(\sigma^2)^{3/2}}. \tag{2.42}$$

This is an improper prior but leads to a proper posterior as long as $n \geq 2$. This prior can be seen as the limit of the product of independent priors $\mu \sim$ Normal$(0, c^2)$ and $\sigma^2 \sim$ InvGamma$(0.5, b)$ as $c \to \infty$ and $b \to 0$. The JP is flat for all values of μ and, similar to the inverse gamma distribution with small shape parameter, the JP peaks at $\sigma^2 = 0$ and drops sharply away from the origin.

Multiple linear regression model: Linear regression (Section 4.2) estimates the linear relationship between p covariates $X_{i1}, ..., X_{ip}$ on a response variable Y_i. The model is $Y_i|\boldsymbol{\beta}, \sigma^2 \stackrel{indep}{\sim}$ Normal$(\sum_{j=1}^{p} X_{ij}\beta_j, \sigma^2)$, where the vector of p regression coefficients is $\boldsymbol{\beta} = (\beta_1, ..., \beta_p)$ and the error variance is σ^2. Appendix A.3 shows that the JP is

$$\pi(\boldsymbol{\beta}, \sigma^2) \propto \frac{1}{(\sigma^2)^{p/2+1}}. \tag{2.43}$$

As in the mean-only model (which is a special case of the regression model), the JP is improper, completely flat in the mean parameters $\boldsymbol{\beta}$, and decays polynomially in the variance with degree that depends on the number of covariates in the model.

For large p, this prior is heavily concentrated at σ^2 near zero, which may be undesirable in some settings. An alternative [89] is to treat $\boldsymbol{\beta}$ and σ^2 as independent *a priori* and take the joint JP as the product of the JP for $\boldsymbol{\beta}$ given σ^2 and the JP for σ^2 given $\boldsymbol{\beta}$. This leads to the prior

$$\pi(\boldsymbol{\beta}, \sigma^2) \propto \frac{1}{\sigma^2} \tag{2.44}$$

that remains flat in $\boldsymbol{\beta}$ but has less mass near zero for σ^2.

2.4.2 Reference priors

Bernardo's reference prior (RP; [17]) is the prior that maximizes the expected difference between the prior and posterior. This is certainly an intuitive definition of an uninformative prior, as it ensures that the prior does not overwhelm the data. To act on this conceptual definition requires a measure of the discrepancy between two density functions. [17] use (ignoring technical problems that require studying sequences of priors) the Kullback–Leibler (KL) divergence between the prior $\pi(\boldsymbol{\theta})$ and the posterior $p(\boldsymbol{\theta}|\mathbf{Y})$, defined as the expected difference in log densities,

$$KL(p, \pi|\mathbf{Y}) = \mathrm{E}\left\{\log[p(\theta|\mathbf{Y})] - \log[\pi(\theta)]\right\}, \tag{2.45}$$

where the expectation is with respect to the posterior $p(\boldsymbol{\theta}|\mathbf{Y})$. The KL divergence depends on the data and thus cannot be used to derive a prior. To remove the dependence on the data, the RP is the prior density function π that maximizes the expected information gain $\int KL(p, \pi|\mathbf{Y})m(\mathbf{Y})d\mathbf{Y}$, where $m(\mathbf{Y})$ is the marginal distribution of the data (Table 1.4).

Finding the RP is a daunting optimization problem because the solution is a prior *distribution* and not a scalar or vector. In many cases, the RP differs from the JP. However, it can be shown that if the posterior is approximately normal then the RP is approximately the JP. The two are also equivalent in non-Gaussian cases; for example, if $Y \sim \text{Binomial}(n, \theta)$, the RP is $\theta \sim \text{Beta}(1/2, 1/2)$.

2.4.3 Maximum entropy priors

A maximum entropy (MaxEnt) prior assumes that a few properties of the prior are known and selects the prior with the most entropy in the class of priors with these properties. As an example, say that θ has prior support $\theta \in \{0, 1, 2, ...\}$, and the prior mean is assumed to be $E(\theta) = 1$. There are infinitely many priors for θ that satisfy the constraint that $\mathrm{E}(\theta) = 1$, e.g., $\theta \sim \text{Poisson}(1)$, $\theta \sim \text{Binomial}(2, 0.5)$ or $\theta \sim \text{Binomial}(5, 0.2)$. The MaxEnt prior is the prior that maximizes entropy over the priors with support $\{0, 1, 2, ...\}$ and mean one.

In general, let the constraints be $\mathrm{E}[g_k(\theta)] = c_k$ for functions g_k (mean, variance, exceedance probabilities, etc.) and constants c_k for $k = 1, ..., K$. The entropy for a discrete parameter with prior $\pi(\theta)$ and discrete support $\mathcal{S} = \{\theta_1, \theta_2, ...\}$ is

$$\mathrm{E}\{-\log[\pi(\theta)]\} = -\sum_{i \in \mathcal{S}} \log[\pi(\theta_i)]\pi(\theta_i). \tag{2.46}$$

The MaxEnt prior is the value of $\{\pi(\theta_1), \pi(\theta_2), ...\}$ that maximizes $-\sum_{i\in\mathcal{S}} \log[\pi(\theta_i)]\pi(\theta_i)$ subject to the K constraints $\sum_{i\in\mathcal{S}} g_k(\theta_i)\pi(\theta_i) = c_k$. MaxEnt priors can be extended to continuous parameters as well.

2.4.4 Empirical Bayes

An empirical Bayesian (EB) analysis uses the data to set the priors. For example, consider the model $Y_i|\theta_i \stackrel{iid}{\sim} \text{Binomial}(N, \theta_i)$ for $i = 1, ..., n$, where each observation's success probability has prior $\theta_i|a, b \stackrel{iid}{\sim} \text{Beta}(a, b)$. The goal of this analysis is to estimate the individual probabilities θ_i, and thus there are two types of parameters: the parameters of interest $\boldsymbol{\theta} = (\theta_1, ..., \theta_n)$ and the nuisance parameters $\boldsymbol{\gamma} = (a, b)$. Without prior knowledge about $\boldsymbol{\gamma}$, a fully Bayesian analysis requires a prior for the nuisance hyperparameters. In contrast, an EB analysis inspects the data to select $\boldsymbol{\gamma}$, and then performs a Bayesian analysis of $\boldsymbol{\theta}$ as if $\boldsymbol{\gamma}$ was known all along. For example, in this problem it can be shown (Appendix A.1) that marginally over $\boldsymbol{\theta}$, $Y_i|\boldsymbol{\gamma} \sim \text{Beta-Binomial}(N, a, b)$. Therefore, an EB analysis might use the MLE from the beta-binomial model to estimate $\boldsymbol{\gamma}$ and plug this estimate into the Bayesian analysis for $\boldsymbol{\theta}$.

More formally, consider the model $\mathbf{Y}|\boldsymbol{\theta} \sim f(\mathbf{Y}|\boldsymbol{\theta})$ and prior $\boldsymbol{\theta}|\boldsymbol{\gamma} \sim \pi(\boldsymbol{\theta}|\boldsymbol{\gamma})$. The marginal maximum likelihood estimator for the hyperparameter is

$$\hat{\boldsymbol{\gamma}} = \underset{\boldsymbol{\gamma}}{\text{argmax}} \int p(\mathbf{Y}|\boldsymbol{\theta})\pi(\boldsymbol{\theta}|\boldsymbol{\gamma})d\boldsymbol{\theta}. \tag{2.47}$$

The EB analysis then conducts a Bayesian analysis of $\boldsymbol{\theta}$ with $\boldsymbol{\gamma}$ set to $\hat{\boldsymbol{\gamma}}$, i.e., $\mathbf{Y}|\boldsymbol{\theta} \sim f(\mathbf{Y}|\boldsymbol{\theta})$ and $\boldsymbol{\theta}|\boldsymbol{\gamma} = \hat{\boldsymbol{\gamma}} \sim \pi(\boldsymbol{\theta}|\hat{\boldsymbol{\gamma}})$. Although other forms of EB are possible (e.g., using $\hat{\boldsymbol{\gamma}}$ as the prior mean for $\boldsymbol{\gamma}$), this plug-in approach provides a systematic and objective way to set priors.

Consider the Gaussian model $Y_i|\mu, \sigma^2 \stackrel{iid}{\sim} \text{Normal}(\mu, \sigma^2)$ where μ has improper prior $\pi(\mu) = 1$ for all μ. The model has two parameters: the parameter of interest $\theta = \mu$ and the nuisance parameter $\gamma = \sigma^2$. It can be shown that (2.47) gives $\hat{\sigma}^2 = \sum_{i=1}^n (Y_i - \bar{Y}_i)^2/n$. An EB analysis proceeds with a univariate study of μ with the variance fixed at $\sigma^2 = \hat{\sigma}^2$. Applying (2.10) with $m = 0$ (and thus $w = 1$) corresponding to the improper prior, the EB posterior is

$$\mu|\mathbf{Y}, \sigma^2 = \hat{\sigma}^2 \sim \text{Normal}(\bar{Y}, \hat{\sigma}^2/n).$$

Similar to the "plug-in" predictor in Figure 1.12, this posterior ignores uncertainty in σ^2, which ultimately underestimates uncertainty in μ. Graphically, the EB analysis would consider only one horizontal slice (row) of Figure 1.11 rather than averaging over rows to compute the marginal posterior of μ. As will be shown in Section 4.1, the marginal posterior of μ averaging over uncertainty in σ^2 (assuming conjugate prior) is a Student's t-distribution that has slightly larger variance than the normal EB posterior.

The disadvantage of EB is obvious: we are using the data twice and ignoring uncertainty about $\boldsymbol{\gamma}$. As a result, an EB analysis of $\boldsymbol{\theta}$ will often have smaller posterior variances and narrower credible intervals than a fully Bayesian analysis that puts a prior on $\boldsymbol{\gamma}$, and thus the statistical properties of the EB analysis can be questioned. Despite this concern, EB remains a useful tool to stabilize computing in high-dimensional problems, especially when it can be argued (e.g., for large data sets) that uncertainty in the nuisance parameters is negligible.

2.4.5 Penalized complexity priors

The penalized complexity (PC) prior [141] is designed to prevent over-fitting caused by using a model that is too complex to be supported by the data. For example, consider the linear regression model

$$\text{Full model}: Y_i|\boldsymbol{\beta}, \sigma^2 \stackrel{indep}{\sim} \text{Normal}\left(\beta_1 + \sum_{j=2}^p X_{ij}\beta_j, \sigma^2\right). \tag{2.48}$$

If the number of predictors p is large then the model becomes very complex. A PC prior might shrink the full model to the base model with $\beta_2 = ... = \beta_p = 0$,

$$\text{Base model} : Y_i|\boldsymbol{\beta}, \sigma^2 \overset{iid}{\sim} \text{Normal}\left(\beta_1, \sigma^2\right). \tag{2.49}$$

Using uninformative priors for the β_j in the full model puts almost no weight on this base model, which could lead to over-fitting if the true model is sparse with many $\beta_j = 0$.

The PC prior is defined using the KL distance between the priors for the full and base models,

$$KL = \int \log\left(\frac{\pi_{full}(\boldsymbol{\theta})}{\pi_{base}(\boldsymbol{\theta}}\right)\pi_{full}(\boldsymbol{\theta})d\boldsymbol{\theta}. \tag{2.50}$$

The prior on the distance between models is then $\sqrt{2KL} \sim \text{Gamma}(1, \lambda)$, i.e., an exponential prior with rate λ which peaks at the base model with $KD = 0$ and decreases in complexity KD. Technically this is not an objective Bayesian approach because the user must specify the base model and either λ or a prior for λ. Nonetheless, the PC prior provides a systematic way to set priors for high-dimensional models.

2.5 Exercises

1. Assume $Y_1, ..., Y_n|\mu \overset{iid}{\sim} \text{Normal}(\mu, \sigma^2)$ where σ^2 is fixed and the unknown mean μ has prior $\mu \sim \text{Normal}(0, \sigma^2/m)$.
   (a) Give a 95% posterior interval for μ.
   (b) Select a value of m, and argue that for this choice your‘ 95% posterior credible interval has frequentist coverage 0.95. That is, if you draw many samples of size n and compute the 95% interval following the formula in (a) for each sample, in the long-run 95% of the intervals will contain the true value of μ.
2. The Major League Baseball player Reggie Jackson is known as “Mr. October” for his outstanding performances in the World Series (which takes place in October). Over his long career he played in 2820 regular-season games and hit 563 home runs in these games (a player can hit 0, 1, 2, ... home runs in a game). He also played in 27 World Series games and hit 10 home runs in these games. Assuming uninformative conjugate priors, summarize the posterior distribution of his home-run rate in the regular season and World Series. Is there sufficient evidence to claim that he performs better in the World Series?
3. Assume that $Y|\theta \sim \text{Exponential}(\theta)$, i.e., $Y|\theta \sim \text{Gamma}(1, \theta)$. Find a conjugate prior distribution for θ, and derive the resulting posterior distribution.
4. A large factory's objective is to keep the injury rate below 4 injuries per month. In the past 6 months there have been 4, 5, 2, 7, 6 and 6 injuries. Conduct a Bayesian analysis of these data and compute the posterior probability that the injury rate exceeds four injuries per month.
5. Assume that $Y|\theta \sim \text{NegBinomial}(\theta, m)$ (see Appendix A.1) and $\theta \sim \text{Beta}(a, b)$.
   (a) Derive the posterior of θ.
   (b) Plot the posterior of θ, and give its 95% credible interval assuming $m = 5$, $Y = 10$, and $a = b = 1$.

6. The distribution of true heart rate across a population is Gaussian with mean 64 beats per minute (bpm) and standard deviation 10 bpm. An individual measurement is unbiased for the true heart rate, but has error standard deviation 5 bpm.

   (a) Given that an individual has a (single) measured heart rate of y bpm, what is the posterior distribution of their true heart rate? Plot the posterior mean as a function of the observed value and comment on the regression to the mean.

   (b) Consider a study designed to test a new method to treat high heart rate. To obtain a sample, they measure randomly-selected individuals from the population and include them in the study if and only if their heart rate exceeds 90 bpm. For these individuals, they apply the treatment and measure their heart rate again. The estimate of the treatment effect is the difference in mean heart rate before and after the treatment for the individuals in the study. Assuming the new method has no actual effect on heart rate, compute (using mathematical derivations or simulations) the expected difference in the mean.

   (c) How could the study in (b) be redesigned to avoid regression to the mean bias?

7. Over the past 50 years California has experienced an average of $\lambda_0 = 75$ large wildfires per year. For the next 10 years you will record the number of large fires in California and then fit a Poisson/gamma model to these data. Let the rate of large fires in this future period, λ, have prior $\lambda \sim \text{Gamma}(a, b)$. Select a and b so that the prior is uninformative with prior variance around 100 and gives prior probability approximately $\text{Prob}(\lambda > \lambda_0) = 0.5$ so that the prior places equal probability on both hypotheses in the test for a change in the rate.

8. An assembly line relies on accurate measurements from an image-recognition algorithm at the first stage of the process. It is known that the algorithm is unbiased, so assume that measurements follow a normal distribution with mean zero, $Y_i|\sigma^2 \stackrel{iid}{\sim} \text{Normal}(0, \sigma^2)$. Some errors are permissible, but if σ exceeds the threshold c, then the algorithm must be replaced. You make $n = 20$ measurements and observe $\sum_{i=1}^n Y_i = -2$ and $\sum_{i=1}^n Y_i^2 = 15$ and conduct a Bayesian analysis with InvGamma(a, b) prior. Compute the posterior probability that $\sigma > c$ for:

   (a) $c = 1$ and $a = b = 0.1$

   (b) $c = 1$ and $a = b = 1.0$

   (c) $c = 2$ and $a = b = 0.1$

   (d) $c = 2$ and $a = b = 1.0$

   For each c, compute the ratio of probabilities for the two priors (i.e. $a = b = 0.1$ and $a = b = 1.0$). Which, if any, of the results are sensitive to the prior?

9. In simple logistic regression, the probability of a success is regressed onto a single covariate X as

$$p(X) = \frac{\exp(\alpha + X\beta)}{1 + \exp(\alpha + X\beta)}. \tag{2.51}$$

   Assuming the covariate is distributed as $X \sim \text{Normal}(0, 1)$ and the parameters have priors $\alpha \sim \text{Normal}(0, c^2)$ and $\beta \sim \text{Normal}(0, c^2)$, find a prior standard deviation c so that the induced prior on the success probability $p(X)$ is roughly Uniform(0,1). That is, for a given c generate S (say $S = 1,000,000$) samples

$(X^{(s)}, \alpha^{(s)}, \beta^{(s)})$ for $s = 1, ..., S$, compute the S success probabilities, and make a histogram of the S probabilities. Repeat this for several c until the histogram is approximately uniform. Report the final value of c and the resulting histogram. Would you call this an uninformative prior for the regression coefficients α and β?

10. The Mayo Clinic conducted a study of $n = 50$ patients followed for one year on a new medication and found that 30 patients experienced no adverse side effects (ASE), 12 experienced one ASE, 6 experienced two ASEs and 2 experienced ten ASEs.

    (a) Derive the posterior of λ for the Poisson/gamma model $Y_1, ..., Y_n|\lambda \stackrel{iid}{\sim}$ Poisson(λ) and $\lambda \sim$ Gamma(a, b).

    (b) Use the Poisson/gamma model with $a = b = 0.01$ to study the rate of adverse events. Plot the posterior and give the posterior mean and 95% credible interval.

    (c) Repeat this analysis with Gamma(0.1,0.1) and Gamma(1,1) priors and discuss sensitivity to the prior.

    (d) Plot the data versus the Poisson($\hat{\lambda}$) PMF, where $\hat{\lambda}$ is the posterior mean of λ from part (b). Does the Poisson likelihood fit well?

    (e) The current medication is thought to have around one adverse side effect per year. What is the posterior probability that this new medication has a higher side effect rate than the previous medication? Are the results sensitive to the prior?

11. A smoker is trying to quit and their goal is to abstain for one week because studies have shown that this is the most difficult period. Denote Y as the number of days before the smoker relapses, and θ as the probability that they relapse on any given day of the study.

    (a) Argue that a negative binomial distribution (Appendix A.1) for $Y|\theta$ is reasonable. What are the most important assumptions being made, and are they valid for this smoking cessation attempt?

    (b) Assuming a negative binomial likelihood and Uniform(0, 1) prior for θ, what is the prior probability the smoker will be smoke-free for at least 7 days?

    (c) Plot the posterior distribution of θ assuming $Y = 5$.

    (d) Given this attempt resulted in $Y = 5$ and that θ is the same in the second attempt, what is the posterior predictive probability the second attempt will result in at least 7 smoke-free days?

12. A smoking cessation study followed $n = 20$ patients until they relapsed to smoking. The trial terminated after six weeks and so many patients did not relapse during the observation period; i.e., their responses was censored. Seven patients relapsed during the observation period at times 0.5, 0.9, 1.1, 2.4, 3.8, 4.0 and 5.2 weeks, and the rest were censored at 6 weeks. Assume the times to event $Y_1, ..., Y_n$ follow the survival analysis model $Y_i|\lambda \stackrel{iid}{\sim}$ Exponential(λ) with prior $\lambda \sim$ Gamma(a, b).

    (a) Show that the posterior distribution of λ is a gamma distribution (hint: the likelihood contribution of a censored observation is Prob($Y_i > 6|\lambda$) = $\exp(-6\lambda)$).

    (b) Compute the posterior mean and 95% interval for λ for several values of a and b. Are the results sensitive to the prior?

13. You are designing a very small experiment to determine the sensitivity of a new security alarm system. You will simulate five robbery attempts and record the number of these attempts that trigger the alarm. Because the dataset will be small, you ask two experts for their opinion. One expects the alarm probability to be 0.95 with standard deviation 0.05, and the other expects 0.80 with standard deviation 0.20.

    (a) Translate these two priors into beta PDFs, and plot the two beta PDFs and the corresponding mixture of experts prior with equal weight given to each expert.
    (b) Now you conduct the experiment, and the alarm is triggered in every simulated robbery. Plot the posterior of the alarm probability under a uniform prior, each experts' prior, and the mixture of experts prior.

14. Let θ be the binomial success probability and $\gamma = \theta/(1-\theta)$ be the corresponding odds of a success. Use Monte Carlo simulation (you can discard samples with $\gamma > 100$) to explore the effects of parameterization on prior choice in this context.

    (a) If $\theta \sim \text{Uniform}(0, 1)$, what is the induced prior on γ?
    (b) If $\theta \sim \text{Beta}(0.5, 0.5)$, what is the induced prior on γ?
    (c) If $\gamma \sim \text{Uniform}(0, 100)$, what is the induced prior on θ?
    (d) If $\gamma \sim \text{Gamma}(1, 1)$, what is the induced prior on θ?
    (e) Would you say any of these priors are simultaneously uninformative for both θ and γ?

15. Say $Y|\mu \sim \text{Normal}(\mu, 1)$ and μ has improper prior $\pi(\mu) = 1$ for all μ. Prove that the posterior distribution of μ is a proper PDF.

16. Suppose that the likelihood $f(\mathbf{Y}|\theta) \geq c(\mathbf{Y})$ for all θ. Show that an improper prior on θ will yield an improper posterior.

17. Consider a sample of size n from the mixture of two normal densities given by $f(y|\theta) = 0.5\phi(y) + 0.5\phi(y - \theta)$ where $\phi(z)$ denotes the standard normal density function. Show that any improper prior for θ leads to an improper posterior.

18. Say a randomized clinical trial is conducted to compare the difference in the mean response between two treatment groups. The effect size b is estimated as the difference between the sample means of the two groups, and b has standard error s, and thus the z-score is $z = b/s$. The true difference between treatment group means is β, and the signal-to-noise ratio is $\delta = \beta/s$. We fit the model $z|\delta \sim \text{Normal}(\delta, 1)$ and δ has mixture prior in (2.33).

    (a) Plot the posterior distribution of δ if we observe $z = 2$.
    (b) Compute the posterior probability that $\delta > 0$ given $z = 2$.
    (c) Compute the posterior probability that $\delta > 0$ for value of $z \in \{0.0, 0.1, ..., 5.0\}$ and plot the probabilities by z.
    (d) Repeat these analyses using prior $\delta \sim \text{Normal}(0, 100^2)$, and comment on the effect of the prior.

19. Say $Y|\lambda \sim \text{Poisson}(\lambda)$.

    (a) Derive and plot the Jeffreys prior for λ.
    (b) Is this prior proper?
    (c) Derive the posterior, and give conditions on Y to ensure it is proper.

20. Say $Y|\lambda \sim \text{Gamma}(1, \lambda)$.

(a) Derive and plot the Jeffreys prior for λ.

(b) Is this prior proper?

(c) Derive the posterior, and give conditions on Y to ensure it is proper.

21. Assume the model $Y|\lambda \sim \text{Poisson}(\lambda)$, and we observe $Y = 10$. Conduct a Bayesian analysis of these data with default prior distribution.

22. The data in the table below are the result of a survey of commuters in 10 counties likely to be affected by a proposed addition of a high occupancy vehicle (HOV) lane.

County	Approve	Disapprove	County	Approve	Disapprove
1	12	50	6	15	8
2	90	150	7	67	56
3	80	63	8	22	19
4	5	10	9	56	63
5	63	63	10	33	19

(a) Analyze the data in each county separately using the Jeffreys prior distribution, and report the posterior 95% credible set for each county.

(b) Let $\hat{p}_i$ be the sample proportion of commuters in county i that approve of the HOV lane (e.g., $\hat{p}_1 = 12/(12+50) = 0.194$). Select a and b so that the mean and variance of the Beta(a, b) distribution match the mean and variance of the sample proportions $\hat{p}_1, ..., \hat{p}_{10}$.

(c) Conduct an empirical Bayesian analysis by computing the 95% posterior credible sets that results from analyzing each county separately using the Beta(a, b) prior you computed in (b).

(d) How do the results from (a) and (c) differ? What are the advantages and disadvantages of these two analyses?

3

Computational approaches

CONTENTS

Computing is a key ingredient to any modern statistical application and a Bayesian analysis is no different. As of the late 1980's, the application of Bayesian methods was limited to small problems where conjugacy (Section 2.1) led to a simple posterior or the number of parameters was low enough to permit numerical integration (e.g., Section 3.1.2). However, with the advent of Markov Chain Monte Carlo (MCMC) methods (which were developed in the 1950s through 1970s), the popularity of Bayesian methods exploded as it became possible to fit realistic models to complex data sets. This chapter is dedicated to the core Bayesian computational tools that led to this resurgence.

Bayesian methods are often associated with heavy computing. This is not an inherent property of the Bayesian paradigm. Bayesian computing is designed to estimate the posterior distribution, which is analogous to a frequentist estimation of the sampling distribution. Both problems are challenging, but frequentists often make use of large-sample normal approximations to simplify the problem. There is nothing to prohibit a Bayesian from making similar approximations, and this is the topic of Section 3.1.3. However, these approximations are not common because MCMC produces much richer output, and the accuracy of the approximation is not limited by the sample size of the data, but rather the MCMC approximation can be made arbitrarily precise even for non-Gaussian posteriors by increasing the computational effort.

Bayesian computation boils down to summarizing a high-dimensional posterior distribution (Section 1.4). In this chapter we cover methods that span the spectrum from summaries that are fast to compute (e.g., MAP estimation) but produce only a point estimate to slower methods (e.g., MCMC) that permit full uncertainty quantification. Just as a user of maximum likelihood analysis need not be an expert on the theory and practice of iteratively reweighted least squares optimization, there is software (Section 3.3) available to implement most Bayesian analyses without the user possessing a deep understanding of the algorithms being used under the hood. However, at a minimum, the user should have sufficient understanding of the underlying algorithms to properly inspect the output to determine if it is sensible. In Sections 3.1 and 3.2.1, we outline the fundamental algorithms used in Bayesian computing and work through a few small examples for illustration. However, for the remainder of the book, we will use the `JAGS` package introduced in Section 3.3 for computation to avoid writing MCMC code.

3.1 Deterministic methods

3.1.1 Maximum a posteriori estimation

Most of the computational methods discussed in this chapter attempt to summarize the entire posterior distribution. This allows for full quantification of uncertainty about the parameters and leads to inferences such as the posterior probability that a hypothesis is true. However, summarizing the full posterior distribution can be difficult and in some settings unnecessary. For example, in machine-learning, statistical estimation is used as a filtering process where the sole purpose is prediction of future values using estimates from the current data; accounting for uncertainty and accessing statistical significance are not high priorities. In this case, a single point estimate may be a sufficient summary of the posterior.

The MAP (Section 1.4.1) point estimator is

$$\hat{\boldsymbol{\theta}}_{MAP} = \underset{\boldsymbol{\theta}}{\arg\max} \log[p(\boldsymbol{\theta}|\mathbf{Y})] = \underset{\boldsymbol{\theta}}{\arg\max} \log[f(\mathbf{Y}|\boldsymbol{\theta})] + \log[\pi(\boldsymbol{\theta})]. \tag{3.1}$$

The MAP estimator requires optimization of the posterior, unlike the posterior mean which requires integration. Typically optimization is faster than integration, especially in high-dimensions, and therefore MAP estimation can be applied even to hard problems.

The MAP solution can be found using calculus or numerical optimization [117]. The simplest optimization algorithm is gradient ascent (or gradient descent to minimize a function), which is an iterative algorithm that begins with an initial value $\boldsymbol{\theta}^{(0)}$ and updates the parameters at iteration t using the rule

$$\boldsymbol{\theta}^{(t)} = \boldsymbol{\theta}^{(t-1)} + \gamma\nabla\left(\boldsymbol{\theta}^{(t-1)}\right),$$

where the step size γ is a tuning parameter and the gradient vector of the log posterior is $\nabla(\boldsymbol{\theta}) = [\nabla_1(\boldsymbol{\theta}), ..., \nabla_p(\boldsymbol{\theta})]^T$ for

$$\nabla_j(\boldsymbol{\theta}) = \frac{\partial}{\partial\theta_j}\left\{\log[f(\mathbf{Y}|\boldsymbol{\theta})] + \log[\pi(\boldsymbol{\theta})]\right\}.$$

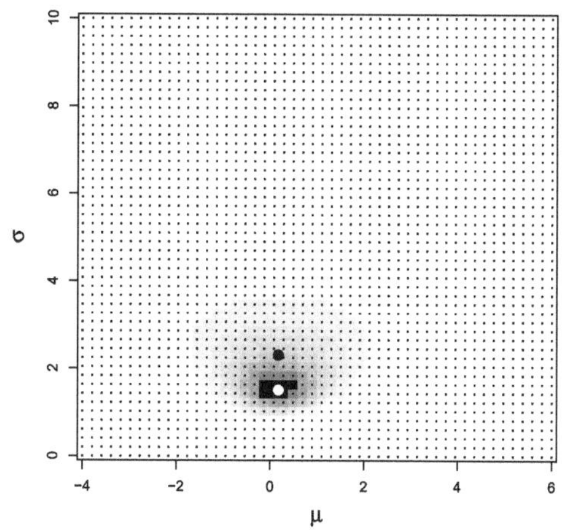

FIGURE 3.1
Numerical integration and optimization. The small points are the grid points $\boldsymbol{\theta}_j = (\mu_j^*, \sigma_j^*)$, and the color of the shading is the posterior density for the model $Y_i|\mu,\sigma \sim \text{Normal}(\mu,\sigma^2)$ with $\mathbf{Y} = (2.68, 1.18, -0.97, -0.98, -1.03)$ and priors $\mu \sim \text{Normal}(0, 100^2)$ and $\sigma \sim \text{Uniform}(0, 10)$. The large black dot is the approximate posterior mean found using numerical integration, and the large white dot is the approximate posterior mode found using numerical optimization.

This step is repeated until convergence, i.e., $\boldsymbol{\theta}^{(t)} \approx \boldsymbol{\theta}^{(t-1)}$. Gradient ascent is most effective when the posterior is convex, and typically requires several starting values to ensure convergence.

There is a rich literature on optimization in statistical computing. Some other useful methods include Newton's method, the expectation-maximization (EM) algorithm, the majorize-minimization (MM) algorithm, genetic algorithms, etc. R has many optimization routines including the general-purpose algorithm `optim`, as illustrated in Listing 3.1 and Figure 3.1 for the model $Y_i|\mu,\sigma \stackrel{iid}{\sim} \text{Normal}(\mu,\sigma^2)$ with priors $\mu \sim \text{Normal}(0, 100^2)$ and $\sigma \sim \text{Unif}(0, 10)$ and data $\mathbf{Y} = (2.68, 1.18, -0.97, -0.98, -1.03)$.

3.1.2 Numerical integration

Many posterior summaries of interest can be written as p-variate integrals over the posterior, including the posterior means, covariances and probabilities of hypotheses. For example, for θ_1, the marginal posterior mean, variance and probability that θ_1 exceeds constant c are

$$\begin{aligned} \text{E}(\theta_1|\mathbf{Y}) &= \int \theta_1 p(\boldsymbol{\theta}|\mathbf{Y}) d\boldsymbol{\theta} \\ \text{Var}(\theta_1|\mathbf{Y}) &= \int [\theta_1 - \text{E}(\theta_1|\mathbf{Y})]^2 p(\boldsymbol{\theta}|\mathbf{Y}) d\boldsymbol{\theta} \\ \text{Prob}(\theta_1 > c|\mathbf{Y}) &= \int_c^\infty \int ... \int p(\boldsymbol{\theta}|\mathbf{Y}) d\theta_1 d\theta_2, ..., d\theta_p. \end{aligned} \tag{3.2}$$

Listing 3.1
Numerical optimization and integration to summarize the posterior.

```
library(cubature)
Y <- c(2.68, 1.18, -0.97, -0.98, -1.03) # Data

# Evaluate the density on the grid for plotting
 m     <- 50
 mu    <- seq(-4,6,length=m)
 sigma <- seq(0,10,length=m)
 theta <- as.matrix(expand.grid(mu,sigma))
 D     <- dnorm(theta[,1],0,100)*dunif(theta[,2],0,10) # Prior
 for(i in 1:length(Y)){ # Likelihood
   D <- D * dnorm(Y[i],theta[,1],theta[,2])
 }
 W     <- matrix(D/sum(D),m,m)

# MAP estimation
 neg_log_post <- function(theta,Y){
   log_like  <- sum(dnorm(Y,theta[1],theta[2],log=TRUE))
   log_prior <- dnorm(theta[1],0,100,log=TRUE)+
                dunif(theta[2],0,10,log=TRUE)
 return(-log_like-log_prior)}

 inits <- c(mean(Y),sd(Y))
 MAP  <- optim(inits,neg_log_post,Y=Y,
               method = "L-BFGS-B", # Since the prior is bounded
               lower = c(-Inf,0), upper = c(Inf,10))$par

# Compute the posterior mean
 post <- function(theta,Y){
   like  <- prod(dnorm(Y,theta[1],theta[2]))
   prior <- dnorm(theta[1],0,100)*dunif(theta[2],0,10)
 return(like*prior)}

 g0 <- function(theta,Y){post(theta,Y)}
 g1 <- function(theta,Y){theta[1]*post(theta,Y)}
 g2 <- function(theta,Y){theta[2]*post(theta,Y)}
 m0 <- adaptIntegrate(g0,c(-5,0.01),c(5,5),Y=Y)$int #constant m(Y)
 m1 <- adaptIntegrate(g1,c(-5,0.01),c(5,5),Y=Y)$int
 m2 <- adaptIntegrate(g2,c(-5,0.01),c(5,5),Y=Y)$int
 pm <- c(m1,m2)/m0

# Make the plot
 image(mu,sigma,W,col=gray.colors(10,1,0),
       xlab=expression(mu),ylab=expression(sigma))
 points(theta,cex=0.1,pch=19)
 points(pm[1],pm[2],pch=19,cex=1.5)
 points(MAP[1],MAP[2],col="white",cex=1.5,pch=19)
 box()
```

If these summaries are sufficient to describe the posterior, then numerical integration can be used for Bayesian computing.

All of the summaries in (3.2) can be written as $\mathrm{E}[g(\boldsymbol{\theta})]$ for some function g, e.g., $\mathrm{Prob}(\theta_1 > c|\mathbf{Y})$ uses $g(\boldsymbol{\theta}) = I(\theta_1 > c)$ where $I(\theta_1 > c) = 1$ if $\theta_1 > c$ and zero otherwise. Assume that a grid of points $\boldsymbol{\theta}_1^*, ..., \boldsymbol{\theta}_m^*$ covers the range of $\boldsymbol{\theta}$ with non-zero posterior density. Then

$$\mathrm{E}[g(\boldsymbol{\theta})] = \int g(\boldsymbol{\theta})p(\boldsymbol{\theta}|Y)d\boldsymbol{\theta} \approx \sum_{j=1}^{m} g(\boldsymbol{\theta}_j)W_j \tag{3.3}$$

where W_j is the weight given to grid point j. To approximate the posterior mean of $g(\boldsymbol{\theta})$ the weights should be related to the posterior PDF $D_j = f(\mathbf{Y}|\boldsymbol{\theta}_j^*)\pi(\boldsymbol{\theta}_j^*)$. The normalizing constant $m(\mathbf{Y})$ cannot be computed and so the posterior is normalized at the grid points as $W_j = D_j / \sum_{l=1}^{m} D_l$.

This provides a very simple approximation to the posterior of $\boldsymbol{\theta}$. Of course, more precise numerical integration can be used, including those in the `R` function `adaptIntegrate` as illustrated in Listing 3.1 and Figure 3.1. However, we do not focus on these methods because they do not scale well with the number of parameters, p. For example, if we use a grid of 20 points for each of the $p = 10$ parameters, then there are $m = 20^{10}$ points in the expanded grid, which is over a trillion points!

3.1.3 Bayesian central limit theorem (CLT)

The central limit theorem states that the sampling distribution of the sample mean for a large sample is approximately normal even for non-Gaussian data. This can be extended to the sampling distribution of the MLE. Many frequentist standard error and p-value computations rely on the approximation

$$\hat{\boldsymbol{\theta}}_{MLE} \sim \mathrm{Normal}(\boldsymbol{\theta}_0, \hat{\boldsymbol{\Sigma}}_{MLE}), \tag{3.4}$$

where $\boldsymbol{\theta}_0$ is the true value and the $p \times p$ covariance matrix $\hat{\boldsymbol{\Sigma}}_{MLE} = (-\mathbf{H})^{-1}$ and the (j,k) element of the Hessian matrix $\mathbf{H}$ is

$$\left.\frac{\partial^2}{\partial\theta_j\partial\theta_k}\log[f(\mathbf{Y}|\boldsymbol{\theta})]\right|_{\boldsymbol{\theta}=\hat{\boldsymbol{\theta}}_{MLE}}. \tag{3.5}$$

To define the analogous Bayesian approximation, suppose that $Y_i|\boldsymbol{\theta} \overset{indep}{\sim} f(Y_i|\boldsymbol{\theta})$ for $i = 1, ..., n$ and $\boldsymbol{\theta} \sim \pi(\boldsymbol{\theta})$. Under general conditions (see Section 9.2.2), for large n the posterior is approximately

$$\boldsymbol{\theta}|\mathbf{Y} \sim \mathrm{Normal}\left(\hat{\boldsymbol{\theta}}_{MAP}, \hat{\boldsymbol{\Sigma}}_{MAP}\right) \tag{3.6}$$

where $\hat{\boldsymbol{\Sigma}}_{MAP} = (-\mathbf{H})^{-1}$ is defined the same way as $\hat{\boldsymbol{\Sigma}}_{MLE}$ except that the Hessian $\mathbf{H}$ is

$$\left.\frac{\partial^2}{\partial\theta_j\partial\theta_k}\log[p(\boldsymbol{\theta}|\mathbf{Y})]\right|_{\boldsymbol{\theta}=\hat{\boldsymbol{\theta}}_{MAP}}, \tag{3.7}$$

which will be similar to $\hat{\boldsymbol{\Sigma}}_{MLE}$ for large samples. Of course, this is not appropriate if the parameters are discrete, but the Bayesian CLT can still be used in some cases where the observations are dependent.

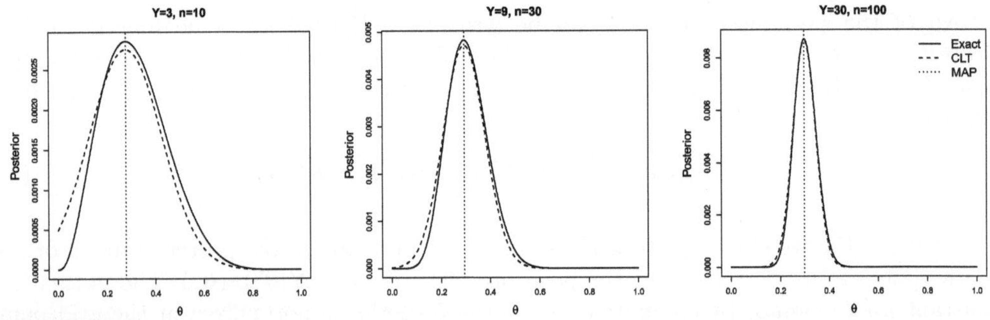

FIGURE 3.2
Illustration of the Bayesian CLT. The exact $\text{Beta}(Y+1/2, n+1/2)$ posterior versus the Gaussian approximation for the model $Y|\theta \sim \text{Binomial}(n, \theta)$ with prior $\theta \sim \text{Beta}(1/2, 1/2)$ for various Y and n.

For example, consider the beta-binomial model $Y|\theta \sim \text{Binomial}(n, \theta)$ with Jeffreys prior $\theta \sim \text{Beta}(1/2, 1/2)$. The MAP estimate is $\hat{\theta}_{MAP} = A/(A+B)$, and the approximate posterior variance is

$$\left[\frac{A}{\hat{\theta}_{MAP}^2} + \frac{B}{(1-\hat{\theta}_{MAP})^2}\right]^{-1},$$

where $A = Y - 0.5$ and $B = n - Y - 0.5$. Of course, there is no need for an approximation in this simple case because the exact posterior is $\theta|Y \sim \text{Beta}(Y + 1/2, n - Y + 1/2)$, but Figure 3.2 illustrates that the Gaussian approximation works well for large n.

As discussed in Section 3.1.1, if the prior is uninformative, then the MAP and MLE estimates are similar, and these approximations suggest that the entire posterior will be similar. This is the first glimpse at a recurring theme: Bayesian and maximum likelihood methods give similar results for large sample sizes irrespective of the prior.

Advantages of the Bayesian CLT approximation are that it replaces integration (Section 3.1.2) with differentiation, and is thus easier to compute, especially for large p, and the method is deterministic rather than stochastic. A disadvantage is that approximation may not be accurate for small or even moderate sample sizes. However, more elaborate approximation methods have been proposed that combine numerical integration with Gaussian approximations, such as the Integrated Nested Laplace Approximation (INLA) of [135] (Appendix A.4). Alternatively, in variational Bayesian inference [85] the user postulates a parametric (potentially non-Gaussian) posterior distribution and then solves for the parameters in the postulated posterior that give the best approximation to the true posterior.

3.1.4 Variational Bayesian (VB) Methods

The normal approximation to the joint posterior distribution motivated by the Bayesian CLT is ideal when the sample size (n) is large and the model size (p) is small and fixed. Under these conditions, the approximation is theoretically justified and computationally efficient. However, for small sizes and/or large models, the posterior may not be well approximated by a normal distribution leading to poor performance. Further, for large models the approximate posterior covariance $\hat{\boldsymbol{\Sigma}}_{MAP}$ is slow to compute because there are on the order of p^2 elements to be computed; if $p = 1{,}000$ then $\hat{\boldsymbol{\Sigma}}_{MAP}$ has 500,500 distinct elements.

Variational Bayesian (VB) approximations address both of these limitations and have thus emerged as one of the leading computational methods for high-dimensional problems. As with the Bayesian CLT approximation, VB presumes the posterior resides in a parametric family of distributions. However, this distribution need not be Gaussian or even the same family for all parameters; e.g., it may be presumed that the posterior of a variance parameter follows a gamma distribution and a mean parameter follows a normal distribution. To avoid computing posterior covariance terms, it can be presumed that some or all parameters are independent in the posterior. These assumptions can lead to massive computational savings but should be applied carefully because gross violations lead to poor performance.

A VB analysis begins by specifying a variational posterior. The variational posterior is the parametric family of distributions that is assumed to include a good approximation to the true posterior. Denote the variational posterior density function as $q(\boldsymbol{\theta}|\mathbf{v})$, where $\mathbf{v} = (v_1, ..., v_q)$ are the variational parameters. For example, if there is a single parameter θ with support $[0, 1]$, then we might select $q(\theta|\mathbf{v})$ to be the Beta(v_1, v_2) PDF and assume that for some value of $\mathbf{v} = (v_1, v_2)$, $p(\theta|\mathbf{Y}) \approx q(\theta|\mathbf{v})$.

The assumption that the posterior resides in a parametric family, q, simplifies the problem down to finding the values of the variational parameters, $\mathbf{v}$, that give a close approximation. These parameters are chosen to minimize the discrepancy between the variational and true posterior distributions. This requires a measure of the similarity between two distributions, and typically the Kullback–Leibler divergence (KLD) is used (Figure 3.3). The KLD is the

$$KLD(\mathbf{v}) = \mathrm{E}_q\left[\log\left\{\frac{q(\boldsymbol{\theta}|\mathbf{v})}{p(\boldsymbol{\theta}|\mathbf{Y})}\right\}\right] = \int \log\left\{\frac{q(\boldsymbol{\theta}|\mathbf{v})}{p(\boldsymbol{\theta}|\mathbf{Y})}\right\} q(\boldsymbol{\theta}|\mathbf{v})d\boldsymbol{\theta}. \tag{3.8}$$

The KLD is non-negative for any $\mathbf{v}$ and equals zero if the approximation is perfect and $q(\boldsymbol{\theta}|\mathbf{v}) = p(\boldsymbol{\theta}|\mathbf{Y})$ for all $\boldsymbol{\theta}$. The VB approximation selects $\mathbf{v}$ to minimize $KLD(\mathbf{v})$.

Similar to selecting a prior distribution, there is no single correct variational posterior family (with the trivial exception of the true posterior distribution), but families with the same support as the parameters are a good starting place. Especially for high-dimensional problems, postulating a variational distribution that captures the dependence between parameters is difficult, and so the mean-field approximation that assumes the posterior distributions of the parameters are independent is appealing,

$$q(\boldsymbol{\theta}|\mathbf{v}) = \prod_{j=1}^{p} q_j(\theta_j|\mathbf{v}_j).$$

Finding the optimal $\mathbf{v}$ requires computing or approximating the integral in (3.8) and then optimizing it with respect to $\mathbf{v}$. In some cases, the integral can be solved analytically using similar techniques to find conjugate priors in Section 2.1, in others numerical approximation is required (using algorithms mentioned in 3.1.2). Optimization can also achieved analytically in small problems and numerically (using algorithms mentioned in Section 3.1.1) in most cases. Conjugate gradient descent which optimizes over single elements of $\mathbf{v}$ in sequence is often used when $\mathbf{v}$ is high-dimensional.

To illustrate this approach, consider the global mean temperature anomaly (difference between current year and the 1951-1980 average) data on the course website (downloaded from https://data.giss.nasa.gov/gistemp/) in Figure 3.4. We analyze only the data in the stable period prior to 1980 using a first-order autoregressive time series model. Since the data are normalized, we assume the mean response is zero. The first observation is $Y_1 \sim \text{Normal}(0, \sigma^2)$, and subsequent observations are $Y_t|Y_{t-1} \sim \text{Normal}(\rho Y_{t-1}, \sqrt{1-\rho^2}\sigma^2)$, where $\rho \in (0, 1)$ controls the temporal correlation and σ^2 controls the variance. In this model, the marginal distribution has mean zero and variance σ^2 for all t and $\text{Cor}(Y_t, Y_{t+h}) =$

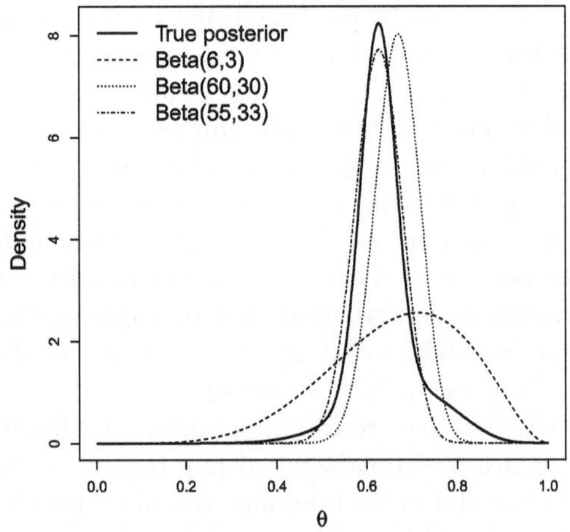

FIGURE 3.3
Illustration of the Kullback–Leibler divergence (KLD) The KLD between the (hypothetical) true posterior (solid line) and the three beta distributions (dashed lines) are 0.74 for Beta(6,3), 0.31 for Beta(60,30) and 0.08 for Beta(55,33).

ρ^h. The Bayesian model is completed by specifying priors that match the parameters' support, $\sigma^2 \sim \text{InvGamma}(a, b)$ and $\rho \sim \text{Beta}(c, d)$ with $a = b = 0.1$ and $c = d = 1$. The posterior is

$$\begin{aligned} p(\sigma^2, \rho|\mathbf{Y}) &\propto (\sigma^2)^{-n/2-a-1}(1-\rho^2)^{-(n-1)} \exp\left(-\frac{S(\rho)}{2\sigma^2}\right) \\ &\times (\sigma^2)^{a-1} \exp\left(-\frac{b}{\sigma^2}\right) \rho^{c-1}(1-\rho)^{d-1} \end{aligned} \tag{3.9}$$

where $S(\rho) = Y_1^2 + \sum_{t=2}^n (Y_t - \rho Y_{t-1})^2/(1-\rho^2)$.

To match the support of the two parameters, we take the variational posterior to be $\sigma^2|\mathbf{Y} \sim \text{InvGamma}(v_1, v_2)$ independent of $\rho|\mathbf{Y} \sim \text{Beta}(v_3, v_4)$. The variational posterior is then

$$\begin{aligned} q(\sigma^2, \rho|\mathbf{v}) &= \frac{v_2^{v_1}}{\Gamma(v_1)}(\sigma^2)^{-(v_1+1)} \exp(-v_2/\sigma^2) \\ &\times \frac{\Gamma(v_3+v_4)}{\Gamma(v_3)\Gamma(v_4)} \rho^{v_3-1}(1-\rho)^{v_4-1}. \end{aligned} \tag{3.10}$$

Summarizing the posterior is then reduced to finding the values of $\mathbf{v} = (v_1, v_2, v_3, v_4)$ so that q in (3.10) approximates p in (3.9). The integral in the KLD is over both σ^2 and ρ, and can be taken analytically for σ^2 but requires univariate numerical integration for ρ.

The true posterior plotted in the upper left panel of Figure 3.5 clearly shows dependence between the two parameters, violating the mean field approximation of posterior independence. As a result, the VB approximation in the top right is poor overall. Surely there are richer variational families that would yield a better fit, but even for this simple mean-field

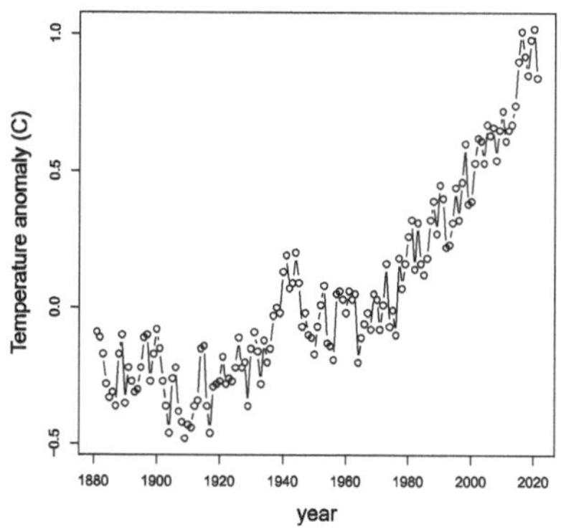

FIGURE 3.4
Global mean temperature data. Global mean temperature anomaly (difference between current year and the 1951-1980 average) by year.

approximation the marginal distributions in the bottom row of Figure 3.5 are centered near the true posterior distribution, and thus the VB point estimates are similar to those from the true posterior.

3.2 Stochastic methods

Monte Carlo (MC) methods are appealing to statisticians because they mirror the fundamental statistical concept of using a sample to make inferences about a population. A canonical problem in statistics is to estimate a summary of the population, such as the population mean or standard deviation. In most cases (i.e., not a census) we cannot observe the entire population to compute the mean or standard deviation directly, so instead we take samples and use them to make inferences about the population. For example, we use the sample mean to approximate the population mean, and, assuming the sample is sufficiently large, the law of large numbers guarantees the approximation is reliable.

MC sampling from the posterior works the same way. In this analogy, the population of interest is the posterior distribution. We would like to summarize the posterior using its mean and variance, but in most cases these posterior summaries cannot be computed directly, and so we take a sample from the posterior and use the MC sample mean to approximate the posterior mean and the MC sample variance to approximate the posterior variance. Assuming the MC sample is sufficiently large, then the approximation is reliable. This holds for virtually any posterior distribution and any summary of the posterior, even those that are high-dimensional integrals.

Assume that we have generated S posterior samples, $\boldsymbol{\theta}^{(1)},...,\boldsymbol{\theta}^{(S)} \sim p(\boldsymbol{\theta}|\mathbf{Y})$, where sample s is $\boldsymbol{\theta}^{(s)} = (\theta_1^{(s)},...,\theta_p^{(s)})$. These draws can be used to approximate any posterior summary discussed in Section 1.4. For a univariate model with $p = 1$, we could approximate the posterior density with a histogram of the S samples, the posterior mean with the MC sample mean $\mathrm{E}(\theta_1|\mathbf{Y}) \approx \sum_{s=1}^{S} \theta_1^{(s)}/S$, the posterior variance with the MC sample variance

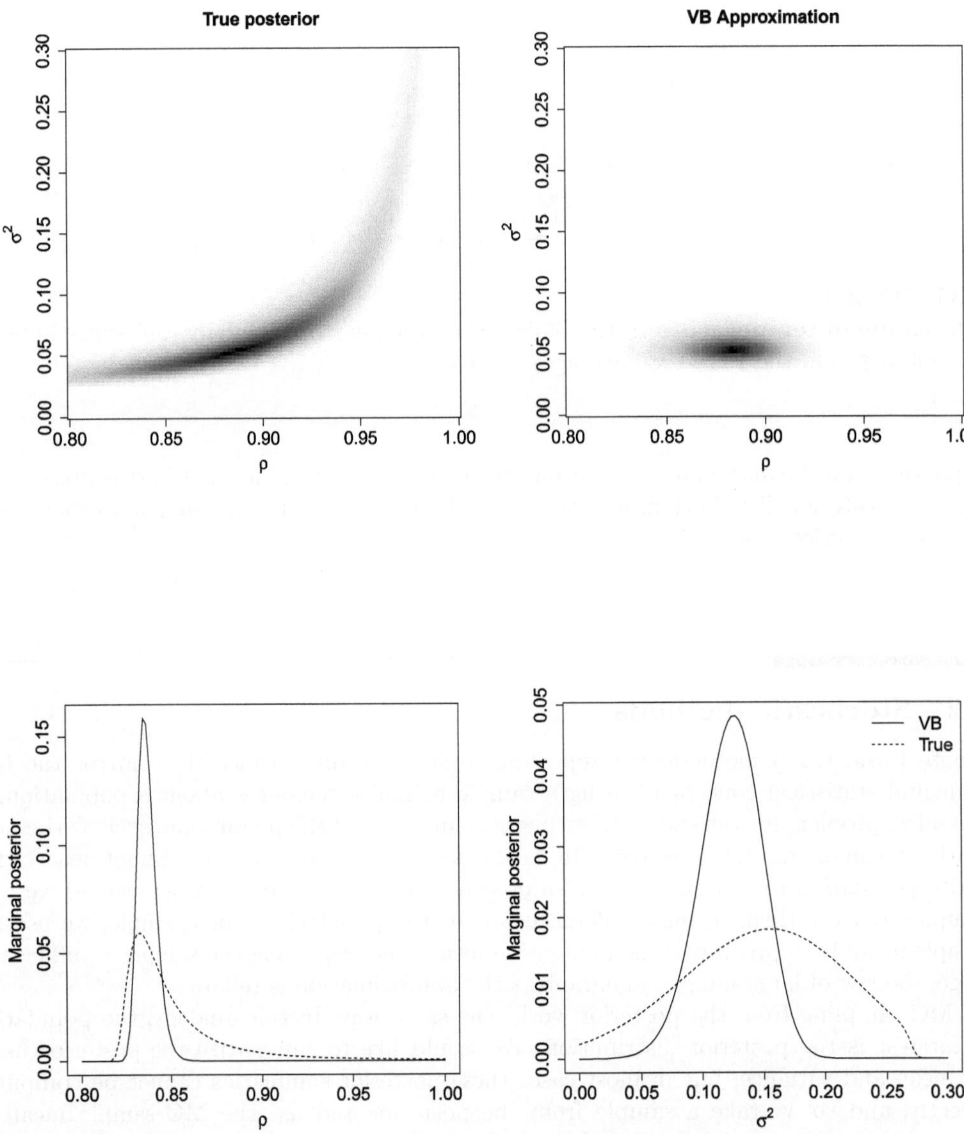

FIGURE 3.5
Variational Bayesian analysis of the global temperature data. The top row shows the exact and VB bivariate posterior distributions, and the bottom row shows the corresponding marginal distributions.

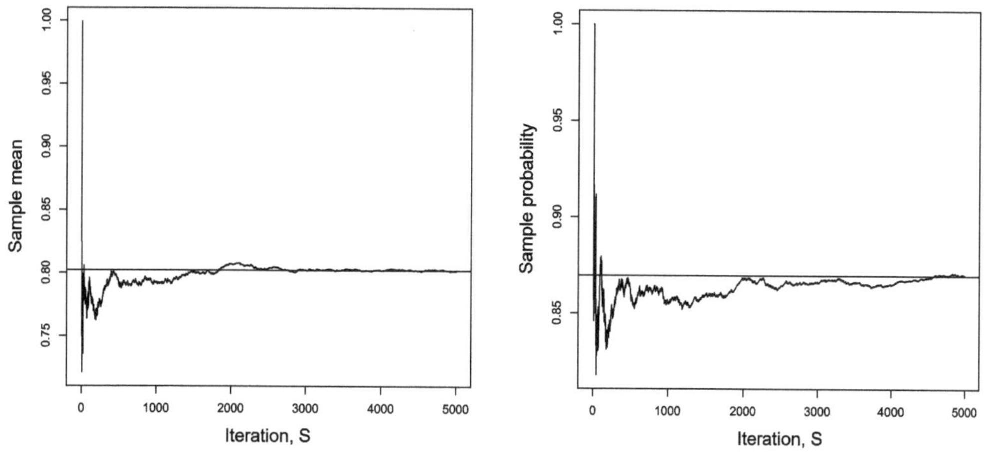

FIGURE 3.6
Convergence of a Monte Carlo approximation. Assuming $\theta^{(s)} \stackrel{iid}{\sim}$ Gamma(8.1, 10.1), the plot below gives the sample mean $\sum_{s=1}^{S} \theta^{(s)}/S$ and sample proportion $\sum_{s=1}^{S} I(\theta^{(s)} > 0.5)/S$ by the number of samples, S; the horizontal lines are the true values.

of the S samples, the probability that $\theta_1 > 0$ with the MC sample proportion with $\theta_1^{(s)} > 0$, etc.

To illustrate convergence of MC sampling, consider the Poisson-gamma model $Y|\theta \sim$ Poisson($N\theta$) with prior $\theta \sim$ Gamma(a, b). The posterior is then $\theta|Y \sim$ Gamma($Y{+}a, N{+}b$). Figure 3.6 assumes $N = 10$, $Y = 8$ and $a = b = 0.1$ and plots the approximate posterior mean and probability that $\theta > 0.5$ as a function of the number of MC iterations. The sample mean and proportion are noisy for a small number of MC samples, but as the number of samples increases, they converge to the true posterior mean and probability that $\theta > 0.5$.

For multivariate models with $p > 1$ parameters, the samples $\theta_j^{(1)}, ..., \theta_j^{(S)}$ follow the marginal posterior distribution of θ_j, $p(\theta_j|\mathbf{Y})$. Critically, we do not need to analytically integrate $p(\theta_j|\mathbf{Y}) = \int f(\boldsymbol{\theta}|\mathbf{Y})d\theta_1...d\theta_{j-1}d\theta_{j+1}...d\theta_p$. Because each sample consists of a random draw from all parameters, MC sampling automatically produces samples from the marginal distribution θ_j accounting for uncertainty in the other parameters.

In addition to the marginal distribution, MC sampling can be used to approximate the posterior of parameters defined as transformations of the original parameters. For example, we can approximate the posterior of $\gamma = g(\theta_1)$ for a function g using the MC samples $\gamma^{(1)}, ..., \gamma^{(S)}$ where $\gamma^{(s)} = g(\theta_1^{(s)})$; that is, we simply transform each sample, and these transformed samples approximate the posterior distribution of γ. The posterior of the transformed parameter is summarized like any other, e.g., $\text{E}(\gamma|\mathbf{Y}) \approx \sum_{s=1}^{S} \gamma^{(s)}/S$. As another example, say we want to test if $\Delta = \theta_1 - \theta_2 > 0$. The posterior of Δ is approximated by the S samples $\Delta^{(s)} = \theta_1^{(s)} - \theta_2^{(s)}$, and thus Prob($\Delta > 0|\mathbf{Y}$) is approximated using the proportion of the S samples for which $\Delta^{(s)} > 0$.

Once we have posterior samples, summarizing the posterior or even complicated functions of the posterior is straightforward, and this is one of the appeals of MC sampling. However, generating valid samples from the posterior distribution is not always straightforward.

3.2.1 MCMC sampling

We will focus on two sampling algorithms: Gibbs sampling and Metropolis–Hastings sampling. Gibbs sampling is preferred when it is possible to directly sample from the conditional distribution of each parameter. Metropolis–Hastings is a generalization to more complicated problems. We also briefly mention other more advanced algorithms in Appendix A.4 including delayed rejection and adaptive Metropolis (DRAM), the Metropolis-adjusted Langevin algorithm, Hamiltonian MC (HMC) and slice sampling.

3.2.1.1 Gibbs sampling

As a motivating example, consider the Gaussian model

$$Y_1, ..., Y_n|\mu, \sigma^2 \stackrel{iid}{\sim} \text{Normal}(\mu, \sigma^2) \tag{3.11}$$

where the priors are $\mu \sim \text{Normal}(\gamma, \tau^2)$ and $\sigma^2 \sim \text{InvGamma}(a, b)$. To study the posterior of $\boldsymbol{\theta} = (\mu, \sigma^2)$ we would like to make S draws from the joint posterior distribution $p(\boldsymbol{\theta}|\mathbf{Y})$. Section 2.1.3 provides a means to sample from the posterior of μ with σ^2 assumed known since $\mu|\sigma^2, \mathbf{Y}$ follows a Gaussian distribution, and Section 2.1.4 provides a means to sample from the posterior of σ^2 with μ assumed known since $\sigma^2|\mu, \mathbf{Y}$ follows an inverse gamma distribution. Gibbs sampling generates draws from the desired joint posterior of $\boldsymbol{\theta}$ using only these univariate conditional distributions.

Gibbs sampling (proposed by [59]) begins with initial values for both parameters and proceeds by alternating between sampling $\mu|\sigma^2, \mathbf{Y}$ and then sampling $\sigma^2|\mu, \mathbf{Y}$ until S samples have been collected (for an extensive discussion about selecting S, see Section 3.4). For example, we might begin by setting μ to the sample mean of $\mathbf{Y}$ and σ^2 to the sample variance of $\mathbf{Y}$ (for an extensive study of initialization, see Section 3.4). Implementing successive steps requires deriving the full conditional posterior distributions, i.e., the distribution of one parameter conditioned on the data and all other parameters. Following Section 2.1.3 and 2.1.4 (with slight modification to the prior distributions), the full conditional posterior distributions are

$$\begin{aligned}
\mu|\sigma^2, \mathbf{Y} &\sim \text{Normal}\left(\frac{\sum_i^n Y_i/\sigma^2 + \gamma/\tau^2}{n/\sigma^2 + 1/\tau^2}, \frac{1}{n/\sigma^2 + 1/\tau^2}\right) \\
\sigma^2|\mu, \mathbf{Y} &\sim \text{InvGamma}\left(n/2 + a, \sum_{i=1}^n (Y_i - \mu)^2/2 + b\right).
\end{aligned}$$

This results in S posterior samples $\boldsymbol{\theta}^{(1)}, ..., \boldsymbol{\theta}^{(S)}$ to summarize the posterior, where $\boldsymbol{\theta}^{(s)} = (\mu^{(s)}, \sigma^{2(s)})$. Listing 3.2 provides `R` code to perform these steps using the same data as in Section 1.1.2. Figure 3.7 (bottom row) plots the draws from the joint posterior and marginal density for μ, which both closely resemble Figure 1.11.

Algorithm 1 provides a general recipe for Gibbs sampling. The algorithm begins with an initial value for the parameter vector, and each iteration consists of a sequence of updates from each parameter's full conditional distribution. Because each step cycles through the parameters and updates them given the current values of all other parameters, the samples are not independent. All sampling is performed conditionally on the previous iteration's value, and so the samples form a specific type of stochastic process called a Markov chain, hence the name Markov chain Monte Carlo (MCMC) sampling. There are MC samplers that attempt to sample independent draws from the posterior (e.g., rejection sampling and approximate Bayesian computing (ABC) [100]), but MCMC is preferred for high-dimensional problems.

Listing 3.2
Gibbs sampling for the Gaussian model with unknown mean and variance.

```
 # Load the data

 Y <- c(2.68,1.18,-0.97,-0.98,-1.03)
 n <- length(Y)

# Create an empty matrix for the MCMC samples

 S                  <- 25000
 samples            <- matrix(NA,S,2)
 colnames(samples) <- c("mu","sigma")

# Initial values

 mu   <- mean(Y)
 sig2 <- var(Y)

# priors: mu ~ N(gamma,tau), sig2 ~ InvG(a,b)

 gamma <- 0
 tau   <- 100^2
 a     <- 0.1
 b     <- 0.1

# Gibbs sampling

 for(s in 1:S){
   P    <- n/sig2 + 1/tau
   M    <- sum(Y)/sig2 + gamma/tau
   mu   <- rnorm(1,M/P,1/sqrt(P))

   A    <- n/2 + a
   B    <- sum((Y-mu)^2)/2 + b
   sig2 <- 1/rgamma(1,A,B)

   samples[s,]<-c(mu,sqrt(sig2))
 }

 # Plot the joint posterior and marginal of mu
 plot(samples,xlab=expression(mu),ylab=expression(sigma))
 hist(samples[,1],xlab=expression(mu))

 # Posterior mean, median and credible intervals
 apply(samples,2,mean)
 apply(samples,2,quantile,c(0.025,0.500,0.975))
```

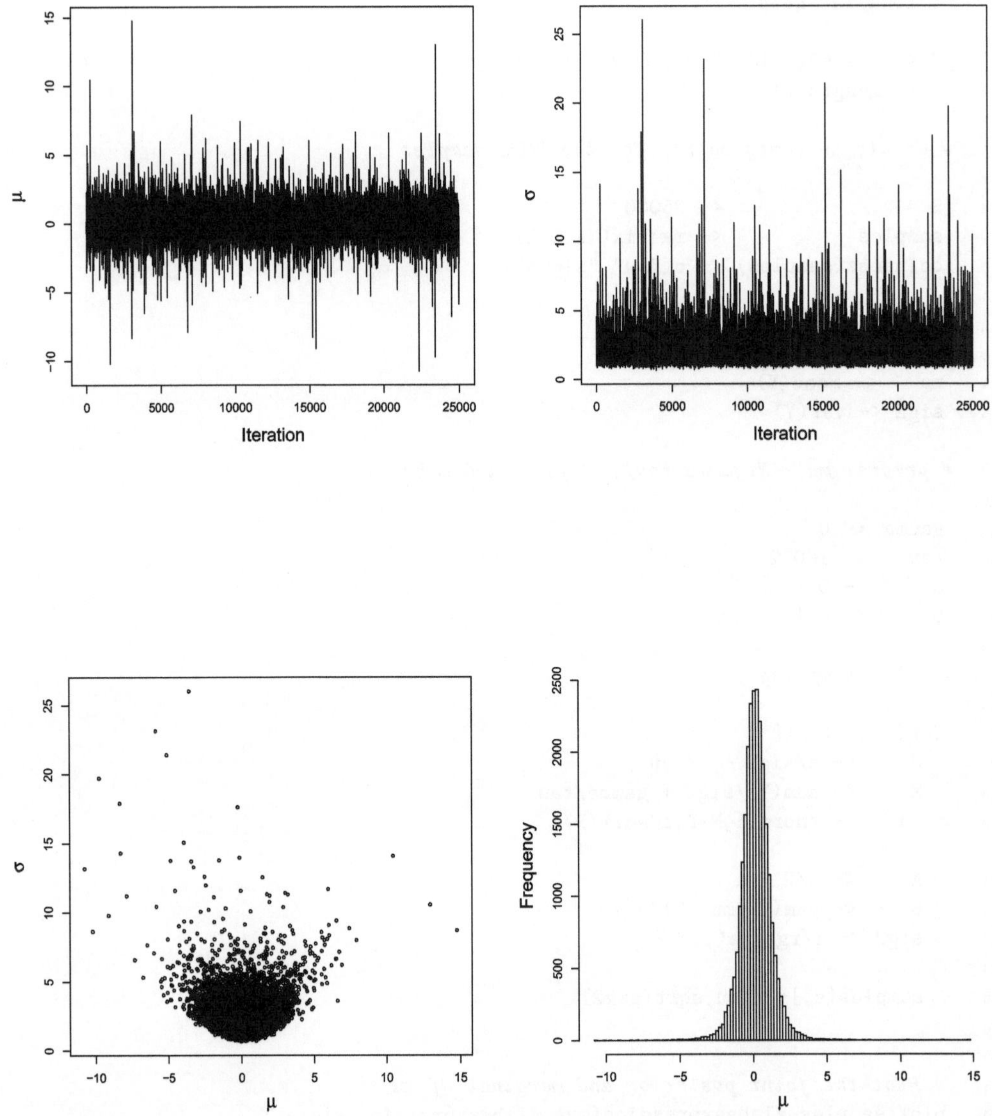

FIGURE 3.7
Summary of posterior samples from MCMC for the Gaussian model with unknown mean and variance. The first row gives trace plots of the samples for μ and σ; the second row shows the samples from the joint posterior of (μ, σ) and the marginal posterior of μ.

Algorithm 1 Gibbs Sampling

Initialize $\boldsymbol{\theta}^{(0)} = (\theta_1^{(0)}, ..., \theta_p^{(0)})$
for $s = 1, ..., S$ **do**
  for $j = 1, ..., p$ **do**
    sample $\theta_j^{(s)} \sim p_j(\theta_j|\theta_1^{(s)}, ..., \theta_{j-1}^{(s)}, \theta_{j+1}^{(s-1)}, ..., \theta_p^{(s-1)}, \mathbf{Y})$
  end for
end for

The beauty of this algorithm is that it reduces the difficult problem of sampling from a multivariate distribution down to a sequence of simpler univariate problems. This assumes that the full conditional distributions are easy to sample from, but as we will see in the following examples, even for high-dimensional problems with large p, the full conditional distributions often follow familiar conjugacy pairs (Section 2.1) which are conducive to sampling.

Before continuing with other examples, we pause and discuss the theoretical properties of this algorithm. Why does it work? That is, why does repeated sampling from full conditional distributions lead to samples from the joint posterior distribution? Surely we cannot trust that the initial value is a sample from the posterior because it is selected subjectively by the user. It would also be dangerous to trust that the first random sample follows the posterior distribution because it depends on the initial values, which might be far from the posterior. Appendix A.3 provides an argument using stochastic process theory that (1) under general conditions, the samples produced by the algorithm converge to the posterior distribution for any choice of initial values, and (2) once a sample is drawn from the posterior, all subsequent samples are also samples from the posterior.

The theoretical convergence arguments dictate how Gibbs sampling is used in practice. The user follows the chain until it has converged, and discards all previous samples from this burn-in period. Convergence is often assessed by visual inspection of the trace plots (i.e., the a plot of the samples by iteration number), but formal measures are available (Section 3.4). All remaining samples are used to summarize the posterior. It is important to remember that unlike other optimization algorithms, we do not expect Gibbs sampling to converge to a single optimal value. Rather, we hope that after burn-in the algorithm produces samples from the posterior distribution. That is, we hope the samples generated by the algorithm converge in distribution to the posterior, and do not get stuck in one place. Figure 3.8 provides idealized output where convergence clearly occurs around iteration 1000 for both parameters and the remaining samples consistently draw from the same distribution after this point and thus the trace plot resembles a "bar code" or "caterpillar."

Bivariate normal example: We will use Gibbs sampling to make draws from posterior distributions, but in fact it can be used to sample from any distribution (so long as we can sample from the full conditional distributions). For example, say $\boldsymbol{\theta} = (U, V)$ is bivariate normal with mean zero and variance 1 for both parameters and correlation ρ between the parameters. This is not a posterior distribution because we are not conditioning on data, but in this toy example we are using Gibbs sampling to draw from this bivariate normal distribution to get a feel for the algorithm and how it converges to the target distribution. As given in (1.26) (see Figure 1.7), the full conditional distributions are $U|V \sim \text{Normal}(\rho V, 1-\rho^2)$ and $V|U \sim \text{Normal}(\rho U, 1-\rho^2)$. Given initial values $(U^{(0)}, V^{(0)})$, the first three iterations are:

1a Draw $U^{(1)}|V^{(0)} \sim \text{Normal}(\rho V^{(0)}, 1-\rho^2)$
1b Draw $V^{(1)}|U^{(1)} \sim \text{Normal}(\rho U^{(1)}, 1-\rho^2)$
2a Draw $U^{(2)}|V^{(1)} \sim \text{Normal}(\rho V^{(1)}, 1-\rho^2)$

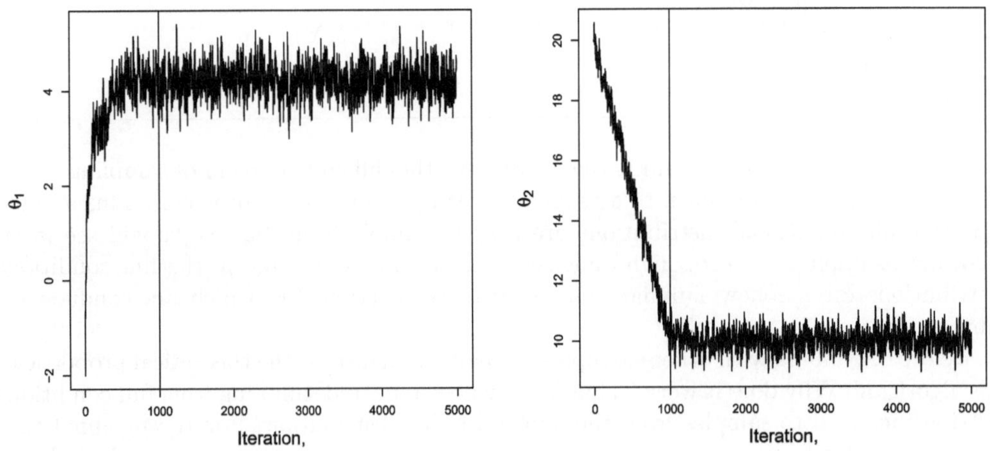

FIGURE 3.8
Gibbs sampling trace plots. These trace plots of the iteration number s by the posterior samples $\theta_j^{(s)}$ for $j = 1, 2$ show convergence at iteration 1000, which is denoted with a vertical line.

2b Draw $V^{(2)}|U^{(2)} \sim \text{Normal}(\rho U^{(2)}, 1-\rho^2)$
3a Draw $U^{(3)}|V^{(2)} \sim \text{Normal}(\rho V^{(2)}, 1-\rho^2)$
3b Draw $V^{(3)}|U^{(3)} \sim \text{Normal}(\rho U^{(3)}, 1-\rho^2)$.

The algorithm continues until S draws have been made.

It can be shown that the distribution after s iterations is

$$U^{(s)} \sim \text{Normal}(\rho^{2s-1}V^{(0)}, 1-\rho^{4s-2}).$$

The true marginal distribution of U is Normal(0, 1), and since $\rho^{2s-1} \approx 0$ and $\rho^{4s-2} \approx 0$ for large s, for any initial value the MCMC posterior is very close but never exactly equal to the true posterior. Convergence is immediate if $\rho = 0$ and slow if $|\rho| \approx 1$, illustrating how cross-correlation between parameters can hinder MCMC convergence (e.g., Figure 3.11 and the discussion of blocked Gibbs sampling below). Nonetheless, for any ρ and large S the approximation is surely sufficient and can be made arbitrarily precise by increasing S.

Constructing a Gibbs sampler for NFL concussion data: In Section 2.1 we analyzed the NFL concussion data. The data consist of the number of concussions in the years 2012–2015, $Y_1 = 171$, $Y_2 = 152$, $Y_3 = 123$, and $Y_4 = 199$, respectively. In Section 2.1 we analyzed the years independently, but here we analyze all years simultaneously with the model

$$Y_i|\lambda_i \overset{indep}{\sim} \text{Poisson}(N\lambda_i) \quad \lambda_i|\gamma \overset{indep}{\sim} \text{Gamma}(1,\gamma) \quad \gamma \sim \text{Gamma}(a,b) \tag{3.12}$$

where $N = 256$ is the number of games per season. The model has five unknown parameters $\boldsymbol{\theta} = (\lambda_1, ..., \lambda_4, \gamma)$ and posterior

$$p(\lambda_1, ..., \lambda_4, \gamma|\mathbf{Y}) \propto \left[\prod_{i=1}^{4} f(Y_i|\lambda_i)\pi(\lambda_i|\gamma)\right]\pi(\gamma), \tag{3.13}$$

where f is the Poisson PMF and π is the gamma PDF.

The Gibbs sampler requires the full conditional posterior for each of the five parameters. We first compute the full conditional of λ_1 given all the other parameters. For this computation, we only need to consider terms that depend on λ_1, and all other terms can be absorbed in the normalizing constant. For λ_1 this gives

$$p(\lambda_1, |\lambda_2, \lambda_3, \lambda_4, \gamma, \mathbf{Y}) \propto f(Y_1|\lambda_1)\pi(\lambda_1|\gamma). \tag{3.14}$$

This is exactly the form of the posterior for the Poisson-gamma conjugacy pair studied in Section 2.1, and so the full conditional of λ_1 is Gamma$(Y_1+1, N+\gamma)$; the full conditionals for the other λ_j are similar.

The terms in the likelihood $f(Y_i|\lambda_i)$ do not depend on γ, and so the full conditional is

$$p(\gamma|\lambda_1, \lambda_2, \lambda_3, \lambda_4, \mathbf{Y}) \propto \left[\prod_{i=1}^{4} \pi(\lambda_i|\gamma)\right] \pi(\gamma). \tag{3.15}$$

The Gamma$(1, \gamma)$ prior is $\pi(\lambda_i|\gamma) = \gamma \exp(-\gamma\lambda_i)$. Therefore, the full conditional reduces to

$$\begin{aligned} p(\gamma|\lambda_1, \lambda_2, \lambda_3, \lambda_4, \mathbf{Y}) &\propto \left[\prod_{i=1}^{4} \gamma \exp(-\gamma\lambda_i)\right] \gamma^{a-1} \exp(-\gamma b) \\ &\propto \gamma^{4+a-1} \exp\left[-\gamma\left(\sum_{i=1}^{4} \lambda_i + b\right)\right], \end{aligned}$$

and thus the update for γ is Gamma$(4+a, \sum_{i=1}^{4} \lambda_i + b)$. Note that the update does not depend on the data, $\mathbf{Y}$. However, the data are informative about γ indirectly because they largely determine the posterior of the λ_j which in turn influence the posterior of γ.

Listing 3.3 provides the code that generates the results in Figure 3.9. Convergence for γ (left panel of Figure 3.9) is immediate (as it is for the λ_i, not shown), and so we do not need a burn-in period, and all samples are used to summarize the posterior. The posterior distribution from this analysis of all years (right panel of Figure 3.9) is actually quite similar to the one-year-at-a-time analysis in Figure 2.2.

Constructing a blocked Gibbs sampler for a T. rex growth chart: [48] studies growth charts for several tyrannosaurid dinosaur species, including the tyrannosaurus rex (T. rex). There $n = 6$ T. rex observations with weights (kg) 29.9, 1761, 1807, 2984, 3230, 5040, and 5654 and corresponding ages (years) 2, 15, 14, 16, 18, 22, and 28, plotted in Figure 3.10 (top left). Non-linear models are likely more appropriate for growth curve data (Section 7.2.2), but we fit a linear model for illustration. We assume that $Y_i = \beta_1 + x_i\beta_2 + \varepsilon_i$ where Y_i is the weight of dinosaur i, x_i is the age, β_1 and β_2 are the regression intercept and slope, respectively, and $\varepsilon_i \overset{iid}{\sim}$ Normal$(0, \sigma^2)$. We select uninformative priors $\beta_j \overset{indep}{\sim}$ Normal$(0, \tau)$ and $\sigma^2 \sim$ InvGamma(a, b) with $\tau = 10{,}000^2$ and $a = b = 0.1$. Figure 3.10 (top right) plots the posterior distribution of the three parameters $\boldsymbol{\theta} = (\beta_1, \beta_2, \sigma)$. The error standard deviation σ is not strongly dependent on the other parameters, but the regression coefficients $\boldsymbol{\beta} = (\beta_1, \beta_2)$ have correlation Cor$(\beta_1, \beta_2) = -0.91$.

Gibbs sampling has trouble when parameters have strong posterior dependence. To see this, consider the hypothetical example plotted in Figure 3.11. The two parameters $\boldsymbol{\beta} = (\beta_1, \beta_2)$ follow a bivariate normal posterior with means equal to zero, variances equal to one, and correlation equal -0.98. The initial values are $\boldsymbol{\beta}^{(0)} = (0, -3)$. The first step is to update β_1 with $\beta_2 = -3$. This corresponds to a draw along the bottom row of Figure 3.11, and so the sample is $\beta_1 \approx 3$. The next update is from the conditional distribution of β_2 given $\beta_1 \approx 3$. Since the two variables are negatively correlated, with $\beta_1 \approx 3$ we must

Listing 3.3
Gibbs sampling for the NFL concussion data.

```
# Load data
 Y <- c(171, 152, 123, 199)
 n <- 4
 N <- 256
# Create an empty matrix for the MCMC samples
 S                  <- 25000
 samples            <- matrix(NA,S,5)
 colnames(samples) <- c("lam1","lam2","lam2","lam4","gamma")
# Initial values
 lambda <- log(Y/N)
 gamma  <- 1/mean(lambda)
# priors: lambda[i]|gamma ~ Gamma(1,gamma), gamma ~ InvG(a,b)
 a <- 0.1
 b <- 0.1
# Gibbs sampling
 for(s in 1:S){
   for(i in 1:n){
     lambda[i] <- rgamma(1,Y[i]+1,N+gamma)
   }
   gamma      <- rgamma(1,4+a,sum(lambda)+b)
   samples[s,] <- c(lambda,gamma)
 }
 boxplot(samples[,1:4],outline=FALSE,
         ylab=expression(lambda),names=2012:2015)
 plot(samples[,5],type="l",xlab="Iteration",
        ylab=expression(gamma))
 # Posterior mean, median and credible interval
 apply(samples,2,mean)
 apply(samples,2,quantile,c(0.025,0.500,0.975))
```

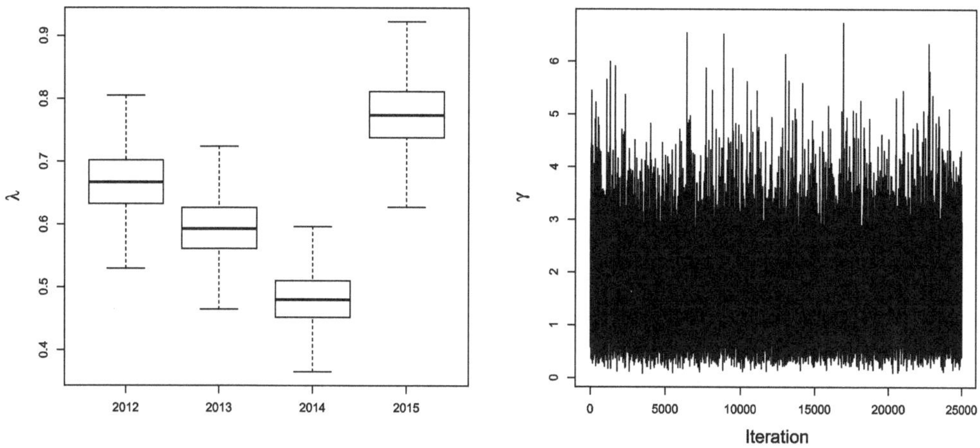

FIGURE 3.9
MCMC analysis of the NFL concussion data. The left panel plots the posterior distribution of the concussion rate λ_i for each year, and the right panel is a trace plot of the hyperparameter γ.

have $\beta_2 \approx -3$, and so β_2 is only slightly changed from its initial value. These small updates continue for the next four iterations, and $\boldsymbol{\beta}$ stays in the lower-right quadrant. This toy example illustrates that with strong posterior dependence between parameters, the one-at-a-time Gibbs sampler will slowly traverse the parameter space, leading to poor convergence.

One way to improve convergence is to update dependent parameters in blocks. Returning to the T. rex example, since β_1 and β_2 are correlated with each other but not σ^2 (Figure 3.10), we might set blocks $\boldsymbol{\theta}_1 = \boldsymbol{\beta}$ and $\theta_2 = \sigma^2$ and apply Algorithm 1 by alternating between sampling from $\boldsymbol{\theta}_1|\theta_2, \mathbf{Y}$ and $\theta_2|\boldsymbol{\theta}_1, \mathbf{Y}$. Since $\boldsymbol{\beta}$ is drawn from its joint posterior, it will not get stuck in a corner as it might in one-at-a-time sampling, and thus convergence is improved.

To derive the full conditional distributions of the blocks and MCMC code we use matrix notation. Let the data vector be $\mathbf{Y} = (Y_1, ..., Y_n)^T$ and $\mathbf{X}$ be an $n \times 2$ covariate matrix with the ith row equal to $(1, x_i)$. The linear regression model can be written $\mathbf{Y}|\boldsymbol{\beta}, \sigma^2 \sim \text{Normal}(\mathbf{X}\boldsymbol{\beta}, \sigma^2\mathbf{I}_n)$. The full conditional distributions (derived in Appendix A.3 and discussed in Section 2.1.6) are

$$
\begin{aligned}
\boldsymbol{\beta}|\sigma^2, \mathbf{Y} &\sim \text{Normal}\left(\mathbf{P}^{-1}\mathbf{W}, \mathbf{P}^{-1}\right) \\
\sigma^2|\boldsymbol{\beta}, \mathbf{Y} &\sim \text{InvGamma}\left(n/2 + a, (\mathbf{Y} - \mathbf{X}\boldsymbol{\beta})^T(\mathbf{Y} - \mathbf{X}\boldsymbol{\beta})/2 + b\right)
\end{aligned}
$$

where $\mathbf{P} = \mathbf{X}^T\mathbf{X}/\sigma^2 + \mathbf{I}/\tau$ and $\mathbf{W} = \mathbf{X}^T\mathbf{Y}/\sigma^2$. Code to implement this model is given in Listing 3.4. Figure 3.10 (bottom row) shows excellent convergence. If there were more than one covariate, then $\mathbf{X}$ would have additional columns and $\boldsymbol{\beta}$ would have additional elements, but the full conditionals and Gibbs sampling steps would be unchanged.

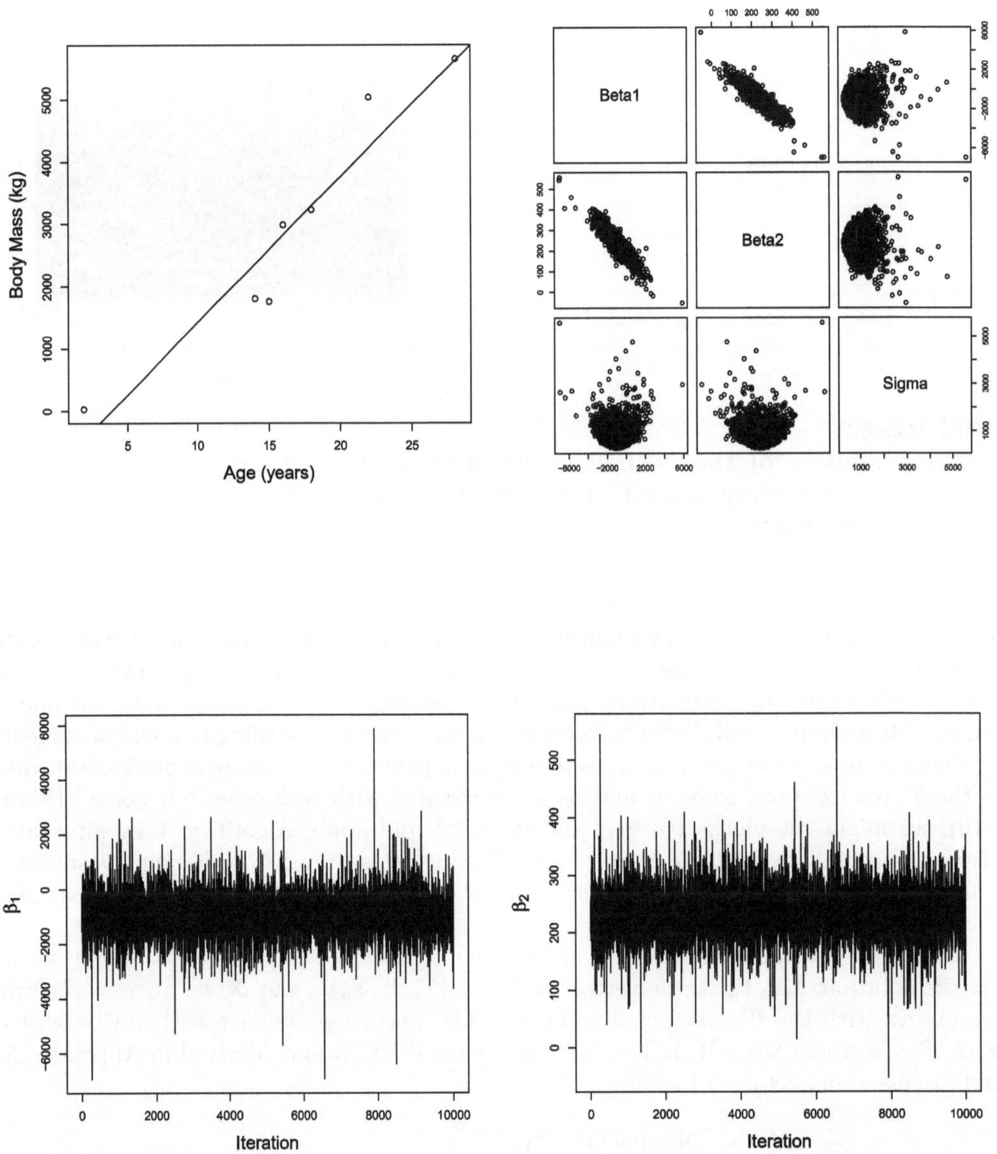

FIGURE 3.10
Analysis of the T. rex growth chart. The data are plotted in the top left panel and samples from the joint posterior of the three model parameters are plotted in the top right. The second row gives trace plots of β_1 and β_2.

Listing 3.4
Gibbs sampling for linear regression applied to the T-rex data.

```

library(mvtnorm)

# Load T-Rex data

 mass <- c(29.9, 1761, 1807, 2984, 3230, 5040, 5654)
 age <- c(2, 15, 14, 16, 18, 22, 28)
 n    <- length(age)
 X    <- cbind(1,age)
 Y    <- mass

# Create an empty matrix for the MCMC samples

 S                  <- 10000
 samples            <- matrix(NA,S,3)
 colnames(samples) <- c("Beta1","Beta2","Sigma")

# Initial values

 beta <- lm(mass~age)$coef
 sig2 <- var(lm(mass~age)$residuals)

# priors: beta ~ N(0,tau I_2), sigma^2 ~ InvG(a,b)

 tau    <- 10000^2
 a      <- 0.1
 b      <- 0.1

# Blocked Gibbs sampling

 V      <- diag(2)/tau
 tXX    <- t(X)%*%X
 tXY    <- t(X)%*%Y

 for(s in 1:S){
   P    <- tXX/sig2 + V
   W    <- tXY/sig2
   beta <- rmvnorm(1,solve(P)%*%W,solve(P))
   beta <- as.vector(beta)

   A    <- n/2 + a
   B    <- sum((Y-X%*%beta)^2)/2 + b
   sig2 <- 1/rgamma(1,A,B)

   samples[s,]<-c(beta,sqrt(sig2))
 }

 pairs(samples)
```

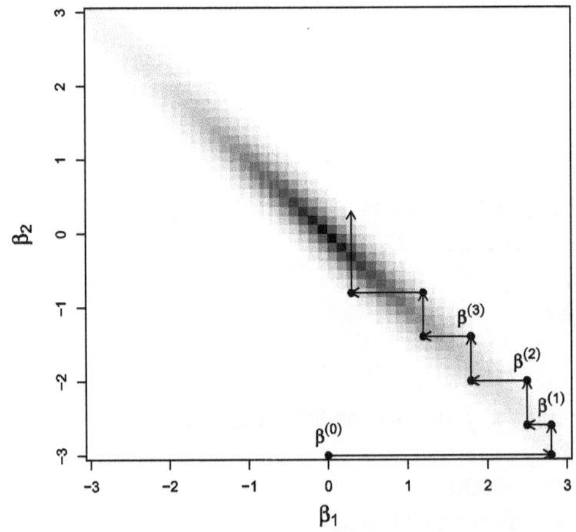

FIGURE 3.11
Gibbs sampling for correlated parameters. The background color represents the posterior PDF (black is high, white is low) of a hypothetical bivariate posterior for $\boldsymbol{\beta} = (\beta_1, \beta_2)$, and dots represent the starting value $\boldsymbol{\beta}^{(0)}$ and five hypothetical Gibbs sampling updates $\boldsymbol{\beta}^{(1)}, ..., \boldsymbol{\beta}^{(5)}$.

3.2.1.2 Metropolis–Hastings (MH) sampling

Each step in Gibbs sampling requires taking a sample from the full conditional distribution of one parameter (or block of parameters) conditional on all other parameters. In the examples in the previous section, all priors were conditionally conjugate, and so we were able to show that the full conditionals were members of familiar families of distributions, and therefore sampling was straightforward. However, it is not always possible to specify a conditionally conjugate prior.

For example, returning to the NFL concussion data (Figure 3.12), suppose we select the model

$$Y_i|\boldsymbol{\beta} \sim \text{Poisson}[N\exp(\beta_1 + \beta_2 i)] \tag{3.16}$$

for year $i = 1, ..., 4$ so that the log concussion rate changes linearly in time. The likelihood is then

$$f(\mathbf{Y}|\boldsymbol{\beta}) \propto \prod_{i=1}^{4} \exp[-N\exp(\beta_1 + \beta_2 i)]\exp[Y_i(\beta_1 + \beta_2 i)]. \tag{3.17}$$

The parameters appear in the exponential within another exponential, and there is no well-known family of distributions for $\boldsymbol{\beta}$ with this feature. Therefore, the posterior will not belong to a known distribution, and it is not clear how to directly sample from the posterior.

MH sampling [74] replaces draws from the exact full conditional distribution with a draw from a proposal/candidate distribution followed by an accept/reject step. That is, the Gibbs update of β_1 is a simple sample $\beta_1^{(s)}|\beta_2^{(s-1)}, \mathbf{Y}$, whereas an MH makes a candidate draw $\beta_1^* \sim q(\beta_1|\beta_1^{(s-1)})$ that is conditioned on the current value of β_1 (and potentially β_2 and/or $\mathbf{Y}$). Of course, the candidate cannot be blindly accepted because the candidate distribution

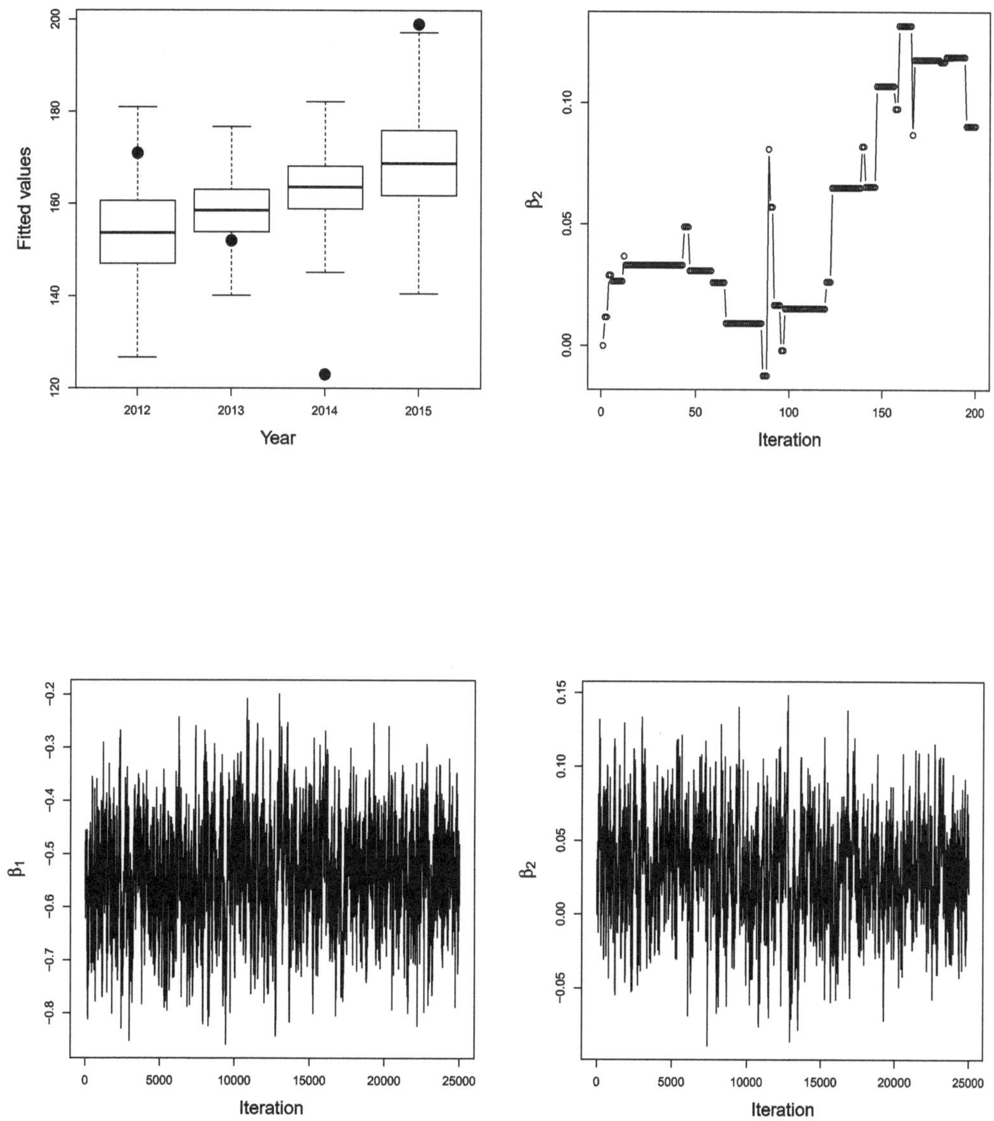

FIGURE 3.12

Poisson regression for the NFL concussion data. The first panel plots the posterior distribution of the mean values $N\exp(\beta_1 + \beta_2 i)$ for each year (boxplots) and the observed number of concussions, Y_i (points); the remaining three panels are trace plots of the regression coefficients β_1 and β_2.

may not be related to the posterior distribution. To correct for this, the candidate is accepted with probability $\min\{1, R\}$ where R is the ratio

$$R = \frac{p(\beta_1^*|\beta_2, \mathbf{Y})}{p(\beta_1^{(s-1)}|\beta_2, \mathbf{Y})} \frac{q(\beta_1^{(s-1)}|\beta_1^*)}{q(\beta_1^*|\beta_1^{(s-1)})}. \tag{3.18}$$

Equivalently, the accept/reject step generates $U \sim \text{Uniform}(0, 1)$ and accepts the candidate if $U < R$. Listing 2 formally describes the MH sampler, which can be justified by following the steps outlined in Appendix A.3 for Gibbs sampling.

Algorithm 2 Metropolis–Hastings Sampling

Initialize $\boldsymbol{\theta}^{(0)} = (\theta_1^{(0)}, ..., \theta_p^{(0)})$
for $s = 1, ..., S$ **do**
    for $j = 1, ..., p$ **do**
        sample $\theta_j^* \sim q_j(\theta_j|\theta_j^{(s-1)})$
        set $\boldsymbol{\theta}^* = \left(\theta_1^{(s)}, ..., \theta_{j-1}^{(s)}, \theta_j^*, \theta_{j+1}^{(s-1)}, ..., \theta_p^{(s-1)}\right)$
        set $R = \frac{f(\mathbf{Y}|\boldsymbol{\theta}^*)\pi(\boldsymbol{\theta}^*)}{f\left(\mathbf{Y}|\boldsymbol{\theta}^{(s-1)}\right)\pi\left(\boldsymbol{\theta}^{(s-1)}\right)} \cdot \frac{q_j\left(\theta_j^{(s-1)}|\theta_j^*\right)}{q_j\left(\theta_j^*|\theta_j^{(s-1)}\right)}$
        sample $U \sim \text{Uniform}(0, 1)$
        if $U < R$ **then**
            set $\theta_j^{(s)} = \theta_j^*$
        else
            set $\theta_j^{(s)} = \theta_j^{(s-1)}$
        end if
    end for
end for

The acceptance ratio R depends on the ratio of the posteriors of the candidate and current values. For updating parameter θ_j, terms in the likelihood or prior that do not include θ_j cancel and can thus be ignored. Crucially, this includes the often intractable normalizing constant $m(\mathbf{Y})$. Other terms can cancel as well; for example, the entire likelihood for the example in Listing 3.3 did not depend on one parameter (γ) and would cancel in the posterior ratio for this parameter.

The MH sampler can be applied more generally than Gibbs sampling, but at the cost of having to select and tune a candidate distribution for each parameter (or block of parameters). A common choice is a random-walk Gaussian candidate distribution

$$\theta_j^*|\theta_j^{(s-1)} \sim \text{Normal}(\theta_j^{(s-1)}, c_j^2) \tag{3.19}$$

where c_j is the candidate standard deviation. This is called a random-walk proposal distribution because it simply adds Gaussian jitter to the current state of the chain. A Gaussian candidate distribution can be used for any continuous parameter, even for those without a Gaussian prior; however, this is not always ideal. For example, a Gaussian candidate for a variance parameter may propose negative values which will automatically be discarded because the prior density and thus acceptance probability will be zero. Alternatively, a parameter with bounded support (e.g., $\sigma > 0$) can be transformed to have support over the whole real line ($\theta = \log(\sigma)$), and a Gaussian can be used in this transformed space. A Gaussian candidate is not appropriate for a discrete parameter, however, because a Gaussian draw would almost surely not be in the support of the discrete parameter's prior.

The standard deviation of the candidate distribution, c_j, is a tuning parameter. A rule of thumb is to tune the algorithm so that it accepts 30-50% of the candidates. It is hard to know which value of c_j will lead to this acceptance rate before starting the sampling, and so it is common to adapt c_j based on batches of samples during the burn-in. For example, if less than 30% of the last 100 candidates were accepted, then you might decrease c_j by 20%, and if more than 50% of the last 100 candidates were accepted then you might increase c_j by 20%. However, once the burn-in has completed, you must fix the candidate distribution unless you consider more advanced methods (Appendix A.4).

Two benefits of the random-walk candidate distribution are that it does not require knowledge about the form of the posterior, and that the acceptance ratio simplifies. If the candidate PDF q is Gaussian,

$$q(\theta_j^{(s-1)}|\theta_j^*) = q_j(\theta_j^*|\theta_j^{(s-1)}) = \frac{1}{\sqrt{2\pi}c_j} \exp\left(-\frac{(\theta_j^{(s-1)} - \theta_j^*)^2}{2c_j^2}\right), \tag{3.20}$$

the ratio of candidate distributions in the MH acceptance probabilities cancels. This is called a symmetric candidate distribution. The MH algorithm reduces to the Metropolis algorithm [104] in Algorithm 3 if the candidate distribution is symmetric.

Algorithm 3 Metropolis Sampling

Initialize $\boldsymbol{\theta}^{(0)} = (\theta_1^{(0)}, ..., \theta_p^{(0)})$
for $s = 1, ..., S$ **do**
    for $j = 1, ..., p$ **do**
        sample $\theta_j^* \sim q_j(\theta_j|\theta_j^{(s-1)})$ for symmetric q_j
        set $\boldsymbol{\theta}^* = \left(\theta_1^{(s)}, ..., \theta_{j-1}^{(s)}, \theta_j^*, \theta_{j+1}^{(s-1)}, ..., \theta_p^{(s-1)}\right)$
        set $R = \dfrac{f(\mathbf{Y}|\boldsymbol{\theta}^*)\pi(\boldsymbol{\theta}^*)}{f\left(\mathbf{Y}|\boldsymbol{\theta}^{(s-1)}\right)\pi\left(\boldsymbol{\theta}^{(s-1)}\right)}$
        sample $U \sim \text{Uniform}(0, 1)$
        if $U < R$ **then**
            set $\theta_j^{(s)} = \theta_j^*$
        else
            set $\theta_j^{(s)} = \theta_j^{(s-1)}$
        end if
    end for
end for

Listing 3.5 gives `R` code for the Poisson regression model of NFL concussions. The Gaussian candidate distributions were set to have standard deviation 0.1. Since the parameters appear in the exponential function in the mean count, adding normal jitter with standard deviation 0.1 is similar to a 10% change in the mean counts, which leads to acceptance probability of 0.42 for β_1 and 0.18 for β_2. Further tuning might reduce the candidate standard deviation for β_2 to increase the acceptance probability, but the trace plots in Figure 3.12 show good convergence. The top right panel of Figure 3.12 zooms in on the first few samples of the Metropolis algorithm and shows how β_2 stays constant for several iterations before jumping to a new value when a candidate is accepted.

For the accept/reject step in Lines 40–41 of Listing 3.5 it is important to perform computation on the log scale. For larger datasets, the ratio of posteriors is numerically unstable because a small value in the denominator can cause the ratio to be numerically infinite. Therefore, the ratio is replaced by the difference of the log scale. The original

Listing 3.5
Metropolis sampling for the NFL concussion data.

```
# Load data

 Y <- c(171, 152, 123, 199)
 t <- 1:4
 n <- 4
 N <- 256

# Create an empty matrix for the MCMC samples

 S                 <- 25000
 samples           <- matrix(NA,S,2)
 colnames(samples) <- c("beta1","beta2")
 fitted            <- matrix(NA,S,4)

# Initial values

 beta   <- c(log(mean(Y/N)),0)

# priors: beta[j] ~ N(0,tau^2)

 tau    <- 10

# Prep for Metropolis sampling

 log_post <- function(Y,N,t,beta,tau){
    mn    <- N*exp(beta[1]+beta[2]*t)
    like  <- sum(dpois(Y,mn,log=TRUE))
    prior <- sum(dnorm(beta,0,tau,log=TRUE))
    post  <- like + prior
 return(post)}

 can_sd <- rep(0.1,2)

# Metropolis sampling

 for(s in 1:S){
   for(j in 1:2){
     can    <- beta
     can[j] <- rnorm(1,beta[j],can_sd[j])
     logR   <- log_post(Y,N,t,can,tau)-log_post(Y,N,t,beta,tau)
     if(log(runif(1))<logR){
       beta <- can
     }
   }
   samples[s,] <- beta
   fitted[s,] <- N*exp(beta[1]+beta[2]*t)
 }

 boxplot(fitted,outline=FALSE,ylim=range(Y),
         xlab="Year",ylab="Fitted values",names=2012:2015)
 points(Y,pch=19)
```

rejection step is to reject if $U < f(\boldsymbol{\theta}^*|\mathbf{Y})/f(\boldsymbol{\theta}^{(s-1)}|\mathbf{Y})$. Taking the log of each side gives the equivalent inequality $\log(U) < \log[f(\boldsymbol{\theta}^*|\mathbf{Y})] - \log[f(\boldsymbol{\theta}^{(s-1)}|\mathbf{Y})]$.

Random-walk Gaussian distributions can also be used for block updates. As with Gibbs, updating dependent parameters simultaneously can improve convergence. Say the p parameters are partitioned into q blocks, $\boldsymbol{\theta} = (\boldsymbol{\theta}_1, ..., \boldsymbol{\theta}_q)$, and the candidate distribution for block j is $\boldsymbol{\theta}_j^* \sim \text{Normal}(\boldsymbol{\theta}^{(s-1)}, c_j\mathbf{V}_j)$. As before, the scalar c_j must be tuned to give a reasonable acceptance probability. The matrix $\mathbf{V}_j$ must also be tuned, which is difficult because it involves the variance of each parameter and the correlation between each pair of parameters, and all of these elements contribute to the single acceptance probability. A reasonable choice is to set $\mathbf{V}_j$ to the posterior covariance of $\boldsymbol{\theta}_j$, but unfortunately this is unknown. A common remedy is to use a short burn-in period and set $\mathbf{V}_j$ to the sample covariance of the burn-in samples, and then tune the scalar c_j to give reasonable acceptance. The proposal distribution can also be adapted based on prior MCMC samples (Appendix A.4)

A drawback of the random walk candidate distribution is that it may be suboptimal if it does not closely approximate the posterior. Using additional information can improve convergence. For example, a Bayesian CLT (Section 3.1.3) approximation $\boldsymbol{\theta}|\mathbf{Y} \approx \text{Normal}(\hat{\boldsymbol{\theta}}_{MAP}, \hat{\boldsymbol{\Sigma}}_{MAP})$ could be used to tune the sampler to match the candidate to the posterior. Taking this to the extreme, if we can tune the candidate to be exactly the full conditional distribution, then the candidate distribution is proportional to the posterior, $q(\theta_j^*|\theta_j^{(s-1)}) \propto f(\mathbf{Y}|\boldsymbol{\theta}^*)\pi(\boldsymbol{\theta}^*)$, and the ratio in Line 6 of Algorithm 2 reduces to 1. This algorithm then cycles through the parameters, samples each from their full conditional distribution, and keeps all draws. This is the Gibbs sampling algorithm in Algorithm 1! This shows that Gibbs sampling is a special case of MH with careful selection of the candidate distributions. It also opens the door for mixing Gibbs and MH updates in the same algorithm because the algorithm can be framed as an MH sampler with some well-tuned candidates that lead to Gibbs steps.

Metropolis-within-Gibbs example: To illustrate this Metropolis-within-Gibbs sampling algorithm, we use logistic regression, which is an extension of linear regression for binary data. For binary data, $\text{E}(Y_i) = \text{Prob}(Y_i = 1)$, and thus we cannot directly model the mean as a linear combination of the covariates $\mathbf{X}_i = (1, X_{i2}..., X_{ip})$ because this linear combination might not be between zero and one. The linear regression model is thus modified to $Y_i|\boldsymbol{\beta} \sim \text{Bernoulli}(p_i)$, where $p_i = 1/[1 + \exp(-\eta_i)]$ for $\eta_i = \beta_1 + \sum_{j=2}^{p} \beta_j X_{ij}$, which ensures that the mean response is between zero and one. We select an uninformative prior for the intercept $\beta_1 \sim \text{Normal}(0, 10^2)$, and for the slopes with $j = 2, ..., p$ the prior is

$$\beta_j|\sigma^2 \sim \text{Normal}(0, \sigma^2) \text{ and } \sigma^2 \sim \text{InvGamma}(a, b). \tag{3.21}$$

The regression coefficients β_j appear in the likelihood inside a non-linear function, and so their normal prior is not conditionally conjugate. However, the prior variance σ^2 has conditionally conjugate inverse gamma prior and thus inverse gamma full conditional distribution. Therefore a Metropolis-within-Gibbs algorithm cycles through the β_j and updates them using Metropolis steps with a Gaussian candidate distribution, and applies a Gibbs step to sample σ^2 from its inverse gamma full conditional distribution.

This model has $p + 1$ parameters, and the algorithm has different types of updates and p tuning parameters. Due to its complexity, it should be interrogated before being used for a real data analysis. A common method for validating code is a simulation study. In a simulation study, we fix the values of the parameters and generate data using these values. The synthetic data are then analyzed, and the posterior distributions are compared with the true values. Unlike a real data analysis, since the true values are known, we can verify that the algorithm is able to recover the true values.

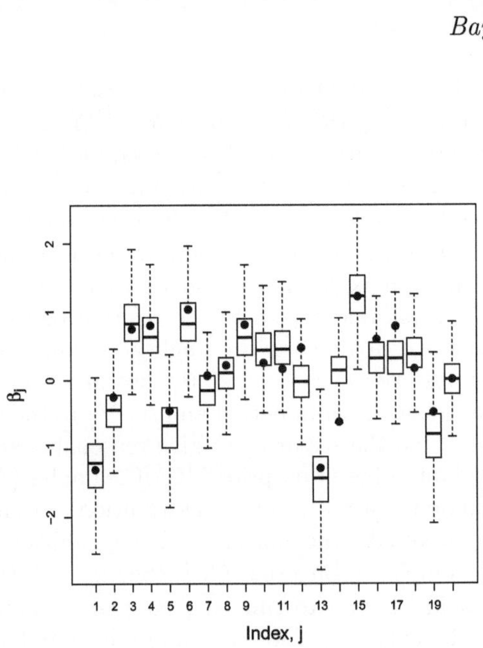

FIGURE 3.13
Logistic regression analysis of simulated data. The boxplots are the posterior distribution of the β_j, and the points are the true values used to simulate the data.

Listing 3.6 provides code to simulate the data, carry out the Metropolis-within-Gibbs algorithm, and summarize the results. The results are plotted in Figure 3.13. The true values of β_j used to generate the data are included by the posterior distribution for all but one or two coefficients, as expected. Therefore, it seems the algorithm is working well.

3.2.2 Approximate Bayesian Computing (ABC)

Metropolis-Hastings sampling is an incredibly versatile algorithm that can be applied to any statistical model where the PDF/PMF of the likelihood and prior can be computed. This covers the vast majority of, but not all, applications. As a counterexample, if the data are assumed to arise as the solution to a stochastic differential equation such as a population-dynamics model, then the likelihood may be too complicated to evaluate, leading to an intractable likelihood. Approximate Bayesian Computing (ABC; [100]) fills this gap. ABC requires only the ability to generate random draws from the model and not that the likelihood function can be evaluated.

As with MCMC, ABC produces S Monte Carlo draws $\boldsymbol{\theta}^{(1)}, ..., \boldsymbol{\theta}^{(S)}$ that can be used to approximate the posterior exactly the same way as MCMC; that is, the sample mean of the S ABC samples approximates the posterior mean, etc. Unlike a Markov chain, for ABC the S samples are independent and identically distributed.

We consider only the most basic version of ABC and refer interested readers to the review of [100] for extensions including combining ABC with MCMC. As sketched in Algorithm 4, ABC begins by sampling from the prior distribution, $\boldsymbol{\theta}^* \sim \pi(\boldsymbol{\theta})$, and then generating a simulated dataset $\mathbf{Y}^*$ given the parameters $\boldsymbol{\theta}^*$, $\mathbf{Y}^*|\boldsymbol{\theta}^* \sim f(\mathbf{Y}|\boldsymbol{\theta}^*)$. The simulated dataset should be the same size and design as the original dataset; e.g., in regression the observed covariate values should be used to simulate data, so that observation i in the simulated data corresponds to observation i in the real data. The simulated dataset is then compared to

Listing 3.6
Metropolis-within-Gibbs sampling for simulated logistic regression data.

```
# Simulate data
n          <- 100
p          <- 20
X          <- cbind(1,matrix(rnorm(n*(p-1)),n,p-1))
beta_true <- rnorm(p,0,.5)
prob       <- 1/(1+exp(-X%*%beta_true))
Y          <- rbinom(n,1,prob)

# Function to compute the log posterior
log_post_beta <- function(Y,X,beta,sigma){
    prob <- 1/(1+exp(-X%*%beta))
    like <- sum(dbinom(Y,1,prob,log=TRUE))
    prior <- dnorm(beta[1],0,10,log=TRUE) +        # Intercept
             sum(dnorm(beta[-1],0,sigma,log=TRUE)) # Slopes
return(like+prior)}

# Create empty matrix for the MCMC samples
S                <- 10000
samples          <- matrix(NA,S,p+1)

# Initial values and priors
beta   <- rep(0,p)
sigma <- 1
a      <- 0.1
b      <- 0.1
can_sd <- 0.1

for(s in 1:S){

  # Metropolis for beta
  for(j in 1:p){
    can    <- beta
    can[j] <- rnorm(1,beta[j],can_sd)
    logR   <- log_post_beta(Y,X,can,sigma)-
              log_post_beta(Y,X,beta,sigma)
    if(log(runif(1))<logR){
      beta <- can
    }
  }

  # Gibbs for sigma
  sigma <- 1/sqrt(rgamma(1,(p-1)/2+a,sum(beta[-1]^2)/2+b))

  samples[s,] <- c(beta,sigma)

 }

 boxplot(samples[,1:p],outline=FALSE,
         xlab="Index, j",ylab=expression(beta[j]))
 points(beta_true,pch=19,cex=1.25)
```

the actual dataset using a discrepancy measure

$$d^* = D(\mathbf{Y}, \mathbf{Y}^*).$$

If d^* is less than the threshold δ, then $\boldsymbol{\theta}^*$ is retained as a posterior sample; if not, then $\boldsymbol{\theta}^*$ is discarded. This process is repeated until S samples have been collected. The intuition is that parameters that lead to data similar to the observations are more likely to be the true value, which is similar to the intuition behind maximum likelihood estimation.

The choice of discrepancy measure and threshold are critical to the accuracy of the algorithm. If samples $\boldsymbol{\theta}^*$ are retained if and only if $\mathbf{Y}^* = \mathbf{Y}$, then the algorithm produces S draws from the actual posterior without approximation. However, if the response is continuous or n is large, then we cannot expect $\mathbf{Y}^*$ to be exactly equal to $\mathbf{Y}$ (actually, all that is required is that the sufficient statistics of $\mathbf{Y}^*$ match those of $\mathbf{Y}$, but this does not resolve the problem). To sample from an approximate posterior distribution, we can retain samples for which $\mathbf{Y}^* \neq \mathbf{Y}$ but are similar as measured by the discrepancy function D. One option is the mean squared differences, $D(\mathbf{Y}, \mathbf{Y}^*) = \sum_{i=1}^{n}(Y_i - Y_i^*)^2/n$. However, for large n it is unlikely that $\mathbf{Y}^*$ will be similar to $\mathbf{Y}$ for all n observations, and a discrepancy measure more targeted at the parameters is more efficient. For example, if the parameters control the mean and standard deviation of the response distribution, then the

$$D(\mathbf{Y}, \mathbf{Y}^*) = (\bar{Y} - \bar{Y}^*)^2 + w(s - s^*)^2$$

targets these parameters, for sample means $\bar{Y}$ and $\bar{Y}^*$, sample standard deviations s and s^*, and weight w chosen to reflect the relative scale and importance of the two parameters. In general, the discrepancy measure should include sufficient information to identify all model parameters. This requires some creativity and may be difficult if there are many parameters, which is a limitation of the method.

Listing 3.7 provides a worked example of ABC for simulated data. The $n = 100$ observations are simulated from the logistic regression model with $Y_i|\boldsymbol{\beta} \sim \text{Bernoulli}(p_i)$ with $p_i = 1/[1 + \exp(-\eta_i)]$ for $\eta_i = \beta_1 + \beta_2 X_i$. The priors are $\beta_1, \beta_2 \stackrel{iid}{\sim} \text{Normal}(0, 1)$. Of course, Metropolis-Hastings sampling could be applied for this model as above, but notice how in the likelihood function and prior distribution need not be computed during ABC. The true values used to generate the data are $\boldsymbol{\beta} = (2, -1)$ and $X_i \sim \text{Normal}(0, 1)$. Since this model tracks the probability that $Y_i = 1$ as a function of X_i, one possibility for D is to bin the observations by X_i and compute the sample proportion of Y_i in each bin. Specifically, we create $L = 4$ bins formed by equally-spaced quantiles of X_i, denoted $q_1, ..., q_{L+1}$. The sample proportion for bin l, $\bar{Y}_l$, is the sample mean of Y_i over observations with $q_l \leq X_i < q_{l+1}$. The discrepancy measure is then the mean squared error between the bin probabilities,

$$D(\mathbf{Y}, \mathbf{Y}^*) = \sum_{l=1}^{L}(\bar{Y}_l - \bar{Y}_l^*)^2/L.$$

Figure 3.14 (left) plots the mean of $D(\mathbf{Y}, \mathbf{Y}^*)$ by $\boldsymbol{\beta}^*$. The discrepancy is clearly lower in areas around the true value of $\boldsymbol{\beta}$, and thus draws with $\boldsymbol{\beta}^*$ near the truth are more likely to be retained. Figure 3.14 also plots the 95% credible sets as a function of the threshold, δ. Smaller values of δ provide a better approximation to the posterior distribution, but the proportion of sample retained drops to 0.02 for the smallest threshold of $\delta = 0.01$.

Algorithm 4 Approximate Bayesian Computing (ABC)

initialize $s = 0$
while $s < S$ **do**
  sample parameters from prior $\boldsymbol{\theta}^* \sim \pi(\boldsymbol{\theta})$
  sample data from likelihood $\mathbf{Y}^*|\boldsymbol{\theta}^* \sim f(\mathbf{y}|\boldsymbol{\theta}^*)$
  compute the discrepancy measure $d^* = D(\mathbf{Y}^*, \mathbf{Y})$
  if $d^* < \delta$ **then**
    set $s = s + 1$
    set $\boldsymbol{\theta}^{(s)} = \boldsymbol{\theta}^*$
  end if
end while

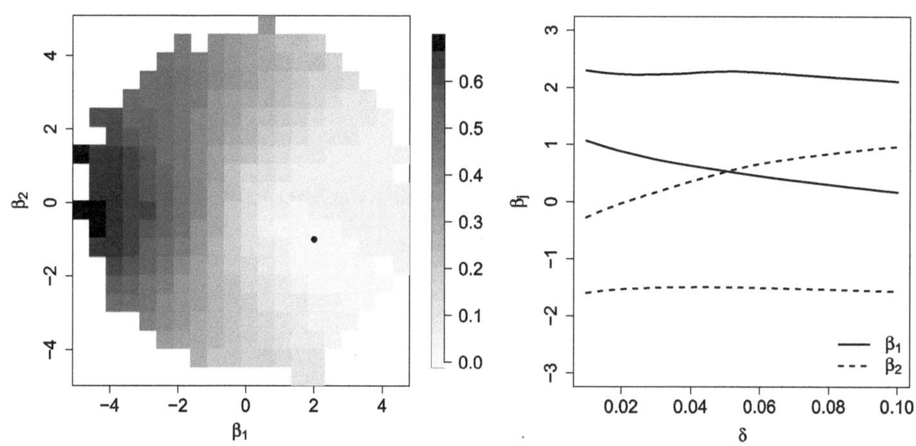

FIGURE 3.14
Illustration of ABC. The mean discrepancy D as a function of the simulated logistic regression parameters $\boldsymbol{\beta}^*$ (left), and the approximate posterior 90% credible intervals for β_1 and β_2 as a function of the threshold δ; the true value of $\boldsymbol{\beta} = (2, -1)$ is represented by the point in the left panel.

Listing 3.7
ABC for logistic regression.

```
# Define the discrepancy measure
D <- function(X,Y_obs,Y_sim,q){
  L <- length(q)-1
  d <- 0
  for(l in 1:L){
    d <- d + (mean(Y_sim[X>=q[l] & X<q[l+1]]) -
              mean(Y_obs[X>=q[l] & X<q[l+1]]))^2
  }
d/L}

# Compute discrepancy
deviation
 S      <- 1000000 # Number of draws from the prior
 b_sim  <- matrix(rnorm(S*2),S,2) # Draw from prior
 d      <- rep(0,S)
 q      <- quantile(X,c(0.00,0.25,0.50,0.75,1.00))

 for(s in 1:S){
   set.seed(s)
   eta   <- b_sim[s,1]+b_sim[s,2]*X
   p     <- 1/(1+exp(-eta))
   Y_sim <- rbinom(n,1,p)
   d[s]  <- D(X,Y,Y_sim,q)
 }

# Plot d as a function of beta
 quilt.plot(b_sim[,1],b_sim[,2],d,
           xlab=expression(beta[1]),ylab=expression(beta[2]))

# Compute posterior approximation by delta
 m     <- 20
 delta <- seq(0.01,0.10,length=m)
 qb1   <- qb2 <- matrix(0,m,3)
 for(i in 1:m){
   qb1[i,] <- quantile(b_sim[d<delta[i],1],c(0.05,0.50,0.95))
   qb2[i,] <- quantile(b_sim[d<delta[i],2],c(0.05,0.50,0.95))
 }
 matplot(delta,qb1,type="l",lty=c(2,1,2),col=1,lwd=2,
         xlab=expression(delta),ylab=expression(beta[1]))
 matplot(delta,qb2,type="l",lty=c(2,1,2),col=1,lwd=2,
         xlab=expression(delta),ylab=expression(beta[2]))
```

3.2.3 Simulation-based inference

Simulation-based inference (SBI; [37]) is closely related to ABC and also useful in settings where the posterior is challenging to evaluate but sampling is straightforward. As with ABC, SBI begins by sampling from the prior and then the data-generating process,

$$\boldsymbol{\theta}^{(s)} \sim \pi(\boldsymbol{\theta}) \quad \text{and} \quad \mathbf{Y}^{(s)}|\boldsymbol{\theta}^{(s)} \sim f(\mathbf{Y}|\boldsymbol{\theta}^{(s)})$$

for $s \in \{1, ..., S\}$. SBI departs from ABC because rather than retaining a subset of the samples $\boldsymbol{\theta}^{(s)}$ to approximate posterior, SBI uses all S samples and machine learning to build an approximation to the posterior $p(\boldsymbol{\theta}|\mathbf{Y})$.

By construction, $(\boldsymbol{\theta}^{(s)}, \mathbf{Y}^{(s)})$ are samples from the joint distribution of the parameters and the data. These samples can thus be used to build a statistical model to regress $\boldsymbol{\theta}^{(s)}$ onto $\mathbf{Y}^{(s)}$ to approximate the posterior, $p(\boldsymbol{\theta}|\mathbf{Y})$. For a univariate analysis with $p = 1$, a very crude approximation is the linear regression

$$\theta^{(s)}|\mathbf{Y}^{(s)} \sim \text{Normal}\left(b_0 + \sum_{i=1}^{n} Y_i^{(s)} b_j, \tau^2\right).$$

With this (unrealistic) assumption, the S training samples are used to estimate $\mathbf{b} = (b_0, ..., b_n)$ and τ^2 using least squares, giving estimates $\widehat{\mathbf{b}}$ and $\widehat{\tau}^2$, and then for the real dataset, $\mathbf{Y}$, the posterior is approximated as $\theta|\mathbf{Y} \sim \text{Normal}\left(\widehat{b}_0 + \sum_{i=1}^{n} Y_i \widehat{b}_j, \widehat{\tau}^2\right)$. Note the role reversal in this regression with the observed outcomes Y_i used as predictors; this illustrates that Bayesian parameter inference is an inverse problem rather than the usual forward prediction problem.

As a toy example, consider Figure 3.15 that illustrates SBI for the model $Y|\theta \sim \text{Normal}(\theta, 1)$ with prior $\theta \sim \text{Normal}(0, 1)$. As derived in Section 2.1, the exact posterior is $\theta|Y \sim \text{Normal}(Y/2, 1/2)$. Figure 3.15 (left) plots $S = 1,000$ samples from the joint density of (θ, Y). To learn the posterior distribution without mathematical derivation, we fit the simple linear regression model $\theta|Y \sim \text{Normal}\left(b_0 + Yb_1, \tau^2\right)$ to these S samples. The least squares analysis gives $\widehat{b}_0 = -0.045$, $\widehat{b}_1 = 0.495$ and $\widehat{\tau}^2 = 0.488$. Figure 3.15 (right) shows that SBI provides a close approximation to the true posterior. This approximation is close largely because it relies on the knowledge that the true posterior is Gaussian and linear in the observations, which are both dubious assumptions in non-trivial problems. Therefore, in practice, more flexible statistical modeling of the posterior and a larger number of samples are required.

This approach is appealing to statisticians because it replaces advanced computing with advanced statistical modeling, facilitated by an effectively unlimited number of samples, S. Given this large dataset and presumed complex non-linear and non-Gaussian relationship between the data and the parameters, machine learning, in particular deep learning (see Section 8.3.4), is typically used to build a model to approximate the posterior rather than the linear regression above. As in ABC or any statistical model-building exercise, summary statistics can replace the full dataset in the approximate posterior and distributions other than the prior can be used to generate the samples (although this must be accounted for later in the analysis).

SBI is an example of amortized inference. That is, while it takes substantial effort to simulate the S samples and build a model relating the parameters and data, once this is completed, inference for a given dataset is immediate. This holds for any dataset generated from the model, not just the dataset that motivated the initial analysis, so the same approximation could be used by any potential user. Thus, by learning the map from the data to the posterior, future users could bypass the challenging computational steps and immediately obtain approximate inference.

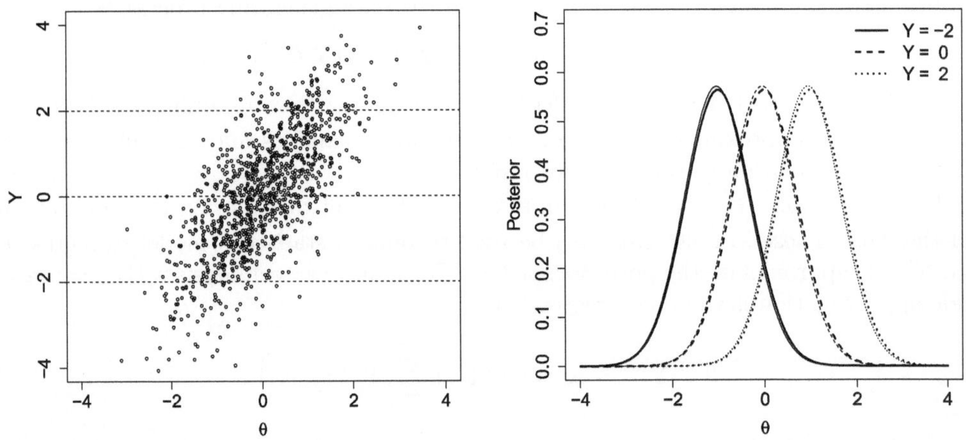

FIGURE 3.15
Illustration of SBI. Samples from the joint distribution of (θ, Y) (left) and the true (thick) and approximate (thin) posterior distributions for three values of Y (right).

3.2.4 Sequential Monte Carlo (SMC) sampling

As with MCMC, ABC and SBI, Sequential Monte Carlo (SMC) sampling (also called particle filtering) approximates the posterior distribution using Monte Carlo sampling. SMC resembles ABC in that samples are made from an initial distribution, often the prior distribution, and evaluated based on their fit to the data. However, SMC differs from ABC in two important ways: the exact, not approximate, likelihood is used, and samples are weighted rather than accepted or rejected. Although SMC can be used to analyze a single static dataset, its name derives from the fact that it is conducive to a sequential analysis of streaming data; e.g., each day a new batch of data is collected, and the Bayesian analysis is updated based on this new information.

As with ABC, we present only the simplest version here; we refer to the rich literature on SMC (e.g., [45]) for more advanced algorithms that include targeted initial particle sampling and replacing errant particles. A simple SMC algorithm begins by drawing S samples/particles from the prior distribution

$$\boldsymbol{\theta}^{(1)}, ..., \boldsymbol{\theta}^{(S)} \stackrel{iid}{\sim} \pi.$$

The particles are then assigned weights based on the likelihood function, $u^{(s)} = f(\mathbf{Y}|\boldsymbol{\theta}^{(s)})$, and normalized to sum to one as

$$w^{(s)} = \frac{u^{(s)}}{\sum_{t=1}^{S} u^{(t)}}.$$

Posterior quantities are then computed using weighted averages, e.g., $\text{E}(\theta_j|\mathbf{Y}) \approx \sum_{s=1}^{S} w^{(s)}\theta_j^{(s)}$ and

$$\text{Prob}(\theta_j > 0|\mathbf{Y}) \approx \sum_{s=1}^{S} w^{(s)} I(\theta_j^{(s)} > 0),$$

i.e., the sum of the weights for particles with $\theta_j^{(s)} > 0$. Note that particles that do not fit the data well will have weight $w^{(s)} \approx 0$ and thus be "filtered out" of the approximate posterior. Therefore, this algorithm will be inefficient if the prior (or another initial distribution) is too diffuse and most particles are filtered out.

The algorithm above approximates the posterior distribution using the entire dataset. However, say the data arrive sequentially and are assumed to be independent over time. Then the unnormalized weights after observing n observations are

$$u_n^{(s)} = \prod_{i=1}^{n} f(Y_i|\boldsymbol{\theta}^{(s)}).$$

When an additional observation arrives, the weights are updated as

$$u_{n+1}^{(s)} = f(Y_{n+1}|\boldsymbol{\theta}^{(s)})u_n^{(s)}$$

and the normalized weights are $w_{n+1}^{(s)} = u_{n+1}^{(s)}/\sum_{t=1}^{S} u_{n+1}^{(t)}$. Compared to reanalyzing the data at each time step, this saves time and storage because we do not need to recompute the likelihood for, or even store, the data from the first n observations to obtain the current particle weights.

To see why this gives a valid posterior approximation, consider a discrete parameter with support $\boldsymbol{\theta} \in \{\boldsymbol{\theta}_1, ..., \boldsymbol{\theta}_K\}$. The true posterior probability that $\boldsymbol{\theta} = \boldsymbol{\theta}_k$ is

$$\text{Prob}(\boldsymbol{\theta} = \boldsymbol{\theta}_k|\mathbf{Y}) = \frac{f(\mathbf{Y}|\boldsymbol{\theta}_k)\pi(\boldsymbol{\theta}_k)}{\sum_{l=1}^{K} f(\mathbf{Y}|\boldsymbol{\theta}_l)\pi(\boldsymbol{\theta}_l)}.$$

The SMC approximation is

$$\text{Prob}(\boldsymbol{\theta} = \boldsymbol{\theta}_k|\mathbf{Y}) \approx \sum_{s=1}^{S} w_s I(\boldsymbol{\theta}^{(s)} = \boldsymbol{\theta}_k) = \frac{\sum_{s=1}^{S} f(\mathbf{Y}|\boldsymbol{\theta}^{(s)}) I(\boldsymbol{\theta}^{(s)} = \boldsymbol{\theta}_k)}{\sum_{s=1}^{S} f(\mathbf{Y}|\boldsymbol{\theta}^{(s)})}.$$

Denoting p_k as the proportion of the S terms with $\boldsymbol{\theta}^{(s)} = \boldsymbol{\theta}_k$ and collecting terms gives

$$\text{Prob}(\boldsymbol{\theta} = \boldsymbol{\theta}_k|\mathbf{Y}) \approx \frac{f(\mathbf{Y}|\boldsymbol{\theta}_k)p_k}{\sum_{l=1}^{K} f(\mathbf{Y}|\boldsymbol{\theta}_l)p_l}.$$

As the sample size S increases, p_k converges to $\pi(\boldsymbol{\theta}_k)$, and the approximate posterior probability converges to true probability. This argument extends to continuous parameters by evaluating the probability that $\boldsymbol{\theta}$ is in a small region.

SMC typically considers a large number of particles, S, but this only provided a good approximation if many of the particles are supported by the posterior. For example, if $S = 10000$ but only two of the weights are non-zero, then the effective sample size for the posterior is two, not S. This intuition is formalized using the effective sample size statistic,

$$ESS = \frac{(\sum_{s=1}^{S} w_s)^2}{\sum_{s=1}^{S} w_s^2}.$$

The ESS is between 1 and S; in the best case the particles are weighted equally, $w_s = w$ for all s, and $ESS = S$; in the worst case $w_s = 1$ for one s and $w_s = 0$ for all other s, and $ESS = 1$. Typically, S is increased until ESS is sufficiently large, say $ESS = 1000$.

Listing 3.8 and Figure 3.16 illustrate SMC for the beta-binomial model $Y|\theta \sim \text{Binomial}(n, \theta)$ and $\theta \sim \text{Beta}(a, b)$ with $Y = 10$, $n = 40$ and $a, b = 1$. First, the posterior is approximated using MCMC sampling (in this case just independent samples from

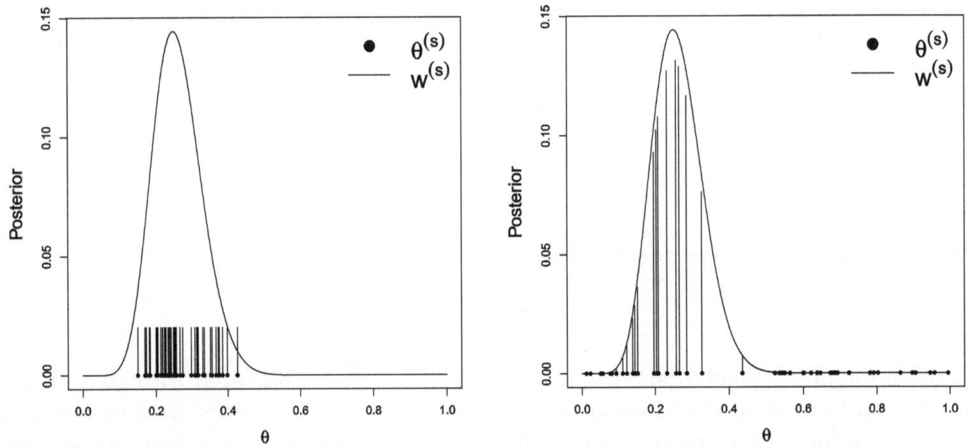

FIGURE 3.16
Illustration of SMC versus MCMC. $S = 50$ equally-weighted samples from MCMC (left) and weighted samples from SMC (right) for a beta-binomial model.

the beta posterior). This produces S samples that are all in the area with high posterior mass and are weighted equally, $w_s = 1/S$, when approximating the posterior mean. The SMC samples, in contrast, are uniformly distributed across the parameter space following the uniform prior and weighted highly for samples with high likelihood. Because many particles have low weight, the effective sample size of the $S = 50$ particles is only $ESS = 9.7$; thus additional particles would be needed to provide a precise approximation.

Algorithm 5 Sequential Monte Carlo sampling

1: Sample particles from the prior, $\theta^{(s)} \stackrel{iid}{\sim} \pi(\boldsymbol{\theta})$
2: Compute weights $u_s = \prod_{i=1}^{n} f(y_i|\boldsymbol{\theta}^{(s)})$ and $w_s = u_s/(\sum_{t=1}^{S} u_t)$
3: Approximate $\text{E}\{g(\boldsymbol{\theta}|\mathbf{Y})\} \approx \sum_{s=1}^{S} w_s g(\boldsymbol{\theta}^{(s)})$.

3.3 MCMC software options in R

Writing MCMC code step-by-step is great practice and the only way to truly understand the algorithms, and tailoring the code to a specific model can improve speed and stability for complex models. However, coding basic MCMC samplers becomes repetitive. For Gibbs sampling, you must derive and encode the full conditional distribution of each parameter, but there are only a few dozen known conjugacy pairs (e.g., Appendix A.2), and so writing a Gibbs sampler is simply a matter of looking up the correct full conditional distribution and entering this into R. If the parameter's full conditional distribution does not appear in the table of conjugacy pairs, you might use a Metropolis update with Gaussian random walk

Listing 3.8
MCMC versus SMC sampling for a beta-binomial model.

```
 Y <- 10; n <- 40            # Data
 a <- b <- 1                 # Prior
 t <- seq(0,1,length=1000) # Posterior
 p <- dbinom(Y,n,t)*dbeta(t,a,b)
 (Y+a)/(n+a+b)               # True post mean
 [1] 0.2619048

# MCMC
 set.seed(919)
 S     <- 50                 # Number of MC samples
 theta <- rbeta(S,Y+a,n-Y+b)
 w     <- rep(1/S,S)

 plot(t,p,type="l",xlab=expression(theta),ylab="Posterior")
 points(theta,rep(0,S),cex=.6,pch=19)
 lines(theta,w,type="h")
 legend("topright",c(expression(theta^{(s)}),expression(w^{(s)})),
        pch=c(19,NA),lty=c(NA,1),bty="n")

mean(theta) # Approximate E(theta|Y)
[1] 0.2634136

# SMC
set.seed(919)
theta <- rbeta(S,a,b)        # Sample from prior
u     <- dbinom(Y,n,theta) # Weight by likelihood
w     <- u/sum(u)

plot(t,p,type="l",xlab=expression(theta),ylab="Posterior")
points(theta,rep(0,S),cex=.6,pch=19)
lines(theta,w,type="h")
legend("topright",c(expression(theta^{(s)}),expression(w^{(s)})),
       pch=c(19,NA),lty=c(NA,1),bty="n")

sum(w*theta) # Approximate E(theta|Y)
[1] 0.2355692
ESS <- (sum(w)^2)/sum(w^2)
ESS
[1] 9.706214
```

candidate distribution. Each Metropolis update consists of a Gaussian proposal followed by computing the acceptance ratio, which requires only tuning the candidate standard deviation and inserting the specific likelihood and prior chosen for the model into the acceptance ratio. But the tuning step is not model specific, and most models are built using only a few dozen distributions for the likelihood and prior (Appendix A.1), and so writing MH code becomes simply a matter of coding the correct distributions.

Fortunately, this process has been automated! There are several general-purpose MCMC packages that can be called from `R`, including Just Another Gibbs Sampler (`JAGS`), `OpenBUGS` ([148]), `STAN` ([28]), and `NIMBLE` ([42]), as well as libraries in `python` including `pymc` ([121]) and the `SAS` procedure `PROC MCMC` ([31]). These packages are not specific to a single model; rather they take a script specifying the likelihood and prior as input and use this information to construct an MCMC sampler for the specified model. The formats of these packages are similar, and so with a solid understanding of MCMC sampling to interpret the results, switching from one to another is fairly straightforward.

For the remainder of the book we will use `JAGS` to carry out MCMC sampling. Compared to other packages, `JAGS` is relatively easy to code and sufficiently fast for the size and complexity of models considered here. `JAGS` is a very general package that can be used to fit all the models discussed in this book and many more. Of course, for a specific application there may be more focused and thus more efficient code. For example, the `BLR` package in `R` surely performs better than `JAGS` for Bayesian linear regression. However, to avoid having to learn dozens of packages for different models we simply use `JAGS` throughout.

`JAGS` must first be downloaded from `http://mcmc-jags.sourceforge.net/`, and installed on your computer. The `R` package `rjags` must be downloaded and installed to communicate with `JAGS` from `R`. An MCMC analysis using `JAGS` has five steps:

(1) Specify the model as a character string in `R`;

(2) Upload the data, and compile the model using the function `jags.model`;

(3) Perform burn-in sampling using the function `update`;

(4) Generate posterior samples using the function `coda.samples`;

(5) Summarize the results using the functions `summarize` and `plot`.

Listing 3.9 contains the `R` code for simple linear regression applied to the T. rex growth chart data, and the output of the `plot` command in the last line is given in Figure 3.17. A few notes on this code:

- All of this code is run in `R`. You must install `JAGS` on your computer, but you never have to open `JAGS` because the `rjags` library passes data and results back and forth between `R` and `JAGS`.

- The model specification format resembles `R` code, but not all `R` commands are applicable in `JAGS`, and some commands with the same syntax have different properties. For example, `dnorm` in `JAGS` specifies the normal distribution via its mean and precision (inverse variance) unlike the `R` command by the same name that uses mean and standard deviation. For a list of `JAGS` commands and their meaning see the user manual in [123].

- The symbol "∼" indicates that a variable follows the distribution given to the right of the symbol, and deterministic operations are denoted by the left arrow"<-".

- In the model definition, the prior is placed on the precision `tau`, but each sample is converted to the standard deviation `sigma <- 1/sqrt(tau)`, and samples of `sigma` are returned. That is, this line tells `JAGS` to compute $\sigma^{(s)} = 1/\sqrt{\tau^{(s)}}$ at each iteration and return the samples $\sigma^{(s)}$.

Listing 3.9
JAGS code for linear regression applied to the T-rex data.

```
library(rjags)
# Load T-Rex data
 mass <- c(29.9, 1761, 1807, 2984, 3230, 5040, 5654)
 age  <- c(2, 15, 14, 16, 18, 22, 28)
 n    <- length(age)
 data <- list(mass=mass,age=age,n=n)

# (1) Define the model as a string
 model_string <- textConnection("model{
   # Likelihood (dnorm uses a precision, not variance)
    for(i in 1:n){
      mass[i] ~ dnorm(beta1 + beta2*age[i],tau)
    }
   # Priors
    tau  ~ dgamma(0.1, 0.1)
    sigma <- 1/sqrt(tau)
    beta1 ~ dnorm(0, 0.001)
    beta2 ~ dnorm(0, 0.001)
 }")

# (2) Load the data, specify initial values and compile the MCMC code
 inits <- list(beta1=rnorm(1),beta2=rnorm(1),tau=rgamma(1,1))
 model <- jags.model(model_string,data = data, inits=inits, n.chains=2)

# (3) Burn-in for 10000 samples
 update(model, 10000, progress.bar="none")

# (4) Generate 20000 post-burn-in samples
 params <- c("beta1","beta2","sigma")
 samples <- coda.samples(model,
            variable.names=params,
            n.iter=20000, progress.bar="none")

# (5) Summarize the output
 summary(samples)
 1. Empirical mean and standard deviation for each variable,
   plus standard error of the mean:

            Mean      SD Naive SE Time-series SE
beta1    2.512   31.61   0.1580         0.1580
beta2   52.763   39.21   0.1961         0.3727
sigma 2792.738 1177.88 5.8894         9.7678

2. Quantiles for each variable:

          2.5%     25%      50%     75%   97.5%
beta1  -59.53  -18.87    2.601   23.57   64.61
beta2  -21.36   25.71   51.531   78.34  134.17
sigma 1083.16 1997.85 2601.864 3361.14 5622.69

 plot(samples)
```

- The object `model` contains the data, the code to update each parameter, and the current state of each parameter in each chain.
- The initial values function can specify either random samples for initial values as is done here, or give specific values (e.g., `beta1 = 0`). If the `inits` argument in `jags.model` is omitted, `JAGS` will set initial values automatically.
- The `update` function modifies the state of each parameter in each chain, but does not store intermediate samples.
- The `coda.samples` function does retain all MCMC samples, but only for those parameters listed in `variable.names`. The function outputs the samples into the object `samples` which are in the format used by the `coda` package that can be used to study convergence.
- `JAGS` has built-in `plot` and `summary` functions that are used in Listing 3.9, but you can make your own plots by extracting the $S \times p$ matrix of samples for chain c, `samples[[c]]`. All the samples can be put in one $2S \times p$ matrix

  ```
  samps_matrix <- rbind(samples[[1]],samples[[2]])
  ```

  and, for example, the posterior quantiles could be computed as

  ```
  apply(samps_matrix,2,quantile,
        c(0.025,0.250,0.500,0.750,0.975))
  ```

- The `summary` function combines the samples across the chains and gives the sample mean, standard deviation, and quantiles of the posterior samples.
- The posterior could be summarized using the posterior mean and 95% intervals of the marginal posterior for each parameter given by the `summarize` function (i.e., the "`Mean`" and "`2.5%`" and "`97.5%`" entries).

Listing 3.10 provides a second example of fitting the Poisson-gamma model to the NFL concussion data. The steps and code are very similar to the previous example, except that instead of using the built-in summary and plot function, Listing 3.10 extracts the samples from the `rjags` object `samples` and combines them into the $2S \times 4$ matrix of samples `samps`. This allows for more flexibility in the posterior summaries used to illustrate the results. For example, Listing 3.10 computes the 90% posterior intervals.

In the remainder of the book we will use `JAGS` for all MCMC coding. We will often show the model statement and summaries of the output, but we will not report the steps of loading the data, generating the samples, etc., because these blocks of code are virtually identical for all models.

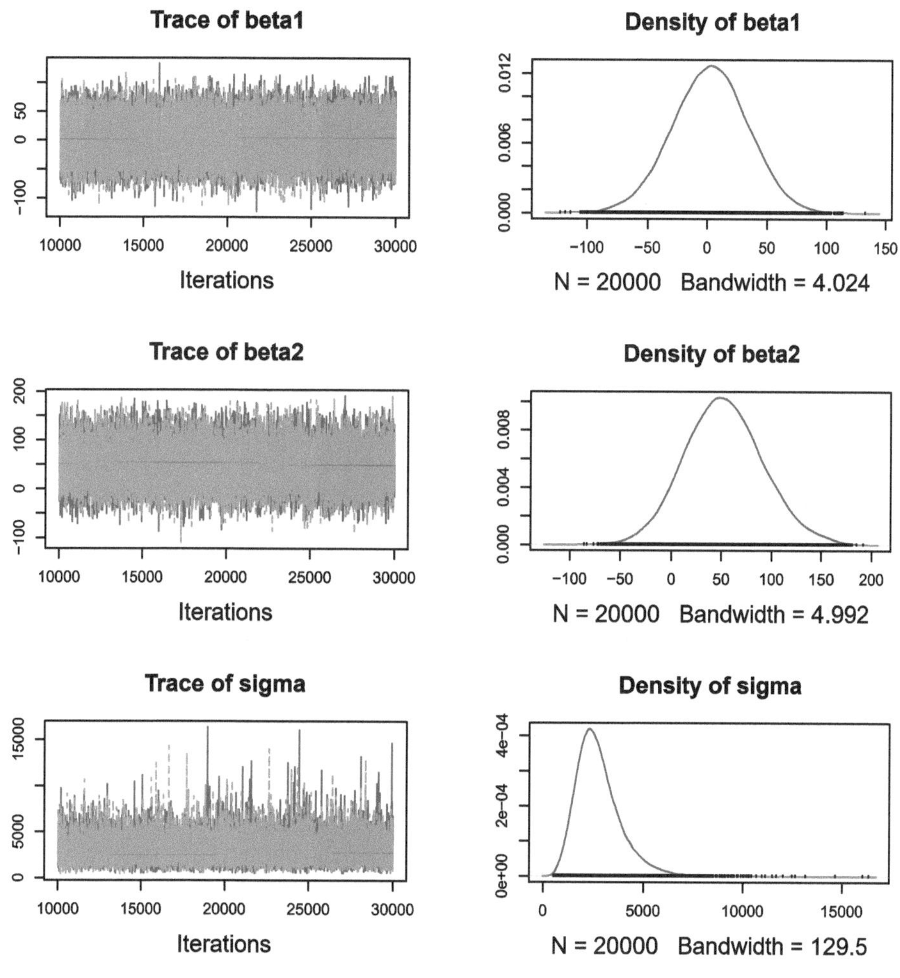

FIGURE 3.17
JAGS output for the T. rex analysis. This is the output of the `plot` function in JAGS for the linear regression of the T. rex growth chart data. The two lines with different shades of gray in the plots in the first columns are the samples from the two chains. The density plots on the right summarize the marginal distribution of each parameter combining samples across chains.

Listing 3.10
JAGS code for the NFL concussion data.

```
library(rjags)
# Load the NFL concussion data
 Y <- c(171, 152, 123, 199)
 n <- 4
 N <- 256

# (1) Define the model as a string
 model_string <- textConnection("model{
   # Likelihood
    for(i in 1:n){
      Y[i] ~ dpois(N*lambda[i])
    }
   # Priors
    for(i in 1:n){
      lambda[i] ~ dgamma(1,gamma)
    }
    gamma  ~  dgamma(a, b)
 }")

# (2) Load the data and compile the MCMC code
 inits <- list(lambda=rgamma(n,1,1),gamma=1)
 data <- list(Y=Y,N=N,n=n,a=0.1,b=0.1)
 model <- jags.model(model_string,data = data, inits=inits, n.chains=2)

# (3) Burn-in for 10000 samples
 update(model, 10000, progress.bar="none")

# (4) Generate 20000 post-burn-in samples
 params <- c("lambda")
 samples <- coda.samples(model,
           variable.names=params,
           n.iter=20000, progress.bar="none")

# (5) Compute 90% credible intervals
 samps <- rbind(samples[[1]],samples[[2]]) #2S x 4 matrix of samples
 apply(samps,2,quantile,c(0.05,0.95))

       lambda[1] lambda[2] lambda[3] lambda[4]
  2.5% 0.5722272 0.5035036 0.4005522 0.6717104
 97.5% 0.7704071 0.6925751 0.5685348 0.8878783
```

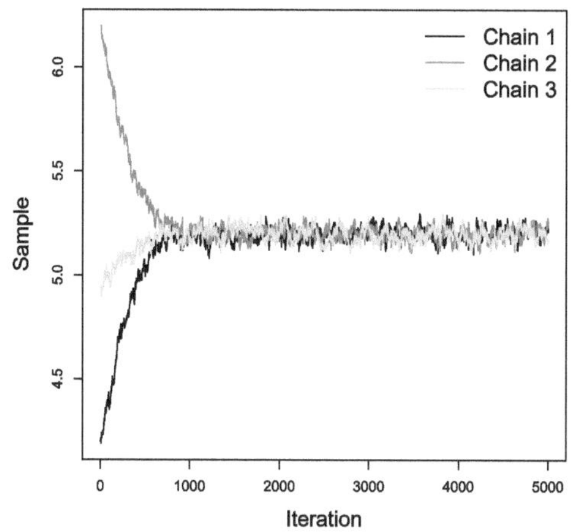

FIGURE 3.18
Convergence of three parallel chains. The three trace plots represent the samples for one parameter from three MCMC chains run in parallel. The chains converge around iteration 1,000.

3.4 Diagnosing and improving MCMC convergence

3.4.1 Selecting initial values

Theoretically, Gibbs and Metropolis–Hastings sampling should converge for any initial values, but in practice the choice of starting values is important. There are two schools of thought on initialization: select initial values close to the posterior mode and run one long chain or run several parallel chains with intentionally diffuse starting values to verify that the algorithm has converged.

Selecting good starting values and running one long chain has the advantage of shortening or eliminating the burn-in period. A common method to select initial values is the MLE or MAP estimates computed using numerical optimization, which is often easier to compute than MCMC sampling. A disadvantage of this approach is that it is hard to rule out the possibility that the chain is stuck in a local mode and is completely missing the bulk of the posterior distribution.

On the other hand, starting multiple (typically 2–5) chains with diffuse starting values requires more burn-in samples, but if all chains give the same results then this provides evidence that the algorithm has properly converged. Figure 3.18 shows idealized convergence of three chains around iteration 500. MCMC is not easily parallelizable because of its sequential nature, but running several independent chains is one way to exploit parallel computing to improve Bayesian computation.

Both single and parallel chains have their relative merits, but given the progress in parallel computing environments, it is preferable to run at least two chains. Care should be taken up front to ensure that the starting values of each chain are sufficiently spread across the posterior distribution so that convergence of the two chains necessarily represents strong evidence of convergence.

3.4.2 Convergence diagnostics

Verifying that the MCMC chains have converged and run long enough to sufficiently explore the posterior is often done informally by visual inspection of trace plots as in Figure 3.18. However, many formal diagnostics have been proposed to diagnose non-convergence. In this chapter, we focus on a few key diagnostics that are produced by `JAGS` via the `coda` package in `R`.

Throughout this section we use two toy examples for illustration:

$$\begin{array}{ll} \text{Poor convergence model} & Y|\boldsymbol{\mu} \sim \text{Poisson}[\exp(\mu_1+\mu_2)] \\ \text{Good convergence model} & Y_1|\boldsymbol{\mu} \sim \text{Poisson}[\exp(\mu_1)] \\ & Y_2|\boldsymbol{\mu} \sim \text{Poisson}[\exp(\mu_2)]. \end{array}$$

In both models the priors are $\mu_j \stackrel{iid}{\sim} \text{Normal}(0, 1000)$. In the first model the two parameters are unidentified, and this leads to poor convergence, as shown in Listing 3.11 and Figure 3.19. In the second model the two parameters are related to separate observations, and so both parameters are identified, leading to good convergence as shown in Listing 3.12 and Figure 3.20. The first model is ridiculous and would never be fit to data, but provides a simple illustration of poor convergence. Of course, not all convergence problems are related to unidentified parameters, but this is a common source of problems.

The first diagnostic is to plot the trace plots and verify that the chains have all reached the same distribution and have mixed properly. Figure 3.20 (top left) provides a good example where the chains for both parameters look like a bar code, while Figure 3.19 (top left) is a problematic case where the chains are mixing slowly and likely need many more iterations to provide a good approximation to the posterior. In these small examples it is possible to plot all chains for all parameters, but for more complex models it may be necessary to inspect the chains for only a representative subset of the parameters.

Autocorrelation of the chains provides a numerical measure of how fast the chains are mixing. Ideally, the iterations would be independent of each other, but the Markovian nature of the sampler induces dependence between successive iterations. The lag-l autocorrelation of the chain for parameter θ_j is defined as

$$\rho_j(l) = \text{Cor}(\theta_j^{(s)}, \theta_j^{(s-l)}), \tag{3.22}$$

and the function $\rho_j(l)$ is called the autocorrelation function. The right panels of Figures 3.19 and 3.20 plot the sample autocorrelation functions for the two examples. Ideally we find that $\rho_j(l) \approx 0$ for all $l > 0$, but some correlation is expected. For example, the lag-1 autocorrelation for μ_1 in Figure 3.20 is around 0.4, but the chain converges nicely.

Rather than reporting the entire autocorrelation function for each parameter, the lag-1 autocorrelation is typically sufficient. Another common one-number summary of the entire function is the effective sample size (ESS). Recall that the sample mean of the MCMC samples is only an estimate of the true posterior mean, and we can quantify the uncertainty in this estimate with the standard error of the sample mean. If the samples are independent, then the standard error of the sample mean is $sd_j/\sqrt{S}$ where sd_j is the sample standard deviation of the draws for θ_j and S is the number of samples. This is labelled "naive SE" in the `JAGS` summary output as in Listing 3.9. However, this underestimates the uncertainty

Listing 3.11
Toy example with poor convergence.

```
# Define the model as a string
> model_string <- textConnection("model{
>   Y     ~ dpois(exp(mu[1]+mu[2]))
>   mu[1] ~ dnorm(0,0.001)
>   mu[2] ~ dnorm(0,0.001)
> }")

# Generate MCMC samples
> inits <- list(mu=rnorm(2,0,5))
> data <- list(Y=1)
> model <- jags.model(model_string,data = data,
>                      inits=inits, n.chains=3)

> update(model, 1000, progress.bar="none")
> samples <- coda.samples(model,
>           variable.names=c("mu"),
>           n.iter=5000, progress.bar="none")

># Apply convergence diagnostics

> # Plots
>  plot(samples)
>  autocorr.plot(samples)

> # Statistics
>  autocorr(samples[[1]],lag=1)
, , mu[1]
          mu[1]      mu[2]
Lag 1 0.9948544 -0.9926385
, , mu[2]
           mu[1]     mu[2]
Lag 1 -0.9960286 0.9947489

>  effectiveSize(samples)
   mu[1]    mu[2]
22.90147 22.71505

>  gelman.diag(samples)
      Point est. Upper C.I.
mu[1]       1.62       2.88
mu[2]       1.62       2.88

Multivariate psrf
1.48

>  geweke.diag(samples[[1]])
  mu[1]  mu[2]
-0.6555 0.6424
```

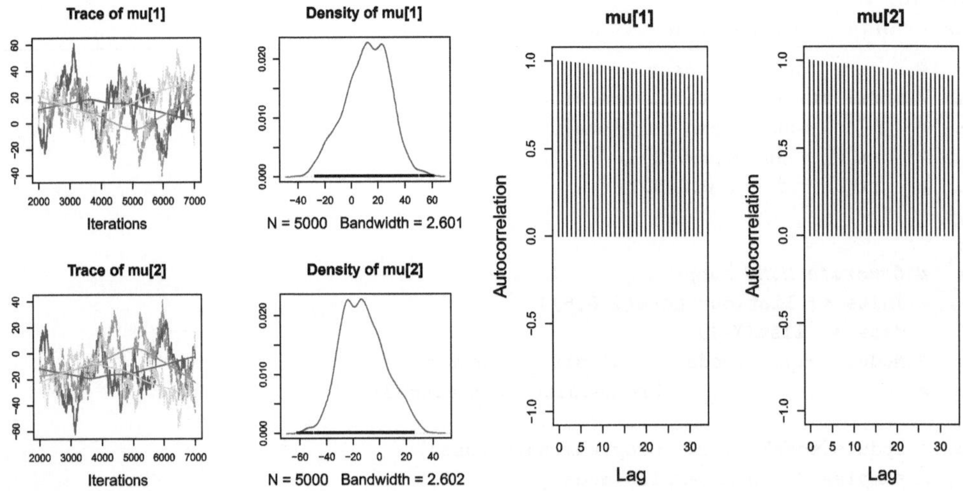

FIGURE 3.19
Convergence diagnostics for a toy example with poor convergence. The left panel gives the trace plot for each parameter and each chain (distinguished by grayscale), and the right panel gives the autocorrelation functions for the first chain.

in the posterior-mean estimate if the samples have autocorrelation. It can be shown that the actual standard error accounting for autocorrelation is $sd_j/\sqrt{ESS_j}$ where

$$ESS_j = \frac{S}{1 + 2\sum_{l=1}^{\infty} \rho_j(l)}. \tag{3.23}$$

In practice the infinite sum must be truncated as $\sum_{l=1}^{L} \rho_j(l)$ for a large L. Assuming the autocorrelation function $\rho_j(l)$ is non-negative for all l, then $ESS_j \le S$ and thus the sample size is discounted to adjust for autocorrelation. The standard error $sd_j/\sqrt{ESS_j}$ is labeled "Time-series SE" in the `JAGS` summary output. One way to determine if the number of samples is sufficient is to increase S until this standard error is acceptably low for all parameters. Another method is to increase S until the effective sample size is acceptably high for all parameters, say $ESS_j > 1000$.

Geweke's diagnostic ([61]) is used to detect non-convergence. It tests for convergence using a two-sample t-test to compare the mean of the chain between batches at the beginning versus the end of the sampler. The default in `coda` is to test that the mean is the same for the first 10% of the samples and the last 50% of the samples. Let $\bar{\theta}_{jb}$ and se_{jb} be the sample mean and standard error, respectively, for θ_j in batch $b = 1, 2$. The standard errors are computed accounting for autocorrelation (as in (3.23)) in the chains. The Geweke statistic is then

$$Z = \frac{\bar{\theta}_{j1} - \bar{\theta}_{j2}}{\sqrt{se_{j1}^2 + se_{j2}^2}}. \tag{3.24}$$

Under the null hypothesis that the means are the same for batches (and assuming each batch has large ESS), Z follows a standard normal distribution, and so $|Z| > 2$ is cause

Listing 3.12
Toy example with good convergence.

```
# Define the model as a string
> model_string <- textConnection("model{
>   Y1    ~ dpois(exp(mu[1]))
>   Y2    ~ dpois(exp(mu[2]))
>   mu[1] ~ dnorm(0,0.001)
>   mu[2] ~ dnorm(0,0.001)
> }")

# Generate MCMC samples
> inits <- list(mu=rnorm(2,0,5))
> data <- list(Y1=1,Y2=10)
> model <- jags.model(model_string,data = data,
>                      inits=inits, n.chains=3)

> update(model, 1000, progress.bar="none")
> samples <- coda.samples(model,
>           variable.names=c("mu"),
>           n.iter=5000, progress.bar="none")

># Apply convergence diagnostics

> # Plots
>  plot(samples)
>  autocorr.plot(samples)

> # Statistics
>  autocorr(samples[[1]],lag=1)
, , mu[1]
         mu[1]      mu[2]
Lag 1 0.359733 0.02112005
, , mu[2]
            mu[1]     mu[2]
Lag 1 0.002213494 0.2776712

>   effectiveSize(samples)
   mu[1]    mu[2]
6494.326 8227.748
>   gelman.diag(samples)

      Point est. Upper C.I.
mu[1]          1          1
mu[2]          1          1
Multivariate psrf
1

>   geweke.diag(samples[[1]])
  mu[1]   mu[2]
-0.5217 -0.2353
```

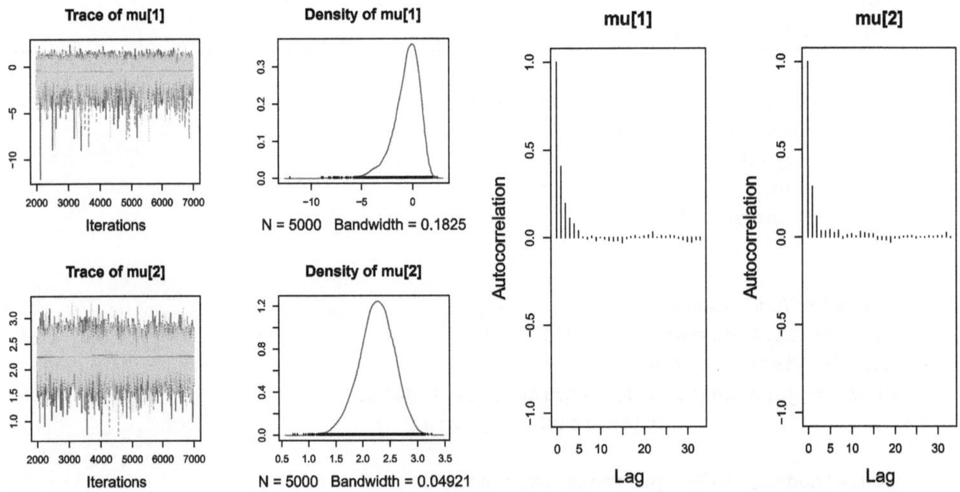

FIGURE 3.20
Convergence diagnostics for a toy example with good convergence. The left panel gives the trace plot for each parameter and each chain (distinguished by grayscale) and the right panel gives the autocorrelation functions for the first chain.

for concern. In the two examples in Listings 3.11 and 3.12, $|Z|$ is less than one, and so this statistic does not detect non-convergence.

Geweke's statistic uses only one chain. A multi-chain extension is the Gelman–Rubin statistic ([57]). The Gelman–Rubin statistic measures agreement in the means between C chains. It is essentially an ANOVA test of whether the chains have the same mean, but the statistic is scaled so that 1.0 indicates perfect convergence and values above 1.1 are questionable. The statistic is

$$R_j = \frac{\frac{S-1}{S}W + \left(\frac{1}{S} + \frac{1}{SC}\right)B}{W} \tag{3.25}$$

where B is S times the variance of the C MCMC sample means for θ_j and W is the average of the C MCMC sample variances for θ_j. The `coda` package plots the Gelman–Rubin statistic as a function of the iteration, and so when the statistic reaches one this indicates good mixing among the parallel chains and that the chains have possibly reached their stationary distribution. The Gelman–Rubin statistics clearly distinguish between the examples in Listing 3.11 and 3.12: $R_j = 1.62$ in the poor-convergence case and $R_j = 1.00$ for the good-convergence case.

3.4.3 Improving convergence

Armed with these diagnostic tools, if the user honestly inspects the sample output, poor convergence can usually be detected. This leaves the more challenging problem of improving convergence! There is no single step that always resolves convergence problems, but the list below offers some suggestions:

1. **Increase the number of iterations**: Theoretically MCMC algorithms should converge and cover the posterior distribution as S goes to infinity. Of course, time limits the number of samples that can be generated, but in some cases increasing S can be faster than searching for other improvements. Nonetheless, this is not the most satisfying solution to poor convergence, specifically when the chains move slowly due to high autocorrelation.
2. **Tune the MH candidate distribution**: The acceptance probability can be tuned during burn-in to be around 30-50%, but as the chain evolves the acceptance probability can change. If groups of parameters have strong cross-correlation then updating them in blocks can dramatically improve convergence. Block steps must be carefully tuned however so the candidate correlation matrix approximates the posterior correlation matrix.
3. **Improve initial values**: Often a sampler works well once it finds the posterior distribution, but the posterior PDF may be too flat away from the mode for the sampler to find its center. One way to improve initial values is to use the maximum likelihood estimates, $\boldsymbol{\theta}^{(0)} = \hat{\boldsymbol{\theta}}_{MLE}$. Another option is to use simulated annealing during burn-in. In simulated annealing, the MH acceptance rate is raised to the power $T_s \in (0, 1]$, i.e., R^{T_s}, where the temperature T_s increases to one throughout the burn-in, e.g., $T_s = s/B$ for $s < B$ and $T_s = 1$ for $s > B$, where B is the number of burn-in samples. The intuition behind this modification is that by raising the acceptance ratio to a power the algorithm is more likely to make big jumps during the burn-in and will slowly settle down in the heart of the posterior as the burn-in period ends.
4. **Use a more advanced algorithm**: Appendix A.4 provides several advanced algorithms that can be used for a particularly vexing problem. For example, Hamiltonian Monte Carlo (HMC) is a Metropolis sampler that uses the gradient of the posterior to intelligently propose candidates and adaptive Metropolis estimates allows the proposal distribution to evolve across iterations.
5. **Simplify the model**: Poor convergence often occurs when models are too complex to be fit with the data at hand. Overly complex models often have unidentified parameters that cannot be estimated. For example, in a model with $Y_i \stackrel{iid}{\sim} \text{Normal}(\theta_1 + \theta_2, \sigma^2)$, the two mean parameters are not identified, meaning that there are infinitely many combinations of the parameters that give the same likelihood (e.g., $\boldsymbol{\theta} = (-10, 10)$ or $\boldsymbol{\theta} = (10, -10)$ give the same mean). Of course, no one would fit such a blatantly unidentified model, but in complex models with dozens of parameters, unidentifiability can be hard to detect. Convergence can be improved by simplifying the model by eliminating covariates, removing non-linear terms, reducing the covariance structure to independence, etc.
6. **Pick more informative priors**: Selecting more informative priors has a similar effect to simplifying the model. For example, even in the silly model $Y_i \stackrel{iid}{\sim} \text{Normal}(\theta_1 + \theta_2, \sigma^2)$, the MCMC algorithm would likely converge quickly with priors $\theta_1 \sim \text{Normal}(-3, 1)$ and $\theta_2 \sim \text{Normal}(3, 1)$. Even a weakly informative prior can stabilize the posterior and improve convergence. As an extreme example, an empirical Bayesian prior (Chapter 2) fixes nuisance parameters at their MAP estimates, which can dramatically improve convergence at the expense of suppressing some uncertainty.
7. **Run a simulation study**: Determining whether a chain has converged can be frustrating for a real data analysis where the true values of the parameters are unknown. Simulating data from the model and then fitting the MCMC algorithm

(as in Listing 3.6) for different sample sizes and parameter values can shed light on the properties of the algorithm and build trust that the sampler is producing reliable output.

3.4.4 Dealing with large datasets

Big data poses computational challenges to all statistical methods, but its effects are acutely felt by MCMC because MCMC requires thousands of passes through the data. Fortunately, recent years have seen a surge in Bayesian computing methods to handle massive datasets. Likely the next few years will see further developments in this area, but below we outline some of the most useful current approaches.

1. **MAP estimation**: MCMC is the gold standard for Bayesian computing because it returns estimates of the entire joint distribution of all of the parameters. However, if the scope of the analysis is limited to prediction, then it is much faster to simply compute the MAP estimate and ignore uncertainty in parameter estimation.
2. **Gaussian approximation**: The Bayesian CLT states that the posterior is approximately normal if the sample size is large. Therefore, running a long MCMC chain for, say, a logistic regression with $n = 1{,}000{,}000$ observations, $p = 50$ covariates and flat priors is unjustifiable; it is much faster and equally accurate to approximate the posterior as Gaussian, centered on the MAP estimate with covariance determined by the information matrix. This computation can often be carried out using MLE software, and thus the posterior will be very similar to the approximate sampling distribution of the MLE, but the interpretation of uncertainty remains Bayesian.
3. **Variational Bayesian computing**: Variational Bayesian approximations (Section 3.1.4) an be even faster to compute than evoking the Bayesian CLT if you target only the marginal distribution of each parameter. That is, if you assume the posterior (not prior) is independent across parameters and use the best approximation to the true posterior that has this form, the posterior can be approximated without computing the joint distribution. This of course raises questions about properly accounting for uncertainty, but does lead to impressive computational savings.
4. **Parallel computing**: MCMC is inherently sequential; i.e., you generally cannot update the second parameter until after updating the first. In special cases, steps of the MCMC routine can be performed simultaneously in parallel on different cores of a CPU or GPU cluster. For example, if parameters are (conditionally) independent, then they can be updated in parallel. Alternatively, if the data are independent and the likelihood factors as the product of n terms, then likelihood calculations within each MCMC step can be done in parallel, e.g., n_1 terms computed on core 1, n_2 terms computed on core 2, etc. While parallelization can theoretically have huge benefits, the overhead associated with passing data across cores can dampen the benefits of using multiple cores unless parallelization is done carefully.
5. **Divide and conquer**: Divide and conquer methods provide an approximation to the posterior that is embarrassingly parallelizable. Say we partition the data into T batches $\mathbf{Y}_1, ..., \mathbf{Y}_T$ that are independent given the model parameters. Then the

posterior can be written

$$p(\boldsymbol{\theta}|\mathbf{Y}) \propto f(\mathbf{Y}|\boldsymbol{\theta})\pi(\boldsymbol{\theta}) = \prod_{t=1}^{T} \left[f(\mathbf{Y}_t|\boldsymbol{\theta})\pi(\boldsymbol{\theta})^{1/T} \right]. \tag{3.26}$$

This decomposition suggests fitting T separate Bayesian analyses (in parallel) with analysis t using data $\mathbf{Y}_t$ and prior $\pi(\boldsymbol{\theta})^{1/T}$ and then combining the results. Each of the T analyses uses the original prior $\pi(\boldsymbol{\theta})$ raised to the $1/T$ power to spread the prior information across batches; e.g., if $\theta \sim \text{Normal}(0, \tau^2)$ then the prior to the $1/T$ power is $\text{Normal}(0, T\tau^2)$. Scott et al. [140] discuss several ways to combine the T posterior distributions. The simplest method is to assume the posteriors are approximately Gaussian, and thus analysis t returns $\boldsymbol{\theta}|\mathbf{Y}_t \approx \text{Normal}(\mathbf{M}_t, \mathbf{V}_t)$. In this case, the posteriors combine as

$$p(\boldsymbol{\theta}|\mathbf{Y}) \approx \text{Normal}(\mathbf{V}\sum_{t=1}^{T} \mathbf{V}_t^{-1}\mathbf{M}_t, \mathbf{V}) \tag{3.27}$$

where $\mathbf{V}^{-1} = \sum_{t=1}^{T} \mathbf{V}_t^{-1}$.

6. **Sequential Monte Carlo**: SMC methods were introduced in Section 3.2.4 for streaming data with each additional new batch of data points modifying the weights of the particles used to approximate the posterior. This approach can also be applied for a large datasets by artificially splitting the data into small (mini) batches and analyzing the batches sequentially. An advantage of this approach for large datasets is that only a single batch and the particle weights need to be stored in memory.
7. **Stochastic gradient MCMC**: Splitting data into minibatchs can be extended to MCMC methods [115]. As described in Section 9.4, Hamiltonian Monte Carlo methods use the derivative of the posterior to optimize the candidate distribution of the Metropolis-Hastings algorithm. Stochastic gradient MCMC uses only a random subset of the dataset to approximate the gradient, which leads to impressive gains for massive datasets.

3.5 Exercises

1. Give an advantage and a disadvantage of the following methods:
   (a) Maximum a posteriori estimation
   (b) Numerical integration
   (c) Bayesian central limit theorem
   (d) Gibbs sampling
   (e) Metropolis–Hastings sampling
2. Assume that $Y_i|\mu \overset{indep}{\sim} \text{Normal}(\mu, \sigma_i^2)$ for $i \in \{1, ..., n\}$, with σ_i known and improper prior distribution $\pi(\mu) = 1$ for all μ.
   (a) Give a formula for the MAP estimator for μ.
   (b) We observe $n = 3$, $Y_1 = 12$, $Y_2 = 10$, $Y_3 = 22$, $\sigma_1 = \sigma_2 = 3$ and $\sigma_3 = 10$; compute the MAP estimate of μ.

(c) Use numerical integration to compute the posterior mean of μ.
(d) Plot the posterior distribution of μ, and indicate the MAP and the posterior mean estimates on the plot.

3. Assume $Y_i|\lambda \overset{indep}{\sim} \text{Poisson}(N_i\lambda)$ for $i = 1, ..., n$.

   (a) Identify a conjugate prior for λ, and derive the posterior that follows from this prior.
   (b) Using the prior $\lambda \sim \text{Uniform}(0, 20)$, derive the MAP estimator of λ.
   (c) Using the prior $\lambda \sim \text{Uniform}(0, 20)$, plot the posterior on a grid of λ assuming $n = 2$, $N_1 = 50$, $N_2 = 100$, $Y_1 = 12$, and $Y_2 = 25$, and show that the MAP estimate is indeed the maximizer.
   (d) Use the Bayesian CLT to approximate the posterior of λ under the setting of (c), plot the approximate posterior, and compare the plot with the plot from (c).

4. Consider the model $Y_i|\sigma_i^2 \overset{indep}{\sim} \text{Normal}(0, \sigma_i^2)$ for $i = 1, ..., n$ where $\sigma_i^2|b \sim \text{InvGamma}(a, b)$ and $b \sim \text{Gamma}(1, 1)$.

   (a) Derive the full conditional posterior distributions for σ_1^2 and b.
   (b) Write pseudocode for Gibbs sampling; i.e., describe in detail each step of the Gibbs sampling algorithm.
   (c) Write your own Gibbs sampling code (not in `JAGS`), and plot the marginal posterior density for each parameter. Assume $n = 10$, $a = 10$ and $Y_i = i$ for $i = 1, ..., 10$.
   (d) Repeat the analysis with $a = 1$, and comment on the convergence of the MCMC chain.
   (e) Implement this model in (c) using `JAGS`, and compare the results with the results in (c).

5. Consider the model $Y_i|\mu, \sigma^2 \sim \text{Normal}(\mu, \sigma^2)$ for $i = 1, ..., n$ and $Y_i|\mu, \delta, \sigma^2 \sim \text{Normal}(\mu + \delta, \sigma^2)$ for $i = n + 1, ..., n + m$, where $\mu, \delta \sim \text{Normal}(0, 100^2)$ and $\sigma^2 \sim \text{InvGamma}(0.01, 0.01)$.

   (a) Give an example of a real experiment for which this would be an appropriate model.
   (b) Derive the full conditional posterior distributions for μ, δ and σ^2.
   (c) Simulate a dataset from this model with $n = m = 50$, $\mu = 10$, $\delta = 1$ and $\sigma = 2$. Write your own Gibbs sampling code (not in `JAGS`) to fit the model above to the simulated data, and plot the marginal posterior density for each parameter. Are you able to recover the true values reasonably well?
   (d) Implement this model using `JAGS`, and compare the results with the results in (c).

6. Consider the model $Y|\theta, b \sim \text{Binomial}(n, \theta)$, $\theta|b \sim \text{Beta}\{\exp(b), \exp(b)\}$ and $b \sim \text{Normal}(0, 1)$.

   (a) Specify initial values for θ and b.
   (b) What is the full conditional distribution of θ?
   (c) The full conditional distribution of b does not have a nice form and thus cannot be updated using Gibbs sampling. Sketch a Metropolis-within-Gibbs sampler for the posterior of (θ, b).

   (d) Implement your Metropolis-within-Gibbs sampler with $n = 100$ and $Y = 10$, and summarize the posterior.

7. Fit the following model to the NBA free throw data in the table in Exercise 22 in Chapter 1:

$$Y_i|\theta_i \sim \text{Binomial}(n_i, \theta_i) \text{ and } \theta_i|m \sim \text{Beta}[\exp(m)q_i, \exp(m)(1-q_i)],$$

   where Y_i is the number of made clutch shots for player $i = 1, ..., 10$, n_i is the number of attempted clutch shots, $q_i \in (0,1)$ is the overall proportion, and $m \sim \text{Normal}(0, 10)$.

   (a) Explain why this is a reasonable prior for θ_i.
   (b) Explain the role of m in the prior.
   (c) Derive the full conditional posterior for θ_1.
   (d) Write your own MCMC algorithm to compute a table of posterior means and 95% credible intervals for all 11 model parameters $(\theta_1, ..., \theta_{10}, m)$. Turn in commented code.
   (e) Fit the same model in `JAGS`. Turn in commented code, and comment on whether the two algorithms returned the same results.
   (f) What are the advantages and disadvantages of writing your own code as opposed to using `JAGS` in this problem and in general?

8. Assume the model $Y|\theta \sim \text{Gamma}(\theta, 1)$ and prior $\theta \sim \text{Uniform}(0, 10)$. Say we observe $Y = 5$. This is not a conjugate prior, and so you will use Metropolis-Hastings sampling to approximate the posterior of θ.

   (a) Plot the posterior of θ for reference, i.e. (in `R`),
```
> theta <- seq(0,10,0.01)
> plot(theta,dgamma(5,theta,1),type="l")
```
   (b) What is a reasonable candidate distribution for θ?
   (c) Give a formula for the acceptance probability (preferably in `R` code). How will you handle candidates outside (0,10)?
   (d) How will you tune the candidate distribution? Be specific.
   (e) Generate and plot samples from the posterior distribution, and compare results with the curve from (a)

9. Open and plot the galaxies data in `R` using the code below,

```
> library(MASS)
> data(galaxies)
> ?galaxies
> Y <- galaxies
> n <- length(Y)
> hist(Y,breaks=25)
```

   Model the observations $Y_1, ..., Y_{82}$ using the Student's t-distribution with location μ, scale σ and degrees of freedom k. Assume prior distributions $\mu \sim \text{Normal}(0, 10000^2)$, $1/\sigma^2 = \tau \sim \text{Gamma}(0.01, 0.01)$ and $k \sim \text{Uniform}(1, 30)$.

   (a) Give reasonable initial values for each of the three parameters.
   (b) Fit the model using `JAGS`. Report trace plots of each parameter, and discuss convergence.

(c) Graphically compare the t distribution with parameters set to their posterior mean with the observed data. Does the model fit the data well?

10. Download the galaxy data

```
> library(MASS)
> data(galaxies)
> Y <- galaxies
```

Suppose $Y_i|\boldsymbol{\theta} \stackrel{iid}{\sim} \text{Laplace}(\mu, \sigma)$ for $i = 1, \ldots, n$ where $\boldsymbol{\theta} = (\mu, \sigma)$.

(a) Assuming the improper prior $\sigma \sim \text{Uniform}(0, 100000)$ and $\pi(\mu) = 1$ for all $\mu \in (-\infty, \infty)$, plot the joint posterior distribution of $\boldsymbol{\theta}$ and the marginal posterior distributions of μ and σ.

(b) Compute the posterior mean of $\boldsymbol{\theta}$ from your analysis in (a), and plot the Laplace PDF with these values against the observed data. Does the model fit the data well?

(c) Plot the posterior predictive distribution (PPD) for a new observation $Y^*|\boldsymbol{\theta} \sim \text{Laplace}(\mu, \sigma)$. How do the mean and variance of the PPD compare to the mean and variance of the "plug-in" distribution from (b)?

11. In Section 2.5 we compared Reggie Jackson's home run rate in the regular season and World Series. He hit 563 home runs in 2820 regular-season games and 10 home runs in 27 World Series games (a player can hit 0, 1, 2, ... home runs in a game). Assuming Uniform(0,10) priors for both home run rates, use **JAGS** to summarize the posterior distribution of (i) his home run rate in the regular season, (ii) his home run rate in the World Series, and (iii) the ratio of these rates. Provide trace plots for all three parameters and discuss convergence of the MCMC sampler including appropriate convergence diagnostics.

12. As discussed in Section 1.6, [47] report that the number of marine bivalve species discovered each year from 2010-2015 was 64, 13, 33, 18, 30 and 20. Denote Y_t as the number of species discovered in year $2009 + t$ (so that $Y_1 = 64$ is the count for 2010). Use **JAGS** to fit the model

$$Y_t|\alpha, \beta \stackrel{indep}{\sim} \text{Poisson}(\lambda_t)$$

where $\lambda_t = \exp(\alpha + \beta t)$ and $\alpha, \beta \stackrel{indep}{\sim} \text{Normal}(0, 10^2)$. Summarize the posterior of α and β and verify that the MCMC sampler has converged. Does this analysis provide evidence that the rate of discovery is changing over time?

13. Write your own Metropolis sampler (i.e., not **JAGS**) for the previous problem. Turn in commented code, report your candidate distribution for each parameters and corresponding acceptance ratio, and use trace plots for each parameter to show the chains have converged.

14. A clinical trial assigned 100 patients each to the placebo, the low dose of a new drug, and the high dose of the new drug (for a total of 300 patients); the data are given in the table below.

Treatment	Positive outcome	Negative outcome
Placebo	52	48
Low dose	60	40
High dose	54	46

Conduct a Bayesian analysis using JAGS with uniform priors on the probability of a patient having a positive outcome under each treatment.

(a) Report the posterior mean and 95% interval for the probability of a positive outcome for each treatment group.

(b) Compute the posterior probability that the low dose is the best of the three treatment options.

15. Consider the normal mixture model

$$Y_i|\theta \stackrel{iid}{\sim} f(y|\theta) = \frac{1}{2}\left[\phi(y-\theta) + \phi(y)\right],$$

for $i = 1, ..., n$ where $\theta \in \mathbb{R}$ and $\phi(z) = \exp\{-z^2/2\}/\sqrt{2\pi}$ denotes the density of a standard normal distribution. We have shown (Section 2.4, problem 14) that an improper prior cannot be used for this likelihood. In this problem we explore the use of vague but proper priors. Analyze data simulated in R as:

```
> set.seed(27695)
> theta_true <- 4
> n          <- 30
> B          <- rbinom(n,1,0.5)
> Y          <- rnorm(n,B*theta_true,1)
```

(a) Argue that the R code above generates samples from $f(y|\theta)$.

(b) Plot the simulated data, and plot $f(y|\theta)$ for $y \in [-3, 10]$ separately for $\theta = \{2, 4, 6\}$.

(c) Assuming prior $\theta \sim \text{Normal}(0, 10^2)$, obtain the MAP estimate of θ and its asymptotic posterior standard deviation using the following R code:

```
> nlp <- function(theta,Y){
>    like <- 0.5*dnorm(Y,0,1)+
>            0.5*dnorm(Y,theta,1)
>    prior <- dnorm(theta,0,10)
>    neg_log_post <- -sum(log(like))-sum(log(prior))
> return(neg_log_post)}

> map     <- optim(mean(Y),nlp,Y=Y,hessian=TRUE)
> map_est <- map$par
> map_sd  <- 1/sqrt(map$hessian)
```

(d) Suppose the prior distribution is $\theta \sim N(0, 10^k)$. Plot the posterior density of θ for $k \in \{0, 1, 2, 3\}$, and compare these posterior densities with the asymptotic normal distribution in (c) (by overlaying all five densities on the same plot).

(e) Use JAGS to fit this model via its mixture representation $Y_i|B_i, \theta \sim \text{Normal}(B_i\theta, 1)$, where $B_i \stackrel{iid}{\sim} \text{Bernoulli}(0.5)$ and $\theta \sim \text{Normal}(0, 10^2)$. Compare the posterior distribution of θ with the results from part (d).

16. Let $Y_1, \ldots, Y_n$ be a random sample from a shifted exponential distribution with density

$$f(y|\alpha, \beta) = \begin{cases} \beta \exp^{-\beta(y-\alpha)} & y \geq \alpha \\ 0 & y < \alpha \end{cases}$$

where $\alpha > 0$ and $\beta > 0$ are the parameters of interest. Assume prior distributions $\alpha \sim \text{Uniform}(0, c)$ and $\beta \sim \text{Gamma}(a, b)$.

(a) Plot $f(y|\alpha, \beta)$ for $\alpha = 2$ and $\beta = 3$ and $y \in [0, 5]$, give interpretations of α and β, and describe a real experiment where this model might be appropriate.

(b) Give the full conditional distributions of α and β, and determine if they are members of a common family of distributions.

(c) Write pseudo code for an MCMC sampler including initial values, full details of each sampling step, and how you plan to assess convergence.

17. Using the data (`mass` and `age`) provided in Listing 3.9, fit the following non-linear regression model:

$$\texttt{mass}_i \sim \text{Normal}\left(\mu_i, \sigma^2\right) \;\; \text{where } \mu_i = \theta_1 + \theta_2 \texttt{age}_i^{\theta_3},$$

with priors $\theta_1 \sim \text{Normal}(0, 100^2)$, $\theta_2 \sim \text{Uniform}(0, 20000)$, $\theta_3 \sim \text{Normal}(0, 1)$, and $\sigma^2 \sim \text{InvGamma}(0.01, 0.01)$.

(a) Fit the model in `JAGS` and plot the data (`age` versus `mass`) along with the posterior mean of μ_i to verify the model fits reasonably well.

(b) Conduct a thorough investigation of convergence.

(c) Give three steps you might take to improve convergence.

18. Assume the (overly complicated) model $Y|n, p \sim \text{Binomial}(n, p)$ with prior distributions $n \sim \text{Poisson}(\lambda)$ and $p \sim \text{Beta}(a, b)$. The observed data is $Y = 10$.

(a) Describe why convergence may be slow for this model.

(b) Fit the model in `JAGS` with $\lambda = 10$ and $a = b = 1$. Check convergence and summarize the posterior of n, p and $\theta = np$.

(c) Repeat the analysis in (b) except with $a = b = 10$. Comment on the effect of the prior distribution of p on convergence for all the three parameters.

19. Consider the (overly complicated) model

$$Y_i|\mu_i, \sigma_i^2 \sim \text{Normal}(\mu_i, \sigma_i^2)$$

where $\mu_i \stackrel{iid}{\sim} \text{Normal}(0, \theta_1^{-1})$, $\sigma_i^2 \sim \text{InvGamma}(\theta_2, \theta_3)$ and $\theta_j \sim \text{Gamma}(\epsilon, \epsilon)$ for $j = 1, 2, 3$. Assume the data are $Y_i = i$ for $i = 1, ..., n$.

(a) Describe why convergence may be slow for this model.

(b) Fit the model in `JAGS` with all four combinations of $n \in \{5, 25\}$ and $\epsilon \in \{0.1, 10\}$, and report the effective sample size for θ_1, θ_2 and θ_3 for all four fits.

(c) Comment on the effect of n and ϵ on convergence.

4

Linear models

CONTENTS

Linear models form the foundation for much of statistical modeling and intuition. In this chapter, we introduce many common statistical models and implement them in the Bayesian framework. We focus primarily on Bayesian aspects of these analyses including selecting priors, computation and comparisons with classical methods. The chapter begins with analyses of the mean of a normal population (Section 4.1.1) and comparison of the means of two normal populations (Section 4.1.2), which are analogous to the classic one-sample and two-sample t-tests. Section 4.2 introduces the more general Bayesian multiple linear regression model including priors that are appropriate for high-dimensional problems. Multiple regression is extended to non-Gaussian data in Section 4.3 via generalized linear models and correlated data in Sections 4.4 and 4.5 via linear mixed models.

4.1 Analysis of normal means

4.1.1 One-sample/paired analysis

In a one-sample study, the n observations are modeled as $Y_i|\mu,\sigma^2 \stackrel{iid}{\sim} \text{Normal}(\mu,\sigma^2)$ and the objective is to determine if $\mu = 0$. This model is often applied to experiments where each unit i actually has two measurements taken under different conditions and Y_i is the difference between the measurements. For example, say student i's math scores before and after a tutorial session are Z_{0i} and Z_{1i}, respectively, then testing if the (population) mean of $Y_i = Z_{1i} - Z_{0i}$ is zero is a way to evaluate the effectiveness of the tutorial sessions.

A Bayesian analysis specifies priors for μ and σ^2 and then summarizes the marginal posterior of μ accounting for uncertainty in σ^2. In this chapter we use the Jeffreys prior, but conjugate normal/inverse gamma priors can also be used. In most cases it is sufficient to plot the posterior density $p(\mu|\mathbf{Y})$ and report the posterior mean and 95% interval. We will also address the problem using a formal hypothesis test. In this chapter we will compute the posterior probabilities of the one-sided hypotheses $\mathcal{H}_1 : \mu \le 0$ and $\mathcal{H}_2 : \mu > 0$; in Chapter 6 we test the point hypothesis $\mathcal{H}_1 : \mu = 0$ versus $\mathcal{H}_2 : \mu \ne 0$.

Known variance: Conditional on σ^2, μ has Jeffreys prior $\pi(\mu) \propto 1$. The posterior is $\mu|\mathbf{Y} \sim \text{Normal}\left(\bar{Y}, \frac{\sigma^2}{n}\right)$, and thus the $100(1-\alpha)\%$ posterior credible interval for μ is

$$\bar{Y} \pm z_{\alpha/2}\frac{\sigma}{\sqrt{n}}, \tag{4.1}$$

where z_τ is the τ quantile of the standard normal distribution (i.e., the normal distribution with mean zero and variance one). This exactly matches the classic $100(1-\alpha)\%$ confidence interval. While the credible interval and confidence interval are numerically identical in this case, they are interpreted differently. The Bayesian credible interval quantifies uncertainty about μ given this dataset, $\mathbf{Y}$, whereas the confidence interval quantifies the anticipated variation in $\bar{Y}$ if we were to repeat the experiment.

For the test of null hypothesis $\mathcal{H}_1 : \mu \le 0$ versus the alternative hypothesis $\mathcal{H}_2 : \mu > 0$, the posterior probability of the null hypothesis is

$$\text{Prob}(\mathcal{H}_1|\mathbf{Y}) = \text{Prob}(\mu < 0|\mathbf{Y}) = \Phi(-Z) \tag{4.2}$$

where Φ is the standard normal cumulative distribution function and $Z = \sqrt{n}\bar{Y}/\sigma$ and thus matches exactly with a frequentist p-value. By definition, $\Phi(z_\tau) = \tau$, and so the decision rule to reject $\mathcal{H}_1$ in favor of $\mathcal{H}_2$ if the posterior probability of $\mathcal{H}_1$ is less than α is equivalent to rejecting $\mathcal{H}_1$ if $-Z < z_\alpha$, or equivalently if $Z > z_{1-\alpha}$ ($-z_\alpha = z_{1-\alpha}$ due to the symmetry of the standard normal PDF). Therefore, the decision rule to reject $\mathcal{H}_1$ in favor of $\mathcal{H}_2$ at significance level α if $Z > z_{1-\alpha}$ is identical to the classic one-sided z-test. However, unlike the classical test, we can quantify our uncertainty using the posterior probability that the hypothesis $\mathcal{H}_1$ (or $\mathcal{H}_2$) is true since we have computed $\text{Prob}(\mathcal{H}_1|\mathbf{Y})$.

Unknown variance: As shown in Section 2.4, the Jeffreys prior for (μ,σ^2) is

$$\pi(\mu,\sigma^2) \propto \left(\frac{1}{\sigma^2}\right)^{3/2}. \tag{4.3}$$

Appendix A.3 shows that the marginal posterior of μ integrating over σ^2 is

$$\mu|\mathbf{Y} \sim t_n\left(\bar{Y}, \hat{\sigma}^2/n\right), \tag{4.4}$$

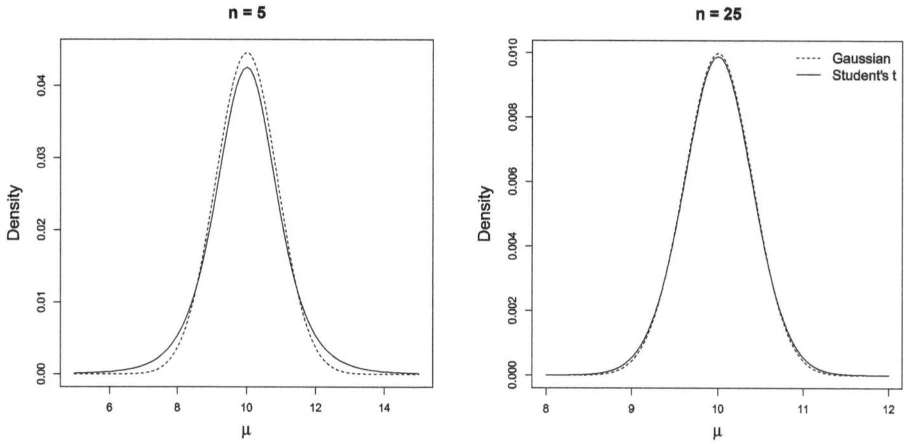

FIGURE 4.1
Comparison of the Gaussian and Student's t-distributions. Below are the Gaussian PDF with mean $\bar{Y}$ and standard deviation $\sigma/\sqrt{n}$ compared to the PDF of Student's t-distribution with location $\bar{Y}$, scale $\hat{\sigma}/\sqrt{n}$, and n degrees of freedom. The plots assume $\bar{Y} = 10$, $\sigma = \hat{\sigma} = 2$, and $n \in \{5, 25\}$.

where $\hat{\sigma}^2 = \sum_{i=1}^{n}(Y_i - \bar{Y})^2/n$' i.e., the posterior is Student's t-distribution with location $\bar{Y}$, variance parameter $\hat{\sigma}^2/n$ and n degrees of freedom. Posterior inference such as credible sets or the posterior probability that μ is positive follow from the quantiles of Student's t-distribution. The credible set is slightly different than the frequentist confidence interval because the degrees of freedom in the classic t-test is $n-1$, whereas the degrees of freedom in the posterior is n; this is the effect of the prior on σ^2.

In classical statistics, when σ^2 is unknown, the Z-test based on the normal distribution is replaced with the t-test based on Student's t-distribution. Similarly, the posterior distribution of the mean changes from a normal distribution given σ^2 to Student's t-distribution when uncertainty about the variance is considered. Figure 4.1 compares the Gaussian and t density functions. The density functions are virtually identical for $n = 25$, but for $n = 5$ the t-distribution has heavier tails than the Gaussian distribution; this is the effect of accounting for uncertainty in σ^2.

As discussed in Section 2.4, an alternative Jeffreys prior is the product of univariate priors, $\pi(\mu, \sigma^2) \propto 1/\sigma^2$. This leads to marginal posterior

$$\mu|\mathbf{Y} \sim t_{n-1}\left(\bar{Y}, \hat{\sigma}^2/n\right). \tag{4.5}$$

With this prior, the posterior distribution matches the frequentist sampling distribution, and thus credible intervals match confidence intervals, etc.

4.1.2 Comparison of two normal means

The two-sample test compares the mean in two groups. For example, an experiment might take the blood pressure of a random sample of n_1 mail carriers that walk their route and n_2 mail carriers that drive their route to determine if these groups have different mean blood pressure. Let $Y_i \stackrel{iid}{\sim} \text{Normal}(\mu, \sigma^2)$ for $i = 1, ..., n_1$ and $Y_i \stackrel{iid}{\sim} \text{Normal}(\mu + \delta, \sigma^2)$ for $i = n_1+1, ..., n_1+n_2 = n$, so that δ is the difference in means and the parameter of interest. Denote the sample mean of the n_j observations in group $j = 1, 2$ as $\bar{Y}_j$ and the group-specific variance estimators as $\hat{s}_1^2 = \sum_{i=1}^{n_1}(Y_i - \bar{Y}_1)^2/n_1$ and $\hat{s}_2^2 = \sum_{i=n_1+1}^{n_1+n_2}(Y_i - \bar{Y}_2)^2/n_2$.

Conditional distribution with the variance fixed: Conditional on the variance and flat prior $\pi(\mu, \delta) \propto 1$ it can be shown that the posterior of the difference in means is

$$\delta|\mathbf{Y} \sim \text{Normal}\left(\bar{Y}_2 - \bar{Y}_1, \sigma^2\left[\frac{1}{n_1} + \frac{1}{n_2}\right]\right). \tag{4.6}$$

As in the one-sample case, posterior intervals and probabilities of hypotheses can be computed using the quantiles of the normal distribution. As in the one-sample case, the credible set and rejection rule match the classic confidence interval and one-sided z-test numerically but have different interpretations.

Unknown variance: The Jeffreys prior for (μ, δ, σ^2) is (Section 2.4)

$$\pi(\mu, \delta, \sigma^2) \propto \frac{1}{(\sigma^2)^2}. \tag{4.7}$$

Appendix A.3 shows (as a special case of multiple linear regression, see Section 4.2) the marginal posterior distribution of δ integrating over both μ and σ^2 is

$$\delta|\mathbf{Y} \sim \text{t}_n\left[\bar{Y}_2 - \bar{Y}_1, \hat{\sigma}^2\left(\frac{1}{n_1} + \frac{1}{n_2}\right)\right], \tag{4.8}$$

where $\hat{\sigma}^2 = (n_1\hat{s}_1^2 + n_2\hat{s}_2^2)/n$ is a pooled variance estimator. As with the one-sample model, the difference between the posterior for the known-variance versus unknown-variance cases is that an estimate of σ^2 is inserted in the posterior, and the Gaussian distribution is replaced with Student's t-distribution with n degrees of freedom. As in the one-sample case, using the alternative Jeffreys prior $\pi(\mu, \sigma^2) \propto 1/\sigma^2$ gives posterior distribution that matches the frequentist sampling distribution,

$$\delta|\mathbf{Y} \sim \text{t}_{n-2}\left[\bar{Y}_2 - \bar{Y}_1, \hat{\sigma}^2\left(\frac{1}{n_1} + \frac{1}{n_2}\right)\right]. \tag{4.9}$$

In the Bayesian analysis we did not "plug in" an estimate of σ^2. Rather, by accounting for its uncertainty and marginalizing over μ and σ^2 the posterior for δ happens to have a natural estimator of σ^2 in δ's scale. Listing 4.1 implements this method.

Unequal variance: If the assumption that the variance is the same for both groups is violated, then the two-sample model can be extended as $Y_i \stackrel{iid}{\sim} \text{Normal}(\mu_1, \sigma_1^2)$ for $i = 1, ..., n_1$ and $Y_i \stackrel{iid}{\sim} \text{Normal}(\mu_2, \sigma_2^2)$ for $i = n_1 + 1, ..., n_1 + n_2$. Since no parameters are shared across the two groups, we can apply the one-sample model separately for each group to obtain

$$\mu_j|\mathbf{Y} \stackrel{indep}{\sim} \text{t}_{n_j}(\bar{Y}_j, s_j^2/n_j) \tag{4.10}$$

for $j = 1, 2$, and the posterior of the difference in means $\delta = \mu_2 - \mu_1$ follows. The posterior of δ is the difference between two Student's t-distributed random variables, which does not in general have a simple form. However, the posterior can be approximated with arbitrary precision using Monte Carlo sampling as in Listing 4.1.

Listing 4.1
R code for comparing two normal means with the Jeffreys prior.

```
# Y1 is the vector of length n1 with data for group 1
# Y2 is the vector of length n2 with data for group 2

# Statistics from group 1
 Ybar1 <- mean(Y1)
 s21   <- mean((Y1-Ybar1)^2)
 n1    <- length(Y1)

# Statistics from group 2
 Ybar2 <- mean(Y2)
 s22   <- mean((Y2-Ybar2)^2)
 n2    <- length(Y2)

# Posterior of the difference assuming equal variance
 delta_hat <- Ybar2-Ybar1
 s2        <- (n1*s21 + n2*s22)/(n1+n2)
 scale     <- sqrt(s2)*sqrt(1/n1+1/n2)
 df        <- n1+n2
 cred_int  <- delta_hat + scale*qt(c(0.025,0.975),df=df)

# Posterior of delta assuming unequal variance using MC sampling
 mu1       <- Ybar1 + sqrt(s21/n1)*rt(1000000,df=n1)
 mu2       <- Ybar2 + sqrt(s22/n2)*rt(1000000,df=n2)
 delta     <- mu2-mu1

 hist(delta,main="Posterior distribution of the difference in means")
 quantile(delta,c(0.025,0.975)) # 95% credible set
```

4.2 Linear regression

The multiple linear regression model with response (also called the dependent variable, outcome or label) Y_i and covariates (also called independent variables, predictors, features or inputs) $X_{i1},...,X_{ip}$ is

$$Y_i = \sum_{j=1}^{p} X_{ij}\beta_j + \varepsilon_i, \tag{4.11}$$

where $\boldsymbol{\beta} = (\beta_1,...,\beta_p)^T$ are the regression coefficients and the errors (also called the residuals) are $\varepsilon_i \stackrel{iid}{\sim} \text{Normal}(0,\sigma^2)$. We assume throughout that $X_{i1} = 1$ for all i so that β_1 is the intercept (i.e., the mean if all other covariates are zero).

The regression model assumes the response Y_i is continuous, but can accommodate either continuous or categorical variables as covariates. Categorical predictors are typically included as binary dummy variables and can include the mean comparisons and Analysis of Variance (ANOVA) models as special cases. With only the intercept term ($p = 1$) and β_1 relabeled as μ, the linear regression model simplifies to the one-sample mean model in Section 4.1.1. The two-sample mean model in Section 4.1.2 is a special case with $p = 2$, X_{i2} set to the binary dummy variable equal to one if observation i is from the second group and zero otherwise, and the parameters relabelled as $\beta_1 = \mu$, and $\beta_2 = \delta$. Similarly, a categorical variable with K levels can be included using an ANOVA framework with a different mean for each level by setting with $X_{i1} = 1$ and including $K - 1$ dummy variables $X_{i2},...,X_{iK}$ with X_{ik} equal 1 if observation i is from group k and zero otherwise.

The coefficient β_j is the slope associated with the jth covariate. For the remainder of this subsection we will assume that all $p-1$ covariates (excluding the intercept term) have been standardized to have mean zero and variance one so the prior can be specified without considering the scales of the covariates. That is, if the original covariate j, $\tilde{X}_{ij}$, had sample mean $\bar{X}_j$ and standard deviation $\hat{s}_j$, then we set $X_{ij} = (\tilde{X}_{ij} - \bar{X}_j)/\hat{s}_j$. After standardization, the slope β_j is interpreted as the change in the mean response corresponding to an increase of one standard deviation unit ($\hat{s}_j$) in the original covariate. Similarly, $\beta_j/\hat{s}_j$ is the expected increase in the mean response associated with an increase of one in the original covariate. The model actually has $p+1$ parameters (p regression coefficients and variance σ^2), but we temporarily use p as the number of regression parameters and not the total number of parameters in the model.

A key concept when selecting a regression model and interpreting results is collinearity. Collinearity generally refers to correlation between the covariates X_{ij}. Recall the interpretation of β_j is the effect of increasing X_{ij} with all other covariates held fixed, but this is hard to envision if X_{ij} is strongly correlated with other covariates. Consider the extreme case with temperature and heat index as covariates. Interpreting the slope for temperature requires envisioning the heat index held fixed while temperature increases, which is counter-intuitive.

When the n observations of one of the covariates is exactly a linear combinations of the n values of the other covariates, then the covariates are said to be linearly dependent. If covariates are not linearly dependent, then they are said to be linearly independent. If the number of predictors exceeds the number of observations, $p > n$, then the covariates will automatically be linearly dependent. As discussed below, the methods that can be applied differ based on whether the covariates are linearly independent.

4.2.1 Classical approaches

Non-Bayesian estimators of $\boldsymbol{\beta}$ range from classical least squares for simple problems to modern penalized regression methods for more complex settings. The Bayesian approaches described in the following sections often give identical point estimators to penalized regression estimators with the prior distribution corresponding to the penalty term. Of course, even in cases with identical point estimators the interpretation of uncertainty is different between the Bayesian and classical approaches.

The classical estimator of $\boldsymbol{\beta}$ is the least squares estimator

$$\widehat{\boldsymbol{\beta}}_{LS} = \arg\min_{\boldsymbol{\beta}} \sum_{i=1}^{n} \left(Y_i - \sum_{j=1}^{p} X_{ij}\beta_j \right)^2. \tag{4.12}$$

This estimator is only valid when the covariates are linearly independent. The sampling distribution of $\widehat{\boldsymbol{\beta}}_{LS}$ follows a normal distribution if σ^2 is known and a Student's t-distribution when σ^2 must be estimated, as is typical.

The likelihood function for the linear model with n observations is the product of n Gaussian PDFs with means $\sum_{j=1}^{p} X_{ij}\beta_j$ and variance σ^2,

$$f(\mathbf{Y}|\boldsymbol{\beta}, \sigma^2) = \prod_{i=1}^{n} \frac{1}{\sqrt{2\pi}\sigma} \exp\left[-\frac{1}{2\sigma^2} \left(Y_i - \sum_{j=1}^{p} X_{ij}\beta_j \right)^2 \right]. \tag{4.13}$$

Since maximizing the likelihood is equivalent to minimizing the negative log-likelihood, the maximum likelihood estimator is identical to the least squares estimator, i.e., $\widehat{\boldsymbol{\beta}}_{MLE} = \widehat{\boldsymbol{\beta}}_{LS}$.

Penalized regression is often used for challenging problems with strong collinearity or a large number of covariates. The ridge regression estimator [79] is

$$\widehat{\boldsymbol{\beta}}_{R} = \arg\min_{\boldsymbol{\beta}} \sum_{i=1}^{n} \left(Y_i - \sum_{j=1}^{p} X_{ij}\beta_j \right)^2 + \lambda \sum_{j=2}^{p} \beta_j^2. \tag{4.14}$$

The addition of the penalty term $\lambda \sum_{j=2}^{p} \beta_j^2$ encourages solutions with estimates near zero and discourages/penalizes implausibly large estimates, thus adding stability. In fact, this (and the LASSO below) solution is valid even if $p > n$ and/or the covariates are linearly dependent. The tuning parameter λ determines how strongly the estimator is shrunk towards zero and can be chosen via cross validation.

While ridge regression adds stability, the elements of $\widehat{\boldsymbol{\beta}}_{R}$ are never exactly zero, and thus all covariates, even those with small effect estimates, are included in the predictive model. The famous LASSO [155] penalized regression estimator simultaneously shrinks estimates to zero and gives some estimates with $\hat{\beta}_j$ set exactly to zero, and thus performs variable selection simultaneously with estimation. The estimator is

$$\widehat{\boldsymbol{\beta}}_{LASSO} = \arg\min_{\boldsymbol{\beta}} \sum_{i=1}^{n} \left(Y_i - \sum_{j=1}^{p} X_{ij}\beta_j \right)^2 + \lambda \sum_{j=2}^{p} |\beta_j|, \tag{4.15}$$

where again the tuning parameter λ can be chosen using cross validation. The LASSO estimator differs from the ridge estimator only in that the squared penalty is replaced by the absolute penalty; however, the discontinuity of the absolute value at the origin gives solutions with exact zeros and thus the properties of the LASSO regressions are quite different than ridge regression.

4.2.2 Prior distributions for Bayesian linear regression

The model assumptions and likelihood in a Bayesian linear regression are identical to classical linear regression, that is, $Y_i = \sum_{j=1}^p X_{ij}\beta_j + \varepsilon_i$ where $\varepsilon_i \stackrel{iid}{\sim} \text{Normal}(0, \sigma^2)$. A Bayesian analysis requires prior distributions for the regression coefficients β_j and variance σ^2. This section provides several options and discusses the scenarios that favor each option. We first introduce the improper Jeffreys prior that, analogous to least squares in classical statistics, is a default choice for regression with a small number of covariates. This is generalized to the Gaussian priors which is useful in the presence of collinearity or for incorporating prior information. Finally we discuss continuous shrinkage priors that are useful when there are many covariates and the prior belief is that many are irrelevant with regression coefficients near zero.

4.2.2.1 Jeffreys prior

Conditioned on σ^2, the Jeffreys prior is $\pi(\boldsymbol{\beta}) \propto 1$. This improper prior leads to a proper posterior under the same conditions required for the least squares estimator to be valid, namely that the covariates are linearly independent. The posterior of $\boldsymbol{\beta}$ is multivariate normal with mean $\widehat{\boldsymbol{\beta}}_{LS}$ and covariance that depends on σ^2 and the covariates X_{ij}. In fact, the posterior distribution of $\boldsymbol{\beta}$ is identical to the sampling distribution of the least squares estimator. Therefore, the posterior credible intervals from this model will numerically match the confidence intervals from a least squares analysis with known error variance.

With σ^2 unknown, the Jeffreys prior is $\pi(\boldsymbol{\beta}, \sigma^2) \propto (\sigma^2)^{-p/2-1}$ (Section 2.4). Under this prior with $n > p$ and linearly independent covariates, the marginal posterior of $\boldsymbol{\beta}$ averaging over σ^2 follows the p-dimensional Student's t-distribution with location $\hat{\boldsymbol{\beta}}_{LS}$ and covariance matrix that depends on $\hat{\sigma}^2$ and the covariates. As described in Section 4.2.3, the estimator of σ^2 and the degrees of freedom differ slightly from the least-squares procedure, but these differences are small if n is much larger than p, and thus the credible intervals approximate the least-squares confidence intervals in most cases. Alternatively, the product of independent Jeffreys prior $\pi(\boldsymbol{\beta}, \sigma^2) \propto 1/\sigma^2$ gives

$$\boldsymbol{\beta}|\mathbf{Y} \sim \text{t}_{n-p}\left(\hat{\boldsymbol{\beta}}_{LS}, \widehat{\Sigma}\right) \tag{4.16}$$

where $\widehat{\Sigma}$ is the estimated covariance of the sampling distribution of $\hat{\boldsymbol{\beta}}_{LS}$ used in least-squares analysis (see (4.23) for details). That is, under this prior, the posterior distribution of $\boldsymbol{\beta}$ matches the sampling distribution of $\hat{\boldsymbol{\beta}}_{LS}$.

4.2.2.2 Gaussian prior

For high-dimensional cases with more predictors than observations, the improper Jeffreys prior does not lead to a valid posterior distribution. Impropriety can be remedied using a proper prior, which always leads to a proper posterior. Even in less extreme situations with moderate p, a proper prior can stabilize the posterior and give better results than the improper prior.

The conjugate prior (conditioned on σ^2) for $\boldsymbol{\beta}$ is the multivariate normal distribution. The most common approach assumes prior independence for the elements of $\boldsymbol{\beta}$ with $\beta_1|\sigma^2 \sim \text{Normal}(\mu_1, \sigma^2\omega^2)$ for the intercept and $\beta_j|\sigma^2 \stackrel{iid}{\sim} \text{Normal}(\mu_j, \sigma^2\tau^2)$ for the slopes with $j > 1$. We include σ^2 in the prior variance to account for the scale of the response, although this is not required. This prior can be informative if ω and τ are small and the μ_j are selected based on prior knowledge. However, typically the prior is uninformative with $\mu_j = 0$ for all j and ω and τ set to be large values or given uninformative prior distributions. Since $\mu_j = 0$ is common, we assume the prior mean is zero unless otherwise noted.

Listing 4.2
JAGS code for multiple linear regression with Gaussian priors.

```
# Likelihood, note that
# inprod(X[i,],beta[]) = X[i,1]*beta[1]+...+X[i,p]*beta[p]
 for(i in 1:n){
    Y[i] ~ dnorm(inprod(X[i,],beta[]),taue)
 }
# Priors
 beta[1] ~ dnorm(0,0.001) #X[i,1]=1 for the intercept
 for(j in 2:p){
    beta[j] ~ dnorm(0,taub*taue)
 }
 taue  ~ dgamma(0.1, 0.1)
 taub  ~ dgamma(0.1, 0.1)
```

Under this prior with $1/\omega^2 = 0$ and $\lambda = 1/\tau^2$, the posterior-mean (and MAP) estimator is the ridge regression estimator [79]

$$\mathrm{E}(\boldsymbol{\beta}|\mathbf{Y}) = \boldsymbol{\beta}_R = \arg\min_{\boldsymbol{\beta}} \sum_{i=1}^{n}\left(Y_i - \sum_{j=1}^{p} X_{ij}\beta_j\right)^2 + \lambda\sum_{j=2}^{p}\beta_j^2. \tag{4.17}$$

Ridge regression is often used to stabilize least squares problems when the number of predictors is large and/or the covariates are collinear. In ridge regression, the tuning parameter λ can be selected based on cross-validation. In a fully Bayesian analysis, τ^2 can either be fixed to a large value to give an uninformative prior, or given a conditionally conjugate inverse gamma prior as in Listing 4.2 to allow the data to determine how much to shrink the coefficients towards zero. If τ^2 (and thus $\lambda = 1/\tau^2$) is given a prior, then the intercept term β_1 should be given a different variance because it plays a different role than the other regression coefficients.

A non-conjugate but more interpretable prior for τ^2 is the R^2 prior of [172] and [168]. They define the proportion of the variance explained by the mean $\eta_i = \sum_{j=1}^{p} X_{ij}\beta_j$ as

$$R^2 = \frac{\mathrm{Var}(\eta_i)}{\mathrm{Var}(Y_i)} = \frac{\mathrm{Var}(\eta_i)}{\mathrm{Var}(\eta_i) + \sigma^2}. \tag{4.18}$$

Assuming the response is standardized and thus $\beta_1 = 0$, and the covariates are standardized to have mean zero and variance one, then (see Appendix) the variance of μ_i, over both the β_j and the distribution of the covariates, is $(p-1)\sigma^2\tau^2$ and thus

$$R^2 = \frac{(p-1)\tau^2}{(p-1)\tau^2 + 1} \text{ so that } \tau^2 = \frac{R^2}{(p-1)(1-R^2)}.$$

The coefficient of determination $R^2 \in (0,1)$ is a measure of model complexity with $R^2 = 0$ and $R^2 = 1$ corresponding to simple and complex models, respectively. Rather than placing a prior on τ^2, they assume a more interpretable prior $R^2 \sim \mathrm{Beta}(a,b)$. Since R^2 is a one-to-one transformation of τ^2, this induces a prior on τ^2. It can be shown that the induced prior of τ^2 is generalized beta prime distribution (defined in the Appendix). A reasonable default choice is $a = b = 1$, which gives a uniform prior to R^2 as in Listing 4.3. An alternative is $a = b = 0.5$ which induces a half-Cauchy prior (see Appendix) on the standard deviation τ as in [54].

Listing 4.3
JAGS code for multiple linear regression with R^2 prior.

```
# Likelihood
for(i in 1:n){
   Y[i] ~ dnorm(inprod(X[i,],beta[]),taue)
}
# Priors
beta[1] ~ dnorm(0,0.001) #X[i,1]=1 for the intercept
for(j in 2:p){
   beta[j] ~ dnorm(0,taub*taue)
}
taue   ~  dgamma(0.1, 0.1)
taub <- (p-1)*(1-R2)/R2
R2     ~ dbeta(1,1)
```

4.2.2.3 Double exponential prior

In regressions with many predictors it is often assumed that most of the p predictors have little effect on the response. The Gaussian prior can reflect this prior belief with a small prior variance, but a small prior variance has a negative side effect; it shrinks the posterior mean of the important regression coefficients toward zero and and induces bias. An alternative [120] is to use double exponential priors for the β_j that have (Figure 4.2) a peak at zero to reflect the prior belief that most predictors have no effect on the response, but a heavy tail to reflect the prior belief that there are a few predictors with strong effects. Listing 4.4 provides `JAGS` code to implement this model (although the `R` package `BLR` [41] is likely faster). This notion of prior distribution with concentration at zero is extended to other distributions in Section 8.1.

Assuming the model $Y_i|\boldsymbol{\beta},\sigma^2 \sim \text{Normal}(\sum_{j=1}^p X_{ij}\beta_j,\sigma^2)$ with improper prior for the intercept and double exponential priors $\beta_j \stackrel{iid}{\sim} \text{DE}(\lambda/\sigma^2)$ with PDF $\pi(\beta_j) \propto \exp\left(-\frac{\lambda}{2\sigma^2}|\beta_j|\right)$, then the maximum a posterior (MAP) estimate is (since maximizing the posterior is equivalent to minimizing twice the negative log posterior) is exactly the LASSO estimator in (4.15) and thus the double exponential prior is often called the Bayesian LASSO prior. An attractive feature of this estimator is that some of the estimates may have $\hat{\beta}_j$ set exactly to zero, and this then performs variable selection simultaneously with estimation. While the posterior mode can be zero, the posterior mean is never zero. Also, because the prior distribution is a continuous PDF, there is zero posterior probability that a coefficient is exactly zero, and so every MCMC sample from this model will have exclusively non-zero values.

4.2.3 Linear regression in matrix notation

The linear regression model is conveniently written in matrix notation. While matrix notation is not necessary to implement Bayesian linear regression, for Gaussian prior distributions this compact notation leads to a closed-form expression for the posterior distribution and thus eliminates the need for MCMC. Let $\mathbf{Y} = (Y_1, ..., Y_n)^T$ be the response vector of length n and the design matrix $\mathbf{X}$ be the $n \times p$ matrix with the first column equal to the vector of ones for the intercept and column j having elements $X_{1j}, ..., X_{nj}$. The linear regression model is then

$$\mathbf{Y} \sim \text{Normal}(\mathbf{X}\boldsymbol{\beta}, \sigma^2\mathbf{I}_n) \tag{4.19}$$

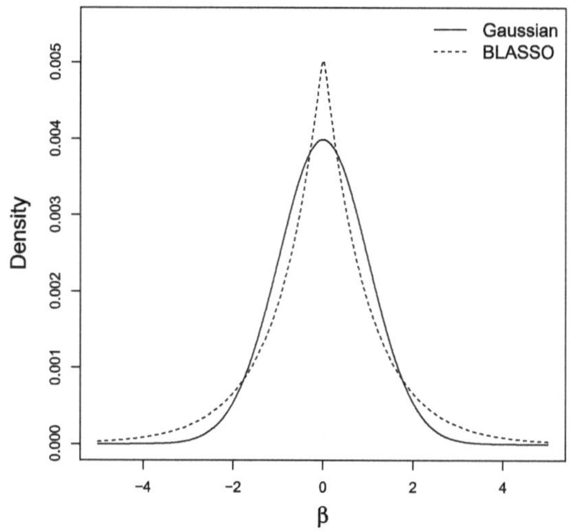

FIGURE 4.2
Comparison of the Gaussian and double exponential priors. Plotted are the standard normal and double exponential/BLASSO PDFs, both set to have the same interquartile range.

Listing 4.4
JAGS code for the Bayesian LASSO.

```
# Likelihood
 for(i in 1:n){
    Y[i] ~ dnorm(inprod(X[i,],beta[]),taue)
 }
# Priors
 beta[1] ~ dnorm(0,0.001)
 for(j in 2:p){
    beta[j] ~ ddexp(0,taub*taue)
 }
 taue  ~ dgamma(0.1, 0.1)
 taub  ~ dgamma(0.1, 0.1)
```

Listing 4.5
R code for Bayesian linear regression under the full Jeffreys prior.

```
# This code assumes:
#   Y is the vector of length n with the observations
#   X is the n x p matrix of covariates
#   The first column of X is all ones for the intercept

# Compute posterior mean and 95% interval
 beta_mean  <- solve(t(X)%*%X)%*%t(X)%*%Y
 sig2_hat   <- mean((Y-X%*%beta_mean)^2)
 beta_cov   <- sig2_hat*solve(t(X)%*%X)
 beta_scale <- sqrt(diag(beta_cov))
 df         <- length(Y)
 beta_025   <- beta_mean + beta_scale*qt(0.025,df=df)
 beta_975   <- beta_mean + beta_scale*qt(0.975,df=df)

# Package the output
 out            <- cbind(beta_mean,beta_025,beta_975)
 rownames(out) <- colnames(X)
 colnames(out) <- c("Mean","Q 0.025","Q 0.975")
```

where $\mathbf{I}_n$ is the $n \times n$ identity matrix with ones on the diagonal (so all responses have variance σ^2) and zeros off the diagonal (so the responses are uncorrelated).

Classical approaches: The usual least squares estimator in matrix notation is

$$\hat{\boldsymbol{\beta}}_{LS} = (\mathbf{X}^T\mathbf{X})^{-1}\mathbf{X}^T\mathbf{Y}. \tag{4.20}$$

Note that the least squares estimator is unique only if $\mathbf{X}^T\mathbf{X}$ is full rank (i.e., $p < n$ and none of $\mathbf{X}$'s columns are redundant) and $(\mathbf{X}^T\mathbf{X})^{-1}$, and thus the estimator is undefined if $\mathbf{X}^T\mathbf{X}$ is not full rank. Assuming $\mathbf{X}$ is full rank and σ^2 is known, the sampling distribution is $\hat{\boldsymbol{\beta}}_{LS} \sim \text{Normal}\left(\boldsymbol{\beta}_0, \sigma^2(\mathbf{X}^T\mathbf{X})^{-1}\right)$, where $\boldsymbol{\beta}_0$ is the true value. Typically σ^2 is estimated based on the residual variance, leading to the Student's t-distribution for the sampling distribution of $\widehat{\boldsymbol{\beta}}_{LS}$ that is used to construct frequentist confidence intervals and p-values.

The ridge regression estimator can also be written in closed form

$$\hat{\boldsymbol{\beta}}_R = (\mathbf{X}^T\mathbf{X} + \lambda\mathbf{D})^{-1}\mathbf{X}^T\mathbf{Y} \tag{4.21}$$

where $\mathbf{D}$ is the $p \times p$ diagonal matrix with zero in the first diagonal element for the intercept and one in the remaining diagonal elements. Given λ, the sampling distribution of $\widehat{\boldsymbol{\beta}}_R$ is also a Student's t-distribution that can be used for uncertainty quantification. Unfortunately, the LASSO estimator $\widehat{\boldsymbol{\beta}}_{LASSO}$ does not have a closed form, and thus a sophisticated numerical optimization routine is required for computation, and the sampling distribution is complex.

Jeffreys prior: Conditioned on σ^2, the Jeffreys prior is $\pi(\boldsymbol{\beta}) \propto 1$. This improper prior only leads to a proper posterior if $\mathbf{X}^T\mathbf{X}$ has full rank, which is the same condition needed for the least squares estimator to be unique. Assuming this condition is met, the posterior is

$$\boldsymbol{\beta}|\mathbf{Y}, \sigma^2 \sim \text{Normal}\left[\hat{\boldsymbol{\beta}}_{LS}, \sigma^2(\mathbf{X}^T\mathbf{X})^{-1}\right]. \tag{4.22}$$

The posterior mean is the least squares solution and the posterior covariance matrix is the covariance matrix of the sampling distribution of the least squares estimator. Therefore, the posterior credible intervals from this model will numerically match the confidence intervals from a least squares analysis with known error variance.

With σ^2 unknown, the Jeffreys prior is $\pi(\boldsymbol{\beta},\sigma^2) \propto (\sigma^2)^{-p/2-1}$ (Section 2.4). Assuming $\mathbf{X}^T\mathbf{X}$ has full rank, Appendix A.3 shows that $\boldsymbol{\beta}|\mathbf{Y} \sim \mathrm{t}_n\left[\hat{\boldsymbol{\beta}}_{LS}, \hat{\sigma}^2(\mathbf{X}^T\mathbf{X})^{-1}\right]$, where $\hat{\sigma}^2 = (\mathbf{Y}-\mathbf{X}\hat{\boldsymbol{\beta}}_{LS})^T(\mathbf{Y}-\mathbf{X}\hat{\boldsymbol{\beta}}_{LS})/n$. That is, the marginal posterior of $\boldsymbol{\beta}$ follows the p-dimensional Student's t-distribution with location $\hat{\boldsymbol{\beta}}_{LS}$, scale matrix $\hat{\sigma}^2(\mathbf{X}^T\mathbf{X})^{-1}$, and n degrees of freedom. Under the alternative Jeffreys prior $\pi(\boldsymbol{\beta},\sigma^2) \propto 1/\sigma^2$ that is often preferred for moderate to large p, the posterior is

$$\boldsymbol{\beta}|\mathbf{Y} \sim \mathrm{t}_{n-p}\left[\hat{\boldsymbol{\beta}}_{LS}, \hat{\sigma}^2(\mathbf{X}^T\mathbf{X})^{-1}\right]. \tag{4.23}$$

A property of the multivariate t distribution is that the marginal distribution of each element is univariate t so that under the alternative Jeffreys prior

$$\beta_j|\mathbf{Y} \sim \mathrm{t}_{n-p}(\hat{\beta}_j, s_j^2) \tag{4.24}$$

where $\hat{\beta}_j$ is the jth element of $\hat{\boldsymbol{\beta}}_{LS}$ and s_j^2 is the jth diagonal element of $\hat{\sigma}^2(\mathbf{X}^T\mathbf{X})^{-1}$ (the degrees of freedom is n under the full Jeffreys prior). Therefore, the joint distribution of all the regression coefficients as well as the marginal distribution of each regression coefficient belong to a known family of distributions, and can thus be computed without MCMC sampling. Listing 4.5 provides `R` code to compute the posterior mean and 95% intervals.

Gaussian priors: For high-dimensional cases with more predictors than observations the improper Jeffreys prior does not lead to a valid posterior distribution, and thus a proper prior is required. The conjugate prior (conditioned on σ^2) for $\boldsymbol{\beta}$ is $\boldsymbol{\beta}|\sigma^2 \sim \mathrm{Normal}(\boldsymbol{\mu}, \sigma^2\boldsymbol{\Omega})$. As above, we include σ^2 in the prior variance to account for the scale of the response. Typically the prior is centered around zero, and so from here on we set $\boldsymbol{\mu} = \mathbf{0}$. This prior, combined with the likelihood $\mathbf{Y}|\boldsymbol{\beta},\sigma^2 \sim \mathrm{Normal}(\mathbf{X}\boldsymbol{\beta}, \sigma^2\mathbf{I}_n)$, gives posterior

$$\boldsymbol{\beta}|\mathbf{Y},\sigma^2 \sim \mathrm{Normal}\left[(\mathbf{X}^T\mathbf{X}+\boldsymbol{\Omega}^{-1})^{-1}\mathbf{X}^T\mathbf{Y}, \sigma^2(\mathbf{X}^T\mathbf{X}+\boldsymbol{\Omega}^{-1})^{-1}\right], \tag{4.25}$$

which is a proper distribution as long as the prior is proper ($\boldsymbol{\Omega}$ is positive definite). As with the Jeffreys prior, if σ^2 is unknown and has a conjugate inverse gamma prior, the the marginal posterior of $\boldsymbol{\beta}$ follows a Student's t-distribution.

There are several choices for the prior covariance matrix $\boldsymbol{\Omega}$. The independent prior in Section 4.2.2.2 is obtained by taking $\boldsymbol{\Omega}$ as the diagonal matrix with diagonal elements $\{\omega^2, \tau^2, ..., \tau^2\}$. The variance component τ^2 can either be fixed at a large value or given a prior distribution, such as the R^2 prior of [172] and [168]. Alternatively, the form of the posterior simplifies under Zellner's g-prior [171] $\boldsymbol{\beta}|\sigma^2 \sim \mathrm{Normal}\left[\mathbf{0}, \frac{\sigma^2}{g}(\mathbf{X}^T\mathbf{X})^{-1}\right]$ for $g > 0$. The conditional posterior is then

$$\boldsymbol{\beta}|\mathbf{Y},\sigma^2 \sim \mathrm{Normal}\left[c\hat{\boldsymbol{\beta}}_{LS}, c\sigma^2(\mathbf{X}^T\mathbf{X})^{-1}\right], \tag{4.26}$$

where $c = 1/(g+1) \in (0,1)$ is the shrinkage factor. This speeds up computation because $\hat{\boldsymbol{\beta}}_{LS}$ and $(\mathbf{X}^T\mathbf{X})^{-1}$ can be computed once outside the MCMC sampler. The shrinkage factor c determines how strongly the posterior mean and covariance are shrunk towards zero. A common choice is $g = 1/n$ and thus $c = n/(n+1)$. Since Fisher's information matrix for the Gaussian distribution is the inverse covariance matrix, the prior contributes $1/n^{th}$ the information as the likelihood, and so this prior is called the unit information prior [88].

4.2.4 Linear regression prediction

One use of linear regression is to make a prediction for a new set of covariates, $\mathbf{X}^{pred} = (X_1^{pred}, ..., X_p^{pred})$. Given the model parameters, the distribution of the new response is

Listing 4.6
JAGS code for linear regression predictions.

```
# Likelihood
for(i in 1:n){
   Y[i] ~ dnorm(inprod(X[i,],beta[]),taue)
}
# Priors
for(j in 1:p){
   beta[j] ~ dnorm(0,0.001)
}
taue  ~  dgamma(0.1, 0.1)

# Predictions
for(i in 1:n_pred){
   Y_pred[i] ~ dnorm(inprod(X_pred[i,],beta[]),taue)
}
# User must pass JAGS the covariates X_pred and integer n_pred
# JAGS returns PPD samples of Y_pred
```

$Y^{pred}|\boldsymbol{\beta},\sigma^2 \sim \text{Normal}(\sum_{j=1}^{p} X_j^{pred}\beta_j,\sigma^2)$. To properly account for parametric uncertainty, we should use the posterior predictive distribution (Section 1.5) that averages over the uncertainty in $\boldsymbol{\beta}$ and σ^2. MCMC provides a means to sample from the PPD by making a sample from the predictive distribution for each of the $s = 1,...,S$ MCMC samples of the parameters, $Y^{(s)}|\boldsymbol{\beta}^{(s)},\sigma^{2(s)} \sim \text{Normal}(\sum_{j=1}^{p} X_j^{pred}\beta_j^{(s)},\sigma^{2(s)})$, and using the S draws $Y^{(1)},...,Y^{(S)}$ to approximate the PPD. The PPD is then summarized the same way as other posterior distributions, such as by the posterior mean and 95% interval. Similar approaches can be used to analyze missing data as in Section 7.1.4.

Listing 4.6 gives `JAGS` code to generate linear regression predictions. This example uses independent Gaussian priors, but the prediction step would be the same for other prior distributions. The matrix of predictors X_{pred} must be passed to `JAGS` and `JAGS` will return the predictions Y_{pred}. Making predictions with `JAGS` can slow the sampler and consume memory, and so it is often more efficient to first perform MCMC sampling for the parameters using `JAGS` and then make predictions in `R` as in Listing 4.7.

4.2.5 Example: Factors that affect a home's microbiome

We use the data from [12], downloaded from `http://figshare.com/articles/1000homes/1270900`. The data are dust samples from the ledges above doorways from $n = 1,059$ homes (after removing samples with missing data; for missing data methods see Section 7.1.4) in the continental US. Bioinformatics processing detects the presence or absence of 763 species (technically operational taxonomic units) of fungi. The response is the log of the number of fungi species present in the sample, which is a measure of species richness. The objective is to determine which factors influence a home's species richness. For each home, eight covariates are included in this example: longitude, latitude, annual mean temperature, annual mean precipitation, net primary productivity (NPP), elevation, the binary indicator that the house is a single-family home, and the number of bedrooms in the home. These covariates are all centered and scaled to have mean zero and variance one.

We apply the Gaussian model in Listing 4.2 first with $\beta_j \stackrel{iid}{\sim} \text{Normal}(0,100^2)$ ("Flat prior") and then with $\beta_j|\sigma^2,\tau^2 \stackrel{iid}{\sim} \text{Normal}(0,\sigma^2\tau^2)$ with $\tau^2 \sim \text{BP}(1,1)$ to give a uniform distribution for the model R^2 ("Gaussian shrinkage prior"), and the Bayesian LASSO prior

Listing 4.7
R code to use JAGS MCMC samples for linear regression predictions.

```
# INPUTS
#   beta_samples := S x p matrix of MCMC samples (from JAGS)
#   taue_samples := S x 1 matrix of MCMC samples (from JAGS)
#   X_pred       := n_pred x p matrix of prediction covariates

S      <- nrow(beta_samples)
n_pred <- nrow(X_pred)
Y_pred <- matrix(NA,S,n_pred)
sigma  <- 1/sqrt(taue_samples)

for(s in 1:S){
  Y_pred[s,] <- X_pred%*%beta_samples[s,]+rnorm(n_pred,0,sigma[s])
}

# OUTPUT
#   Y_pred       := S x n_pred matrix of PPD samples
```

in Listing 4.4. For each of the three models we ran two MCMC chains with 10,000 samples in the burn-in and 20,000 samples after burn-in. Trace plots (not shown) showed excellent convergence, and the effective sample size exceeded 1,000 for all parameters and all models.

The results are fairly similar for the three priors (Figure 4.3). In all three models, temperature, NPP, elevation, and single-family home are the most important predictors, with the most richness estimated to occur in single-family homes with low temperature, NPP, and higher elevation. In all three models, the sample with largest fitted value (i.e., the posterior mean of $\mathbf{X}\boldsymbol{\beta}$) is a single-family home with three bedrooms in Montpelier, VT, and the sample with the smallest fitted value is a multiple-family home with two bedrooms in Tempe, AZ.

Although the results are not that sensitive to the prior in this analysis, there are some notable differences. For example, compared to the posterior under a flat prior, the posterior of the slope for latitude (top right panel in Figure 4.3) is shrunk towards zero by the Gaussian shrinkage model, and the posterior density concentrates even more around the origin for the Bayesian LASSO prior. However, it is not clear from this plot which of these three fits is preferred; model comparison is discussed in Chapter 6.

4.3 Generalized linear models

Multiple linear regression assumes that the response variable is Gaussian and thus the mean response can be any real number. Many analyses do not conform to this assumption. For example, in Section 4.3.1 we analyze binary responses with support $\{0, 1\}$ and in Section 4.3.3 we analyze count data with support $\{0, 1, 2, ...\}$. Clearly these data are not Gaussian, and their mean cannot be any real number because the mean must be between zero and one for binary data and positive for count data. The generalized linear model (GLM) extends linear regression concepts to handle these non-Gaussian outcomes. There is a deep theory of GLMs (exponential families, canonical links, etc; see [103]), but here we focus only on casting the GLM in the Bayesian framework through a few examples.

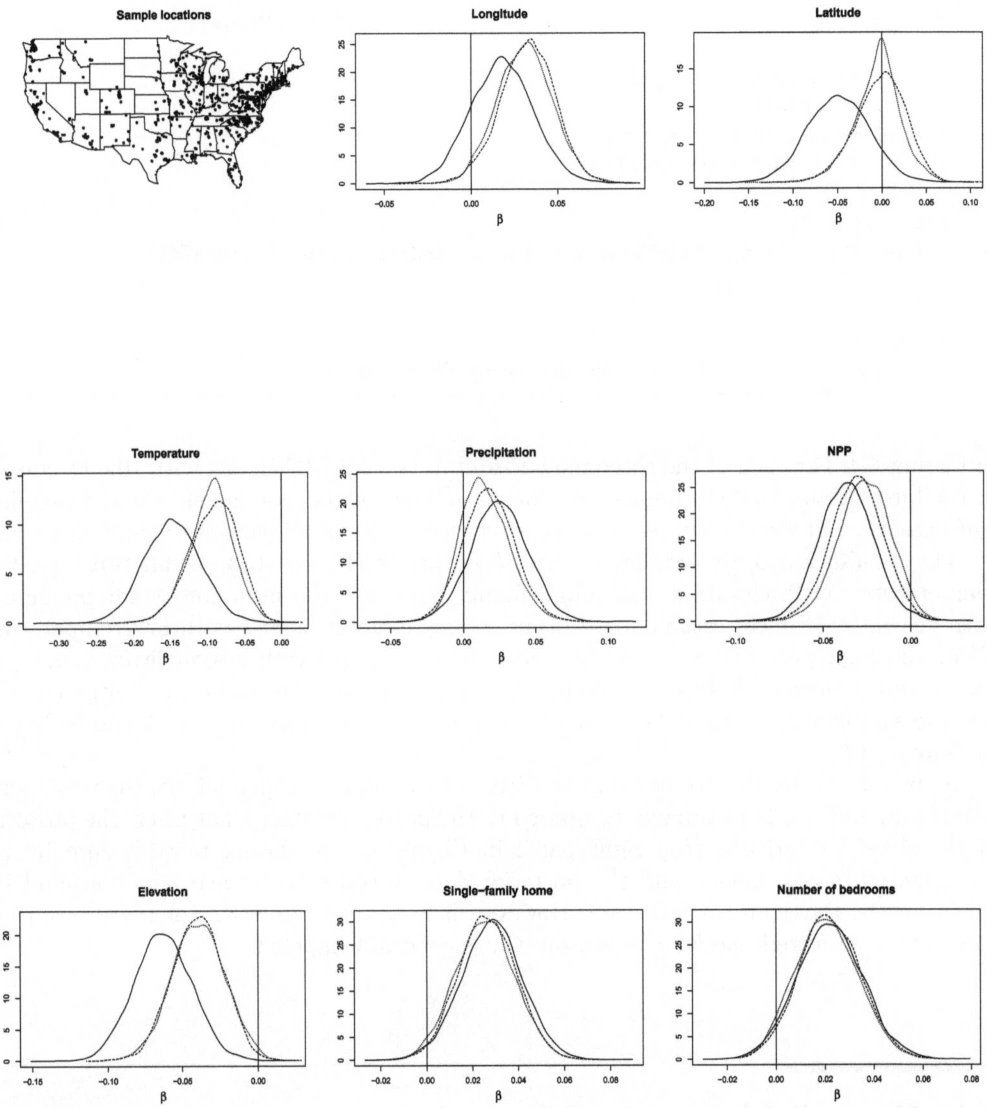

FIGURE 4.3
Regression analysis of the richness of a home's microbiome. The first panel shows the sample locations and the remaining panels plot the posterior distributions of the regression coefficients, β_j. The three models are distinguished by their priors for β_j: the flat prior is $\beta_j \sim \text{Normal}(0, 100^2)$ (solid line), the Gaussian shrinkage prior is $\beta_j \overset{iid}{\sim} \text{Normal}(0, \sigma^2\tau^2)$ with $\tau^2 \sim \text{BP}(1, 1)$ (dashed line) to give a uniform distribution for the model R^2, and the Bayesian LASSO is $\beta_j \overset{iid}{\sim} \text{DE}(0, \sigma^2\tau^2)$ with $\tau^2 \sim \text{InvGamma}(0.1, 0.1)$ (dotted line).

The basic steps in selecting a GLM are (1) determine the support of the response and select an appropriate parametric family and (2) link the covariates to the parameters that define this family. As an example, consider the Gaussian linear regression model in Section 4.2. If the support of the response is $(-\infty, \infty)$, a natural parametric family is the Gaussian distribution. Of course, there are other families with this support, and the fit of the Gaussian family to the data should be verified empirically. Once the Gaussian family has been selected, the covariates must be linked to one or both of the parameters, i.e., the mean or the variance. Let the linear combination of the covariates be

$$\eta_i = \sum_{j=1}^{p} X_{ij}\beta_j. \tag{4.27}$$

The linear predictor η_i can take any value in $(-\infty, \infty)$ depending on X_{ij}. Therefore, to link the covariates with the mean, we can simply set $\text{E}(Y_i) = \eta_i$ as in standard linear regression. To complete the standard model we elect not to link the covariates with the variance, and simply set $\text{Var}(Y_i) = \sigma^2$ for all i.

The function that links the linear predictor with a parameter is called the link function. Say that the parameter in the likelihood for the response is θ_i (e.g., $\text{E}(Y_i) = \theta_i$ or $\text{Var}(Y_i) = \theta_i$); then the link function g is

$$g(\theta_i) = \eta_i. \tag{4.28}$$

The link function must be an invertible function that is well-defined for all permissible values of the parameter. For example, in the Gaussian case the link function for the mean is the identity function $g(x) = x$ so that the mean can be any real number. To link the covariates to the variance we must ensure that the variance is positive, and so natural-log function $g(x) = \log(x)$ is more appropriate. Link functions are not unique, and must be selected by the user. For example, the link function for the mean could be replaced by $g(x) = x^3$, and the link function for the variance could be replaced with $g(x) = \log_{10}(x)$.

Bayesian fitting of a GLM requires selecting the prior and computing the posterior distribution. The priors for the regression coefficients discussed for Gaussian data in Section 4.2 can be applied for GLMs. If the number of observations is much larger than the number of covariates, uninformative normal prior distributions can be selected for the β_j. Shrinkage prior distributions $\beta_j|\tau^2 \sim \text{Normal}(0, \tau^2)$ with prior distribution assigned to τ^2 are more stable with a large number of covariates. The prior for τ^2 can be the conditionally conjugate inverse gamma distribution or the more flexible generalized beta prime prior in [168] (via the `R` package `r2d2glmm`) that leads to a beta distribution for the model R^2 as in Section 4.2.2.2. The posterior distributions for GLMs are usually too complicated to derive the posterior in closed-form and prove that, say, a particular prior distribution leads to a Student's t-posterior distribution with interpretable mean and covariance. However, much of the intuition developed with Gaussian linear models carries over to GLMs.

With non-Gaussian responses the full conditional distributions for the β_j are usually not conjugate, and Metropolis sampling must be used. Maximum likelihood estimates can be used as initial values, and the corresponding standard errors can suggest appropriate candidate distributions. In the examples in this chapter we use `JAGS` to carry out the MCMC sampling. `R` also has dedicated packages for Bayesian GLMs, such as the `MCMClogit` package for logistic regression (Section 4.3.1) that are likely faster than `JAGS`. With flat priors and a large sample it is even more efficient to invoke the Bayesian central limit theorem (Section 3.1.3) and approximate the posterior as Gaussian (e.g., using the `glm` function in `R`) and avoid MCMC altogether.

Listing 4.8
Model statements for several GLMs in JAGS.

```

# (a) Logistic regression
  for(i in 1:n){
    Y[i]          ~ dbern(q[i])
    logit(q[i]) <- inprod(X[i,],beta[])
  }
  for(j in 1:p){beta[j] ~ dnorm(0,0.01)}

# (b) Probit regression
  for(i in 1:n){
    Y[i]          ~ dbern(q[i])
    probit(q[i]) <- inprod(X[i,],beta[])
  }
  for(j in 1:p){beta[j] ~ dnorm(0,0.01)}

# (c) Poisson regression
  for(i in 1:n){
    Y[i]               ~ dpois(lambda[i])
    log(lambda[i]) <- inprod(X[i,],beta[])
  }
  for(j in 1:p){beta[j] ~ dnorm(0,0.01)}

# (d) Negative binomial regression
  for(i in 1:n){
    Y[i]               ~ dnegbin(q[i],m)
    q[i]              <- m/(m + lambda[i])
    log(lambda[i]) <- inprod(X[i,],beta[])
  }
  for(j in 1:p){beta[j] ~ dnorm(0,0.01)}
  m ~ dgamma(0.1,0.1)

# (e) Zero-inflated Poisson
  for(i in 1:n){
    Y[i]               ~ dpois(q[i])
    q[i]              <- Z[i]*lambda[i]
    Z[i]               ~ dbern(p[i])
    log(lambda[i]) <- inprod(X[i,],beta[])
    logit(p[i])     <- inprod(X[i,],alpha[])
  }
  for(j in 1:p){beta[j] ~ dnorm(0,0.01)}
  for(j in 1:p){alpha[j] ~ dnorm(0,0.01)}

# (f) Beta regression
  for(i in 1:n){
    Y[i]          ~ dbeta(r*q[i],r*(1-q[i]))
    logit(q[i]) <- inprod(X[i,],beta[])
  }
  for(j in 1:p){beta[j] ~ dnorm(0,0.01)}
  r ~ dgamma(0.1,0.1)
```

4.3.1 Binary outcome

Binary outcomes $Y_i \in \{0,1\}$ occur frequently when the result of an experiment is recorded only as an indicator of a success or failure. For example, in Section 1.2 the response was the binary indicator that a test for HIV was positive. Binary variables must follow a Bernoulli distribution. The Bernoulli distribution has one parameter, the success probability $\text{Prob}(Y_i = 1) = q_i \in [0,1]$. Since the parameter is a probability, the link function must have input range $[0,1]$ and output range $[-\infty,\infty]$. Below we discuss two such link functions: logistic and probit.

Logistic regression: The logistic link function is

$$g(q) = \text{logit}(q) = \log\left(\frac{q}{1-q}\right). \tag{4.29}$$

This link function converts the event probability q first to the odds of the event, $q/(1-q) > 0$, and then to the log odds, which can be any real number. The logistic regression model is written

$$Y_i \overset{indep}{\sim} \text{Bernoulli}(q_i) \text{ and } \text{logit}(q_i) = \eta_i = \sum_{j=1}^{J} X_{ij}\beta_j. \tag{4.30}$$

The inverse logistic (expit) function is $g^{-1}(x) = \exp(x)/[1+\exp(x)]$ and so the model can also be expressed as $Y_i \overset{indep}{\sim} \text{Bernoulli}\left(\exp(\eta_i)/[1+\exp(\eta_i)]\right)$. `JAGS` code for this model is in Listing 4.8a.

Because the log odds of the event that $Y_i = 1$ are linear in the covariates, β_j is interpreted as the increase in the log odds corresponding to an increase of one in X_j with all other covariates held fixed. Similarly, with all other covariates held fixed, increasing X_j by one multiplies the odds by $\exp(\beta_j)$. Therefore if $\beta_j = 2.3$, increasing X_j by one multiplies the odds by ten, and if $\beta_j = -2.3$, increasing X_j by one divides the odds by ten. This interpretation is convenient for communicating the results and specifying priors. If a change of one in the covariate is deemed a large change, then a standard normal prior may have sufficient spread to represent an uninformative prior. The prior for the regression coefficients can also be evaluated on the induced prior for the success probabilities, q_i. Figure 4.4 shows that prior standard deviation 10 gives a prior for q_i that has most of its mass on zero and one, which may be unrealistic in many problems. This is an example of prior predictive checks, which will be discussed in Section 6.5

Probit regression: There are many possible link functions from $[0,1]$ to $[-\infty,\infty]$. In fact, the quantile function (inverse CDF) of any continuous random variable with support $[-\infty,\infty]$ would suffice. The link function in logistic regression is the quantile function of the logistic distribution. In probit regression, the link function is the Gaussian quantile function

$$Y_i \overset{indep}{\sim} \text{Bernoulli}(q_i) \text{ and } q_i = \Phi(\eta_i), \tag{4.31}$$

where Φ is the standard normal CDF (Listing 4.8b).

Unfortunately, the regression coefficients β_j in probit regression do not have a nice interpretation such as their log-odds interpretation in logistic regression. However, probit regression is useful because it leads to Gibbs sampling and can be used to model dependence between binary variables. Probit regression is equivalent to specifying latent variables $Z_i \overset{indep}{\sim} \text{Normal}(\eta_i, \sigma^2)$ and assuming that observing $Y_i = 1$ indicates that Z_i exceeds the threshold z. For example, Z_i might represent a patient's blood pressure, and Y_i is the corresponding binary indicator that the patient has high blood pressure (defined as exceeding

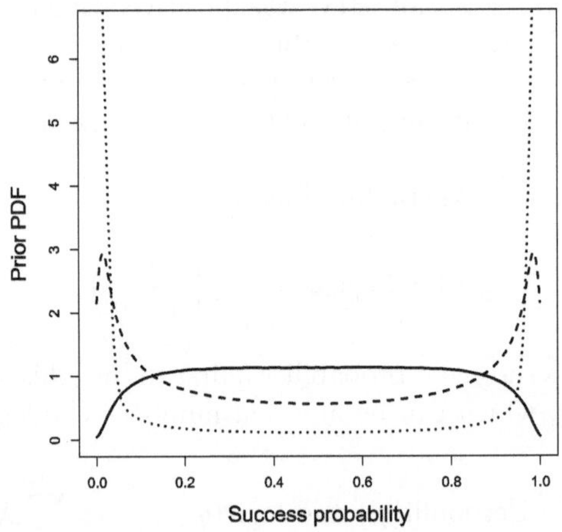

FIGURE 4.4
Logistic regression prior. Prior distribution for success probability q_i with $\text{logit}(q_i) = \beta_1 + X_i\beta_2$ and $X_i \sim \text{Normal}(0, 1)$ and $\beta_1, \beta_2 \sim \text{Normal}(0, \tau^2)$; the prior is plotted for $\tau = 1$ (solid), $\tau = 2$ (dashed) and $\tau = 10$ (dotted).

z). In this example, the probability that the patient has high blood pressure is

$$\begin{aligned}\text{Prob}(Y_i = 1) &= \text{Prob}(Z_i > z)\\ &= \text{Prob}[(Z_i - \eta_i)/\sigma > (z - \eta_i)/\sigma]\\ &= 1 - \Phi[(z - \eta_i)/\sigma]\\ &= \Phi\left[(\eta_i - z)/\sigma\right]\\ &= \Phi\left[\left(\sum_{j=1}^{J} X_{ij}\beta_j - z\right)/\sigma\right].\end{aligned}$$

Since we never observe the latent variables Z_i, we cannot estimate the threshold z or the variance σ^2. The threshold z is coupled with the intercept because adding a constant to the threshold and subtracting the same constant from the intercept does not affect the event probabilities q_i. Therefore, the threshold is typically set to $z = 0$. Similarly, multiplying by σ and dividing the slopes β_j by the same constant do not affect the event probabilities, and so the variance is typically set to $\sigma^2 = 1$. This gives the usual probit regression model $q_i = \Phi(\eta_i)$.

In this formulation, the regression coefficients $\boldsymbol{\beta}$ have conjugate full conditional distributions conditioned on the latent Z_i. This leads to Gibbs sampling if the Z_i are imputed at each MCMC iteration [5]. Also, dependence between binary outcomes Y_i can be induced by a multivariate normal model for the latent variables Z_i.

4.3.2 Categorical outcomes

Logistic and probit regression can be generalized to model categorical data with more than two possible outcomes. Categorical data might arise if a survey asks customers which of $R = 4$ products they prefer, and their response is coded as $Y_i \in \{1, ..., R\}$, or how strongly they agree with a statement (e.g., "strongly", "very strongly", etc.), again coded as $Y_i \in \{1, ..., R\}$. Models differ depending on whether the response levels are ordered (strength of agreement) or unordered (product choice), but in either case the main objectives are to determine how covariates $\mathbf{X}_i$ affect the likelihood of each response and make predictions.

A natural model for unordered responses is the multinomial logistic regression model,

$$\text{Prob}(Y_i = r|\boldsymbol{\beta}) = \frac{\exp(\mathbf{X}_i\boldsymbol{\beta}_r)}{\sum_{k=1}^{R} \exp(\mathbf{X}_i\boldsymbol{\beta}_k)} \tag{4.32}$$

for $\boldsymbol{\beta} = (\boldsymbol{\beta}_1, ..., \boldsymbol{\beta}_R)$. In this model, each of the R levels has a different set of regression parameters, $\boldsymbol{\beta}_r$, that determine how the covariates influence the probability of response r. The linear predictors $\mathbf{X}_i\boldsymbol{\beta}_r$ are exponentiated to ensure the probabilities are positive and the sum in the denominator ensures that $\sum_{r=1}^{R} \text{Prob}(Y_i = r|\boldsymbol{\beta}) = 1$, giving a valid PMF for all $\mathbf{X}_i$ and $\boldsymbol{\beta}$.

This formulation maintains the log-odds interpretation that

$$\log\left\{\frac{\text{Prob}(Y_i = r|\boldsymbol{\beta})}{\text{Prob}(Y_i = s|\boldsymbol{\beta})}\right\} = \mathbf{X}_i(\boldsymbol{\beta}_r - \boldsymbol{\beta}_s).$$

As demonstrated by the log-odds above, only the differences between the regression coefficients are estimable. To avoid overparameterization, one of the R responses, say $r = 1$, is set as the baseline with $\boldsymbol{\beta}_1 = 0$. With this restriction, the log-odds of category r compared to the baseline is $\log\left\{\frac{\text{Prob}(Y_i=r|\boldsymbol{\beta})}{\text{Prob}(Y_i=1|\boldsymbol{\beta})}\right\} = \mathbf{X}_i\boldsymbol{\beta}_r$, and thus $\boldsymbol{\beta}_r$ is interpreted as in logistic regression with the caveat that all comparisons are with respect to the baseline response (often chosen as the most common response). In fact, if there are only $R = 2$ categories with $Y_i = 1$ a failure and $Y_i = 2$ a success, then the multinomial logistic regression model reduces to the standard logistic regression model for binary data.

Ordinal data where $Y_i = r$ (agree) is in some sense between $Y_i = r - 1$ (disagree) and $Y_i = r + 1$ (strongly agree) can be modeled as a sequence of logistic regressions. The sequence of logistic regressions mimics the respondent first deciding whether to give the first response; if not they decide whether to give the second, etc. Mathematically, the first logistic regression is the probability of selecting the first category versus all others $\log\{\text{Prob}(Y_i = 1|\boldsymbol{\beta})/\text{Prob}(Y_i > 1|\boldsymbol{\beta})\} = \mathbf{X}_i\boldsymbol{\beta}_1$. The second logistic regression is fit only using data from observations with $Y_i > 1$ and is $\log\{\text{Prob}(Y_i = 2|\boldsymbol{\beta})/\text{Prob}(Y_i > 2|\boldsymbol{\beta})\} = \mathbf{X}_i\boldsymbol{\beta}_2$. This is repeated until all $R-1$ decisions have been made giving parameters $\boldsymbol{\beta}_1, ..., \boldsymbol{\beta}_{R-1}$. For both the ordered and unordered logistic regression models, any of the priors for standard logistic regression discussed in Section 4.3.1 can be applied, independently across the $\boldsymbol{\beta}_r$. If R is large, then it may be beneficial to treat the β_r as random effects following the methods in Section 4.4.

Probit regression can also be applied to categorical data. One approach for ordinal data is to assume that there is a latent continuous variable $Z_i|\boldsymbol{\beta} \sim \text{Normal}(\mathbf{X}_i\boldsymbol{\beta}, 1)$ for each observation, and the response is determined by this latent value. In this model, if there are R categories, there are $R+1$ breakpoints $L_0 < L_1 < ... < L_R$ with $L_0 = -\infty$ and $L_R = \infty$, and $Y_i = k$ if $L_{k-1} < Z_i < L_k$. This is visualized in Figure 4.5 where $R = 4$ and the shaded probability is $\text{Prob}(Y_i = 3|\boldsymbol{\beta}) = \text{Prob}(L_2 < Z_i < L_3|\boldsymbol{\beta})$.

The appeal of this model is that it connects categorical data analysis back to the well-understood Gaussian linear regression. For instance, say Z_i is a continuous measure of a respondent's view on a topic; then $\boldsymbol{\beta}$ summarizes the covariate effects on this continuous measure. However, the Z_i are not observed; rather the respondent is asked to rate the topic as strongly unfavorable, unfavorable, favorable or strongly favorable, and these responses are coded as $Y_i \in \{1, ..., 4\}$.

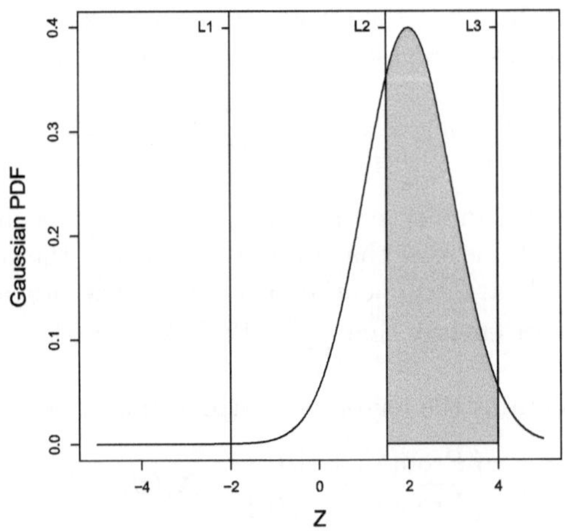

FIGURE 4.5
Probit regression for categorical data. The curve is the PDF of $Z|\boldsymbol{\beta} \sim \text{Normal}(X\beta, 1)$ with $X\beta = 2$. The vertical lines are at the cutpoints L_1, L_2 and L_3, and the area under the curve is the probability that $Y = 3$.

The probit regression model for categorical data is conducive to MCMC computing. As with the probit regression model for binary data, the Z_i can be imputed at each iteration, and then $\boldsymbol{\beta}$ can be updated using Gibbs sampling. A complication is selecting the cutpoints, L_k. They should be given priors so that $L_k < L_{k+1}$ for all k, and one of the cutpoints must be fixed, e.g., $L_1 = 0$, because increasing the intercept in $\boldsymbol{\beta}$ and increasing all the cutpoints by the same amount gives the same bin probabilities (i.e., in Figure 4.5 the x-axis would be shifted but the area of the shaded region would not be affected). An example is $L_1 = 0$ and $L_k = L_{k-1} + \delta_k$ for $k > 1$ with gamma prior distributions for the δ_k so the cutpoints maintain proper ordering.

4.3.3 Count outcomes

Random variables with support $Y_i \in \{0, 1, 2, ...\}$ often arise as the number of events that occur in a time interval or spatial region. For example, in Section 2.1 we analyze the number of NFL concussions by season. In this chapter, we will focus on modeling the mean as a function of the covariates. The link between the linear predictor and the mean must ensure that $\text{E}(Y_i) = \lambda_i \geq 0$. A natural link function is the log link,

$$\log(\lambda_i) = \sum_{j=1}^{p} X_{ij}\beta_j. \tag{4.33}$$

In this model, the mean is multiplied by $\exp(\beta_j)$ if X_j increases by one with all other covariates remaining fixed. Unlike binary data, specifying the mean does not completely determine the likelihood for count data. We discuss below two families of distributions for the likelihood function: Poisson and negative binomial.

Poisson regression: The Poisson regression model is

$$Y_i|\lambda_i \stackrel{indep}{\sim} \text{Poisson}(\lambda_i) \text{ where } \lambda_i = \exp\left(\sum_{j=1}^{p} X_{ij}\beta_j\right). \tag{4.34}$$

As with logistic regression, the regression coefficients $\beta_1, ..., \beta_p$ can be given the priors discussed in Section 4.2, and Metropolis sampling can be used to explore the posterior (Listing 4.8c). A critical assumption of the Poisson model is that the mean and variance are equal, that is,

$$\text{E}(Y_i) = \text{Var}(Y_i) = \lambda_i. \tag{4.35}$$

The distribution of a count is over-dispersed (under-dispersed) if its variance is greater than (less than) its mean. If over-dispersion is present, then the Poisson model is inappropriate and another model should be considered so that the posterior accurately reflects all sources of variability.

Negative binomial regression: One approach to accommodating over-dispersion is to incorporate gamma random variables $e_i \stackrel{iid}{\sim} \text{Gamma}(m, m)$ with mean 1 and variance $1/m$. Then

$$Y_i|e_i, \lambda_i, m \stackrel{indep}{\sim} \text{Poisson}(\lambda_i e_i). \tag{4.36}$$

Marginally over e_i, Y_i follows the negative binomial distribution,

$$Y_i|\lambda_i, m \sim \text{NegBinomial}(q_i, m) \tag{4.37}$$

with probability $q_i = m/(\lambda_i + m)$ and size m. The size m need not be an integer, but if it is and we envision a sequence of independent Bernoulli trials each with success probability q_i, then Y_i can be interpreted as the number of failures that occur before the mth success. More importantly for handling over-dispersion is that $\text{E}(Y_i) = \lambda_i$ and $\text{Var}(Y_i) = \lambda_i + \lambda_i^2/m > \lambda_i$. The parameter m controls over-dispersion. If m is large, then $e_i \approx 1$, and the model reduces to the Poisson regression model with $\text{Var}(Y_i) \approx \text{E}(Y_i)$; if m is close to zero, then e_i has large variance, and thus $\text{Var}(Y_i) > \text{E}(Y_i)$. The over-dispersion parameter can be given a gamma prior, as in Listing 4.8d.

Zero-inflated Poisson: Another common deviation from the Poisson distribution is an excess of zeros. For example, say that Y_i is the number of fish caught by visitor i to a state park. It may be that most of the Y_i are zero because only a small proportion (say p) of the visitors fished, but the distribution of Y_i for those that did fish is Poisson with mean λ. The probability of an observation being zero is then $1 - p$ for the non-fishers plus p times the Poisson probability at zero for the fishers. The PMF corresponding to this scenario is

$$f(y|p,\lambda) = \begin{cases} (1-p) + pf_P(0|\lambda) & \text{if } y = 0 \\ pf_P(y|\lambda) & \text{if } y > 0 \end{cases} \tag{4.38}$$

where f_P is the Poisson PMF. This is equivalent to the two-stage model

$$Y_i|Z_i, \lambda \sim \text{Poisson}(Z_i\lambda) \text{ and } Z_i \sim \text{Bernoulli}(p)$$

where Z_i is the latent indicator that visitor i fished and thus the mean of Y_i is zero for non-fishers with $Z_i = 0$. In this scenario, we do not observe the Z_i, but this two-stage model gives the equivalent model to (4.38) but uses only standard distributions. As a result, the model can be coded in `JAGS` with covariates included in the mass at zero and the Poisson rate as in Listing 4.8e.

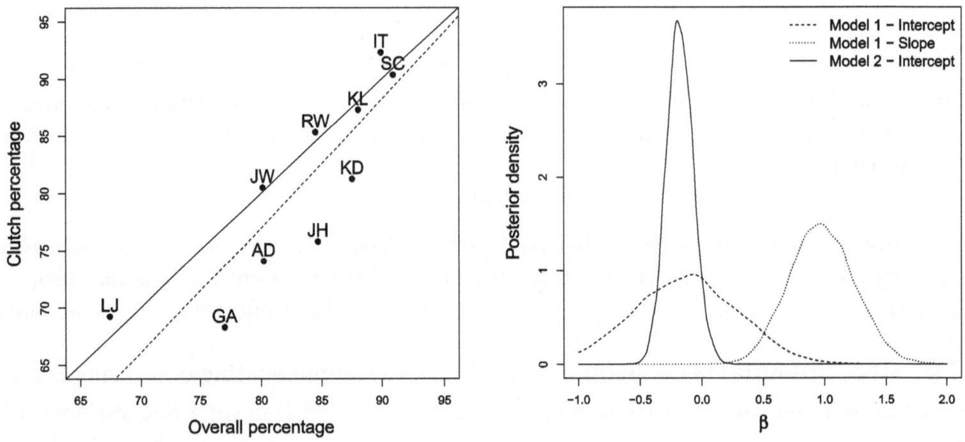

FIGURE 4.6
Logistic regression analysis of NBA free throws. The left panel shows the overall percentage of free throws made versus the percentage for clutch shots made for each player (denoted by the player's initials). The solid line is the $x = y$ lines and the dashed line is the fitted value from Model 2. The right plot is the posterior density for the slope and intercept from Model 1 and the intercept from Model 2.

4.3.4 Example: Logistic regression for NBA clutch free throws

The table in Problem 17 of Section 1.6 gives the overall free-throw proportion ($q_i \in [0,1]$), the number of clutch shots made (Y_i), and the number of attempted clutch shots (n_i) for players $i = 1, ..., 10$. Figure 4.6 shows that most players have similar success rates in clutch and non-clutch situations, but some players appear to have less success in pressure situations. We fit logistic regression models to formally explore this relationship.

The support of the response is $Y_i \in \{0, 1, ..., n_i\}$, and so we select a binomial likelihood $Y_i|p_i \sim \text{Binomial}(n_i, p_i)$ where p_i is the probability of player i making a clutch shot. The number of shots n_i is known, and so we link the covariate to the success probability p_i. The two models are

1. $\text{logit}(p_i) = \beta_1 + \beta_2 X_i$
2. $\text{logit}(p_i) = \beta_1 + X_i$

where $X_i = \text{logit}(q_i)$ is the log odds of making a regular free throw. If $\beta_1 = 0$ and $\beta_2 = 1$ in Model 1 or $\beta_1 = 0$ in Model 2, then the clutch performance equals the overall performance. To determine if this is the case, we fit the model using JAGS with two chains with burn-in 10,000 and 20,000 additional samples and with uninformative Normal(0, 100) priors for all parameters. The model specification for Model 1 is

```
for(i in 1:10){
  Y[i]          ~ dbinom(p[i],n[i])
  logit(p[i]) <- beta[1] + beta[2]*X[i]
}
beta[1] ~ dnorm(0,0.01)
```

TABLE 4.1
Beta regression of microbiome data. Posterior median and 95% intervals for the regression coefficients β_j and concentration parameter r.

	Median	95% Interval
Intercept	−1.01	(−1.05, −0.96)
Longitude	−0.21	(−0.28, −0.14)
Latitude	−0.08	(−0.22, 0.05)
Temperature	−0.15	(−0.30, −0.01)
Precipitation	0.03	(−0.04, 0.11)
NPP	−0.04	(−0.10, 0.02)
Elevation	−0.02	(−0.09, 0.05)
Single-family home	0.07	(0.02, 0.13)
Number of bedrooms	−0.08	(−0.13, −0.03)
r	7.97	(7.34, 8.66)

```
beta[2] ~ dnorm(0,0.01)
```

The results are plotted in Figure 4.6. For Model 1, the slope is centered squarely on one, and the intercept is centered slightly below zero, but both parameters have considerable uncertainty, and so the model with $p_i = q_i$ remains plausible. However, the intercept in Model 2 is negative with posterior probability 0.96, so there is some evidence that even the best players in the NBA underperform in pressure situations. The fitted curve (dashed line) in Figure 4.6 (left panel) is $1/[1 + \exp(-\bar{\beta}_1 - X_i)]$, where $\bar{\beta}_1$ is the posterior mean, and shows that the clutch performance can be several percentage points lower than the overall percentage.

4.3.5 Example: Beta regression for microbiome data

In Section 4.2.5 we regressed the diversity of a sample's microbiome onto features of the home. Diversity was measured as the log of the number of the $L = 763$ species present in the sample. However, this measure fails to account for the relative abundance of the species. Let $A_{il} \geq 0$ be the abundance of species l in sample i. Another measure of the diversity is the proportion of the total abundance attributed to the most abundant species in sample i,

$$Y_i = \frac{\max\{A_{i1}, ..., A_{iL}\}}{\sum_{l=1}^{L} A_{il}} \in [0, 1]. \tag{4.39}$$

This measure is plotted in Figure 4.7 (top left).

Since Y_i is between 0 and 1, Gaussian linear regression is inappropriate. One option is to transform the response from support $[0, 1]$ to $(-\infty, \infty)$, e.g., $Y_i^* = \text{logit}(Y_i)$, and model the transformed data Y_i^* using a Gaussian linear model. Another option is to model the data directly using a non-Gaussian model. A natural model for a continuous variable with support $[0, 1]$ is the beta distribution. Since the mean must also be in $[0, 1]$, a logistic link can be used to relate the covariates to the mean response. Therefore, we fit the beta regression model

$$Y_i|\boldsymbol{\beta}, r \sim \text{Beta}[rq_i, r(1 - q_i)] \text{ and } \text{logit}(q_i) = \mathbf{X}_i^T\boldsymbol{\beta}. \tag{4.40}$$

$Y_i|\boldsymbol{\beta}, r$ has mean q_i and variance $q_i(1-q_i)/(r+1)$, and so $r > 0$ determines the concentration of the beta distribution around the mean q_i.

Using priors $\beta_j \sim \text{Normal}(0, 100)$ and $r \sim \text{Gamma}(0.1, 0.1)$, we fit this model in JAGS using the code in Listing 4.8e (although sampling would likely be faster using the betareg package in R). Before fitting the model, the covariates are standardized to have mean zero and variance one. Convergence was excellent for all parameters (the bottom left panel of Figure 4.7 shows the trace plot for r). The posterior distributions in Table 4.1 indicate that there is more diversity (smaller Y_i) on average in homes in cool regions in the east and in multiple-family homes with many bedrooms.

4.4 Random effects

The standard linear regression model assumes all observations are independent. This assumption is tenuous if data are collected in groups. For example, Figure 4.8 plots the jaw bone density measurements of $n = 20$ children measured over the course of $m = 4$ visits. Analyzing data from this repeated measures design requires accounting for likely correlation between subsequent observations on the same child. This section introduces the concept of a random effect to account for this and other settings with potential dependence between observations. The random effects methods introduced here are a preview of hierarchical modeling (Section 7.1), which is one of the most powerful applications of the Bayesian framework.

These 20 children represent only a random sample from a much larger population, but we can use these samples to make an inference about the larger population. If we let Y_{ij} be the jth measurement for patient i, then (ignoring age as in the top right panel of Figure 4.8) the one-way random effects model is

$$Y_{ij}|\alpha_i \overset{indep}{\sim} \text{Normal}(\alpha_i, \sigma^2) \text{ where } \alpha_i \overset{iid}{\sim} \text{Normal}(\mu, \tau^2). \tag{4.41}$$

The random effect α_i is the true mean for patient i, and the observations for patient i vary around α_i with variance σ^2. The α_i are called random effects because if we repeated the experiment with a new sample of 20 children, the α_i would change. In this model, the population of patient-specific means is assumed to follow a normal distribution with mean μ and variance τ^2.

For the one-way random effects model, marginally over the random effects, the mean is $\text{E}(Y_{ij}) = \mu$ for all observations, and the variance of the response is the sum of the random effect and error variances, $\text{Var}(Y_{ij}) = \tau^2 + \sigma^2$. The proportion of the variance attributed to the random effects is

$$R^2 = \frac{\text{Var}(\alpha_i)}{\text{Var}(Y_{ij})} = \frac{\tau^2}{\tau^2 + \sigma^2}.$$

Marginalizing over the random effects induces correlation between observations from the same group. The marginal covariance is

$$\text{Cov}(Y_{ij}, Y_{uv}) = \begin{cases} \tau^2 + \sigma^2 & i = u \text{ and } j = v \\ \tau^2 & i = u \text{ and } j \neq v \\ 0 & i \neq u. \end{cases} \tag{4.42}$$

Since two observations from the same patient share a common random effect, they have covariance τ^2 and thus correlation $\tau^2/(\tau^2 + \sigma^2)$. Observations from different patients have no common source of variability and are thus uncorrelated.

A Bayesian analysis of a random effects model requires priors for the population parameters. Unlike analyses such as linear regression where the main focus is on the mean and

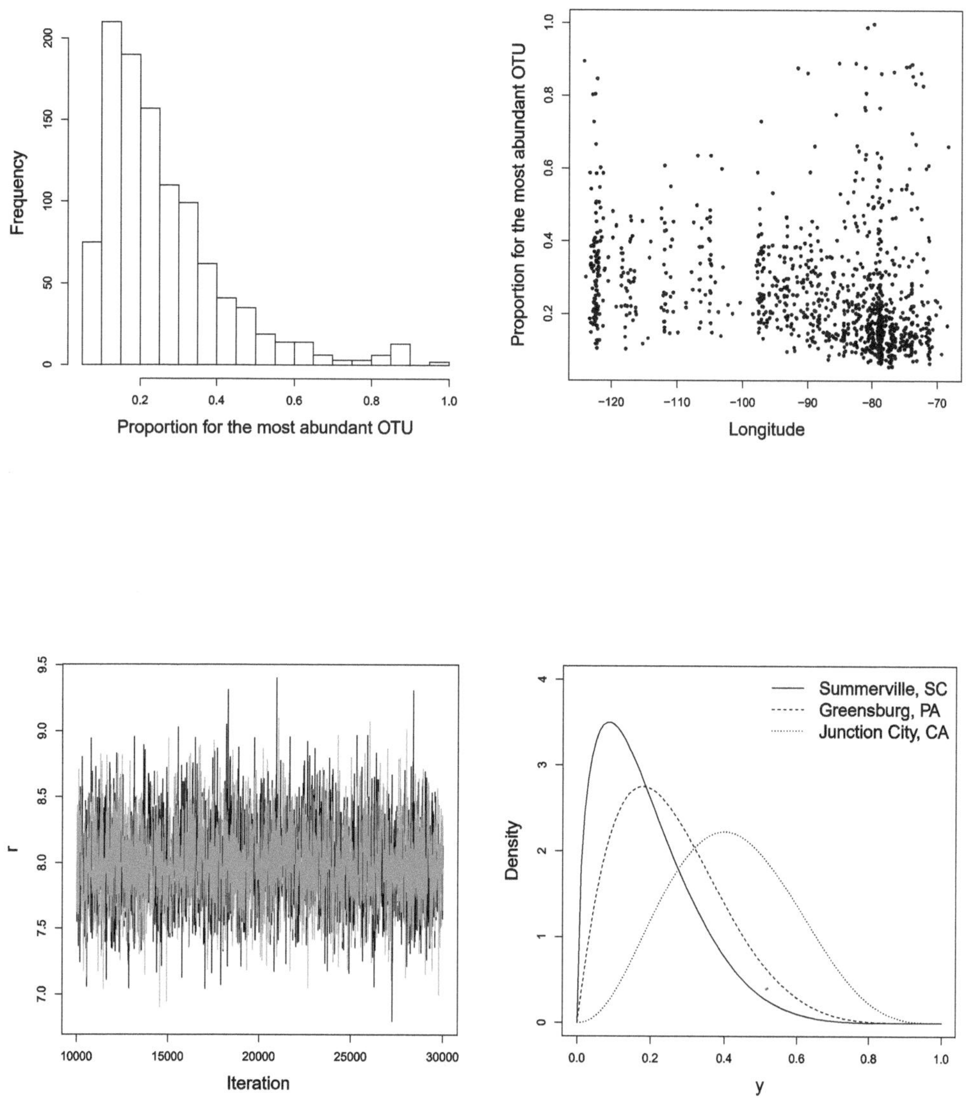

FIGURE 4.7
Beta regression for microbiome data. The top left panel shows the histogram of the observed proportions of abundance allocated to the most abundant OTU, and the top right panel plots this variable against the sample's longitude. The second row gives the trace plots (the two chains are different shades of gray) of the concentration parameter, r, and the fitted Beta$[\hat{r}\hat{q}, \hat{r}(1-\hat{q})]$ density for three samples evaluated at the posterior mean for all parameters.

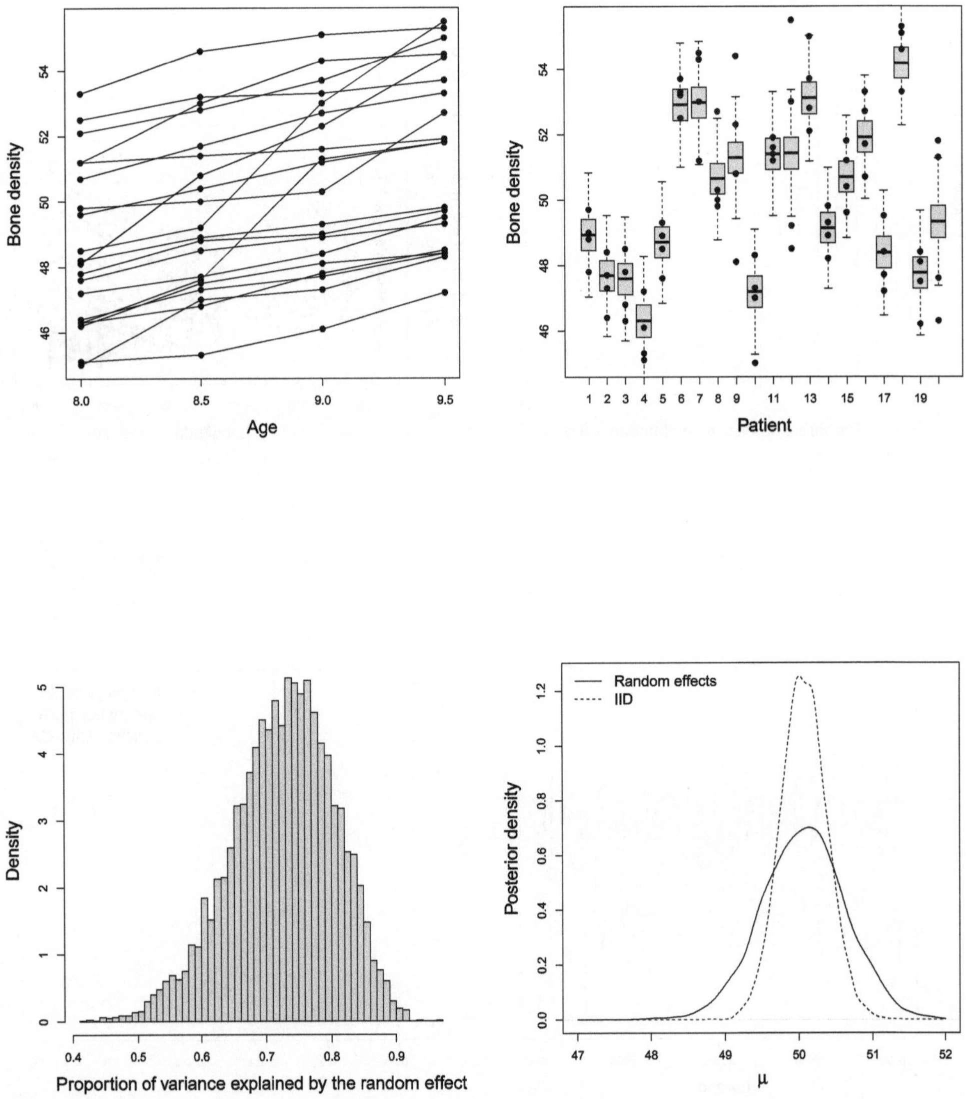

FIGURE 4.8

One-way random effect analysis of the jaw data. The dots in the top left panel show bone density at the four visits for each patient (connected by lines), the top right panel compares the observations (dots) and the posterior distribution of the subject random effect α_i (boxplot), the bottom left panel plots the posterior of the variance ratio $\tau^2/(\tau^2+\sigma^2)$, and the final panel compares the posterior density of the mean μ from the random effects model versus the independence model $Y_{ij} \stackrel{iid}{\sim} \text{Normal}(\mu, \sigma^2)$.

Listing 4.9
The one-way random effects model in JAGS.

```
# Likelihood
 for(i in 1:n){for(j in 1:m){
   Y[i,j] ~ dnorm(alpha[i],sig2_inv)
 }}

# Random effects
 for(i in 1:n){alpha[i] ~ dnorm(mu,tau2_inv)}

# Priors
 mu       ~ dnorm(0,0.0001)
 sig2_inv ~ dgamma(0.1,0.1)
 tau2_inv ~ dgamma(0.1,0.1)
```

prediction of new observations, in random effects models the variance components (e.g., σ^2 and τ^2) are often the main focus of the analysis. Therefore it is important to scrutinize the priors used for these parameters. The inverse gamma prior (Listing 4.9) is conjugate for the variance parameters which often leads to simple Gibbs updates. However, as shown in Figure 4.9, the inverse gamma prior with small shape parameter for a variance induces a prior for the standard deviation with prior PDF equal to zero at the origin. This excludes the possibility that there is no random effect variance, which may not be suitable in many problems.

While the inverse gamma distribution is the only conjugate prior, there are many possibilities including:

1. Inverse gamma prior for the variance
2. Uniform prior for the standard deviation
3. Half-Cauchy prior for the standard deviation
4. Generalized beta-prime prior for the standard deviation/variance

As discussed in Section 2.4.1, selecting the parameterization is a key factor in determining a prior distribution. The main choices here are the variance, τ^2, or the standard deviation, τ. Eliciting prior information is arguably easier for the standard deviation because it is on the scale of the observations. The prior information that the standard deviation cannot exceed τ_{max} is encoded in the uniform prior $\tau \sim \text{Uniform}(0, \tau_{max})$. If the upper bound is justified, then this is a reasonable uninformative prior that adheres to this constraint; however, since the prior probability above τ_{max} is zero, the posterior probability above τ_{max} will also be zero regardless of the data, and so the choice of upper bound is critical.

As an alternative, [54] endorses a half-Cauchy (HC) prior for the standard deviation. The HC distribution (see Appendix) is the Student's t-distribution with one degree of freedom restricted to be positive and has a flat PDF at the origin (Figure 4.9), which is usually a more accurate expression of prior belief than the inverse gamma distribution. Listing 4.10 gives JAGS code for this prior. In this code the HC distribution is assigned directly to the standard deviation, i.e., $\tau \sim \text{HC}(1)$, which breaks the conjugacy relationships for the variance components (this is easily handled by JAGS); conjugacy can be restored using a two-stage model (Section 9.4). If we denote $\omega = 1/\sigma^2$ as the precision, then the scale-mixture of gammas distribution

$$\omega|b \sim \text{Gamma}(1/2, b) \text{ and } b \sim \text{Gamma}(1/2, 1) \tag{4.43}$$

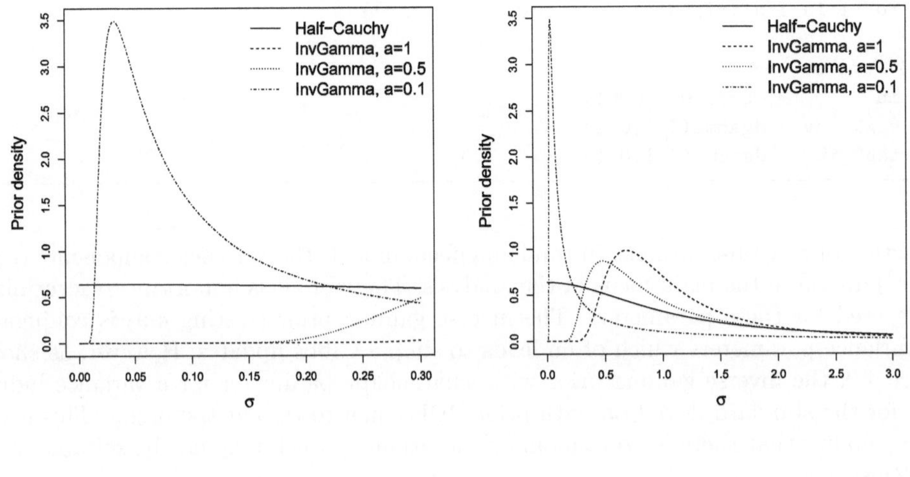

FIGURE 4.9
Priors for a standard deviation. The half-Cauchy prior for σ and the prior induced for σ by inverse gamma priors on σ^2 with different shape parameters a (and scale parameter b set to give median 1) and beta prime (BP) with different shape parameters. The left and right panels differ by the range of σ being plotted.

Listing 4.10
The one-way random effects model with half-Cauchy priors.

```
# Likelihood
 for(i in 1:n){for(j in 1:m){
   Y[i,j] ~ dnorm(alpha[i],sig2_inv)
 }}

# Random effects
 for(i in 1:n){alpha[i] ~ dnorm(mu,tau2_inv)}

# Priors
 mu          ~ dnorm(0,0.0001)
 sig2_inv <- pow(sigma,-2)
 tau2_inv <- pow(tau,-2)
 sigma       ~ dt(0, 1, 1)T(0,) # Half-Cauchy priors with
 tau         ~ dt(0, 1, 1)T(0,) # location 0 and scale 1
```

Listing 4.11
The one-way random effects model with half-Cauchy priors with a mixture of gammas representation.

```
# Likelihood
 for(i in 1:n){for(j in 1:m){
   Y[i,j] ~ dnorm(alpha[i],sig2_inv)
 }}

# Random effects
 for(i in 1:n){alpha[i] ~ dnorm(mu,tau2_inv)}

# Priors
 mu          ~ dnorm(0,0.0001)
 sig2_inv ~ dgamma(0.5,b1); b1 ~ dgamma(0.5,1) # sigma ~ HC(1)
 tau2_inv ~ dgamma(0.5,b2); b2 ~ dgamma(0.5,1) # tau ~ HC(1)
```

induces a HC(1) prior on the standard deviation σ. This model is in Listing 4.11 which is equivalent to Listing 4.10. This code assumes the HC scale parameter is fixed at one. Because the Cauchy prior has a very heavy tail, this gives 0.99 prior quantile equal to 63. Despite this wide prior range, the scale of the HC prior could be adjusted to the scale of the data.

The generalized beta-prime (GBP) distribution has been proposed as a generalization of the half-Cauchy prior [168, 10, 91, 172]. In Section 4.2, the beta-prime (BP) prior $\tau^2 \sim \text{BP}(a, b)$ was used in linear regression to induce a Beta(a, b) prior on the model R^2. With small a and b this gives an uninformative prior for the overall model fit. As discussed below, the BP prior for random effects variance has a similarly appealing interpretation.

The GBP distribution extends the BP distribution by including a power parameter, c, and scale parameter, d. If $X \sim \text{BP}(a, b)$ then

$$X^* = dX^{1/c} \sim \text{GBP}(a, b, c, d)$$

with $c = d = 1$ reducing to the beta-prime distribution $X^* \sim \text{BP}(a, b)$. The flexibility of the power parameter allows the variance and standard deviation parameterizations to stay in the same family because $\tau^2 \sim \text{GBP}(a, b, c, d)$ is equivalent to $\tau \sim \text{GPD}(a, b, 2c, d^{1/2})$. The Half-Cauchy prior is a special case with $a = b = 1/2$ and $c = 1$; that is, $\tau \sim \text{HC}(\sigma)$ is

Listing 4.12
The one-way random effects model with uniform prior on R^2.

```
# Likelihood
for(i in 1:n){for(j in 1:m){
  Y[i,j] ~ dnorm(alpha[i],sig2_inv)
}}

# Random effects
for(i in 1:n){alpha[i] ~ dnorm(mu,tau2_inv)}

# Priors
mu         ~ dnorm(0,0.0001)
sig2_inv   ~ dgamma(0.1,0.1)
R2         ~ dunif(0,1)
tau2_inv <- sig2_inv*(1-R2)/R2
```

equivalent to $\tau^2 \sim \text{GBP}(1/2, 1/2, 1, \sigma)$. However, as shown in Figure 4.9 the BP prior, and thus the GBP prior, includes a wider range of shapes. For example, the PDF at zero can be zero like the inverse gamma, finite like the half-Cauchy or infinite.

If we select GBP prior $\tau^2|\sigma^2 \sim \text{GBP}(a, b, 1, \sigma^2)$, then the proportion of variance explained by the random effects in the one-way random effects model, $R^2 = \tau^2/(\tau^2 + \sigma^2)$, follows a Beta$(a, b)$ prior distribution. Setting $a = b = 1$ as in Listing 4.12 gives an uninformative uniform prior on the proportion of variance explained by the random effects, while selecting a half-Cauchy prior with $a = b = 1/2$ is fairly diffuse, but has more prior mass for R^2 near zero and one. In contrast, the priors $\sigma^2, \tau^2 \sim \text{InvGamma}(\epsilon, \epsilon)$ for small ϵ, induce an unrealistic prior for R^2 with half its mass at $R^2 \approx 0$ and the other half at $R^2 \approx 1$.

Returning to the bone density data, Figure 4.8 plots the posterior approximated with two chains each with 30,000 samples and the first 10,000 discarded as burn-in assuming the model in Listing 4.9. The posterior distribution of the proportion of the variance attributed to the random effect $\tau^2/(\tau^2+\sigma^2)$ ranges from 0.1 to 0.5 (bottom left panel). Accounting for this correlation between observations from the same patient affects the posterior of the fixed effect, μ (bottom right). The posterior variance of μ is larger for the random effects model compared to the posterior of μ for the independence model $Y_{ij} \overset{iid}{\sim} \text{Normal}(\mu, \sigma^2)$. This is expected because there is less information about the mean in 4 repeated measurements for 20 patients than 80 measurements from 80 different patients.

To test for prior sensitivity, we refit the model using a half-Cauchy prior in Listing 4.10 and uniform R^2 prior in Listing 4.12. In this case the prior for the variance components had little effect on the results. The 95% posterior credible sets σ and τ are (1.24, 1.78) and (1.75, 3.51) respectively using inverse gamma priors compared to (1.24, 1.77) and (1.72, 3.45) for the half-Cauchy priors and (1.25, 1.80) and (1.69, 3.35) for the R^2 prior.

Marginal models: Inducing correlation by conditioning on random effects is equivalent to a marginal model (Section 4.6) that does not include random effects but directly specifies correlation between observations in the same group. For example, the one-way random effects model

$$Y_{ij}|\alpha_i \overset{indep}{\sim} \text{Normal}(\alpha_i, \sigma^2) \text{ where } \alpha_i \overset{iid}{\sim} \text{Normal}(\mu, \tau^2) \tag{4.44}$$

is equivalent to the marginal model

$$\mathbf{Y}_i \sim \text{Normal}(\boldsymbol{\mu}, \boldsymbol{\Sigma}), \tag{4.45}$$

where $\mathbf{Y}_i = (Y_{i1}, ..., Y_{im})^T$ is the data vector for group i, $\boldsymbol{\mu} = (\mu, ..., \mu)^T$ is the mean vector, and $\boldsymbol{\Sigma}$ is the covariance matrix with $\tau^2+\sigma^2$ on the diagonals and τ^2 elsewhere. An advantage of the marginal approach is that we no longer have to estimate the random effects α_i; a disadvantage is that for large data sets and complex correlation structure the mean vector and especially the covariance matrix can be large, which slows computation.

In the hierarchical representation, the elements of $\mathbf{Y}_i$ are independent and identically distributed given α_i; in the marginal model the m observations from group i are no longer independent, but they remain exchangeable; i.e., their distribution is invariant to permuting their order. The concept of exchangeability plays a fundamental role in constructing hierarchical models. The representation theorem by Bruno de Finetti states that any infinite sequence of exchangeable variables can be written as independent and identically distributed conditioned on some latent distribution. Therefore, this important type of dependent data can be modeled using a simpler hierarchical model.

Bayesian random effects models can also describe structured (non-exchangeable) correlation as in Section 4.6. For example, observations made over space and time may exhibit correlation that decays with distance in space or lag in time.

4.5 Generalized linear mixed models

Multiple linear regression (Section 4.2) estimates the relationships between variables under the assumption of independence between observations. The regression coefficients ($\boldsymbol{\beta}$) are the parameters of interest and apply to all observations in the analysis and for prediction for other observations from the same population. In this sense, these parameters are *fixed effects*. Random effects (Section 4.4) allow for different parameters (α_i) in different subpopulations and can be used to capture dependence. Fixed and random effects can of course be combined in common model, which is then referred to as a linear mixed model. In this chapter we introduce the linear mixed model for Gaussian data and the generalized linear mixed model for non-Gaussian data.

As we will see, the same algorithms (e.g., Gibbs sampling) can be used for mixed models as for the other models we have considered. In fact, computationally there is no need to distinguish between fixed and random effects (which can lead to confusion, [76]). Conceptually, however, fixed and random effects are distinct because fixed effects describe the population, and a random effect describes an individual from the population. A common source of confusion is that fixed effects are treated as random variables in a Bayesian analysis. However, as with all parameters, the prior and posterior distributions for fixed effects reflect subjective uncertainty about the true values of these fixed but unknown parameters, and not that they are a random sample from a larger population.

Linear mixed model: The linear mixed model extends the multiple regression model (Section 4.2) by adding random effects to account for clustering of the observations. The linear mixed model has Gaussian responses and both fixed and random effects. Returning to the bone density example in Figure 4.8, the response Y_{ij} is the bone density for child $i \in \{1, ..., 20\}$ at visit $j \in \{1, ..., 4\}$. There is an increasing trend in age, and so we may add a fixed effect for the child's age at visit j, X_{ij}. There is also clearly clustering/dependence between the four observations for each child, so we include a random effect for child, giving a linear mixed model with the one-way random effect,

$$Y_{ij}|\alpha_i \stackrel{indep}{\sim} \text{Normal}(\mu + X_{ij}\beta + \alpha_i, \sigma^2) \text{ where } \alpha_i \stackrel{iid}{\sim} \text{Normal}(0, \tau^2), \tag{4.46}$$

where μ is the intercept and β is the fixed age trend. This model has the mean of the random-effect distribution equal zero, which is equivalent to removing μ from the mean of Y_{ij} and assuming the mean of α_k is μ.

In this model, the age trend is fixed in that it applies to all children in this dataset and to all children in the population not included in the sample. The random effects α_i each apply to only one child. If the objective of the study is to learn about the population, then the primary interest is not in the random effects for the children in the random sample, but rather to generalized to the distribution across the population, parameterized by τ^2.

A more general formulation of the linear mixed model with one type of random effect is

$$Y_i|\boldsymbol{\beta},\boldsymbol{\alpha} \stackrel{indep}{\sim} \text{Normal}\left(\sum_{l=1}^{p} X_{il}\beta_l + \sum_{k=1}^{K} Z_{ik}\alpha_k, \sigma^2\right) \tag{4.47}$$

where $\alpha_k \stackrel{iid}{\sim} \text{Normal}(0,\tau^2)$. The fixed component $\sum_{l=1}^{p} X_{il}\beta_l$ is identical to multiple linear regression. The random component is now determined by the Z_{ik}, which are often binary cluster indicators. The bone density model fits this format with $Y_1, ..., Y_{80}$ being the observations for the 20 children and 4 visits in a single vector and $Z_{ik} = 1$ if observation i comes from child k and $Z_{ik} = 0$ otherwise. Writing the model this way allows it to be expressed in matrix notation, extending (4.19) by including the $n \times K$ matrix $\mathbf{Z}$ to give $\mathbf{Y}|\boldsymbol{\beta},\boldsymbol{\alpha} \sim \text{Normal}(\mathbf{X}\boldsymbol{\beta} + \mathbf{Z}\boldsymbol{\alpha}, \sigma^2\mathbf{I}_n)$.

The Bayesian linear mixed model is completed with priors for $\boldsymbol{\beta}$, τ^2 and σ^2. Priors from previous sections can be adapted for the linear mixed model; e.g., the priors $\beta_j \stackrel{iid}{\sim}$ $\text{Normal}(0, c^2)$ and $\sigma^2, \tau^2 \stackrel{iid}{\sim} \text{InvGamma}(a, b)$ for large c and small a and b are conjugate and uninformative. The inverse gamma distributions can be replaced with Half-Cauchy or beta-prime distributions (Section 4.4). As discussed in [168], the beta-prime prior can induce a beta distribution on the overall model fit,

$$R^2 = \frac{\text{Var}\left(\sum_{j=1}^{p} X_{ij}\beta_j + \sum_{k=1}^{K} Z_{ik}\alpha_k\right)}{\text{Var}(Y_i)}.$$

If the covariates are standardized and $\beta_j \sim \text{Normal}\{0, q_1\sigma^2 W/(p-1)\}$ and $\tau^2 = q_2\sigma^2 W$ with $q_1 + q_2 = 1$, then the prior $W \sim \text{BP}(a, b)$ gives the prior $R^2 \sim \text{Beta}(a, b)$. The proportions (q_1, q_2) determine the relative contribution of the fixed and random effects and can be fixed or given a Dirichlet prior.

Two-way random effects: The model above assumes a linear time trend, which may be overly restrictive. A more flexible model includes a separate parameter for the mean response at each of the four time points, $\alpha_{21}, ..., \alpha_{24}$,

$$Y_{ij}|\alpha_i \stackrel{indep}{\sim} \text{Normal}(\mu + X_{ij}\beta + \alpha_{1i} + \alpha_{2j}, \sigma^2)$$

where as before the child random effects are modeled as $\alpha_{1i} \stackrel{iid}{\sim} \text{Normal}(0, \tau_1^2)$. Treating age as a random effect is questionable, as the four ages under consideration are not a random sample from a larger population of ages. One approach is to treat the α_{2j} as fixed effects with uninformative prior, effectively including four dummy variables for visit number as covariates. This model is overparameterized as there are six parameters (μ, β, α_{21}, ..., α_{j4}) to explain the population mean at the four ages, and so at least two parameters must be removed.

An alternative is to model the age indicators as $\alpha_{2j} \stackrel{iid}{\sim} \text{Normal}(0, \tau_2^2)$. In this model, the α_{2j} represent deviations from the linear trend, and the Gaussian priors add stability to avoid overfitting; i.e., small τ_2 shrinks $\alpha_{2j} \approx 0$, and the time trend is approximately linear with slope β. This example illustrates how random effects can be used as a convenient modeling device in Bayesian models that do not have the classical interpretation of a random effects models [77].

Listing 4.13
Random slopes model in JAGS.

```
# Likelihood
 for(i in 1:n){for(j in 1:m){
   Y[i,j] ~ dnorm(alpha[i,1]+alpha[i,2]*age[j],tau)
 }}

# Random effects
 for(i in 1:n){
    alpha[i,1:2] ~ dmnorm(beta[1:2],OmegaInv[1:2,1:2])
 }

# Priors
 tau ~ dgamma(0.1,0.1)
 for(j in 1:2){beta[j] ~ dnorm(0,0.0001)}

 nvar   <- 2 # Set the Wishart parameters so E(Omega) = I_nvar
 nu     <- (nvar+1) + 0.1
 R[1,1] <- nu - (nvar+1)
 R[1,2] <- 0
 R[2,1] <- 0
 R[2,2] <- nu - (nvar+1)

 OmegaInv[1:2,1:2] ~ dwish(R[,],nu)
```

Random effects can also be used to capture complex dependence structure between observations. For the two-way random effects model with $\alpha_{2j} \stackrel{iid}{\sim} \text{Normal}(0, \tau_2^2)$, the covariance is

$$\text{Cov}(Y_{ij}, Y_{uv}) = \begin{cases} \sigma^2 + \tau_1^2 + \tau_2^2 & i = u \text{ and } j = v \\ \tau_1^2 & i = u \text{ and } j \neq v \\ \tau_2^2 & i \neq u \text{ and } j = v \\ 0 & i \neq u \text{ and } j \neq v. \end{cases} \tag{4.48}$$

Therefore, the random effects induce correlation $\tau_1^2/(\sigma^2 + \tau_1^2 + \tau_2^2)$ between observations for the same child at different visits and correlation $\tau_2^2/(\sigma^2 + \tau_1^2 + \tau_2^2)$ between observations for the different children at the same visit time.

Random slopes model: The previous models for the bone growth data allow each child to have a different intercept but assume the same trajectory across time for all children. However, Figure 4.8 indicates that the rate of increase over time varies by child. Therefore we could model the child-specific slopes as random effects

$$Y_{ij}|\boldsymbol{\alpha}_i \stackrel{indep}{\sim} \text{Normal}(\alpha_{i1} + \alpha_{i2}X_j, \sigma^2) \text{ where } \boldsymbol{\alpha}_i \stackrel{iid}{\sim} \text{Normal}(\boldsymbol{\beta}, \boldsymbol{\Omega}), \tag{4.49}$$

and the random intercept and slope for patient i is $\boldsymbol{\alpha}_i = (\alpha_{i1}, \alpha_{i2})^T$. The mean vector $\boldsymbol{\beta}$ includes the population mean intercept and slope, and is thus a fixed effect. The 2×2 population covariance matrix $\boldsymbol{\Omega}$ determines the variation of the random effects over the population. To complete the Bayesian model we specify prior $\sigma^2 \sim \text{InvGamma}(0.1, 0.1)$, $\boldsymbol{\beta} \sim \text{Normal}(\mathbf{0}, 100^2\mathbf{I}_2)$, and $\boldsymbol{\Omega} \sim \text{InvWishart}(3.1, \mathbf{I}_2/3.1)$. The inverse Wishart prior for the covariance matrix has prior mean (which exists because we take $\nu = 3.1 > 2 + 1$) $\mathbf{I}_2$, the 2×2 identify matrix (see Appendix A.1). `JAGS` code for this random-slopes model is in Listing 4.13.

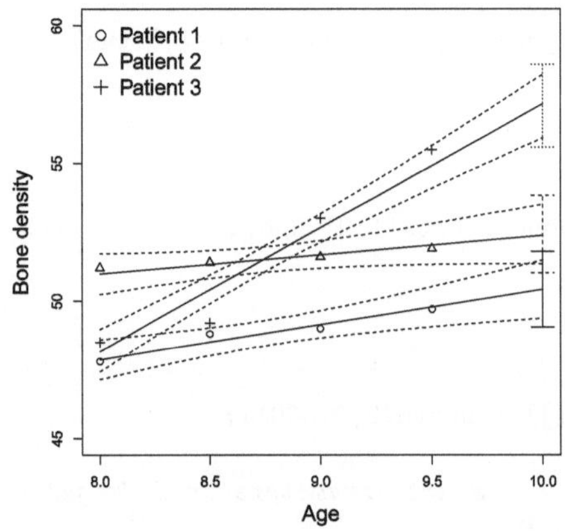

FIGURE 4.10
Mixed effects analysis of the jaw data. The observed bone density (points) for three subjects versus the posterior median (solid lines) and 95% intervals (dashed lines) of the fitted value $\alpha_{i1} + \alpha_{i2}X$ for X ranging from 8–10 years, and 95% credible intervals (vertical lines at Age=10) of the posterior predictive distributions for the measured response at age $X = 10$.

The posterior mean of the population covariance matrix is

$$E(\mathbf{\Omega}|\mathbf{Y}) = \begin{bmatrix} 84.83 & -9.36 \\ -9.36 & 1.11 \end{bmatrix} \tag{4.50}$$

and the posterior 95% interval for the correlation $\text{Cor}(\alpha_{i1}, \alpha_{i2}) = \mathbf{\Omega}_{12}/\sqrt{\mathbf{\Omega}_{11}\mathbf{\Omega}_{22}}$ is $(-0.986, -0.904)$. Therefore there is a strong negative dependence between the intercept and slope, indicating that bone density increases rapidly for children with low bone density at age 8, and vice versa.

Figure 4.10 plots the posterior distribution of the fitted values $\alpha_{i1} + \alpha_{i2}X$ for X between 8 and 10 years for three patients. For each patient and each age, we compute the 95% interval using the quantiles of the S posterior samples $\alpha_{i1}^{(s)} + \alpha_{i2}^{(s)}X$. We also plot the Posterior Predictive Distribution (PPD) for the measured bone density at age 10. The PPD is approximated by sampling $Y_i^{*(s)} \sim \text{Normal}(\alpha_{i1}^{(s)} + \alpha_{i2}^{(s)}10, \sigma^{2(s)})$ at each iteration and then computing the quantiles of the S predictions. The PPD accounts for both uncertainty in the patient's random effect $\boldsymbol{\alpha}_i$ and measurement error with variance σ^2. The intervals in Figure 4.10 suggest that uncertainty in the random effects is the dominant source of variation.

Generalized linear mixed models: Generalized linear mixed models (GLMMs) include random effects to the GLMs of Section 4.3 in the natural way. Recall that for GLMs the mean (or other features of the distribution) response $\text{E}(Y_i)$ is related to the linear predictor η_i via link function g as $g\{\text{E}(Y_i)\} = \eta_i$. For instance, in logistic regression the binary response variable Y_i was regression onto covariates $\mathbf{X}_i$ using the logistic link function $\text{logit}\{\text{E}(Y_i)\} = \eta_i = \sum_{l=1}^{p} X_{il}\beta_l$. This link function ensures that the mean response is between zero and one for η_i.

The GLMM adds random effects to the linear predictor. For example, consider the one-way GLMM model where Y_{ij} is the ith observation in group j. The linear predictor

$g\{E(Y_{ij})\} = \eta_{ij}$ includes both the fixed effect covariate X_{ij} and random effect α_j

$$\eta_{ij} = \mu + X_{ij}\beta + \alpha_j$$

with random effects distribution $\alpha_j \stackrel{iid}{\sim} \text{Normal}(0, \tau^2)$. As with GLMs, different link function, are appropriate for different response types; e.g., if Y_{ij} is binary then we might use a logistic link function or if Y_{ij} is a count we might use a Poisson regression with log link function. Also, the linear mixed model is a special case of the GLMM with identity link function and Gaussian assumption for the response distribution.

The motivation for and interpretation of the random effects is the same for GLMMs as for Gaussian linear mixed models. That is, the distribution of random effects is interpreted as representing a variation across a population of sampling units. The statistical interest is either in the population parameter τ^2, or sometimes in identifying extreme units in the sample with large or small α_j.

Random effects are also used to account for dependence between observations in the same cluster. Unlike in the Gaussian case, the correlation cannot be computed analytically, but the insight that shared random effects leads to dependence remains. To see this, consider the logistic model for binary data without fixed effects so $\text{logit}\{\text{Prob}(Y_{ij} = 1|\alpha_j)\} = \alpha_j$. The mean is $\text{Prob}(Y_{ij} = 1) = 0.5$ marginally over the random effect for all τ. The table below gives the joint PMF for two observations from the same cluster. These probabilities are approximated using Monte Carlo sampling by first drawing $\alpha_j \sim \text{Normal}(0, \tau^2)$ and then generating Y_{1j} and Y_{2j} independently given the draw of α_j. With $\tau = 0$, the observations are independent, and the joint probabilities are the product of marginal probabilities. For large τ the two observations are dependent with probability that they are equal converging to one. This happens because for draws with large α_j, both observations are likely one, and for draws with small α_j both observations are likely zero. Therefore, the random effect induces positive dependence between these binary responses.

τ	0	5	10	15	20
$\text{Prob}(Y_{1j} = 0, Y_{2j} = 0)$	0.25	0.42	0.46	0.47	0.48
$\text{Prob}(Y_{1j} = 0, Y_{2j} = 1)$	0.25	0.08	0.04	0.03	0.02
$\text{Prob}(Y_{1j} = 1, Y_{2j} = 0)$	0.25	0.08	0.04	0.03	0.02
$\text{Prob}(Y_{1j} = 1, Y_{2j} = 1)$	0.25	0.42	0.46	0.47	0.48

In the example below, we use `JAGS` to fit a GLMM with binary outcomes and random effects. `JAGS` is sufficiently fast for small and moderately-sized analyses such as this. However, we note that for larger analyses, packages dedicated for GLMMs such as `brms` [26] may be preferable.

Childhood malaria example: Diggle et al. [44] study factors associated with childhood malaria. The data are available in the `geoR` package and consist of $n = 1,332$ children from $L = 65$ villages (points in Figure 4.11) in the Gambia. The response variable is $Y_i = 1$ if child i tested positive for malaria, and $Y_i = 0$ if child i tested negative. Let $s_i \in \{1, ..., L\}$ be the village number for child i. In addition to the village, observation i has seven covariates ($p = 8$ including the intercept): the x-coordinate of the village, the y-coordinate of the village, the child's age, an indicator of bed net use, an indicator of whether the bed net is treated, the greenness of the village and an indicator of the health center in the village. All covariates are standardized to have mean zero and variance one.

The objectives of the analysis are to estimate the covariate effects and to identify villages with excess risk of malaria, i.e., high malaria rates that cannot be explained by the covariates. We use the generalized logistic regression model with fixed effects for the covariates

Listing 4.14
Random effects logistic regression for The Gambia data.

```
for(i in 1:n){
  Y[i]         ~ dbern(q[i])
  logit(q[i]) <- X[i,1]*beta[1] + X[i,2]*beta[2] +
                 X[i,3]*beta[3] + X[i,4]*beta[4] +
                 X[i,5]*beta[5] + X[i,5]*beta[6] +
                 alpha[s[i]]
}
for(j in 1:6){ # Priors for the fixed effects
  beta[j] <- dnorm(0,0.1)
}
for(k in 1:65){ # Village random effects
  alpha[k] <- dnorm(0,tau2_inv)
}
tau2_inv <- (1-R2)/R2 # Beta-prime/Half-Cauchy prior
R2        ~ beta(0.5,0.5)
```

and a random effect for the village:

$$q_i = \text{Prob}(Y_i = 1|\boldsymbol{\beta}, \boldsymbol{\alpha}) \quad \text{and} \quad \text{logit}(q_i) = \sum_{j=1}^{p} X_{ij}\beta_j + \alpha_{s_i} \tag{4.51}$$

with random effects distribution $\alpha_k \stackrel{iid}{\sim} \text{Normal}(0, \tau^2)$. The random effects are coded using the village indicators s_i, so that if child i is from village l, then $s_i = k$ and $\alpha_{s_i} = \alpha_k$. The random effects both capture correlation between two children from the same village, and also can be used to identify village with excess risk of malaria and $\alpha_k > 0$. Listing 4.14 gives `JAGS` code to fit this model with uninformative priors $\beta_j \sim \text{Normal}(0, 10)$ and $\tau^2 \sim \text{BP}(0.5, 0.5)$, i.e., $\tau \sim \text{HC}(1)$.

Table 4.2 summarizes the posterior of the covariate effects. As expected, bed net use and proximity to a health care center reduce malaria prevalence. The posterior mean of the village effect standard deviation τ is 0.88. Figure 4.2b shows that the posterior mean of the village random effects α_k range from roughly -1.5 to 1.5. These values indicate considerable variation in the positivity rate across villages even after accounting for the fixed effects. The odds of testing positive for a child from a village with $\alpha_k = 1.5$ is $\exp(1.5) = 4.48$ times higher than a child from a village with the same covariates but $\alpha_k = 0$. Investigation of the villages in the clusters with large α_k in Figure 4.2b could be used to generate hypotheses about missing factors that might explain disease prevalence in these areas.

To test for prior sensitivity, we refit the model with $\tau^2 \sim \text{BP}(a, b)$ for (a, b) set to $(1, 1)$, $(5, 1)$, $(5, 1)$ or $(5, 5)$. The posterior mean of the random effect standard deviation τ ranged from 0.84 to 0.94. Therefore, large changes to the prior produce moderate changes to the posterior and there is slight sensitivity to the prior.

4.6 Linear models with correlated data

Section 4.4, specifically (4.42), demonstrates that including random effects can capture dependence between observations. In this approach, we included random effects to the model, assumed conditional independence between the observations given the random effects, and

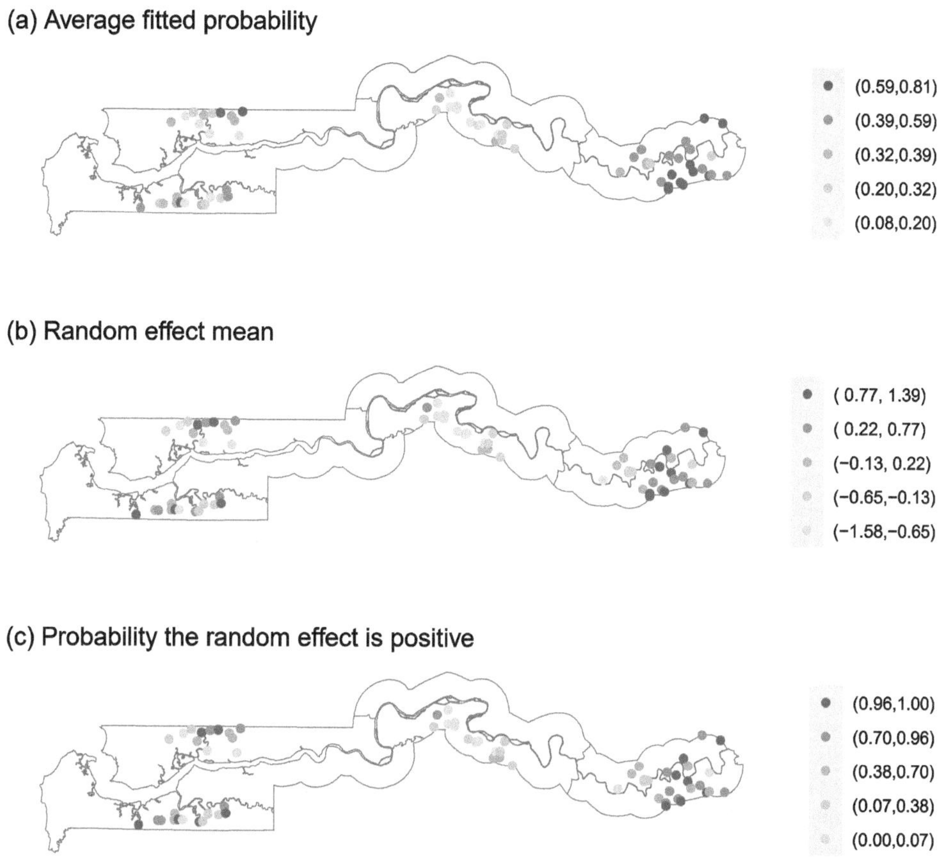

FIGURE 4.11
Fitted values for the Gambia analysis. (a) Average of the posterior means of fitted probabilities q_i for each village, (b) posterior mean of the village random effects α_k and (c) the posterior probability that $\alpha_k > 0$.

TABLE 4.2
Summary of the Gambia analysis. Posterior summaries for the fixed effects coefficients (β_j) and random effect standard deviation (τ).

	Mean	SD	95% Interval
Intercept	−0.64	0.13	(−0.89, −0.39)
Age	0.29	0.05	(0.18, 0.39)
Net use	−0.20	0.07	(−0.33, −0.05)
Treated	−0.18	0.10	(−0.37, 0.01)
Greenness	0.32	0.12	(0.09, 0.55)
Health center	−0.16	0.12	(−0.40, 0.08)
Random effect SD	0.88	0.11	(0.68, 1.12)

studied the correlation induced by marginalizing over the random effects. Alternatively, correlation between observations can be modeled directly without including random effects. For example, rather than including random effects to capture dependence, the covariance model in (4.48) could be used directly, and equivalently, to model dependence. For data with a natural ordering such as spatial or temporal data, modeling the marginal correlation directly as a function of distance in space or time is often the more intuitive approach. As with random effects, there are many correlation models, and finding a reasonable one is important to obtain valid inference and make accurate predictions.

The Bayesian linear model with correlated errors is

$$\mathbf{Y}|\boldsymbol{\beta} \sim \text{Normal}(\mathbf{X}\boldsymbol{\beta}, \boldsymbol{\Sigma}). \tag{4.52}$$

The mean structure $\mathbf{X}\boldsymbol{\beta}$ is interpreted the same as in a linear model with independent errors; i.e., with all other covariates held fixed, increasing a covariate by one increases the mean by the corresponding element in $\boldsymbol{\beta}$; however, the posterior of $\boldsymbol{\beta}$ can be dramatically affected by the choice of correlation (e.g., [78]). A Bayesian analysis of correlated data hinges on correctly specifying the correlation structure to capture, say, spatial or temporal correlation. Say observation i is taken at spatial location (longitude/latitude) $\mathbf{s}_i$. Then we might assume that the correlation between observations at locations $\mathbf{s}_i$ and $\mathbf{s}_j$ decays with distance between locations, $d_{ij} = ||\mathbf{s}_i - \mathbf{s}_j||$, as

$$\text{Cor}(Y_i, Y_j) = C(d_{ij}) = \exp(-\phi d_{ij})$$

where $\phi > 0$ determines the rate of decay and is a parameter to be estimated. Then the (i, j) element of $\boldsymbol{\Sigma}$ is $\sigma^2 C(d_{ij})$, so σ^2 is the variance of $\mathbf{Y}$.

Rather than explicitly modeling the correlation between each pair of observations, the correlation is determined by the correlation function, C, which inputs a distance and outputs a correlation. The exponential correlation function is just one simple example, e.g., the squared exponential correlation function is $C(d) = \exp(-\phi d^2)$, which is popular in machine learning (Section 8.3.2). The correlation function is typically selected from a list of valid correlation functions that ensure than when used to populate $\boldsymbol{\Sigma}$, $\boldsymbol{\Sigma}$ is guaranteed to by symmetric and positive definite. [38] provides a list of common spatial and temporal correlation functions.

Given the correlation structure and priors for the correlation parameters (ϕ in the previous example), standard Bayesian computational tools can be used to summarize the posterior. Correlation parameters usually will not have conjugate priors, and so Metropolis–Hastings sampling is used. An advantage of the Bayesian approach for correlated data is that using MCMC sampling, we can account for uncertainty in the correlation parameters for prediction or inference on other parameters, whereas maximum likelihood analysis often uses plug-in estimates of the correlation parameters and thus underestimates uncertainty.

Gun control example: The data for this analysis come from Kalesan et. al. (2016) [86]. The response variable, Y_i, is the log firearm-related death rate per 10,000 people in 2010 in state i (excluding Alaska and Hawaii). This is regressed onto five potential confounders: log 2009 firearm death rate per 10,000 people; firearm ownership rate quartile; unemployment rate quartile; non-firearm homicide rate quartile; and firearm export rate quartile. The covariate of interest is the number of gun control laws in effect in the state. This gives $p = 6$ covariates.

We first fit the usual Bayesian linear regression model

$$Y_i = \beta_0 + \sum_{j=1}^{p} X_i \beta_j + \varepsilon_i \tag{4.53}$$

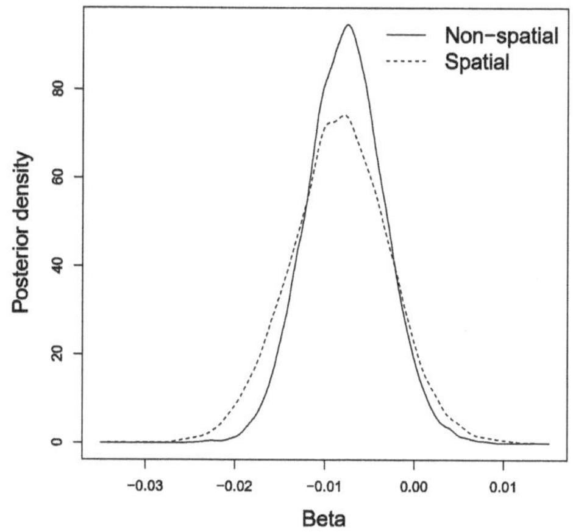

FIGURE 4.12
Effect of gun-control legislation on firearm-related death rate. Posterior distribution of the coefficient associated with the number of gun-control laws in a state from the spatial and non-spatial model of the states' firearm-related death rates.

with independent errors $\varepsilon_i \overset{iid}{\sim} \text{Normal}(0, \sigma^2)$ and uninformative priors. The posterior density of the regression coefficient corresponding to the number of gun laws is plotted in Figure 4.12. The posterior probability that the coefficient is negative is 0.96, suggesting a negative relationship between the number of gun laws and the firearm-related death rate.

The assumption of independent residuals is questionable because neighboring states may be correlated. Spatial correlation may stem from guns being brought across state borders or from missing covariates (e.g., attitudes about and use of guns) that vary spatially. Research has shown that accounting for residual dependence can have a dramatic effect on regression coefficient estimates [78].

We decompose the residual covariance $\text{Cov}[(\varepsilon_1, ..., \varepsilon_n)^T] = \Sigma$ as

$$\Sigma = \tau^2 S + \sigma^2 I_n, \tag{4.54}$$

where $\tau^2 S$ is the spatial covariance and $\sigma^2 I_n$ is the non-spatial covariance. There are many spatial correlation models (e.g., [11]) that allow the correlation between two states to decay with the distance between the states. For example, a common model is to assume the correlation between states decays exponentially with the distance between them. However, quantifying the distance between irregularly shaped states is challenging, and so we model spatial dependence using adjacencies. Let $A_{ij} = 1$ if states i and j share a border and $A_{ij} = 0$ if $i = j$ or the states are not neighbors. The spatial covariance follows the conditionally autoregressive model $S = (M - \rho A)^{-1}$, where A is the adjacency matrix with (i, j) element A_{ij} and M is the diagonal matrix with the ith diagonal element equal to the number of states that neighbor state i. The parameter $\rho \in (0, 1)$ is not the correlation between adjacent sites, but determines the strength of spatial dependence with $\rho = 0$ corresponding to independence.

Listing 4.15
Random slopes model with autoregressive dependence in JAGS.

```
# Likelihood
  for(i in 1:n){
    Y[i,1:m] ~ dmnorm(mn[i,1:m],SigmaInv)
    for(j in 1:m){mn[i,j] <- alpha[i,1]+alpha[i,2]*age[j]}
  }
  SigmaInv[1:m,1:m] <- inverse(Sigma[1:m,1:m])
  for(j in 1:m){for(k in 1:m){
     Sigma[j,k] <- pow(rho,abs(k-j))/tau
  }}

# Random effects
  for(i in 1:n){alpha[i,1:2] ~ dmnorm(beta[1:2],Omega[1:2,1:2])}

# Priors
  tau ~ dgamma(0.1,0.1)
  for(j in 1:2){beta[j] ~ dnorm(0,0.0001)}
  rho ~ dunif(0,1)
  Omega[1:2,1:2] ~ dwish(R[,],2.1)

  R[1,1]<-1/2.1
  R[1,2]<-0
  R[2,1]<-0
  R[2,2]<-1/2.1
```

The posterior mean (standard deviation) of the spatial dependence parameter ρ is 0.38 (0.25), and so the residual spatial dependence in these data is not strong. However, the posterior of the regression coefficient of interest in Figure 4.12 is noticeably wider for the spatial model than the non-spatial model. The posterior probability that the coefficient is negative lowers from 0.96 under the non-spatial model to 0.93 for the spatial model. Therefore, while accounting for residual dependence did not qualitatively change the results, this example illustrates that the chosen model for the residuals can affect the posterior of the regression coefficients.

Jaw bone density example: A possible correlation structure for the longitudinal data in Figure 4.8 (top left) is to assume that correlation decays with the time between visits. A first-order autoregression correlation structure is $\text{Cor}(Y_{ij}, Y_{ik}) = \rho^{|j-k|}$. Denoting the vector of m observations for patient i as $\mathbf{Y}_i = (Y_{i1}, ..., Y_{im})^T$, the $m \times m$ covariance matrix $\mathbf{\Sigma}$ has (j, k) element equal to $\sigma^2\rho^{|j-k|}$. The random slope model for the mean in matrix notation is $E(\mathbf{Y}_i|\boldsymbol{\alpha}_i) = \mathbf{X}\boldsymbol{\alpha}_i$, where $\mathbf{X}$ is the $m \times 2$ matrix with the first column equal to the vector of ones for the intercept and the second column equal to the ages $X_1, ..., X_m$ for the slope. The likelihood is then

$$\mathbf{Y}_i|\alpha_i \overset{indep}{\sim} \text{Normal}(\mathbf{X}\boldsymbol{\alpha}_i, \mathbf{\Sigma}) \tag{4.55}$$

with random effects distribution $\boldsymbol{\alpha}_i \overset{iid}{\sim} \text{Normal}(\boldsymbol{\beta}, \mathbf{\Omega})$. The correlation parameter is given prior $\rho \sim \text{Uniform}(0, 1)$ and all other priors are the same as other fits. JAGS code is given in Listing 4.15.

The posterior median of the correlation parameter ρ is 0.85, and the posterior 95% interval is (0.46, 0.96), so there is evidence of correlation that cannot be explained by the patient-specific linear trend. Including autoregressive correlation has only a modest effect on the posterior distribution of the fixed effects: the 95% posterior intervals for the random effect model with independent errors are (29.9, 38.3) for β_1 and (1.33, 2.38) for β_2 compared to (30.0, 37.3) for β_1 and (1.45, 2.31) for β_2 for the autoregressive model.

This model with random intercept, random slope and autoregressive correlation structure is now very complex. There are almost as many parameters as observations and multiple explanations of dependence (random effects and residual correlation). Perhaps in this case all of these terms are necessary and can be estimated from this relatively small data set, but a simpler yet adequate model is preferred for computational purposes and because simpler models are easier to explain and defend. Model comparisons and tests of model adequacy are the topics of Chapter 6.

4.7 Exercises

1. A clinical trial gave six subjects a placebo and six subjects a new weight loss medication. The response variable is the change in weight (pounds) from baseline (so −2.0 means the subject lost 2 pounds). The data for the 12 subjects are:

Placebo	Treatment
2.0	−3.5
−3.1	−1.6
−1.0	−4.6
0.2	−0.9
0.3	−5.1
0.4	0.1

Conduct a Bayesian analysis to compare the means of these two groups. Would you say the treatment is effective? Is your conclusion sensitive to the prior?

2. Load the classic Boston Housing Data in R:

```
> library(MASS)
> data(Boston)
> ?Boston
```

The response variable is medv, the median value of owner-occupied homes (in \$1,000s), and the other 13 variables are covariates that describe the neighborhood.

(a) Fit a Bayesian linear regression model with uninformative Gaussian priors for the regression coefficients. Verify the MCMC sampler has converged, and summarize the posterior distribution of all regression coefficients.

(b) Perform a classic least squares analysis (e.g., using the lm function in R). Compare the results numerically and conceptually with the Bayesian results.

(c) Refit the Bayesian model with double exponential priors for the regression coefficients, and discuss how the results differ from the analysis with uninformative priors.

(d) Fit a Bayesian linear regression model in (a) using only the first 500 observations and compute the posterior predictive distribution for the final 6 observations. Plot the posterior predictive distribution versus the actual value for these 6 observations, and comment on whether the predictions are reasonable.

3. Download the 2016 Presidential Election data from the book's website. Perform Bayesian linear regression with the response variable for county i being the difference between the percentage of the vote for the Republican candidate in 2016 minus 2012 and all variables in the object $\mathbf{X}$ as covariates.

   (a) Fit a Bayesian linear regression model with uninformative Gaussian priors for the regression coefficients, and summarize the posterior distribution of all regression coefficients.

   (b) Compute the residuals $R_i = Y_i - \mathbf{X}_i\hat{\boldsymbol{\beta}}$ where $\hat{\boldsymbol{\beta}}$ is the posterior mean of the regression coefficients. Are the residuals Gaussian? Which counties have the largest and smallest residuals, and what might this say about these counties?

   (c) Include a random effect for the state, that is, for a county in state $l = 1, ..., 50$,

$$Y_i|\alpha_l \sim \text{Normal}(\mathbf{X}_i\boldsymbol{\beta} + \alpha_l, \sigma^2)$$

   where $\alpha_l \overset{iid}{\sim} \text{Normal}(0, \tau^2)$ and τ^2 has an uninformative prior. Why might adding random effects be necessary? How does adding random effects affect the posterior of the regression coefficients? Which states have the highest and lowest posterior mean random effect, and what might this imply about these states?

4. Download the US gun control data from the book's website. These data are taken from the cross-sectional study in [86]. For state i, let Y_i be the number of homicides and N_i be the population.

   (a) Fit the model $Y_i|\boldsymbol{\beta} \sim \text{Poisson}(N_i\lambda_i)$ where $\log(\lambda_i) = \mathbf{X}_i\boldsymbol{\beta}$. Use uninformative priors and $p = 7$ covariates in $\mathbf{X}_i$: the intercept, the five confounders $\mathbf{Z}_i$, and the total number of gun laws in state i. Provide justification that the MCMC sampler has converged and sufficiently explored the posterior distribution and summarize the posterior of $\boldsymbol{\beta}$.

   (b) Fit a negative binomial regression model, and compare with the results from Poisson regression.

   (c) For the Poisson model in (a), compute the posterior predictive distribution for each state with the number of gun laws set to zero. Repeat this with the number of gun laws set to 25 (the maximum number). According to these calculations, how would the number of deaths nationwide be affected by these policy changes? Do you trust these projections?

5. Download the titanic dataset from R,

```
library("titanic")
dat <- titanic_train
?titanic_train
```

   Let $Y_i = 1$ if passenger i survived and $Y_i = 0$ otherwise. Perform a Bayesian logistic regression of the survival probability onto the passenger's age, gender (dummy variable) and class (two dummy variables). Summarize the effect of each covariate.

6. The T. rex growth chart data plotted in Figure 3.10 has $n = 6$ observations with weights (kg) 29.9, 1761, 1807, 2984, 3230, 5040, and 5654 and corresponding ages (years) 2, 15, 14, 16, 18, 22 and 28. Since weight must be positive, the gamma family of distributions is a reasonable model for these data. Describe a model with gamma likelihood and log mean that increases linearly with age.

Approximate the posterior using MCMC, summarize the posterior distribution of all model parameters, and plot the data versus the fitted mean curve.

7. Consider the one-way random effects model $Y_{ij}|\alpha_i, \sigma^2 \sim \text{Normal}(\alpha_i, \sigma^2)$ and $\alpha_i \sim \text{Normal}(0, \tau^2)$ for $i = 1, ..., n$ and $j = 1, ..., m$. Assuming conjugate priors $\sigma^2, \tau^2 \sim \text{InvGamma}(a, b)$, derive the full conditional distributions of α_1, σ^2 and τ^2, and outline (but do not code) an MCMC algorithm to sample from the posterior.

8. Load the Gambia data in R:

```
> library(geoR)
> data(gambia)
> ?gambia
```

The response variable Y_i is the binary indicator that child i tested positive for malaria (pos), and the remaining seven variables are covariates.

(a) Fit the logistic regression model

$$\text{logit}[\text{Prob}(Y_i = 1)] = \sum_{j=1}^{p} X_{ij}\beta_j$$

with uninformative priors for the β_j. Verify that the MCMC sampler has converged, and summarize the effects of the covariates.

(b) In this dataset, the 2,035 children reside in $L = 65$ unique locations (defined by the x and y coordinates in the dataset). Let $s_i \in \{1, ..., L\}$ be the label of the location for observation i. Fit the random effects logistic regression model

$$\text{logit}[\text{Prob}(Y_i = 1)] = \sum_{j=1}^{p} X_{ij}\beta_j + \alpha_{s_i} \text{ where } \alpha_l \stackrel{iid}{\sim} \text{Normal}(0, \tau^2)$$

and the β_j and τ^2 have uninformative priors. Verify that the MCMC sampler has converged; explain why random effects might be needed here; discuss and explain any differences in the posteriors of the regression coefficients that occur when random effects are added to the model; plot the posterior means of the α_l by their spatial locations, and suggest how this map might be useful to malaria researchers.

9. Download the babynames data in R, and compute the log odds of a baby being named "Sophia" each year after 1950:

```
library(babynames)
dat <- babynames
dat <- dat[dat$name=="Sophia" &
             dat$sex=="F" &
             dat$year>1950,]
yr  <- dat$year
p   <- dat$prop
t   <- dat$year - 1950
Y   <- log(p/(1-p))
```

Let Y_t denote the sample log-odds in year $t + 1950$. Fit the following time series (auto-regressive order 1) model to these data:

$$Y_t = \mu_t + \rho(Y_{t-1} - \mu_{t-1}) + \varepsilon_t$$

where $\mu_t = \alpha + \beta t$ and $\varepsilon_t \stackrel{iid}{\sim} \text{Normal}(0, \sigma^2)$. The priors are $\alpha, \beta \sim \text{Normal}(0, 100^2)$, $\rho \sim \text{Uniform}(-1, 1)$, and $\sigma^2 \sim \text{InvGamma}(0.1, 0.1)$.

(a) Give an interpretation of each of the four model parameters: α, β, ρ, and σ^2.

(b) Fit the model using `JAGS` for $t > 1$, verify convergence, and report the posterior mean and 95% interval for each parameter.

(c) Plot the posterior predictive distribution for Y_t in the year 2020.

10. The global mean temperature anomaly (difference between current year and the 1951-1980 average) data on the course website and plotted in Figure 3.4. Let X_i and Y_i be the year and temperature anomaly (C) for observation i. We will analyze these data using a nonparametric regression model $Y_i = g(X_i) + \varepsilon_i$ as in Section 8.3, where g is modeled as a linear combination of $J = 15$ B-spline basis functions.

    (a) Fit the model, and plot the data along with the posterior mean and 95% posterior interval of $g(X_i)$ for each value X_i.

    (b) Summarize and interpret the posterior distribution of $\Delta = g(2020) - g(1980)$.

    (c) Summarize and interpret the posterior distribution for $\delta_i = g(X_i) - g(X_{i-1})$ for each $i > 1$.

11. Consider an ecological study where the response, Y_i, is a measure of marine biodiversity at location $i = 1, ..., n$. The covariates $X_{i1}, ..., X_{ip}$ are summaries of the environment of location i, such as water depth, distance to shore, fishing regulations, etc. Assume the multiple linear regression model $Y_i = \beta_0 + X_{i1}\beta_1 + ... + X_{ip}\beta_p + \varepsilon_i$. Give a scenario where the following priors for the β_j might be useful:

    (a) Jeffreys prior

    (b) Bayesian LASSO prior

    (c) An informative prior $\beta_j \sim \text{Normal}(m_j, v)$ for small v

    Your scenarios should include both numerical summaries such as n and p and scientific objectives.

12. In a study of the genetic determinants of depression, researchers gathered information from 1000 patients. The response variable is the (continuous) CES-D survey score and the predictors 10,000 genetic markers. The objective is to determine if any of the markers are associated with depression.

    (a) Describe and justify a model and prior for these data

    (b) How would you summarize the results?

13. Open and plot the galaxies data in `R` using the code below,

```
> library(MASS)
> data(galaxies)
> ?galaxies
> Y <- galaxies
> hist(Y,breaks=25)
```

Model the observations $Y_1, ..., Y_{82}$ using a mixture of $K = 3$ normal distributions. For each of the S MCMC iterations evaluate the density function on the grid $y \in \{5000, 5100, ..., 40000\}$ (351 points in the grid), giving an $S \times 351$ matrix of posterior samples. Plot the posterior median and 95% credible set of the density function at each of the 351 grid values. Does this mixture model fit the data well?

14. A group of 10 ecologists is surveying a forest for Red Cockaded Woodpeckers (RCP). Each ecologist will walk along a different path and make 25 stops. At each stop, they will record local conditions (tree density, elevation, etc.) and whether they see or hear an RCP. The objective is to build a model for the types of habitat that are the most favorable to the RCP.

    (a) Describe and justify a model and prior for these data
    (b) How would you summarize the results?

5

Hypothesis testing

CONTENTS

Statistical hypothesis testing is a key ingredient of the scientific method. Hypothesis testing provides a rigorous comparison of experimental conditions or populations. Thus far, we have performed tests informally. For example, in Figure 1.10 we compared two interval hypotheses for a binomial proportion θ. The null hypothesis was $\mathcal{H}_1 : \theta < 0.5$, and the alternative hypothesis was $\mathcal{H}_2 : \theta > 0.5$. We used the posterior probability of each hypothesis as a measure of evidence for each, but we did not discuss how to use these probabilities to arrive at a decision. As another example, for multiple regression in Section 4.2 we computed 95% intervals for regression coefficient β_j and reached conclusions based on whether the interval included zero. A more formal test would set the null hypothesis to be that $\mathcal{H}_1 : \beta_j = 0$, i.e., a point hypothesis versus $\mathcal{H}_2 : \beta_j \neq 0$. However, under a continuous prior distribution such as the Gaussian distribution, the posterior probability of a single point such as $\beta_j = 0$ is always zero, and thus new methods are needed to formalized this hypothesis test.

In this chapter, we introduce fundamental Bayesian hypothesis testing concepts such as decision theory and the Bayes factors. We also cover more advanced topics including multiple, A/B and sequential testing, and sample size calculations.

5.1 Bayes factors

Bayes factors [87] provide a formal summary of the evidence that the data support one statistical model over another. In hypotheses testing, the two models correspond to the null hypothesis $\mathcal{H}_1$ and the alternative hypothesis $\mathcal{H}_2$; therefore, below we use the terms hypotheses and models exchangeably. Although not necessary for the computation of Bayes factors, most testing problems can be framed so that both hypotheses are nested in a common model and distinguished by the model parameters. The full model might be $Y_i|\mu \stackrel{iid}{\sim}$ normal(μ, σ^2), with $\mathcal{H}_1$ defined by $\mu \leq 0$ and $\mathcal{H}_2$ defined by $\mu > 0$.

In a Bayesian analysis, the unknown parameters are treated as random variables. If the hypotheses are stated as functions of the parameters, then they are also random variables; i.e., the posterior probability $\text{Prob}(\mu \leq 0|\mathbf{Y}) = \text{Prob}(\mathcal{H}_1|\mathbf{Y})$ is the posterior probability of the null hypothesis and $\text{Prob}(\mu > 0|\mathbf{Y}) = \text{Prob}(\mathcal{H}_2|\mathbf{Y})$. These posterior probabilities are the most intuitive summaries of uncertainty.

A common mistake in a frequentist hypothesis testing is to confuse the p-value with the probability that the null hypothesis is true; from a frequentist perspective, the parameters and thus the hypotheses are fixed quantities and not random variables, and therefore it is not sensible to assign them probabilities. From the Bayesian perspective however giving the posterior probability of each hypothesis is a legitimate summary of uncertainty.

Posterior probabilities of hypotheses incorporate information from both the data and the prior. Bayes factors remove the effect of the prior and quantify the data's support of the hypotheses. The Bayes factor (BF) of $\mathcal{H}_2$ relative to $\mathcal{H}_1$ is the ratio of posterior odds to prior odds,

$$BF = \frac{\text{Prob}(\mathcal{H}_2|\mathbf{Y})/\text{Prob}(\mathcal{H}_1|\mathbf{Y})}{\text{Prob}(\mathcal{H}_2)/\text{Prob}(\mathcal{H}_1)}. \tag{5.1}$$

If the prior probability of the two hypotheses are equal, then the BF is simply the posterior odds $\text{Prob}(\mathcal{H}_2|\mathbf{Y})/\text{Prob}(\mathcal{H}_1|\mathbf{Y})$. In this case, rejecting $\mathcal{H}_1$ if $BF > c$ is equivalent to rejecting $\mathcal{H}_1$ if the posterior probability of the alternative hypothesis is high, $\text{Prob}(\mathcal{H}_2|\mathbf{Y}) > c/(c+1)$.

Hypothesis tests are usually designed to be conservative so that the null model is rejected in favor of the alternative only if the data strongly support this model. A rule of thumb [87] is that $BF > 10$ provides strong evidence of the alternative hypothesis compared to the null hypothesis, and $BF > 100$ is decisive evidence.

Bayesian decision theory provides a way to use posterior probabilities of hypotheses to make decisions. We defer a full discussion of decision theory to Section 9.1, but introduce the concepts needed for hypothesis testing here. Decision theory requires specifying the loss incurred for each type of error. For hypothesis testing there are two types of losses: a Type I error rejects the null hypothesis when it is in fact true, and a Type II error fails to reject the null hypothesis when it is in fact false. Define the loss of a Type I error as λ_I. To be conservative, we might assume the loss of a Type I error is c times the loss of a Type II error, so $\lambda_I = c\lambda_{II}$ with c set to a large value, say $c = 10$ or $c = 100$.

The risk of a particular decision is defined as the expected value of the loss under the decision, which is the product of the probability of the decision being wrong and the loss associated with this error. The risk for the decision to reject the null hypothesis is

$$R_2 = \text{Prob}(\mathcal{H}_1|\mathbf{Y})\lambda_I = \text{Prob}(\mathcal{H}_1|\mathbf{Y})c\lambda_{II},$$

and the risk for the decision to not reject the null hypothesis is

$$R_1 = \text{Prob}(\mathcal{H}_2|\mathbf{Y})\lambda_{II}.$$

Risk is minimized by rejecting the null hypothesis if $R_2/R_1 < 1$, i.e.,

$$\frac{\text{Prob}(\mathcal{H}_2|\mathbf{Y})}{\text{Prob}(\mathcal{H}_1|\mathbf{Y})} > c,$$

or equivalently rejecting the null hypothesis if the posterior probability of the alternative hypothesis exceeds $\text{Prob}(\mathcal{H}_2|\mathbf{Y}) > c/(1+c)$. To be concrete, if we assume a Type I error is $c = 9$ times as costly as a Type II error, then we reject the null if the posterior probability of the alternative exceeds 0.9. If the prior probabilities of the hypotheses are equal, then the posterior odds are the Bayes factors, and we should reject if $BF > c$.

This formulation using decision theory shows that labelling hypotheses as "null" versus "alternative" has no special role in Bayesian testing. That is, the results for the test of $\mathcal{H}_1 : \theta < 0$ versus $\mathcal{H}_2 : \theta > 0$ with relative loss c are identical to the test of $\mathcal{H}_1 : \theta > 0$ versus $\mathcal{H}_2 : \theta < 0$ with relative loss $1/c$. This symmetry is in contrast with classical testing, where the p-value is computing as if the null hypothesis is true and so changing the labels of the hypotheses can change the results.

Computing BFs for problems with many parameters can be challenging. In general, model selection can be framed as treating the model as an unknown random variable $\mathcal{H} \in \{\mathcal{H}_1, \mathcal{H}_2\}$ with prior probability $\text{Prob}(\mathcal{H} = \mathcal{H}_j) = q_j$. Conditioned on model j, i.e., $\mathcal{H} = \mathcal{H}_j$, the remainder of the Bayesian model is

$$\mathbf{Y} \sim f(\mathbf{Y}|\boldsymbol{\theta}; \mathcal{H}_j) \text{ and } \boldsymbol{\theta} \sim \pi(\boldsymbol{\theta}|\mathcal{H}_j). \tag{5.2}$$

The BF requires the marginal posterior of the model integrating over uncertainty in the parameters,

$$\text{Prob}(\mathcal{H} = \mathcal{H}_j|\mathbf{Y}) = \int p(\boldsymbol{\theta}, \mathcal{H} = \mathcal{H}_j|\mathbf{Y})d\boldsymbol{\theta}. \tag{5.3}$$

Unfortunately, marginalizing over the parameters (as in m in Table 1.4) is rarely possible. Also, the marginal distribution of $\mathcal{H}$ is not defined for improper priors. Therefore, *BFs cannot be used with improper priors.* Even with proper priors, BFs can be very sensitive to the choice of hyperparameters as shown by the examples below. Partial or fractional Bayes factors [118] adapt the Bayes factor for improper priors. These methods have two stages. Part of the data's information is analyzed in the first stage, and the posterior from the first stage is used as a proper prior for the second-stage analysis of the remaining data and computing the Bayes factor.

MCMC provides a means of approximating BFs in special cases. The BF for nested models defined by intervals of the parameters is straightforward to compute using MCMC. For example, in the model above with $Y_i|\mu \overset{iid}{\sim} \text{Normal}(\mu, \sigma^2)$ and $\mu \sim \text{Normal}(0, 10^2)$, and models $\mathcal{H}_1$ defined by $\mu \leq 0$ and $\mathcal{H}_2$ defined by $\mu > 0$, the posterior probability of Model 1 (2) can be approximated by generating samples from the posterior of the full model and recording the proportion of the samples for which μ is negative (positive). Section 6.2 provides a more general computational strategy for computing model probabilities.

MCMC can also be used for point-null hypotheses. If the hypotheses to be tested are $\mathcal{H}_1 : \mu = 0$ and $\mathcal{H}_2 : \mu \neq 0$, then the Bayes factor can be approximated by using MCMC by fitting the model $Y_i|\mu \overset{iid}{\sim} \text{normal}(\mu, \sigma^2)$ with $\mu = \gamma\delta$ and prior distributions $\gamma \sim \text{Bernoulli}(0.5)$ and $\delta \sim \text{Normal}(0, 10^2)$. MCMC samples with $\delta = 0$ correspond to $\mathcal{H}_1$, and samples with $\delta = 1$ correspond to $\mathcal{H}_2$, therefore the posterior probability that $\delta = 1$ approximates the posterior probability of $\mathcal{H}_2$ which can be used to compute the Bayes factor. Section 6.2 expands on this idea in the context of multiple regression.

A final note on Bayes factors is that they resemble the likelihood ratio statistic, which is commonly used in frequentist hypothesis testing. Assuming equal priors, the Bayes factor is

$$BF = \frac{\int f(\mathbf{Y}|\boldsymbol{\theta}; \mathcal{H}_2)\pi(\boldsymbol{\theta}|\mathcal{H}_2)d\boldsymbol{\theta}}{\int f(\mathbf{Y}|\boldsymbol{\theta}; \mathcal{H}_1)\pi(\boldsymbol{\theta}|\mathcal{H}_1)d\boldsymbol{\theta}}, \tag{5.4}$$

i.e., the ratio of marginal distribution of the data under the two models. This resembles the likelihood ratio from classical hypothesis testing,

$$LR = \frac{f(\mathbf{Y}|\hat{\boldsymbol{\theta}}_2)}{f(\mathbf{Y}|\hat{\boldsymbol{\theta}}_1)}, \tag{5.5}$$

where $\hat{\boldsymbol{\theta}}_j$ is the MLE under hypothesis $j = 1, 2$. Therefore, both measures compare hypotheses based on the ratio of their likelihood, but the BF integrates over posterior uncertainty in the parameters, whereas the likelihood ratio statistic plugs in point estimates.

Beta-binomial example: To build intuition about BFs, we begin with the univariate binomial model and test whether the success probability equals 0.5. Let $Y|\theta \sim \text{Binomial}(n, \theta)$, and consider two hypotheses:

$$\mathcal{H}_1 : \theta = 0.5 \text{ versus } \mathcal{H}_2 : \theta \neq 0.5 \text{ and } \theta \sim \text{Beta}(a, b). \tag{5.6}$$

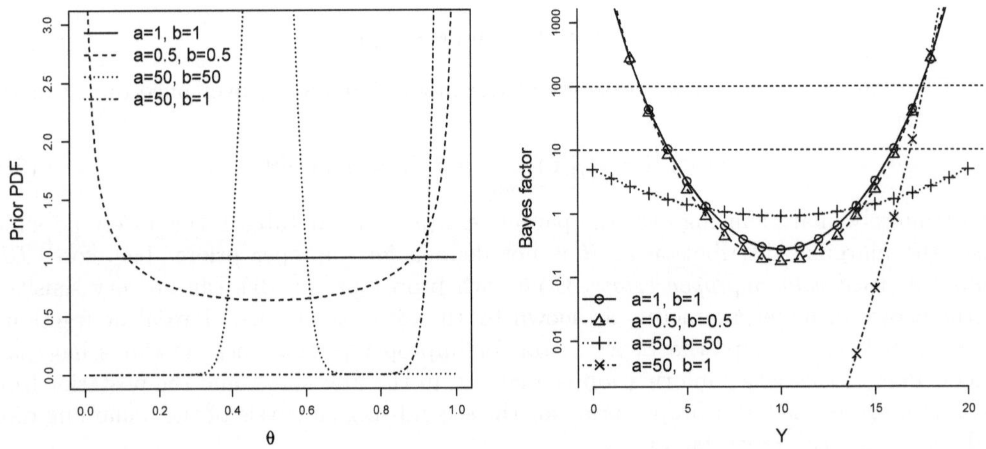

FIGURE 5.1
Bayes factor for the beta-binomial model. (left) Beta(a, b) prior PDF for several combinations of a and b and (right) observed data Y versus the Bayes factor comparing the beta-binomial model $Y|\theta \sim \text{Binomial}(n, \theta)$ and $\theta \sim \text{Beta}(a, b)$ versus the null model $Y \sim \text{Binomial}(n, 0.5)$ for $n = 20$ and several combinations of a and b.

The null hypothesis has no unknown parameters, and for $\mathcal{H}_2$ the parameter θ can be integrated out giving the beta-binomial model for Y (see Appendix A.1). Therefore, these hypotheses about θ correspond to two different models for the data

$$\mathcal{H}_1 : Y \sim \text{Binomial}(n, 0.5) \text{ versus } \mathcal{H}_2 : Y \sim \text{Beta-Binomial}(n, a, b). \tag{5.7}$$

Assuming priors $\text{Prob}(\mathcal{H}_j) = q_j$ and observed data $Y = y$,

$$BF(y) = \frac{f_{BB}(y; n, a, b)}{f_B(y; n, 0.5)} \tag{5.8}$$

where f_{BB} and f_B are the beta-binomial and binomial PMFs, respectively.

Figure 5.1 plots $BF(y)$ for $n = 20$ and different hyperparameters a and b. Assuming the uniform prior ($a = b = 1$) or Jeffreys prior ($a = b = 0.5$), the BF exceeds 10 for $y < 5$ or $y > 15$ successes, and exceeds 100 for $y < 4$ or $y > 16$ successes; these scenarios give strong and decisive evidence, respectively, against the null hypothesis that $\theta = 0.5$ in favor of the alternative that $\theta \neq 0.5$. The BF is less than 10 for all possible y for the strong prior centered on 0.5 ($a = b = 50$). Under this prior the two hypotheses are similar, and it is difficult to distinguish between them. Finally, under the strong prior that θ is close to one ($a = 50$ and $b = 1$), the BF exceeds 10 only for $y > 16$ successes; in this case y near zero is even less likely under the alternative than under the null that $\theta = 0.5$.

Normal-mean example: Say there is a single observation $Y|\mu \sim \text{Normal}(\mu, 1)$, and the objective is to test whether $\mu = 0$. The competing hypotheses are

$$\mathcal{H}_1 : \mu = 0 \text{ versus } \mathcal{H}_2 : \mu \neq 0 \text{ and } \mu \sim \text{Normal}(0, \tau^2). \tag{5.9}$$

Given that we observe $Y = y$, it can be shown that the BF of $\mathcal{H}_2$ relative to $\mathcal{H}_1$ is

$$BF(y) = \left(1+\tau^2\right)^{-1/2} \exp\left[\frac{y^2}{2}\left(\frac{\tau^2}{1+\tau^2}\right)\right]. \tag{5.10}$$

For fixed τ, $BF(y)$ increases to infinity as y^2 increases as expected because data far from zero contradict $\mathcal{H}_1 : \mu = 0$. However, for any fixed y, $BF(y)$ converges to zero as the prior variance τ^2 increases. Therefore, even if the observation is 10 standard deviation units above zero, the BF favors the null model with $\mu = 0$ if the prior variance is sufficiently large. This is an example of Lindley's paradox [97] where Bayesian tests can perform poorly depending on the prior. For the normal means problem, this odd result occurs because with large τ these data are unlikely under both models. For example, the probability of $|Y| \in [9, 11]$ is very small under the null hypothesis that $\mu = 0$, but under the alternative hypothesis this probability converges to zero as τ increases because the marginal distribution of Y (averaging over μ) becomes increasingly diffuse.

BFs are not defined with improper priors, and this example suggests that they can have strange properties for uninformative priors. It has been argued [16] that the standard half-Cauchy prior $\mu \sim t_1(0,1)$ is preferred to the large-variance normal prior because this prior is diffuse without having a variance parameter to tune. In general, prior selection has more impact for hypothesis testing than estimation, making it imperative to report Bayes factors for multiple priors to illuminate this sensitivity.

Lindley's paradox is more pronounced for tests of point null hypotheses such as $\mathcal{H}_1 : \mu = 0$ than for one-sided tests such as

$$\mathcal{H}_1 : \mu \leq 0 \text{ versus } \mathcal{H}_2 : \mu > 0.$$

To compute the BF for these hypotheses we simply fit the Bayesian model $Y|\mu \sim \text{Normal}(\mu, 1)$ and $\mu \sim \text{Normal}(0, \tau^2)$ and compute $\text{Prob}(\mathcal{H}_2|Y) = \text{Prob}(\mu > 0|Y)$ using the results in Section 2.1.3. This test is stable as the prior variance increases because the BF converges to $\Phi(y)/[1 - \Phi(y)] > 0$, where Φ is the standard normal CDF. Since $1 - \Phi(y)$ is the p-value for the classical one-sided z-test, the BF test with large variance is equivalent to the frequentist test.

Tests based on credible sets are another remedy to Lindley's paradox. That is, we simply fit a Bayesian model $Y|\mu \sim \text{Normal}(\mu, 1)$ and $\mu \sim \text{Normal}(0, \tau^2)$ as in Section 4.1 and reject the null hypothesis that $\mu = 0$ if the posterior credible set for μ excludes zero. As the prior variance increases, this rule has the same frequentist operating characteristics as the classic two-sided z-test (or t-test with unknown error variance). Therefore, this approach is not sensitive to the prior and has appealing frequentist properties including controlling Type I error.

5.2 Multiple testing

Traditional experiments are designed to test a single hypothesis. However, modern data analyses often present many variables and thus many possible tests. Consider a genetic study where m genetic markers are used as predictors of a patient's health outcome. We might fit the multiple regression (Section 4.2) $Y_i = \sum_{j=1}^{m} X_{ij}\beta_j + \varepsilon_i$ where X_{ij} is the expression at marker j for subject i and β_j is the effect of marker j on the health outcome. This leads to m potential tests,

$$\mathcal{H}_{1j} : \beta_j = 0 \text{ versus } \mathcal{H}_{2j} : \beta_j \neq 0,$$

for $j \in \{1, ..., m\}$. For genetic studies, the number of tests m is often in the thousands, and thus new testing procedures are required to avoid being overwhelmed with spurious results. As an extreme case, if $m = 10,000$ and in fact none of the markers have any effect on the health response, then a method that incorrectly rejects $\mathcal{H}_{1j}$ with probability 0.05 will result in an average of 500 false positives, which is clearly unacceptable.

A global test is a potential remedy for this multiple testing problem. The global test replaces the m individual tests with the single test of hypotheses

$$\mathcal{H}_1 : \beta_j = 0 \text{ for all } j \text{ versus } \mathcal{H}_2 : \beta_j \neq 0 \text{ for at least one } j.$$

Since this reduces the problem to a single test comparing two models, standard methods can be used such as cross-validation or computing the posterior probability of $\mathcal{H}_1$ and rejecting $\mathcal{H}_1$ if its probability is small.

If we assume the global null hypothesis is false and return to multiple testing, the challenge is identifying the subset of the m tests that should be rejected. To formalize this problem, we establish notation for our decision (r_j) and the true status (δ_j) of test j. Let $r_j = 1$ if we reject $\mathcal{H}_{1j}$ in favor of $\mathcal{H}_{2j}$ and $r_j = 0$ if we do not reject. Ideally, the decisions r_j align with the true status of the hypotheses $\delta_j = 0$ if $\mathcal{H}_{1j}$ is true and $\delta_j = 1$ if $\mathcal{H}_{2j}$ is true. For a set of decisions $r_1, ..., r_m$, the number of "discoveries" is $\sum_{j=1}^m r_j$. However, some of these discoveries may be false, i.e., $r_j = 1$ but in reality $\delta_j = 0$ or equivalently $1 - \delta_j = 1$. The number of false discoveries is then $\sum_{j=1}^m r_j(1 - \delta_j)$ and the false discovery proportion is

$$FDP = \frac{\sum_{j=1}^m r_j(1 - \delta_j)}{\sum_{j=1}^m r_j}.$$

A multiple testing procedure might want to control this proportion to be small, say $FDP \approx \alpha = 0.1$.

Unfortunately, FDP cannot be computed because we do not know the true status of the hypotheses, δ_j. The Bayesian false discovery rate [147] evaluates the posterior mean of FDP averaging over the posterior distribution of the δ_j. The posterior mean is

$$BFDR = \frac{\sum_{j=1}^m r_j(1 - q_j)}{\sum_{j=1}^m r_j}$$

where q_j is the posterior probability of $\mathcal{H}_{2j}$ (as computed by MCMC for example). For any given set of decision $r_1, ..., r_m$, we can compute BFDR and thus select the decision that gives the most discoveries with $BFDR \approx 0.1$.

To make this search practical for large problems, the decision rule can be simplified to thresholding the posterior probabilities. That is, we reject the null hypothesis $\mathcal{H}_{1j}$ if the posterior probability of the alternative is high and let $r_j(t) = 1$ if $q_j > t$ and $r_j(t) = 0$ otherwise. Then the BFDR is a function of t (as in the right panel of Figure 5.2), and we can search for the t so that

$$BFDR(t) = \frac{\sum_{j=1}^m r_j(t)(1 - q_j)}{\sum_{j=1}^m r_j(t)} \approx \alpha,$$

which requires a search over only $t \in (0, 1)$ (Listing 5.1 and Figure 5.2).

Extreme streamflow application: To illustrate this method, we analyze the data in [99]. The data are posted on the course website and originate from the USGS Hydro Climatic Data Network (`https://water.usgs.gov/GIS/metadata/usgswrd/XML/hcdn.xml`), which is designed to measure changes in flood risk not due to human development (damns, etc). Let Y_{ij} be the log annual maximum (i.e., the largest daily value for the year) streamflow in

Listing 5.1
R code to identity the Bayesian false discovery rate threshold.

```
# q is a vector of the posterior probabilities
# of the alternative hypothesis, H_2
alpha <- 0.1
t     <- seq(0,1,0.01)
BFDR  <- t
for(i in 1:length(t)){
  rej_t   <- q>t[i]
  BFDR[i] <- sum(rej_t*(1-q))/sum(rej_t)
}
t_alpha <- t[which.min(abs(BFDR-alpha))]

plot(t,BFDR,type="l",xlab="t",ylab="BFDR(t)")
abline(alpha,0,lty=2) # The desired FDR
abline(v=t_alpha)     # The selected threshold
```

year $i \in \{1, ..., m\}$ at monitoring station $j \in \{1, ..., n\}$. After removing missing observations there are $n = 236$ stations and $m = 72$ years. The objective is to determine whether (and if so where) the likelihood of extreme streamflow is increasing over time so we regress Y_{ij} onto $X_i = (t_i - 1985)/10$ where t_i is the year of observation Y_{ij}. The simple linear regression model is

$$Y_{ij} = \beta_{0j} + X_i\beta_{1j} + \varepsilon_{ij}$$

where β_{1j} is the increase in the mean log streamflow per decade at station j. For illustration, we ignore spatial and temporal correlation; for a spatial treatment of multiple testing, see [151].

The competing hypotheses for station j are

$$\mathcal{H}_{1j} : \beta_{1j} \leq 0 \text{ versus } \mathcal{H}_{2j} : \beta_{1j} > 0.$$

The regressions return the posterior probabilities of the alternative hypotheses $q_j = \text{Prob}(\beta_{1j} > 0|\mathbf{Y})$ for $j = 1, ..., n$. We declare that flood risk is increasing in time if $q_j > t$ where the threshold t is selected so that the BFDR is controlled at $\alpha = 0.1$. Figure 5.2 plots the posterior probabilities of an increase, q_j, for each station and the estimated BFDR as a function of the threshold, t. This analysis suggests that rejecting the null hypothesis if $q_j > t = 0.74$ controls the false discovery rate at 0.1. Using this rejection rule, the locations with a detected change are primarily in Washington and the Midwest (Missouri, Illinois and Indiana).

5.3 A/B testing and the Bayesian bandit

The term A/B testing was popularized by the technology sector, but the testing methods apply to many other settings. The task is to conduct an experiment to compare several experimental conditions and identify which gives the best performance. For example, a marketing company may run three versions of an advertising campaign to determine which is optimal. This task does not fit the classic frequentist hypothesis testing framework since there is no natural control group or null hypothesis. The Bayesian approach is intuitively

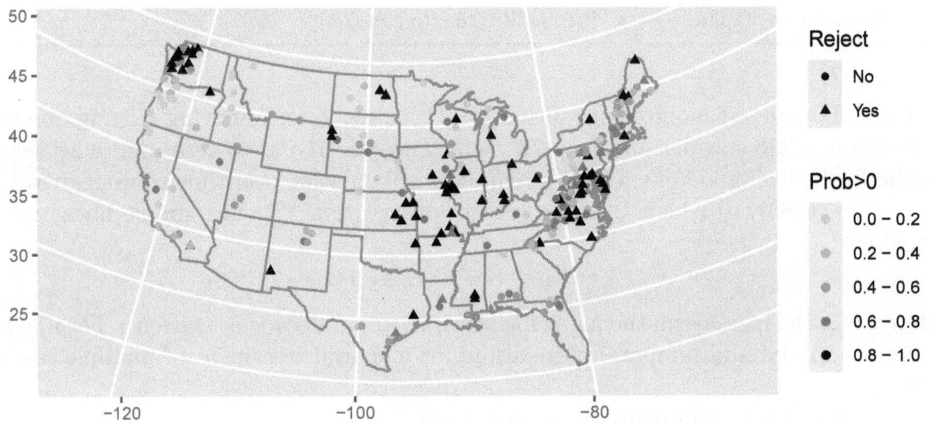

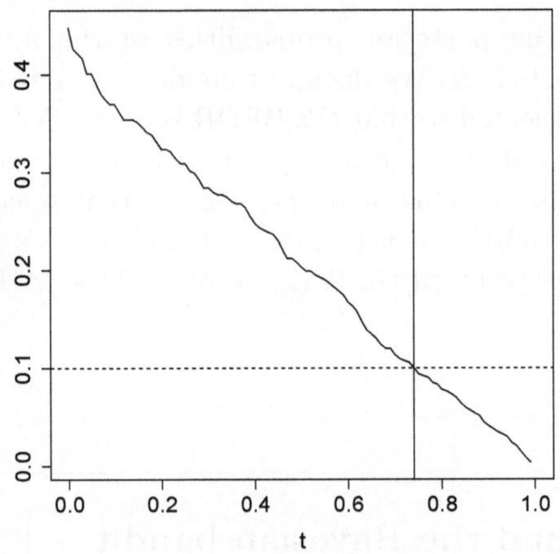

FIGURE 5.2
Bayesian false discovery rate control for multiple testing. (top) Posterior probability (q_j) of an increase in extreme streamflow (grayscale) and decision to reject $\mathcal{H}_{1j}$ (shape) based on BFDR control at each station and (bottom) Bayesian false discovery rate as a function of the threshold (t).

appealing because it returns the posterior probability that each condition is optimal and thus directly addresses the study objectives.

Let n_j be the number of observations made under condition $j \in \{1, ..., J\}$ and Y_{ij} be the response for observation $i \in \{1, ..., n_j\}$ in group j. A/B testing can be applied to continuous (time spent on a website), count (number of clicks) or binary (purchase or not) data. The most basic models for these data types are Gaussian, Poisson and Bernoulli, respectively. Assuming independence, for Poisson and Bernoulli observations we can aggregate across observations giving $Y_j = \sum_{i=1}^{n_j} Y_{ij}$. The likelihoods for these models are

- Continuous response: $Y_{ij}|\boldsymbol{\theta} \stackrel{indep}{\sim} \text{Normal}(\theta_j, \sigma_j^2)$,
- Count response: $Y_j|\boldsymbol{\theta} \stackrel{indep}{\sim} \text{Poisson}(n_j\theta_j)$,
- Binary response: $Y_j|\boldsymbol{\theta} \stackrel{indep}{\sim} \text{Binomial}(n_j, \theta_j)$.

A default prior in this case in the objective Jeffreys prior (Section 4.2.2.1), but informative priors can also be used if this information is available.

The objective of the analysis is to compare θ_j across the J conditions and compute the posterior probability that each has the largest mean, i.e.,

$$p_j = \text{Prob}(\theta_j = \max\{\theta_1, ..., \theta_J\}|\mathbf{Y}),$$

which can often be computed analytically for two groups and approximated using Monte Carlo sampling in other cases. The groups can also be weighted by cost. For example, if profit of a sale is P and advertising campaign j costs C_j units per user and gives probability of purchase θ_j, then the value of campaign j is $V_j = P\theta_j - C_j$, and we might compute

$$p_j = \text{Prob}(V_j = \max\{V_1, ..., V_J\}|\mathbf{Y}),$$

which is also straightforward to approximate using Monte Carlo sampling if P and C_j are known.

These two posterior distributions are plotted in Figure 5.3. Although the retention rates are similar for both groups, there is a slight preference for Level 30 with (computed via Monte Carlo sampling from the posterior beta distributions above) $\text{Prob}(\theta_1 > \theta_2|Y_1, Y_2) = 0.96$. With such a large dataset, the results are not sensitive to the prior with $\text{Prob}(\theta_1 > \theta_2|Y_1, Y_2)$ remaining 0.96 for $a = b$ equal 0.1, 1 or 10.

Bayesian gambit: A/B testing is used to identify the optimal strategy to be used in practice. Conducting an experiment and analyzing the data are crucial to decision making, but time spent experimenting is time lost implementing the optimal strategy. This is a basis of the so-called multi-armed bandit problem [142], where the objective is take actions sequentially to optimize overall results while accounting for uncertainty in the reward of the potential actions. Thompson sampling [154, 136] is a Bayesian approach that balances experimentation and optimization. Thompson sampling randomly assigns the strategy to use at a given time point based on posterior probabilities determined using all information collected up to the decision point.

Consider the following application. We are designing a mobile heath app that sends reminders to users to take an action. There are $J = 3$ treatment strategies under consideration: (1) text message, (2) email and (3) both text message and email. Let θ_j be the probability that a user takes the desired action when given a reminder following strategy $j \in \{1, 2, 3\}$. An initial A/B test is conducted with $n_j = 100$ observations in each group. The model is $Y_j|\theta_j \sim \text{Binomial}(n_j, \theta_j)$ and prior $\theta_j \sim \text{Beta}(0.5, 0.5)$. The posterior distributions of this test are plotted in Figure 5.4 (top left); it appears that the text and email combination is preferred, but the probability it is optimal is only 0.68.

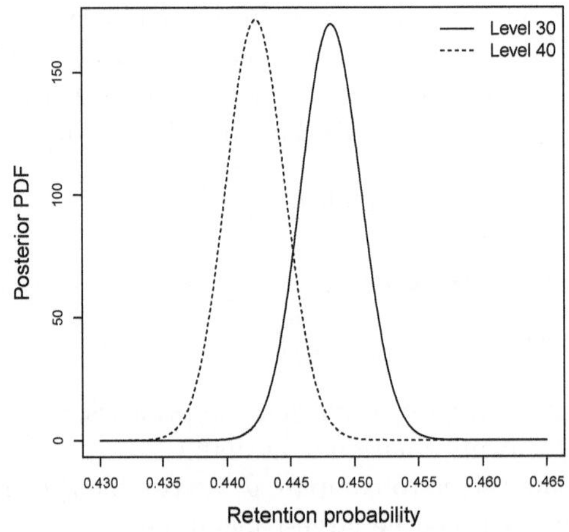

FIGURE 5.3
A/B testing of the Cookie Cats data. Posterior density of the retention rates of the two designs.

To continue accumulating data while favoring the seemingly optimal strategy, the team conducts Thompson sampling for 1000 additional users. At any point in the experiment, denote n_j as the (cumulative) number of users that have been given treatment j and Y_j be the number of these users that took the desired action. At a given time, the posterior distributions are $\theta_j|Y_j \sim \text{Beta}(Y_j + 0.5, n_j - Y_j + 0.5)$. Each day, the treatment is assigned to treatment j with probability equal to the posterior probability that $\theta_j = \max\{\theta_1, ..., \theta_J\}$, which can be implemented by taking a sample from θ_1, θ_2 and θ_3 from their current posterior distributions and assigning the strategy corresponding to the largest sample. The user's action is recorded, and the posteriors are updated for the next step. The algorithm is given in Algorithm 5.4.

Algorithm 6 Thompson sampling for binary data

Initialize $Y_j = n_j = 0$ for $j \in \{1, ..., J\}$ and $a = b = 0.5$
for $t = 1, ..., T$ **do**
  Sample $\theta_j^* \stackrel{indep}{\sim} \text{Beta}(Y_j + a, n_j - Y_j + b)$ for $j \in \{1, ..., J\}$
  Assign the next observation treatment k if $\theta_k^* = \max\{\theta_1^*, ..., \theta_J^*\}$
  Set $n_k = n_k + 1$
  If the observation is a success set $Y_k = Y_k + 1$
end for

The intuition behind this algorithm is that early in the experiment all treatment options are plausibly optimal, and all are included in the randomization. This experimentation accrues information about the effectiveness of the strategies. As the experiment continues, it becomes clear which strategy is optimal, and it is assigned with increasing probability until eventually it is the only one assigned to optimize performance. As shown in Figure 5.4 (bottom left), by the end of the 1000 additional samples the vast majority of users are

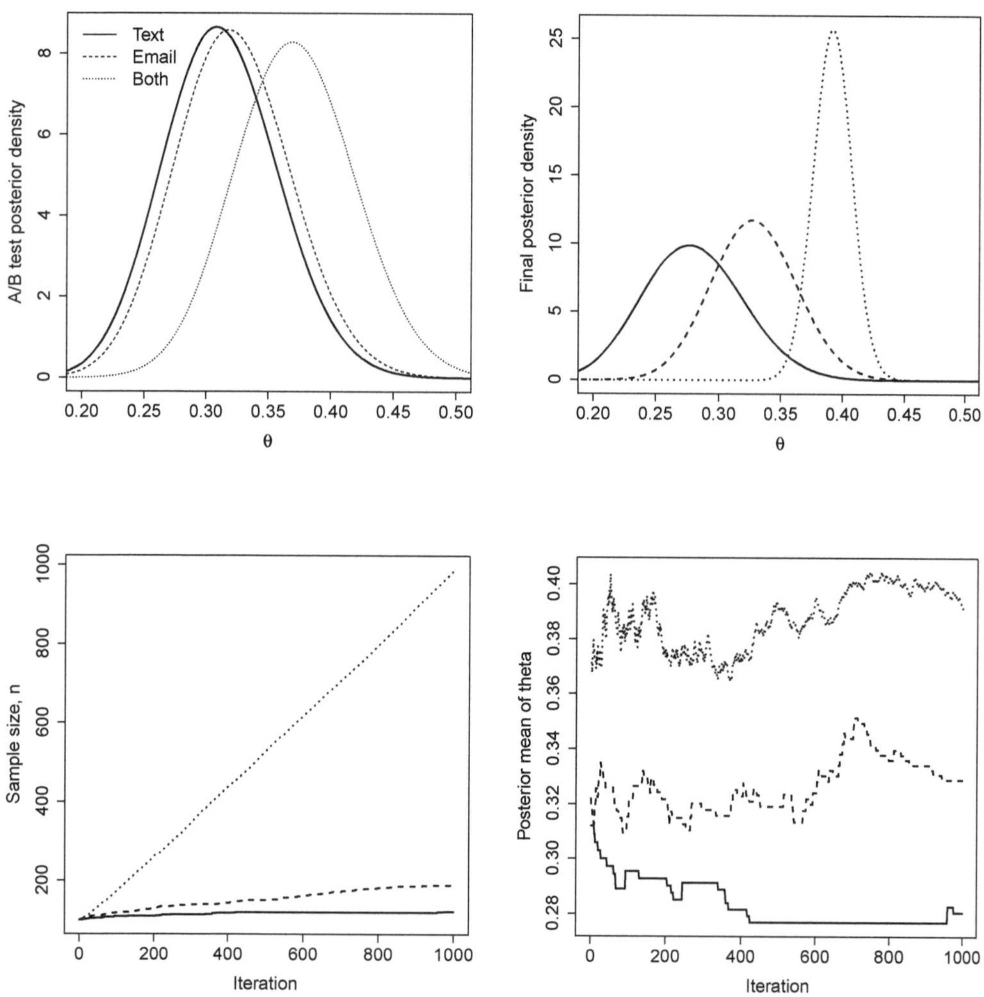

FIGURE 5.4
Illustration of Thompson sampling. Posterior distribution of the θ_j after the initial A/B test (top left) and after 1000 iterations of Thompson sampling (top right), and the sample size n_j (bottom left) and posterior mean of θ_j (bottom right) at each iteration of Thompson sampling.

being given the combination strategy as the posterior of θ_3 separates from θ_1 and θ_2 (top right).

5.4 Sample size and power calculations

Thus far we have considered methods to test hypotheses given an observed dataset, but have ignored decisions in data collection. Although other decisions can be considered (see [125] for a review), a critical design criterion is selecting the sample size, n. Selecting the sample size poses a trade-off between gathering sufficient information to reliably conduct inference with the expense of collecting data. In this section, we discuss methods to select the sample size that acknowledge this trade-off.

To fix ideas, consider the simple experiment where n independent trials will be conducted, each with success probability θ. The data is then generated as $Y|\theta \sim \text{Binomial}(n, \theta)$. Below we refer to the prior distribution that will be used to analyze the data as the *analysis prior*, which for this example is taken to be the conjugate Jeffreys prior $\theta \sim \text{Beta}(a, b)$ with $a = b = 0.5$. The posterior is then $\theta|Y \sim \text{Beta}(Y + a, n - Y + b)$. The Bayes factor, $BF(Y)$, for testing the null hypotheses that $\theta = \theta_0$ versus the alternative that $\theta \neq \theta_0$ and follows prior $\theta \sim \text{Beta}(a, b)$ is given in (5.8), and we reject the null hypothesis in favor of the alternative if $BF(Y) > c$.

There is nothing in this analysis that requires frequentist notions of probability because it is conducted conditioned on an observed dataset. However, in the design stage before the data is collected, we must necessarily consider the data to be random and thus both Bayesian (random θ) and frequentist (random Y) notions of probability and uncertainty. To illustrate this, say the sample size n is selected so that the posterior variance is sufficiently small. For the binomial model, the posterior variance is

$$V(Y) = \frac{(Y + a)(n - Y + b)}{(n + a + b)^2(n + a + b + 1)},$$

which depends on Y and can thus not be computed at the design stage. To account for this uncertainty about Y, we could view Y as random and select n so the expected value (over Y) of $V(Y)$ is sufficiently small.

Treating Y as random to compute the mean of $V(Y)$ requires specifying its distribution. It is reasonable to use the binomial model, but to complete the specification requires selecting θ or its distribution, referred to as the *design prior* [137]. The design prior need not be the same as the analysis prior, which is often set to be diffuse or even improper to reflect prior ignorance, and is often chosen to be a range of values that are plausible and reflect scientifically-meaningful differences. For example, if the study is designed to determine if θ differs from 0.50, then the design prior might be $\theta \sim \text{Uniform}(0.55, 0.65)$An advantage of the Bayesian approach to sample size calculation is the ability to incorporate results from preliminary data or other sources into the design prior distribution.

If the analysis will be a hypothesis test of the null hypothesis that $\theta = \theta_0$ versus the alternative that $\theta \neq \theta_0$, then the design is typically based on frequentist properties of the test including:

- The **Type-I error rate** is the probability of mistakenly rejecting the null hypothesis assuming data are generated under the null hypothesis with $\theta = \theta_0$ and a Bayesian test is performed on the observed data.

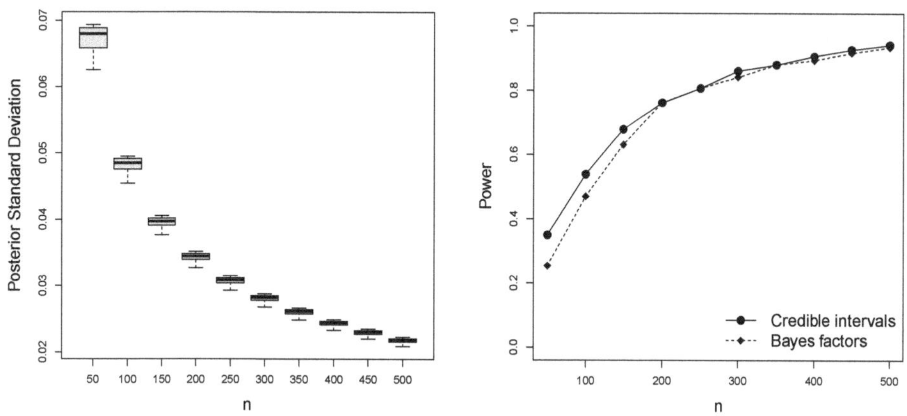

FIGURE 5.5
Illustration of sample size computations. The model is $Y|\theta \sim \text{Binomial}(n, \theta)$, the analysis prior is $\theta \sim \text{Beta}(0.5, 0.5)$ and the design prior is $\theta \sim \text{Uniform}(0.55, 0.65)$. For a sequence of samples sizes, n, the plots give the sampling distribution of the posterior standard deviation (left) and probability of rejecting the null hypothesis $\mathcal{H}_1 : \theta = 0.5$ using credible intervals and Bayes factors (right).

- The **power** is the probability of correctly rejecting the null hypothesis assuming data are generated under the alternative hypothesis/design prior and a Bayesian test is performed on the observed data.

A Bayesian experimental design for this test selects the sample size, n, and other design factors including the Bayes-factor threshold to reject the null hypothesis, c, to control Type I error and/or power.

Listing 5.2 provides three ways to select n for the beta-binomial example via Monte Carlo simulation. Many of these quantities could be computed mathematically without sampling for this simple case [34], but in general, Monte Carlo sampling provides a flexible and reliable way to approximate design quantities of interest [160]. The first approach selects n so that the posterior standard deviation is sufficiently small. Many datasets are generated from the design prior, and for each dataset the posterior standard deviation is computing. Figure 5.5 (left) plots the sampling distribution of the posterior standard deviation as a function of the sample size. If the objective is to have the posterior standard deviation no larger than 0.05, then the sample size should be at least $n = 100$.

The remaining two approaches select n to control the power of the test of the null hypothesis that $\theta = 0.5$ at 0.80. The first test simply computes the posterior 95% credible interval and rejects the null if θ_0 is outside the interval. The Type I error of this test is controlled between 0.049 and 0.066 depending on n, and $n = 300$ gives power exceeding 0.8 (Figure 5.5, right). The final design rejects the null hypothesis if $BF(Y) > c$. The threshold c (which depends on n) is selected to fix the Type I error at 0.05 by simulating data from the null hypothesis and setting c to the 0.95 quantile of the sampling distribution of $BF(Y)$. With c fixed, the power is computed and also results in a sample size of $n = 300$ to control power at 0.80.

Listing 5.2
R code for sample size computations.

```
library(VGAM)
a1     <- 0.55             # Design prior, Unif(a1,b1)
b1     <- 0.65
a2     <- 0.50             # Analysis prior, Beta(a2,b2)
b2     <- 0.50
t0     <- 0.50             # Null values, theta0

n      <- seq(50,500,50)
S      <- 100000           # Number of Monte Carlo simulations
SD     <- matrix(0,S,length(n))
PowCI <- TypeICI <- rep(0,length(n))
PowBF <- TypeIBF <- rep(0,length(n))

for(i in 1:length(n)){

  # Generate data under the null
   Y1        <- rbinom(S,n[i],t0)     # Gen data under the null
   A1        <- Y1 + a2               # Post theta|Y1 ~ Beta(A1,B1)
   B1        <- n[i] - Y1 + b2

   theta     <- runif(S,a1,b1)        # Gen from design prior
   Y2        <- rbinom(S,n[i],theta)  # Gen data under the alt
   A2        <- Y2 + a2               # Post theta|Y2 ~ Beta(A2,B2)
   B2        <- n[i] - Y2 + b2

  # Posterior standard deviation
   SD[,i]    <- sqrt(A2*B2)/((A2+B2)*sqrt(A2+B2+1))

  # Test via credible intervals
   L1         <- qbeta(0.025,A1,B1)
   U1         <- qbeta(0.975,A1,B1)
   TypeICI[i] <- mean((L1>t0) | (U1<t0))

   L2         <- qbeta(0.025,A2,B2)
   U2         <- qbeta(0.975,A2,B2)
   PowCI[i]   <- mean((L2>t0) | (U2<t0))

  # Test via Bayes factors
   BF1        <- dbetabinom.ab(Y1,n[i],a2,b2)/dbinom(Y1,n[i],t0)
   BF2        <- dbetabinom.ab(Y2,n[i],a2,b2)/dbinom(Y2,n[i],t0)
   c          <- quantile(BF1,0.95)
   PowBF[i]   <- mean(BF2>c)
}

boxplot(t(SD)~n,ylab="Posterior Standard Deviation")

plot(n,PowCI,type="l",ylab="Power",ylim=0:1)
lines(n,PowBF,lty=2)
```

5.5 Exercises

1. Assume that $Y|\lambda \sim \text{Poisson}(N\lambda)$ and $\lambda \sim \text{Gamma}(0.1, b)$. The null hypothesis is that $\lambda \leq 1$, and the alternative is that $\lambda > 1$. Select b so that the prior probability of the null hypothesis is 0.5. Using this prior, compute the posterior probability of the alternative hypothesis and the Bayes factor for the alternative relative to the null hypothesis for (a) $N = 10$ and $Y = 12$; (b) $N = 20$ and $Y = 24$; (c) $N = 50$ and $Y = 60$; (d) $N = 100$ and $Y = 120$. For which N and Y is there definitive evidence in favor of the alternative?

2. Data from the initial Moderna vaccine trial in [9] are given in the table below.

Group	Infected (Y)	Total participants (n)
Placebo	185	14073
Vaccine	11	14134

Using a Bayes factor, test the hypotheses that $\mathcal{H}_1$: the infection rate is lower in the placebo group versus $\mathcal{H}_2$: the infection rate is lower in the vaccine group. Describe and justify your model and prior and test for sensitivity to the prior choice.

3. Use the "Mr. October" data (Section 2.5) $Y_1 = 563$, $N_1 = 2820$, $Y_2 = 10$, and $N_2 = 27$. Compare the two models using Bayes factors:

$$\mathcal{H}_1\text{: } Y_1|\lambda_1 \sim \text{Poisson}(N_1\lambda_1) \text{ and } Y_2|\lambda_2 \sim \text{Poisson}(N_2\lambda_2)$$
$$\mathcal{H}_2\text{: } Y_1|\lambda_0 \sim \text{Poisson}(N_1\lambda_0) \text{ and } Y_2|\lambda_0 \sim \text{Poisson}(N_2\lambda_0).$$

Assume the Uniform(0,c) prior for all λ_j and compare the results for $c = 1$ and $c = 10$.

4. Researchers in Brazil are studying the effect of two intervention strategies to reduce deforestation. Their observational data consists of $n_1 = 7$ regions under strategy 1 and $n_2 = 9$ regions under strategy 2. Each observation was matched with a similar control region and the response is the difference in deforestation percent between the treated and control regions. The 16 differences are given in the table below.

Strategy	Differences (%)								
1	−11.5	−8.4	−6.8	−5.6	−0.4	1.9	2.8		
2	−7.0	−4.3	−3.0	−3.0	−0.1	0.0	2.2	4.3	10.8

(a) Let Y_{ij} be the ith observation under strategy j. The model is $Y_{ij}|\mu_j, \sigma_j \overset{indep}{\sim} \text{Normal}(\mu_j, \sigma_j^2)$. Select and justify uninformative priors, compute the posterior probabilities of the hypotheses $\mathcal{H}_1 : \mu_1 < \mu_2$ versus $\mathcal{H}_2 : \mu_2 < \mu_1$, and summarize the results of the test.

(b) Compare strategies using a simple linear regression for all the $i = 1, ..., n = 16$ observations. The likelihood is $Y_i|\boldsymbol{\beta}, \sigma^2 \overset{indep}{\sim} \text{Normal}(\beta_1 + \beta_2 X_i, \sigma^2)$ where $X_i = 0$ if the observation came from strategy 1 and $X_i = 1$ if it came from strategy 2. Select and justify uninformative priors, compute the 95% credible interval for β_2, and use this to compare strategies.

(c) Assume the regression model in (b) except set $\beta_2 = \gamma\delta$ with priors $\gamma \sim \text{Bernoulli}(0.5)$ and $\delta \sim \text{Normal}(0, W\sigma^2)$. Select and justify uninformative priors (try 2-3 values of W), compute the posterior probabilities of the hypotheses $\mathcal{H}_1 : \mu_1 = \mu_2$ versus $\mathcal{H}_2 : \mu_1 \neq \mu_2$, and summarize the results of the test.

(d) Comment on the pros and cons of these three tests.

5. Equation 5.10 gives the Bayes factor for a test of a normal mean.

   (a) For $\tau = 1$, plot the Bayes factor as a function of the observed value y and give a range of values of y that will lead to rejecting the null hypothesis.

   (b) Repeat (a) with $\tau = 10$, and comment on the differences.

6. You are on the design team for a mobile health app. Your team is considering three new designs labelled A, B and C. To identify the optimal design, $n = 100$ users were randomly selected to be given each design and followed for a month. The table below gives the number of users in each treatment arm that opened the app y times for y between 0 and 6 (the maximum number of times the app was opened). For example, 100 users were given design A, and 40 of the 100 opened the app exactly once.

Design	0	1	2	3	4	5	6	Total
A	37	40	15	7	1	0	0	100
B	12	29	28	16	8	6	1	100
C	27	49	18	5	0	1	0	100

   (a) Assuming the observations are Poisson counts, specify a likelihood and an objective Bayesian prior.

   (b) Perform an A/B test. Plot the posterior distribution of the Poisson rate parameter (i.e., the mean response) for each group, and compute the posterior probability that each of the three designs is optimal.

7. The national average rate of new thyroid cancer cases is around 13 per 100,000 people. An epidemiological study is monitoring certain regions for rates exceeding the national rate. The population size (N) and the number of new cases (Y) for each region for the most recent year are:

Region	N	Y	Region	N	Y
1	179611	20	11	30557	5
2	125453	18	12	81858	15
3	41646	7	13	103475	12
4	98826	22	14	166203	14
5	68642	10	15	52032	12
6	71844	6	16	85767	15
7	67702	9	17	145591	20
8	45954	9	18	139176	15
9	96325	18	19	87842	24
10	200201	27	20	97930	11

   Separately for each region i, fit the model $Y_i|\lambda_i \sim \text{Poisson}(N_i\lambda_i)$ with prior $\lambda \sim \text{Gamma}(a, b)$, and compute the posterior probability that the region's rate λ_i exceeds the national average $\lambda_0 = 13/100000$, denoted $q_i = \text{Prob}(\lambda_i > \lambda_0|Y_i)$.

   (a) Select a and b so that (approximately) the prior probability that $\lambda_1 > \lambda_0$ is 0.5, and the prior 0.95 quantile is $10\lambda_0$.

(b) Assuming that you will conclude the rate in region i exceeds the national average if $q_i > t$, select t to control the Bayesian false discovery rate at 10%. With this value, summarize the results.

8. Assume the model $Y_i|\lambda \sim \text{Poisson}(\lambda)$ for $i = 1, ..., n$ and improper prior $\pi(\lambda) \propto 1$. The objective is to test the null hypothesis $\mathcal{H}_1 : \lambda \leq 1$ versus the alternative hypothesis $\mathcal{H}_2 : \lambda > 1$, and the test is to reject the null hypothesis if $\text{Prob}(\lambda > 1|Y_1, ..., Y_n) > 0.95$.

   (a) Assuming n is fixed at 10, compute the probability of rejecting the null hypothesis for true λ ranging from 1 to 2.

   (b) Assuming the true value of λ is 1, compute the Type I error of the test for n ranging from 10 to 100.

   (c) Assuming the true value of λ is 1.5, compute the power of the test for n ranging from 10 to 100.

   (d) How might you modify the test to guarantee that the Type I error is exactly 0.05?

9. Download the Boston Housing Data in `R` from the `Boston` dataset. The response is `medv`, the median value of owner-occupied homes, and the other 13 variables are covariates that describe the neighborhood. Use stochastic search variable selection (SSVS) to compute the most likely subset of the 13 covariates to include in the model and the marginal probability that each variable is included in the model. Clearly describe the model you fit including all prior distributions.

6

Model selection and diagnostics

CONTENTS

A statistical model is a mathematical representation of a system that includes errors and biases in the observation process, and therefore lays bare the assumptions being made in the analysis. Of course, no statistical model is absolutely correct; in reality, functional relationships are not linear, errors are not exactly Gaussian, residuals are not independent, etc. Nonetheless, slight deviations in fit are a small price to pay for a simple and interpretable representation of reality that allows us to probe for important relationships, test scientific hypotheses and make predictions. On the other hand, if a model's assumptions are blatantly violated then the results of the analysis cannot be trusted. Therefore, selecting an appropriate model that is as simple as possible while fitting the data reasonably well is a key step in any parametric statistical analysis.

In this chapter we discuss Bayesian model selection and goodness-of-fit measures. For model selection, we assume there is a finite collection of candidate models denoted as $\mathcal{M}_1, ..., \mathcal{M}_M$. For example, we might compare Gaussian and Student's t-distribution models,

$$\mathcal{M}_1 : Y_i \stackrel{iid}{\sim} \text{Normal}(\mu, \sigma^2) \text{ versus } \mathcal{M}_2 : Y_i \stackrel{iid}{\sim} \text{t}_\nu(\mu, \sigma^2) \tag{6.1}$$

or whether or not to include a covariate

$$\mathcal{M}_1 : Y_i \stackrel{indep}{\sim} \text{Normal}(\beta_1, \sigma^2) \text{ versus } \mathcal{M}_2 : Y_i \stackrel{indep}{\sim} \text{Normal}(\beta_1 + X_i\beta_2, \sigma^2). \tag{6.2}$$

The methods we discuss are general and can be used for other model-selection tasks, including selecting a random effects structure, a link function, etc.

Sections 6.1–6.4 introduce techniques for comparing and selecting statistical models. Section 6.1 begins with cross validation, which is a flexible and intuitive way to compare methods. For a more formal Bayesian treatment of model selection and assessment of model uncertainty, Section 6.2 provides computational tools to approximate Bayes factors when many linear models are under consideration. An attractive feature of Bayes factors is that rather than selecting a single model, uncertainty about the model is captured using posterior probabilities. Section 6.3 uses these posterior probabilities to make predictions that appropriately average over model uncertainty. While Bayes factors are appealing, they typically require extensive derivation or computation and are sensitive to the prior, and so Section 6.4 provides less formal but more broadly applicable model selection criteria. In virtually any analysis, none of the M models will be "right;" rather we are simply searching for the one that fits the "best." Therefore, Section 6.5 provides goodness-of-fit tools to verify that a selected model captures the important features of a dataset.

6.1 Cross validation

Arguably the most intuitive way to compare models is based on out-of-sample prediction performance. Ideally an independent validation set is available and used to evaluate performance. For example, for streaming data one might train the M models on data available at the time of the analysis and use data collected after the analysis to measure predictive performance. This is often infeasible as data are not collected sequentially, and so internal cross validation (CV) is used instead. In K-fold CV, each observation is randomly assigned to one of the K folds, with $g_i \in \{1, ..., K\}$ denoting the group assignment for observation i. Denote the subset of the data in fold k as $\mathbf{Y}_k = \{Y_i; g_i = k\}$ and the data in all other folds as $\mathbf{Y}_{(k)} = \{Y_i; g_i \neq k\}$. The model is then fit K times, with the kth model fit using $\mathbf{Y}_{(k)}$ to train the model and make predictions for $\mathbf{Y}_k$. In this way, a prediction is made for each observation using an analysis that excludes the observation, approximating out of sample prediction performance.

Bayesian prediction is based on the Posterior Predictive Distribution (PPD) of the test set observations $\mathbf{Y}_k$ given the training data $\mathbf{Y}_{(k)}$ (Section 1.5). An advantage of this method of prediction is that it naturally averages over uncertainty in the model parameters. Generating samples from the PPD averaging over parameter uncertainty is straightforward using MCMC. Let $Y_i^{(s)}$ be the prediction made at MCMC iteration s for test set observation Y_i; then $Y_i^{(1)}, ..., Y_i^{(S)}$ can be used to approximate the PPD. For example, we might compute the Monte Carlo sample mean $\hat{Y}_i = \sum_{s=1}^{S} Y_i^{(s)}/S$, median $\tilde{Y}_i$, and τ-quantile $q_i(\tau)$ to approximate the posterior predictive mean, median and quantiles respectively.

Many metrics are available to summarize the accuracy of the predictions from each model [64]. The most common measures of point prediction are bias, mean squared error and mean absolute deviation,

$$\begin{aligned} BIAS &= \frac{1}{n}\sum_{i=1}^{n}(\hat{Y}_i - Y_i) \\ MSE &= \frac{1}{n}\sum_{i=1}^{n}(\hat{Y}_i - Y_i)^2 \\ MAD &= \frac{1}{n}\sum_{i=1}^{n}|\tilde{Y}_i - Y_i|, \end{aligned}$$

with MSE being more sensitive to large errors than MAD. While MSE provides a natural ranking of models, it can be difficult to interpret without knowledge of the scale of the data. A scale-free measure of prediction performance is the correlation between Y_i and $\tilde{Y}_i$,

$$R = \frac{\sum_{i=1}^{n}(\tilde{Y}_i - \tilde{M})(Y_i - M)}{\sqrt{\tilde{V}V}}$$

where $\tilde{M}$ and $\tilde{V}$ are the mean and variance of $\tilde{Y}_1, ..., \tilde{Y}_n$ and M and V are the mean and variance of $Y_1, ..., Y_n$. The squared correlation, R^2, can also be reported.

Performance of credible intervals can be summarized using empirical coverage and average width of prediction intervals

$$\begin{aligned} COV &= \frac{1}{n}\sum_{i=1}^{n} I\left[q_i(\alpha/2) \le Y_i \le q_i(1-\alpha/2)\right] \\ WIDTH &= \frac{1}{n}\sum_{i=1}^{n} \left[q_i(1-\alpha/2) - q_i(\alpha/2)\right], \end{aligned}$$

where $I(A) = 1$ if the statement A is true and zero otherwise. A measure of fit to the entire distribution is the log score,

$$LS = \frac{1}{n}\sum_{i=1}^{n} \log[f(Y_i|\hat{\boldsymbol{\theta}}_i)] \tag{6.3}$$

where $\hat{\boldsymbol{\theta}}_i$ is the parameter estimate based on the fold that excludes observation i. The log score is the average log likelihood of the test set observations given the parameter estimates from the training data.

Based on these measures, we might discard models with COV far below the nominal $1-\alpha$ level and from the remaining models, choose the one with small MSE, MAD and $WIDTH$ and large LS. It is essential that the evaluation is based on out-of-sample predictions rather than within-sample fit. Overly complicated models (e.g., a linear model with too many predictors) may replicate the data used to fit the model but be too unstable to predict well under new conditions.

Cross validation can be motivated by information theory. Suppose the "true" data generating model has PDF f_0 so that in reality $Y_i \stackrel{iid}{\sim} f_0$. Of course, we cannot know the true model and so we choose between M models with PDFs $f_1, ..., f_M$. Our objective is to select the model that is in some sense the closest to the true model. A reasonable measure of the difference between the true and postulated model is the Kullback–Leibler divergence

$$KL(f_0, f_j) = \mathrm{E}\left\{\log\left[\frac{f_j(Y)}{f_0(Y)}\right]\right\} = \mathrm{E}\left\{\log[f_0(Y)]\right\} - \mathrm{E}\left\{\log[f_j(Y)]\right\},$$

where the expectation is with respect to the true model $Y \sim f_0$. The term $\mathrm{E}\{\log[f_0(Y)]\}$ is the same for all $j = 1, ..., M$ and therefore ranking models based on $KL(f_0, f_j)$ is equivalent to ranking models based on their log score $LS_j = \mathrm{E}\{\log[f_j(Y)]\}$. Since the data are generated from f_0, the cross validation log score in (6.3) is a Monte Carlo estimate of the true log score LS_j, and therefore ranking models based on their cross validation log score is an attempt to rank them based on similarity to the true data-generating model.

6.2 Stochastic search variable selection

In Section 5.1, Bayes factors were used to summarize the data's support for two hypotheses. This concept can be generalized to many models based on the inclusion of different subsets of covariates. For example, in the linear regression model with $p = 2$ covariates, $Y_i = \beta_0 + X_{1i}\beta_1 + X_{i2}\beta_2 + \varepsilon_i$, there are four models: $\mathcal{M}_1 : \beta_1 = \beta_2 = 0$, $\mathcal{M}_2 : \beta_1 = 0, \beta_2 \neq 0$, $\mathcal{M}_3 : \beta_1 \neq 0, \beta_2 = 0$ and $\mathcal{M}_4 : \beta_1 \neq 0, \beta_2 \neq 0$. The posterior probability of each model can be used to determine which model to select. However, in many applications, the number of models under consideration is large, and enumerating all models and computing each posterior probability is unfeasible. For example, in linear regression with p parameters

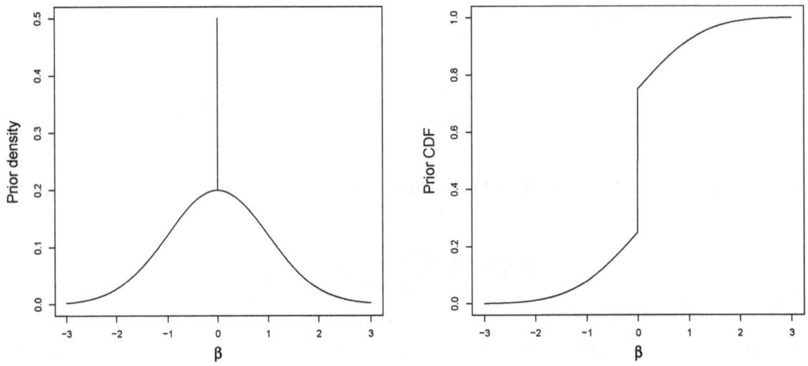

FIGURE 6.1
Spike and slab prior. PDF (left) and CDF of β under the spike and slab prior $\beta = \gamma\delta$ where $\gamma \sim \text{Bernoulli}(0.5)$ and $\delta \sim \text{Normal}(0,1)$.

there are $M = 2^p$ potential models formed by including subsets of the predictors; with $p = 30$ this is over a billion potential models. A classical way to overcome searching over all combinations of covariates is to employ a systematic search such as forward, backward or stepwise selection. In this section we discuss a stochastic alternative that randomly visits models according to their posterior probability.

Stochastic search variable selection (SSVS; [60]) approximates model probabilities using MCMC. SSVS introduces dummy variables to encode the model and then computes posterior probabilities of the dummy variables to approximate the posterior model probabilities. For example, consider the linear regression model

$$Y_i|\boldsymbol{\beta}, \sigma^2 \overset{indep}{\sim} \text{Normal}\left(\sum_{j=1}^{p} X_{ij}\beta_j, \sigma^2\right). \tag{6.4}$$

The $M = 2^p$ models formed by subsets of the predictors can be encoded by $\boldsymbol{\gamma} = (\gamma_1, ..., \gamma_p)$ where $\gamma_j = 1$ if covariate j is included in the model and $\gamma_j = 0$ if covariate j is excluded from the model, so that $\boldsymbol{\gamma} = c(1, 0, 1, 0, 0, ...)$ corresponds to the model $\text{E}(Y_i) = X_{i1}\beta_1 + X_{i3}\beta_3$.

A common prior over models is to fix $\gamma_1 = 1$ for the intercept and $\gamma_j|q \overset{iid}{\sim} \text{Bernoulli}(q)$ for $j > 1$, where q is the prior inclusion probability. Given the model $\boldsymbol{\gamma}$, the regression coefficients that are included in the model can have independent normal priors (other priors are possible, [131]) $\beta_j|\gamma_j = 1 \sim \text{Normal}(0, \sigma^2\tau^2)$. The inclusion probability and regression coefficient variance can be fixed or have a prior such as $q \sim \text{Beta}(a, b)$ and $\tau^2 \sim \text{InvGamma}(\epsilon, \epsilon)$. As with Bayes factors (Section 5.1), the posterior for this model can be sensitive to the prior, and multiple priors should be compared to understand sensitivity.

The prior for β_j induced by this model is plotted in Figure 6.1. The prior is a mixture of two components: a peak at $\beta_j = 0$ corresponding to samples that exclude ($\gamma_j = \beta_j = 0$) covariate j and a Gaussian curve corresponding to samples that include ($\gamma_j = 1$ so $\beta_j = \delta_j$) β_j. Because of this distinct shape, this prior is often called the spike-and-slab prior.

All M models can simultaneously be written as the supermodel

$$\begin{aligned} Y_i|\boldsymbol{\beta},\sigma^2 &\stackrel{indep}{\sim} \text{Normal}\left(\sum_{j=1}^p X_{ij}\beta_j,\sigma^2\right) \\ \beta_j &= \gamma_j\delta_j \\ \gamma_j &\sim \text{Bernoulli}(q) \\ \delta_j &\sim \text{Normal}(0,\tau^2\sigma^2). \end{aligned}$$

MCMC samples from this model include different subsets of the covariates. Posterior samples with $\gamma_j = 0$ have $\beta_j = 0$ and thus covariate j excluded from the model. This supermodel can be fit a once and gives approximations for all M posterior model probabilities. Since the search over models is done within MCMC, this is a stochastic search as opposed to systematic searches such as forward or backward regression. An advantage of a stochastic search is that models with low probability are rarely sampled and so more of the computing time is spent on high-probability models.

With many possible models, no single model is likely to emerge as having high probability. Therefore, with large M extremely long chains can be needed to give accurate estimates of model probabilities. More stable summaries of the model space are the marginal inclusion probabilities $MIP_j = \text{E}(\gamma_j = 1|\mathbf{Y}) = \text{Prob}(\beta_j \neq 0|\mathbf{Y})$. MIP_j is the probability that covariate j is included in the model, averaging over uncertainty in the subset of the other variables that are included in the model, and can be used to compute the Bayes factor for the models that do and do not include covariate j, $BF_j = [MIP_j/(1-MIP_j)]/[q/(1-q)]$. In fact, [13] shows that if a single model is required for prediction, the model that includes covariates with MIP_j greater than 0.5 is preferred to the highest probability model.

Childhood malaria example: Diggle et al. [44] analyze data from $n = 1,332$ children from the Gambia. The binary response Y_i is the indictor that child i tested positive for malaria. We use five covariates in X_{ij}:

- Age: Age of the child, in days
- Net use: Indicator variable denoting whether (1) or not (0) the child regularly sleeps under a bed-net (coded 1 if yes, 0 otherwise)
- Treated: Indicator variable denoting whether the bed-net is treated (coded 1 if yes, 0 otherwise)
- Green: Satellite-derived measure of the greenness of vegetation in the immediate vicinity of the village (arbitrary units)
- PCH: Indicator variable denoting the presence (1) or absence (0) of a health center in the village

All five covariates are standardized to have mean zero and variance one. We use the logit regression model

$$\text{logit}[\text{Prob}(Y_i = 1)] = \alpha + \sum_{j=1}^p X_{ij}\beta_j. \tag{6.5}$$

The spike-and-slab prior for β_j is $\beta_j = \gamma_j\delta_j$ where $\gamma_j \sim \text{Bernoulli}(0.5)$ and $\delta_j \sim \text{Normal}(0,\tau^2)$. Listing 6.1 gives `JAGS` code to fit this model.

Table 6.1 gives the model probabilities, i.e., the proportion of the MCMC samples with γ corresponding to each model. Only three models have posterior probability greater than 0.01. All three models include age, net use and greenness, and differ based on whether they

Listing 6.1
JAGS code for SSVS.

```
for(i in 1:n){
  Y[i]          ~ dbern(pi[i])
  logit(pi[i]) <- alpha           + X[i,1]*beta[1] +
                  X[i,2]*beta[2] + X[i,3]*beta[3] +
                  X[i,4]*beta[4] + X[i,5]*beta[5]
}
for(j in 1:5){
  beta[j] <- gamma[j]*delta[j]
  gamma[j] ~ dbern(0.5)
  delta[j] ~ dnorm(0,tau)
}
alpha ~ dnorm(0,0.01)
tau   ~ dgamma(0.1,0.1)
```

TABLE 6.1
Posterior model probabilities for the Gambia analysis. All other models have posterior probability less than 0.01.

Covariates	Probability
Age, Net use, Greenness, Treated	0.42
Age, Net use, Greenness, Treated, Health center	0.37
Age, Net use, Greenness, Health center	0.20

TABLE 6.2
Marginal posteriors for the Gambia analysis. Posterior inclusion probabilities (i.e., Prob($\beta_j \neq 0|\mathbf{Y}$)) and posterior median and 90% intervals for the β_j.

Covariate	Inclusion Prob	Median	95% Interval
Age	1.00	0.26	(0.19, 0.34)
Net use	1.00	−0.25	(−0.34, −0.17)
Greenness	1.00	0.29	(0.21, 0.37)
Treated	0.79	−0.13	(−0.24, 0.00)
Health center	0.56	−0.05	(−0.19, 0.00)

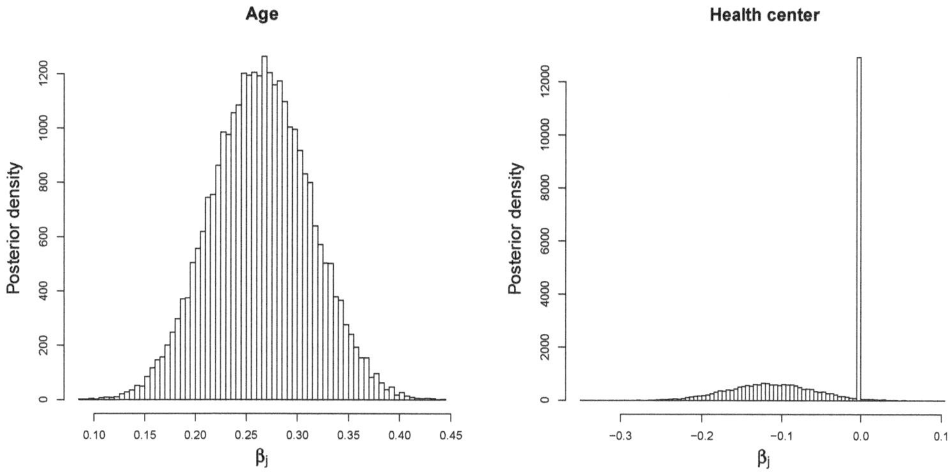

FIGURE 6.2
Posterior distribution for the SSVS analysis. Posterior distribution for the regression coefficients β_j for age and proximity to a health center.

include bed-net treatment, the health center indicator, or both. The marginal inclusion probabilities MIP_j in Table 6.2 exceed 0.5 for all covariates, and therefore the best single model for prediction likely includes all covariates [13]. The posterior density for the regression coefficient corresponding to age (Figure 6.2) is bell-shaped because it escapes the prior spike at zero; however the posterior for proximity to a health center retains considerable mass at zero and thus the spike-and-slab shape.

6.3 Bayesian model averaging

A main advantage of the Bayesian approach is the ability to properly handle uncertainty in model parameters when making statistical inference and predictions. In Section 1.5 (Figure 1.12) we explored the effect of accounting for parameter uncertainty in prediction. The posterior predictive distribution (PPD) for observation Y^* given the observed data $\mathbf{Y}$ averages over uncertainty in the parameters according to their posterior distribution,

$$p(Y^*|\mathbf{Y}) = \int p(Y^*, \boldsymbol{\theta}|\mathbf{Y})d\boldsymbol{\theta} = \int p(Y^*|\boldsymbol{\theta}, \mathbf{Y})p(\boldsymbol{\theta}|\mathbf{Y})d\boldsymbol{\theta}. \tag{6.6}$$

In addition to parametric uncertainty, with many potential models no single model is likely to emerge as the only viable option, and it is important to account for model uncertainty in prediction. Denoting the posterior probability of model m as $\text{Prob}(\mathcal{M} = \mathcal{M}_m|\mathbf{Y}) = w_m(\mathbf{Y})$, the PPD of Y^* averaging over posterior model uncertainty is

$$p(Y^*|\mathbf{Y}) = \sum_{m=1}^{M} w_m(\mathbf{Y})p(Y^*|\mathbf{Y}, \mathcal{M}_m) \tag{6.7}$$

where $p(Y^*|\mathbf{Y}, \mathcal{M}_m)$ is the PPD from model $\mathcal{M}_m$. Making an inference that accounts for model uncertainty using posterior model probabilities is called Bayesian model averaging (BMA; [80]).

SSVS (Section 6.2) via MCMC provides a convenient way to perform model-averaged predictions. Draws from the SSVS model include dummy indicators for different models as unknown parameters, and thus naturally average over models according to their posterior probability. Therefore, if a sample of Y^* is made at each MCMC iteration, then these samples follow the Bayesian model averaged PPD, as desired.

In addition to prediction, BMA can be used for inference on parameters common to all models. For example, if parameter β_j has posterior distribution $p(\beta_j|\mathbf{Y}, \mathcal{M}_m)$ under model $\mathcal{M}_m$, then the BMA posterior is $p(\beta_j|\mathbf{Y}) = \sum_{m=1}^{M} w_m(\mathbf{Y})p(\beta_j|\mathbf{Y}, \mathcal{M}_m)$. However, BMA results for parameters must be carefully scrutinized because the interpretation of parameters can change considerably across models, and so it is not always clear how to interpret an average over models. As an extreme case, in a regression analysis with collinearity the sign of β_j might change depending on the other covariates that are included, and so it is not obvious that results should be combined across models.

6.4 Model selection criteria

Cross validation and Bayes factors are both difficult to compute for large datasets and/or complicated models; cross validation requires fitting each model K times, and Bayes factors require difficult integration. Model-fit criteria provide a useful alternative. The criteria considered in this chapter are defined via the deviance (twice the negative log likelihood) of the data given the parameters,

$$D(\mathbf{Y}|\boldsymbol{\theta}) = -2\log[f(\mathbf{Y}|\boldsymbol{\theta})]. \tag{6.8}$$

In this chapter, we compare only models with the same deviance function but different models/priors for the parameters so that the deviance is a comparable measure of fit across models. For some models, e.g., the random effect models in Section 4.4 which have equivalent representations conditional on and marginal over the random effects, the definition of the likelihood and thus deviance are not unique, but for most models for independent data the definition of the deviance is clear.

The Bayesian Information Criteria (BIC) is one such criteria, defined as

$$BIC = D(\mathbf{Y}|\hat{\boldsymbol{\theta}}_{MLE}) + \log(n)p \tag{6.9}$$

where $\hat{\boldsymbol{\theta}}_{MLE}$ is the maximum likelihood estimate, n is the sample size, and p is the number of parameters in the model. BIC is split into two terms: $D(\mathbf{Y}|\hat{\boldsymbol{\theta}}_{MLE})$ is small for models that fit the data well and $\log(n)p$ is small for simple low-dimensional models. Since simple models with good fit are desirable, models with smaller BIC are preferred.

BIC was originally motivated as an approximation to the Bayes factor. However, BIC has undesirable features from the Bayesian perspective. First, estimating $\boldsymbol{\theta}$ using the MLE does not use prior information and is not available from MCMC output. Second, quantifying model complexity with the number of parameters p does not account for informative priors. For example, if two models have the same number of parameters, but one has very informative priors, and the other does not, then the model with informative priors should be regarded as simpler because it has fewer effective degrees of freedom.

The deviance Information Criteria (DIC; [145]) resolves these issues. From a Bayesian perspective, the deviance $D(\mathbf{Y}|\boldsymbol{\theta})$ is a random variable since it is a function of the random variables $\boldsymbol{\theta}$, therefore $D(\mathbf{Y}|\boldsymbol{\theta})$ has a posterior distribution that can be summarized using its posterior mean $\bar{D}$. Model complexity is summarized by the effective number of parameters,

$$p_D = \bar{D} - \hat{D}, \tag{6.10}$$

where $\hat{D} = D(\mathbf{Y}|\hat{\boldsymbol{\theta}})$ and $\hat{\boldsymbol{\theta}}$ is the posterior mean of $\boldsymbol{\theta}$. As shown by example below, p_D is typically not an integer because prior shrinkage can result in partial degrees of freedom. Both $\bar{D}$ and $\hat{\boldsymbol{\theta}}$ can be approximated using MCMC output by computing $D\left(\mathbf{Y}|\boldsymbol{\theta}\right)$ and $\boldsymbol{\theta}$ at each iteration and taking the mean over iterations. Therefore, DIC is straightforward to compute given MCMC output.

The criteria is then

$$DIC = \bar{D} + p_D = D(\mathbf{Y}|\hat{\boldsymbol{\theta}}) + 2p_D. \tag{6.11}$$

This resembles the Akaike Information Criterion (AIC) [4] $AIC = D(\mathbf{Y}|\hat{\boldsymbol{\theta}}_{MLE}) + 2p$ but with the posterior mean and effective degrees of freedom replacing the MLE and the total number of parameters, respectively. As with AIC and BIC, the actual value of the criteria is hard to interpret, but it can be used to rank models with models having small DIC being preferred (a loose rule of thumb is that a difference of 5 is substantial, and 10 is definitive, but it is difficult to establish statistical significance using DIC). The intuition is that models with small DIC are simple (small p_D) and fit well (small $\bar{D}$). The effective number of parameters p_D is generally less than the number of parameters p if the prior are strong, as desired. Unfortunately p_D can be outside $[0,p]$ in pathological cases, typically where the posterior mean of $\boldsymbol{\theta}$ is not a good summary of the posterior as is the case in mixture models with multimodal priors and posteriors.

The motivation of p_D as a measure of model size is complex, but intuition can be built using a few examples. In multiple linear regression with $\mathbf{Y}|\boldsymbol{\beta} \sim \text{Normal}(\mathbf{X}\boldsymbol{\beta}, \sigma^2\mathbf{I}_n)$ and Zellner's prior $\boldsymbol{\beta} \sim \text{Normal}(0, c\sigma^2(\mathbf{X}^T\mathbf{X})^{-1})$, the effective number of parameters is $p_D = \frac{c}{c+1}p$. In this case, the effective number of parameters increases from zero with tight prior ($c = 0$) to p with uninformative prior ($c = \infty$). Also, in the one-way random effects model (Section 4.4)

$$Y_{ij}|\mu_j \sim \text{Normal}(\mu_j, \sigma^2) \text{ and } \mu_j \sim \text{Normal}(0, c\sigma^2), \tag{6.12}$$

for $i = 1, ..., n$ replications within each of the $j = 1, ..., p$ groups and fixed variance components σ^2 and c. The effective number of parameters is $p_d = \frac{c}{c+1/n}p$, which also increases from 0 to p with the prior variance c.

The Watanabe–Akaike (also known as the widely applicable) information criteria ($WAIC$; [55]) is an alternative to DIC. $WAIC$ is proposed as an approximation to n-fold (i.e., leave-one-out) cross validation. Rather than the posterior mean of the deviance, $WAIC$ computes the posterior mean and variance of the likelihood and log likelihood. The criterion is

$$WAIC = -2\sum_{i=1} \log\{\bar{f}_i\} + 2p_W \tag{6.13}$$

where fit for observation i is measured by the posterior mean of the likelihood,

$$\bar{f}_i = \text{E}[f(Y_i|\boldsymbol{\theta})|\mathbf{Y}] \tag{6.14}$$

and model complexity is measured by

$$p_W = \sum_{i=1}^{n} \text{Var}[\log(f(Y_i|\boldsymbol{\theta}))|\mathbf{Y}], \tag{6.15}$$

defined as the sum of posterior variance of the log likelihood functions. Therefore, as with BIC and DIC, models with small $WAIC$ are preferred because they are simple and fit well.

Selecting a random effect model for the Gambia data: As in Section 6.2, the binary response Y_i indicates that child i tested positive for malaria, and we consider covariates for age, bed-net use, bed-net treatment, greenness and health center. In this analysis we also consider child i's village $v_i \in \{1, ..., 65\}$ to account for dependence in the malaria status of children from the same village (Figure 6.3 plots the location and number of children sampled from each village). We use the random effects logistic regression model

$$\text{logit}[\text{Prob}(Y_i = 1)] = \alpha + \sum_{j=1}^{p} X_{ij}\beta_j + \theta_{v_i}, \tag{6.16}$$

where θ_v is the random effect for village v. We compare three models for the village random effects via DIC and $WAIC$:

1. No random effects: $\theta_v = 0$
2. Gaussian random effects: $\theta_v \sim \text{Normal}(0, \tau^2)$
3. Double-exponential random effects: $\theta_v \sim \text{DE}(0, \tau^2)$

In all models, the priors are $\alpha, \beta_j \sim \text{Normal}(0, 100)$ and $\tau^2 \sim \text{InvGamma}(0.1, 0.1)$.

Code for Model 3 is given in Listing 6.2. DIC is computed using the `dic.samples` function in `JAGS`, although this unfortunately requires extra MCMC sampling. There is no analogous function for $WAIC$ in `JAGS` and so it must be computed outside of `JAGS`. In Listing 6.2 the extra line in the likelihood `like[i]` instructs `JAGS` to return posterior samples of the likelihood function $f(Y_i|\boldsymbol{\theta})$ which in this model is the binomial PMF (`dbin` in `JAGS`).

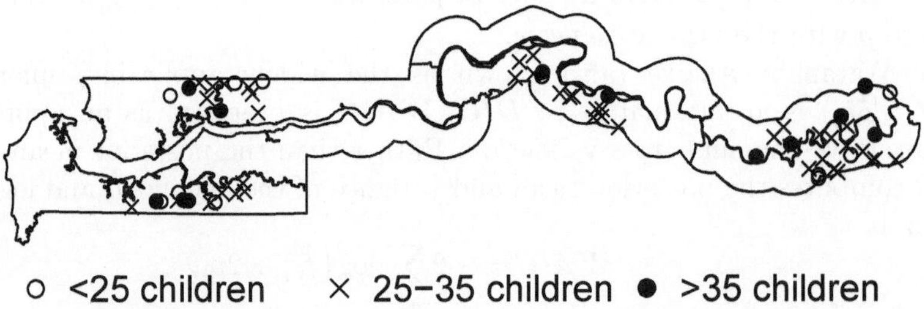

FIGURE 6.3
Gambia data. The location and number of children sampled from each village.

Listing 6.2

JAGS code to compute WAIC and DIC for the random effects model.

```
mod <- textConnection("model{
  for(i in 1:n){
    Y[i]          ~ dbern(pi[i])
    logit(pi[i]) <- beta[1] + X[i,1]*beta[2]
                    + X[i,2]*beta[3] + X[i,3]*beta[4]
                    + X[i,4]*beta[5] + X[i,5]*beta[6]
                    + theta[village[i]]
    like[i]      <- dbin(Y[i],pi[i],1) # For WAIC computation
  }
  for(j in 1:6){beta[j] ~ dnorm(0,0.01)}
  for(j in 1:65){theta[j] ~ ddexp(0,tau)}
  tau   ~ dgamma(0.1,0.1)
}")

data   <- list(Y=Y,X=X,n=n,village=village)
model  <- jags.model(mod,data = data, n.chains=2,quiet=TRUE)
update(model, 10000, progress.bar="none")
samps  <- coda.samples(model, variable.names=c("like"),
                       n.iter=50000, progress.bar="none")

# Compute DIC
DIC    <- dic.samples(model,n.iter=50000,progress.bar="none")

# Compute WAIC
like   <- rbind(samps[[1]],samps[[2]]) # Combine the two chains
fbar   <- colMeans(like)
Pw     <- sum(apply(log(like),2,var))
WAIC   <- -2*sum(log(fbar))+2*Pw
```

After MCMC sampling, the posterior mean of `like`[i] is computed as the approximation to $\bar{f}_i$ and the posterior variance of `log(like`[i]) is computed to approximate p_W.

The *WAIC* and *DIC* results are in Table 6.3. Both measures show strong support for including village random effects, but cannot distinguish between Gaussian and double-exponential random-effect distributions. Since the Gaussian model is more familiar, this is probably the preferred model for these data.

2016 Presidential Election example: The data for this analysis come from Tony McGovern's very useful data repository [1]. The response variable, Y_i, is the percent increase in Republican (GOP) support from 2012 to 2016, i.e.,

$$100\left(\frac{\%\text{ in } 2016}{\%\text{ in } 2012} - 1\right), \tag{6.17}$$

in county $i = 1, ..., n$ (Figure 6.4a). The election data are matched with $p = 10$ county-level census variables (X_{ij}) obtained from Kaggle via Ben Hamner[2]:

1. Population, percent change – April 1, 2010 to July 1, 2014
2. Persons 65 years and over, percent, 2014
3. Black or African American alone, percent, 2014

[1] `https://github.com/tonmcg/County_Level_Election_Results_12-16`
[2] `https://www.kaggle.com/benhamner/2016-us-election`

TABLE 6.3
Model selection criteria for the Gambia data. DIC (p_D) and $WAIC$ (p_W) for the three random effects models.

Random-effects model	DIC (p_D)	$WAIC$ (p_W)
None	2526 (6.0)	2525 (6.0)
Gaussian	2333 (55.1)	2333 (53.4)
Double-exponential	2334 (57.1)	2333 (54.3)

4. Hispanic or Latino, percent, 2014
5. High school graduate or higher, percent of persons age 25+, 2009–2013
6. Bachelor's degree or higher, percent of persons age 25+, 2009–2013
7. Homeownership rate, 2009–2013
8. Median value of owner-occupied housing units, 2009–2013
9. Median household income, 2009–2013
10. Persons below poverty level, percent, 2009–2013.

All covariates are centered and scaled (e.g., Figure 6.4b). The objective is to determine the factors that are associated with an increase in GOP support. Also, following the adage that "all politics is local," we explore the possibility that the factors related to GOP support vary by state.

For a county in state s, we assume the linear model

$$Y_i = \beta_{0s} + \sum_{j=1}^{p} X_i\beta_{sj} + \varepsilon_i, \tag{6.18}$$

where β_{js} is the effect of covariate j in state s and $\varepsilon_i \overset{iid}{\sim} \text{Normal}(0, \sigma^2)$. We compare three models for the β_{js}

1. Constant slopes: $\beta_{js} \equiv \beta_j$ for all counties
2. Varying slopes, uninformative prior: $\beta_{js} \overset{iid}{\sim} \text{Normal}(0, 10^2)$
3. Varying slopes, informative prior: $\beta_{js} \overset{indep}{\sim} \text{Normal}(\mu_j, \sigma_j^2)$

In all models, the prior for the error variance is $\sigma^2 \sim \text{InvGamma}(0.1, 0.1)$. In the first model the slopes have uninformative priors $\beta_j \sim \text{Normal}(0, 10^2)$. In the final model, the mean μ_j and variances σ_j^2 are given priors $\mu_j \sim \text{Normal}(0, 10^2)$ and $\sigma_j^2 \sim \text{InvGamma}(0.1, 0.1)$ and estimated from the data so that information is pooled across states via the prior. The three methods are compared using DIC and WAIC as in Listing 6.3 for the constant slopes model.

We first study the results from Model 1 with the same slopes in all states. Table 6.4 shows that all covariates other than home-ownership rate are associated with the election results. GOP support tended to increase in counties with decreasing population, high proportion of seniors and high school graduates, low proportions of African Americans and Hispanics, high income but low home value and high poverty rate.

The DIC ($\bar{D}$, p_D) for the three models are 21312 (21300, 12) for Model 1 with constant slopes, 18939 (18483, 455) for Model 2 with varying slope and uninformative priors, and 18842 (18604, 238) for Model 3 with varying slopes and informative priors. The first model is not rich enough to capture the important trends in data and thus has high $\bar{D}$ and DIC.

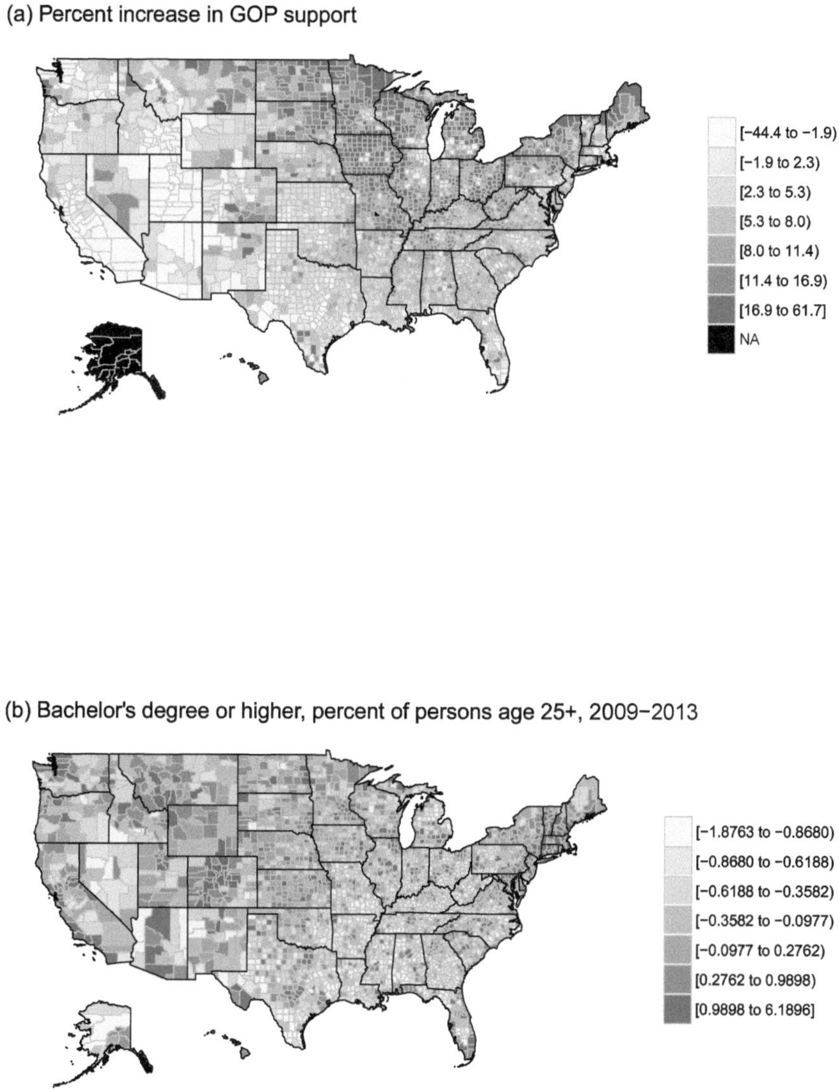

FIGURE 6.4
2016 Presidential Election data. Panel (a) plots the percentage change in Republican (GOP) support from 2012 to 2016 (Y_i) and Panel (b) plots the percent of counties with a bachelor's degree or higher (standardized to have mean zero and variance one; X_{i7}).

Listing 6.3
JAGS code to compute DIC for the constant slopes model.

```
model_string <- "model{
  for(i in 1:n){
     Y[i]      ~ dnorm(Xb[i],taue)
     Xb[i]    <- inprod(X[i,],beta[])
     like[i] <- dnorm(Y[i],Xb[i],taue) # For WAIC
  }
  for(j in 1:p){beta[j] ~ dnorm(0,0.01)}
  taue ~ dgamma(0.1,0.1)
}"

# Put the data and model statement JAGS format
dat   <- list(Y=Y,n=n,X=X,p=p)
init  <- list(beta=rep(0,p),tau=1)
model <- jags.model(textConnection(model_string),n.chains=2,
                    inits=init,data = dat)

# Burn-in samples
update(model, 10000, progress.bar="none")

# Compute DIC
dic <- dic.samples(model1,n.iter=50000)

# Compute WAIC
samp <- coda.samples(model, variable.names=c("like"),
        n.iter=50000)
like <- rbind(samps[[1]],samps[[2]]) # Combine the two chains
fbar <- colMeans(like)
Pw   <- sum(apply(log(like),2,var))
WAIC <- -2*sum(log(fbar)) + 2*Pw
```

TABLE 6.4
2016 Presidential Election multiple regression analysis. Posterior mean (95% interval) for the slopes β_j for the model with the same slopes in each state.

Covariate	Median	95% interval
Population change	−1.14	(−1.46, −0.81)
Percent over 65	0.93	(0.54, 1.32)
Percent African American	−1.56	(−1.89, −1.23)
Percent Hispanic	−2.06	(−2.40, −1.72)
Percent HS graduate	1.75	(1.25, 2.26)
Percent bachelor's degree	−6.19	(−6.71, −5.67)
Home-ownership rate	0.01	(−0.39, 0.41)
Median home value	−1.52	(−1.99, −1.06)
Median income	1.88	(1.13, 2.61)
Percent below poverty	1.48	(0.91, 2.04)

The second model has the best fit to the observed data (smallest $\bar{D}$), but is too complicated (large p_D). The final model balances model complexity and fit and has the smallest DIC. WAIC gives similar results, with $WAIC$ (p_W) equal to 21335 (20), 18971 (406) and 18909 (259) for Models 1–3, respectively.

Inspection of the posterior of the variances σ_j^2 from Model 3 shows that the covariate effect that varies the most across states is the proportion of the county with a dachelor's degree. Figure 6.5 maps the posterior mean and standard deviation of the state-level slopes, β_{s7}. The association between the proportion of the population with a bachelor's degree and change in GOP support is the strongest (most negative) in New England and the Midwest. The posterior standard deviation is the smallest in Colorado and Texas, possibly because these states have high variation in the covariate across counties.

Simulation study to evaluate the performance of DIC **and** $WAIC$: In these examples DIC and $WAIC$ gave similar results. However, in practice there will obviously be cases where they differ, and the user will have to choose between them and defend their choice. Also, when applied to real data as above the "correct" model is unknown, and so we cannot say for certain that either method selected the right model. One way to build trust in these criteria (or any other statistical method) is to evaluate their performance for simulated data where the correct model is known (see Section 9.3 for more discussion of simulation studies).

The data for this simulation experiment are generated to mimic the Gambia data. The response Y_i is binary and generated from the random effects logistic regression model

$$\text{logit}[\text{Prob}(Y_i = 1)] = \alpha + X_i\beta + \theta_{v_i},$$

where v_i is the village index of observation i and $\theta_v \sim \text{Normal}(0, \sigma^2)$ is the random effect for village v. Data are generated with $n = 100$ observations, ten villages each with ten observations, $\alpha = 0$, $\beta = 1$ and $X_i \stackrel{iid}{\sim} \text{Normal}(0, 1)$. We vary the random effect variance σ^2 to determine how large it must be before the model selection criteria consistently favor the random effects model.

For each value of σ we generate $N = 100$ datasets. For each simulated data set we fit two models: (1) the model without random effects (i.e., $\theta_v = 0$) and (2) the full model with random effects and $\theta_v \sim \text{Normal}(0, \sigma^2)$. The priors are $\alpha, \beta \sim \text{Normal}(0, 100)$ and $\sigma^2 \sim \text{InvGamma}(0.1, 0.1)$. For each model and each simulated data set we generate two MCMC chains of length 10,000 (after a burn-in of 1,000) iterations and compute both DIC and $WAIC$. Say DIC_{mj} and $WAIC_{mj}$ are the criteria for model $m \in \{1, 2\}$ for simulated dataset $j \in \{1, ..., N\}$. Figure 6.6 plots the N samples of $DIC_{2j} - DIC_{1j}$ (left) and $WAIC_{2j} - WAIC_{1j}$ (right). The random effects model is selected if these differences are negative, and so the proportion of samples that are negative is the estimated probability of selecting the random effects model. The percentage of the datasets for which the random effects model is selected is given above each boxplot in Figure 6.6.

When data are generated with $\sigma = 0$, then the correct model is the simple model without random effects. In this case, both metrics on average have larger values for the overly complex random effects model and select the random effects model for only 10–20% of the simulated datasets. For data generated with $\sigma > 0$ the random effects model is correct, and both methods reliably select it with probability that increases with σ. Therefore, both methods reliably select the correct model for data generated to mimic the Gambia data. For this particular setting DIC returns the correct model with slightly higher probability than $WAIC$, but of course we cannot generalize this result to other settings.

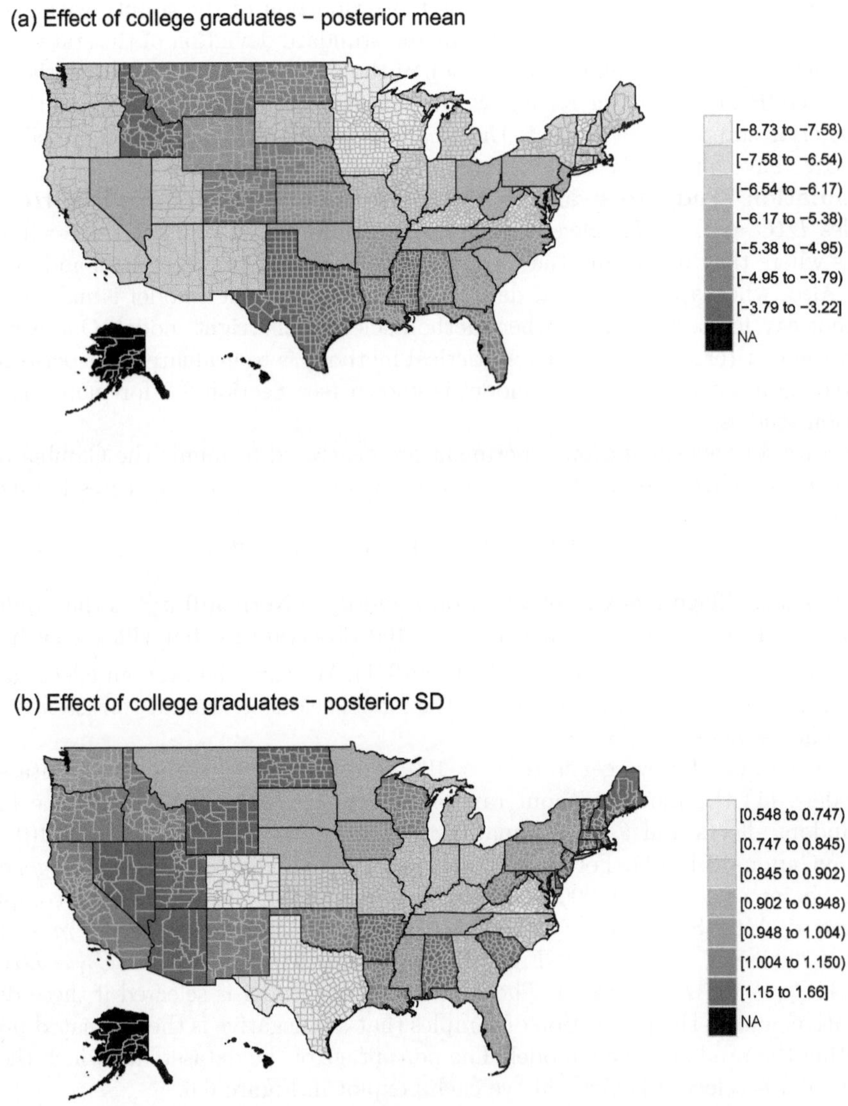

FIGURE 6.5
Results of the 2016 Presidential Election analysis. Posterior mean and standard deviation of the effect of the bachelor's degree rate on GOP support (β_{s7}) for the model with a different slope in each state and informative prior.

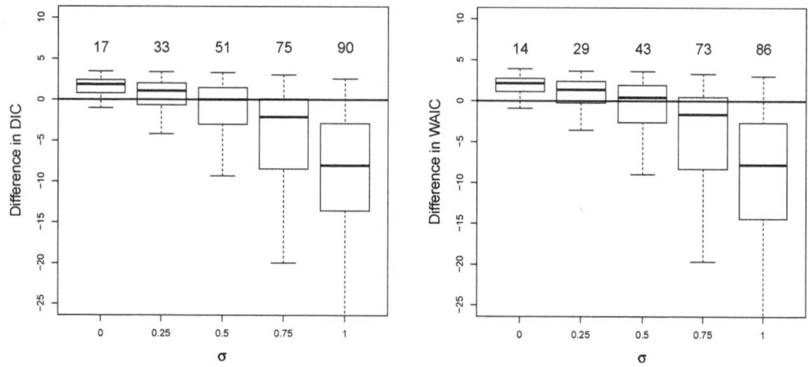

FIGURE 6.6
Simulation to evaluate selection criteria. Boxplots of the difference in DIC (left) and $WAIC$ (right) comparing random effects logistic regression with simple logistic regression for $N = 100$ simulated datasets generated with random effect standard deviation σ. Each boxplot represents the distribution of the difference over N datasets simulated with $\sigma \in \{0.00, 0.25, 0.50, 0.75, 1.00\}$, and the numbers above the boxplots are the percentage of the N datasets for which the difference was negative, and thus the criteria favored the random effects model.

6.5 Goodness-of-fit checks

Thus far we have discussed methods to choose from a prespecified set of models. This model selection step is crucial but cannot guarantee that the selected model actually fits the data well. For example, if all models considered are Gaussian but the data are not, then even the best fitting model is inappropriate. Therefore, in addition to comparing models, diagnostics should be performed to determine if the models capture the important features of the data.

Standard diagnostic tools are equally important to a Bayesian or non-Bayesian analysis. For example, in a linear regression, normality (e.g., residual qq-plot), linearity (e.g., added variable plots), influential points (e.g., Cook's D), etc., should be scrutinized. Many of these classic tools are based on least-squares residuals and therefore are not purely Bayesian, but they remain valuable as informal goodness-of-fit measures.

Another way to critique a model is out-of-sample prediction performance. Say the data are split into a training set $\mathbf{Y}$ and a test set $\mathbf{Y}^* = (Y_1^*, ..., Y_m^*)$. Section 1.5 discusses the Posterior Predictive Distribution (PPD) of the test set observation i given the training data (and averaging over uncertainty in model parameters), $f_i^*(y|\mathbf{Y})$. Comparing the test-set data to the PPD is a way to verify the model fits well. The PPD evaluated at the observed test observation,

$$CPO_i = f_i^*(Y_i^*|\mathbf{Y}), \tag{6.19}$$

is called the conditional predictive ordinate (CPO) [52, 122]. Test set observations with small CPO do not fit the model well and are potentially outliers.

A more interpretable diagnostic is to check that roughly 95% of the test-set observations fall in the 95% posterior prediction intervals. For continuous data, the Probability Integral Transform (PIT) statistic [40] provides a measure of fit for the entire predictive distribution rather than just the 95% intervals. The PIT is the posterior predictive probability below the test set value, Y_i^*,

$$PIT_i = \int_{-\infty}^{Y_i^*} f_i^*(y|\mathbf{Y})dy. \tag{6.20}$$

This integral looks daunting, but can be easily approximated as the proportion of the MCMC samples from the PPD that are below the test-set value. Typically PIT_i is computed for each test-set observation, and these statistics are plotted in a histogram. If the model fits well, then the PIT statistics should follow a Uniform(0,1) distribution and the PIT histogram should be flat.

An important consideration when interpreting diagnostic measures is that even a model that appears to be perfectly calibrated (say with uniform PIT statistics) is not necessarily the true model (if there is such a thing). For example, say the data are generated as $Y|X \sim$ Normal$(X, 1)$ with $X \sim$ Normal$(0, 1)$; then the model $Y \sim$ Normal$(0, 2)$ will fit the data perfectly well, but is clearly inferior to a model that includes X as a predictor. Therefore, both model selection and goodness-of-fit testing are important.

Posterior predictive checks: Rather than focusing on predicting individual test set observations, posterior predictive checks (e.g., [58]) evaluate fit using summaries of the dataset. Let $\tilde{\boldsymbol{\theta}}$ be a posterior sample of the model parameters and $\tilde{\mathbf{Y}}$ be a replicate dataset drawn from the model given $\tilde{\boldsymbol{\theta}}$. To facilitate comparisons, the replicate dataset should have the same dimensions as the observed data. For example, in the linear regression model $\mathbf{Y} \sim$ Normal$(\mathbf{X}\boldsymbol{\beta}, \sigma^2\mathbf{I})$, the parameters are $\tilde{\boldsymbol{\theta}} = (\tilde{\boldsymbol{\beta}}, \tilde{\sigma}^2)$, and we would sample

$$\tilde{\mathbf{Y}}|\tilde{\boldsymbol{\theta}} \sim \text{Normal}(\mathbf{X}\tilde{\boldsymbol{\beta}}, \tilde{\sigma}^2\mathbf{I}) \tag{6.21}$$

using the same covariates $\mathbf{X}$ as in the original dataset.

Assume that MCMC produces S posterior samples $\boldsymbol{\theta}^{(1)}, ..., \boldsymbol{\theta}^{(S)}$ and we generate S replicate datasets $\tilde{\mathbf{Y}}_1, ..., \tilde{\mathbf{Y}}_S$ where $\tilde{\mathbf{Y}}_s|\boldsymbol{\theta}^{(s)} \sim p(\mathbf{y}|\boldsymbol{\theta}^{(s)})$. If the model fits well, then $\mathbf{Y}$ and the $\tilde{\mathbf{Y}}_s$ should follow the same distribution, and thus comparing $\mathbf{Y}$ to the distribution of $\tilde{\mathbf{Y}}_s$ provides an evaluation of whether the proposed data-generating model is valid. Summarizing fit for an entire multivariate distribution such as the distribution of $\mathbf{Y}$ is challenging, and so we restrict comparisons to one-number summaries of the data set, $D(\mathbf{Y})$. For example, $D(\mathbf{Y})$ could be the mean or maximum value of the dataset, or any other measure of interest. A visual goodness-of-fit check plots the predictive distribution of the summary statistics, $D(\tilde{\mathbf{Y}}_1), ..., D(\tilde{\mathbf{Y}}_S)$, as a histogram and compares the observed measure $D(\mathbf{Y})$ to this distribution; if the observed value falls far from the center of the distribution, then this is evidence the model does not fit well.

Selecting the summary measure D is clearly important. The most effective summary measures are those that verify modeling assumptions. For example, when fitting the model $Y \sim$ Normal(μ, σ^2), the mean and variance of the data should fall in the predictive distribution of the mean and variance because there are parameters in the model for these summaries. Therefore, selecting D to be the sample mean or variance is not informative. However, the normal model assumes the distribution is symmetric, and there are no parameters in the model to capture asymmetry. Therefore, taking D to be the skewness provides a useful verification that this modeling assumption holds.

The Bayesian p-value is a more formal summary of the posterior predictive check. The Bayesian p-value is the probability under repeated sampling from the fitted model of observing a summary statistic at least as large as the observed statistic. This probability can be approximated using MCMC output as the proportion of the S draws $D(\tilde{\mathbf{Y}}_1), ..., D(\tilde{\mathbf{Y}}_S)$ that are greater than $D(\mathbf{Y})$. A Bayesian p-value near zero or one indicates that in at least one aspect the model does not fit well.

The Bayesian p-value resembles the p-value from classical hypothesis testing in that they both quantify the probability under repeated sampling of observing a statistic at least as large as the observed statistic. An important difference is that the classical p-value assumes repeated sampling under the null hypothesis, whereas the Bayesian p-value assumes repeated sampling from the fitted Bayesian model. Another important difference is that for the Bayesian p-value, probabilities near either zero or one provide evidence against the fitted model.

Gun control example: Kalesan et al. (2016) [86] study the relationship between state gun laws and firearm-related death rates. The response variable, Y_i, is the number of firearm-related deaths in 2010 in state i. The analysis includes five potential confounders (Z_{ij}): 2009 firearm death rate per 10,000 people; firearm ownership rate quartile; unemployment rate quartile; non-firearm homicide rate quartile and firearm export rate quartile. The covariates of interest in the study are status of gun laws in the state. Let X_{il} indicate that state i has law l. In this example, we simply use the number of laws $X_i = \sum_l X_{il}$ as the covariate. Setting aside correlation versus causation issues, the objective of the analysis is to determine if there is a relationship between the number of gun laws in the state and its firearm-related death rate. Our objective in this section is to illustrate the use of posterior predictive checks to verify that the model fits well.

We check the fit of two models. The first is the usual Poisson regression model

$$Y_i \sim \text{Poisson}(\lambda_i) \text{ where } \lambda_i = N_i \exp\left(\alpha + \sum_{j=1}^{5} Z_{ij}\beta_j + X_i\beta_6\right) \tag{6.22}$$

and N_i is the state's population. A concern with the Poisson model is that because the mean equals the variance, it may not be flexible enough to capture large counts. Therefore, we also consider the negative binomial model with mean λ_i and over-dispersion parameter m

$$Y_i \sim \text{NB}\left(\frac{m}{\lambda_i + m}, m\right). \tag{6.23}$$

The priors for the fixed effects are $\alpha, \beta_j \sim \text{Normal}(0, 10)$ and for the negative-binomial model $m \sim \text{Gamma}(0.1, 0.1)$. Code to fit the negative-binomial model is given in Listing 6.4.

We interrogate the models using posterior predictive checks with the following six test statistics:

1. Minimum count: $D_1(\mathbf{Y}) = \min\{Y_1, ..., Y_n\}$
2. Maximum count: $D_2(\mathbf{Y}) = \max\{Y_1, ..., Y_n\}$
3. Range of counts: $D_3(\mathbf{Y}) = \max\{Y_1, ..., Y_n\} - \min\{Y_1, ..., Y_n\}$
4. Minimum rate: $D_4(\mathbf{Y}) = \min\{Y_1/N_1, ..., Y_n/N_n\}$
5. Maximum rate: $D_5(\mathbf{Y}) = \max\{Y_1/N_1, ..., Y_n/N_n\}$
6. Range of rates: $D_6(\mathbf{Y}) = \max\{Y_1/N_1, ..., Y_n/N_n\} - \min\{Y_1/N_1, ..., Y_n/N_n\}$

We use test statistics related to the range of the counts and rates because the main concern here is properly accounting for large counts. Figure 6.7 plots the PPD for both models and all six test statistics. The PPD from the Poisson model does not capture the largest counts or the range of counts observed in the dataset. The Bayesian p-value is close to zero or one for all four test statistics involving either the maximum or the range. As expected, the negative-binomial model gives much wider prediction intervals and thus less extreme

Listing 6.4
JAGS code for the over-dispersed Poisson regression model.

```
# Likelihood
for(i in 1:n){
  Y[i]               ~ dnegbin(q[i],m)
  q[i]              <- m/(m+N[i]*lambda[i])
  log(lambda[i]) <- alpha + inprod(Z[i,],beta[1:5]) + X[i]*beta[6]
}

#Priors
 for(j in 1:6){
     beta[j] ~ dnorm(0,0.1)
 }
 alpha ~ dnorm(0,0.1)
 m     ~ dgamma(0.1,0.1)

# Posterior predictive checks
for(i in 1:n){
  Y2[i]     ~ dnegbin(q[i],m)
  rate[i] <- Y2[i]/N[i]
}

D[1] <- min(Yp[])
D[2] <- max(Yp[])
D[3] <- max(Yp[])-min(Yp[])
D[4] <- min(rate[])
D[5] <- max(rate[])
D[6] <- max(rate[])-min(rate[])
```

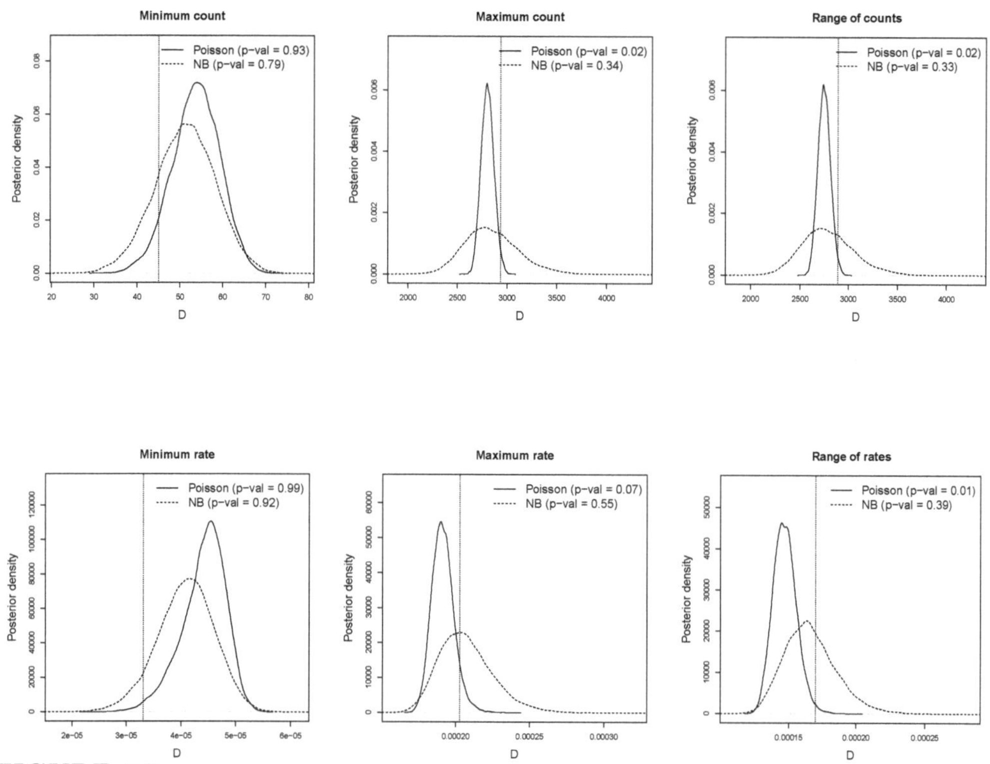

FIGURE 6.7

Bayesian p-values for the gun control example. The density curves show the posterior predictive distribution of the six test statistics for the Poisson and negative-binomial (NB) models, and the vertical lines are the test statistics for the observed data. The Bayesian p-value is given in the legend's parentheses.

Bayesian p-values. Therefore, while the Poisson model is not appropriate for this analysis, the negative-binomial model appears to be adequate based on these (limited) tests. However, in both analyses the coefficient associated with the number of gun laws (β_6) is negative with high probability (95% interval (−0.017,−0.011) for the Poisson model and (−0.026, −0.008) for the negative-binomial model), and so the conclusion that there is a negative association between the number of gun laws in a state and its firearm-related mortality is robust to this modeling assumption.

6.6 Exercises

1. Download the `airquality` dataset in `R`. Compare the following two models using 5-fold cross validation:

   $\mathcal{M}_1$: $\text{ozone}_i \sim \text{Normal}(\beta_1 + \beta_2 \text{solar.R}_i, \sigma^2)$

   $\mathcal{M}_2$: $\text{ozone}_i \sim \text{Normal}(\beta_1 + \beta_2 \text{solar.R}_i + \beta_2 \text{temp}_i + \beta_3 \text{wind}_i, \sigma^2)$.

Specify and justify the priors you select for both models.

2. Fit model $\mathcal{M}_2$ to the `airquality` data from the previous problem, and use posterior predictive checks to verify that the model fits well. If you find model misspecification, suggest (but do not fit) alternatives.

3. Use the "Mr. October" data (Section 2.5) $Y_1 = 563$, $N_1 = 2820$, $Y_2 = 10$, and $N_2 = 27$. Compare the two models:

$$\mathcal{M}_1\text{: } Y_1|\lambda_1 \sim \text{Poisson}(N_1\lambda_1) \text{ and } Y_2|\lambda_2 \sim \text{Poisson}(N_2\lambda_2)$$
$$\mathcal{M}_2\text{: } Y_1|\lambda_0 \sim \text{Poisson}(N_1\lambda_0) \text{ and } Y_2|\lambda_0 \sim \text{Poisson}(N_2\lambda_0).$$

using *DIC* and *WAIC*. Assume the Uniform(0,c) prior for all λ_j, and compare the results for $c = 1$ and $c = 10$.

4. Consider the regression of the (continuous) CES-D survey score of depression onto p genetic markers. The p covariates for subject i, $X_{i1}, ..., X_{ip}$, are aligned along a chromosome giving strong collinearity with the correlation between markers j and k being $\rho^{|j-k|}$ for $\rho = 0.99$.

   (a) Say that $p = 3$ and that these generic markers are in fact very important predictors of depression. What do you anticipate the results of Bayesian linear regression with uninformative priors will be?

   (b) Write a stochastic search variable selection model for these data, and describe how you would summarize the results.

   (c) Assuming the scenario in (a), what do you anticipate the posterior probability of the 2^p models and the p marginal inclusion probabilities will be?

   (d) Repeat (d) except with $p = 1,000$. How would you summarize these results?

5. Use *DIC* and *WAIC* to compare logistic and probit links for the `gambia` data in the `R` package `geoR` using the five covariates in Listing 6.2 and no random effects.

6. Fit logistic regression model to the `gambia` data in the previous question and use posterior predictive checks to verify the model fits well. If you find model misspecification, suggest (but do not fit) alternatives.

7. For the NBA free throw data in Section 1.6, assume that for player i, $Y_i|p_i \sim \text{Binomial}(n_i, p_i)$ where Y_i is the number of clutch makes, n_i is the number of clutch attempts, and p_i is the clutch make probability. Use leave-one-out cross validation to compare the models:

$$\mathcal{M}_1\text{: } \text{logit}(p_i) = \beta_1 + \text{logit}(q_i)$$
$$\mathcal{M}_2\text{: } \text{logit}(p_i) = \beta_1 + \beta_2\text{logit}(q_i),$$

where q_i is the overall free throw percentage.

8. Fit model $\mathcal{M}_2$ to the NBA data in the previous question, and use posterior predictive checks to verify the model fits well. If you find model misspecification, suggest (but do not fit) alternatives.

9. Download the `WWWusage` dataset in `R`. Using data from times $t = 5, ..., 100$ as outcomes (earlier times may be used as covariates), fit the autoregressive model

$$Y_t|Y_{t-1}, ..., Y_1 \sim \text{Normal}(\beta_0 + \beta_1 Y_{t-1} + ... + \beta_L Y_{t-L}, \sigma^2)$$

where Y_t is the WWW usage at time t. Compare the models with $L = 1, 2, 3, 4$, and select the best time lag L.

10. Using the `WWWusage` dataset in the previous problem, fit the model with $L = 2$ and use posterior predictive checks to verify that the model fits well. If you find model misspecification, suggest (but do not fit) alternatives.
11. For the one-way random effects model in (6.12), explain how you would select a prior for c so the p_D has roughly a uniform prior on p_D.
12. The global mean temperature anomaly (difference between current year and the 1951-1980 average) data on the course website are downloaded from `https://data.giss.nasa.gov/gistemp/` and plotted below.

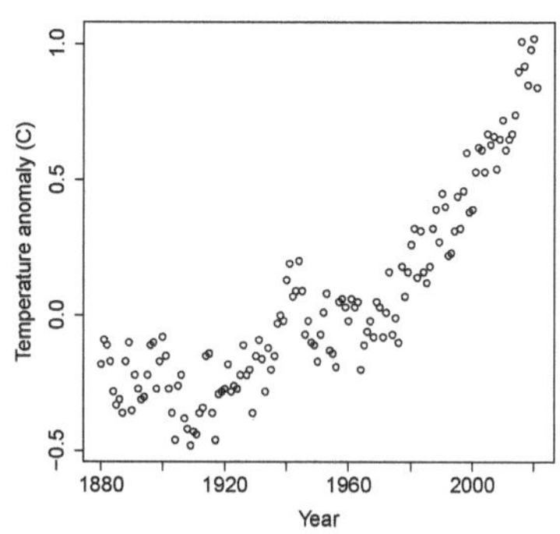

Let X_i and Y_i be the year and temperature anomaly (C) for observation i. We will analyze these data using a nonparametric regression model $Y_i = g(X_i) + \varepsilon_i$ as in Section 8.3, where g is modeled as a linear combination of J B-spline basis functions.

(a) Select J using DIC and WAIC.

(b) Select J using cross-validation.

(c) For the selected value of J, check model fit using posterior predictive checks.

13. Open and plot the galaxies data in `R` using the code below,

```
> library(MASS)
> data(galaxies)
> ?galaxies
> Y <- galaxies
> n <- length(Y)
> hist(Y,breaks=25)
```

Model the observations $Y_1, ..., Y_{82}$ using the Student's t-distribution with location μ, scale σ and degrees of freedom k. Assume prior distributions $\mu \sim \text{Normal}(0, 10000^2)$, $1/\sigma^2 = \tau \sim \text{Gamma}(0.01, 0.01)$ and $k \sim \text{Uniform}(1, 30)$.

(a) Use posterior predictive checks to evaluate whether the t distribution captures the mean, variance, skewness and kurtosis of the data.

(b) Repeat this analysis with a Gaussian model.

14. For each model selection approach below, give an advantage and a limitation of the approach and a hypothetical situation where it would be a good option.

    (a) Bayes factors
    (b) Stochastic search variable selection
    (c) Cross validation
    (d) DIC/WAIC

7

Case studies using hierarchical modeling

CONTENTS

7.1 Overview of hierarchical modeling

Thus far we have introduced Bayesian ideas in the context of standard statistical models. However, one of the primary benefits of Bayesian methodology is flexibility to handle non-standard cases with irregularities such as missing values, censored data, variables measured with error, multiple data sources each with distinct biases and errors, subpopulations with different properties, etc. Incorporating all of these features in an analysis may seem daunting, but can often be accomplished by breaking the large model into manageable layers and combining the layers in a hierarchical model. Hierarchical modeling (also known as a multilevel modeling) is thus an essential model-building tool.

To see how building a model for complex data can be simplified by thinking hierarchically, say our objective is to specify the joint distribution of three variables X, Y and Z. Directly fitting a multivariate joint distribution can be challenging, especially if the three variables have different supports. However, any trivariate distribution can be written

$$f(x,y,z) = f(x)f(y|x)f(z|x,y). \tag{7.1}$$

By ordering the three variables and specifying the univariate marginal distribution of X and then the univariate conditional distributions for $Y|X$ and $Z|X,Y$, the multivariate problem is reduced to three univariate problems. Because the variables are ordered and each conditional distribution depends only on the previous variables in the ordering, the resulting joint distribution is guaranteed to be valid. Also, since any multivariate distribution can be decomposed this way, there is no loss of flexibility by taking this approach.

FIGURE 7.1
Directed acyclic graphs (DAGs). Panel (a) shows the DAG for the model $f(X, Y, Z) = f(X)f(Y|X)f(Z|X, Y)$ and Panel (b) shows the DAG for the model $f(X, Y, Z) = f(X)f(Y|X)f(Z|Y)$.

The trivariate model is represented as a Directed Acyclic Graph (DAG) in Figure 7.1a. A DAG (also called a Bayesian network) represents the model as a graph with each observation and parameter as a node (i.e., the points that define the graph) and edges (i.e., connections between the nodes) to denote conditional dependence. To define a valid stochastic model the graph must be directed and acyclic. A directed graph associates each edge with a direction; an arrow from X to Y indicates that the hierarchical model is defined by modeling the conditional distribution of Y as a function of X. The absence of an arrow from X to Z in Figure 7.1b conveys the choice that conditioned on Y, Z does not depend on X, i.e., $f(z|x, y) = f(z|y)$; in contrast, Figure 7.1a is the DAG for the model with conditional dependence between X and Z given Y. The graph must also be acyclic, meaning that it is impossible to follow directed edges from a node through the graph and return to the original node. These two conditions rule out building models such as $p(x, y, z) = p(x|y, z)p(y|z)p(z|y)$, which may not be a valid joint distribution.

Hierarchical models can take many forms, but a general way to build a model is through creating a data layer, a process layer and a prior layer. Model building should begin with the *process layer* that contains the underlying scientific processes of interest and the unknown parameters. Building this layer is ideally done in consultation with domain experts. Once this layer is defined, the statistical objectives can be articulated, for example, to estimate a particular parameter or test a specific hypothesis. Ideally these objectives dictate the data to be collected for the analysis. The *data layer* relates (via the likelihood function) the data to the process and encodes bias and error in the data collection procedure, which requires knowledge of how the data were collected. Finally the *prior layer* quantifies uncertainty about the model parameters at the onset of the analysis.

Building a model hierarchically is convenient, but not fundamentally different than models we have considered previously. In fact, we have already encountered many hierarchical models such as the random effects models in Section 4.4. This means that the computational methods to approximate the posterior described in Chapter 3 apply to hierarchical models, as do the graphical and numerical methods used to summarize the posterior and check fit. Below we discuss several categories of hierarchical models and the application of Bayesian methods in each case. We then illustrate these concepts with several case studies in Section 7.2.

7.1.1 Mathematical/statistical models

In many scientific disciplines, the process of interest is described mathematically as a set of differential equations. Differential-equation models use the current state of all variables and the physical or biological laws to determine the dynamics of the system. Mathematical models are fundamental to many applications, including the

- track of a hurricane through the Atlantic ocean,
- strength of a bridge under a given design,
- formation and spread of cancer cells,
- population dynamics of an endangered species.

Ideally, it can be assumed that the mathematical model is a reasonable representation of the system, but is imperfect due to discretization or simplification, and there is uncertainty because the initial and current states of the system and model parameters are not known precisely.

Consider modeling the spread of a wildland fire [49]. The mathematical/differential equations model would describe the dispersion of the fire as a function of the topology, terrain, meteorology, etc. There are many reasons forecasts from the model will have error. Even if the equations perfectly describe the system's dynamics, which is unlikely, in practice they would need to be discretized because of computational limitations or the spatial resolution of model inputs such as satellite imagery. Also, the equations are contingent on the current state of the system including the location and intensity of the fire, the condition of the surrounding vegetation and the current wind speed and direction. The model also depends on unknown parameters, such as the likelihood of spread into a high-elevation pine-dominant forest. Neither the current state nor model parameters can be known exactly.

Statistical analyses that account for model bias, unknowns and error are called uncertainty quantification (UQ; [143]). A UQ analysis can improve predictions by calibrating mathematical models with real-world data, communicate uncertainty to stakeholders to improve decision making, improve efficiency by optimizing future data collection for model refinement and formally test for differences between future scenarios while accounting for all sources of uncertainty. A Bayesian approach [90] is natural for these tasks because it is able to seamlessly incorporate prior knowledge when available and account for multiple sources of uncertainty.

As an introduction, consider the simple disease progression model below to illustrate the model-building process. Let S_t and I_t be the number of susceptible and infected individuals in a population, respectively, at time t. Scientific understanding of the disease is used to model disease propagation. The change of S_t and I_t over time can be described by differential equations [23]. Since our data are discretized to daily counts, we consider a difference equation. Many compartmental models have been proposed including the SIER model, but in consultation with an epidemiologist, we might select the simple Reed–Frost model [3]:

$$\begin{aligned}\text{Process layer:}\qquad & I_{t+1}|S_t, I_t \sim \text{Binomial}\left[S_t, 1-(1-q)^{I_t}\right]\\ & S_{t+1} = S_t - I_{t+1}\end{aligned}$$

where it is assumed that all infected individuals are removed from the population before the next time step and q is the probability of a non-infected person coming into contact with and contracting the disease from an infected individual.

The epidemiological process-layer model expresses the disease dynamics up to a few unknown state variables, I_1 and S_1, and an unknown parameter, q. To estimate these

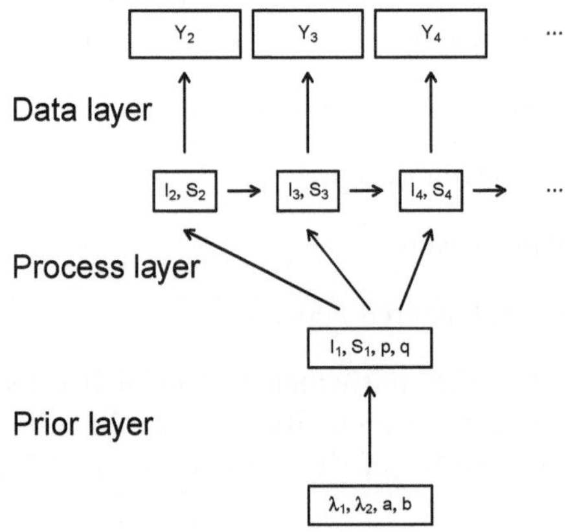

FIGURE 7.2
Directed acyclic graph for the Reed–Frost infectious disease model.

unknowns, the number of cases at time t, denoted Y_t, is collected. The data layer models our ability to measure the process I_t. For example, after discussing the data collection procedure with domain experts, we might assume there are no false positives (uninfected people counted as infected) but potentially false negatives (uncounted infected individuals) and thus

$$\text{Data layer: } Y_t|I_t \sim \text{Binomial}(I_t, p) \tag{7.2}$$

where p is the probability of detecting an infected individual. The Bayesian model is completed using priors for the unknown state variables and parameters

$$\begin{aligned} \text{Prior layer:} \quad & I_1 \sim \text{Poisson}(\lambda_1), S_1 \sim \text{Poisson}(\lambda_2) \\ & p, q \sim \text{beta}(a, b). \end{aligned}$$

Figure 7.2 plots the DAG corresponding to this model.

The Reed-Frost model is very simple and unlikely to provide a perfect representation of the system's dynamics. If diagnostics call the fit of the model into question, a remedy is to select a more sophisticated model, such as the SIER model mentioned above. However, we can almost never be certain that the chosen model has complete fidelity to the true dynamics; thus we should allow for this bias in the hierarchical model. Discrepancy terms can enter either in the data or process layers. Here we add the discrepancy term, $d_t > 0$, to the data layer

$$\text{Data layer with discrepancy: } Y_t|I_t, d_t \sim \text{Binomial}(I_t d_t, p). \tag{7.3}$$

In this parameterization, $d_t = 1.1$ means that the number of infected individuals is 10% higher at time t in reality than in the current state of the model. Of course, if the discrepancy terms are allowed to vary freely across t then it will absorb all the signal in the data and true number of infected individuals cannot be estimated. Identifiability concerns can be lessened by taking $\log(d_t) = \sum_{k=0}^{K} \alpha_k t^k$ as a low-order polynomial function of t and estimating the discrepancy parameters α_k in the Bayesian hierarchical model.

The Reed-Frost model is simple, and an analysis of these data using Bayesian methods can utilize standard MCMC methods. However, more complex mathematical models that are defined as the solution to a set of differential equations pose computational challenges because evaluating the model even a single time can take several hours. Section 8.6.5 shows how machine learning tools can be used to improve computation and thus facilitate a Bayesian analysis of complex mathematical models. This section also discusses Bayesian machine learning tools to learn the equations that govern the process from data rather than specifying models based on domain knowledge.

7.1.2 State-space models

State-space models (SSMs) are a class of hierarchical models that are used to study the evolution of processes over time and make short-term forecasts. The process layer defines the evolution of latent (unobserved) states that are related to the observations in the data layer. Below we discuss two examples of SSMs: the Dynamic Linear Model (DLM) and the Hidden Markov Model (HMM). Both models are described assuming time is discretized, so that the process is defined only at times $t \in \{1, \ldots, n\}$, and the latent space variables follow a Markov process where the distributions of the states are independent of their history given the previous value. Extensions to continuous time and richer temporal dynamics are available, but more complex. The nature of the latent states distinguishes the two models with the DLM modeling them as continuous random variables whereas in HMM models the latent states as discrete random variables.

Dynamic linear models: In the DLM [163], the q latent states at time t, $\mathbf{X}_t = (X_{t1}, ..., X_{tq})$, evolve following the linear model

$$\mathbf{X}_t = \mathbf{A}\mathbf{X}_{t-1} + \mathbf{u}_t$$

where $\mathbf{A}$ $(q \times q)$ determines the evolution of the latent states and $\mathbf{u}_t \overset{iid}{\sim} \text{Normal}(0, \mathbf{\Sigma}_u)$. The latent variables are related to the p response variables, $\mathbf{Y}_t = (Y_{t1}, ..., Y_{tp})$, as

$$\mathbf{Y}_t = \mathbf{B}\mathbf{X}_t + \mathbf{v}_t$$

where $\mathbf{B}$ $(p \times q)$ links the latent states with the outcome and $\mathbf{v}_t \overset{iid}{\sim} \text{Normal}(0, \mathbf{\Sigma}_v)$.

As a concrete example, say the responses are p stock prices that are thought to be driven by q latent economic factors. As another example, say the outcomes are a patient's vital signs (heart rate, blood pressure, etc.) and the latent factors are facets of heart function. In both cases, it is assumed that the responses are correlated over time and with each other, and most of this correlation can be explained by variation in the latent factors. An advantage of modeling dependence between the outcomes using the latent variables is the efficiency gain that comes if a small number of latent factors $(q < p)$ can explain the dependence.

If data are available through time n, sampling from the posterior prediction distribution (PPD) of $\mathbf{Y}_{n+1}$ is straightforward using MCMC. Let the values for iteration s be denoted with a superscript, e.g., $\mathbf{X}_t^{(s)}$. Then a sample from the PPD is generated as $\mathbf{X}_{t+1}^{(s)} \sim \text{Normal}\left(\mathbf{A}^{(s)}\mathbf{X}_t^{(s)}, \mathbf{\Sigma}_u^{(s)}\right)$ and then $\mathbf{Y}_{t+1}^{(s)} \sim \text{Normal}\left(\mathbf{B}^{(s)}\mathbf{X}_{t+1}^{(s)}, \mathbf{\Sigma}_v^{(s)}\right)$. The distribution of the samples $\mathbf{Y}_{t+1}^{(1)}, ..., \mathbf{Y}_{t+1}^{(S)}$ account for uncertainty in the latent states as well as the parameters ($\mathbf{A}$, $\mathbf{B}$, etc). While MCMC provides a straightforward method for short-term prediction, it can be slow because sampling requires updating the latent process $\mathbf{X}_t$ for all n training observations. More efficient methods are available, including the Kalman filter [32].

Label switching [83] poses a challenge when fitting and interpreting SSMs (both DLMs and HMMs). Unless there is some prior knowledge to distinguish the latent factors, changing the labels of the model has no effect on fit. As an example, say $q = 2$ and $\mathbf{X}_t = (X_{t1}, X_{t2})$ is changed to $\mathbf{X}_t = (X_{t2}, X_{t1})$ and the corresponding rows of $\mathbf{A}$, $\mathbf{B}$ and $\mathbf{\Sigma}_u$ are swapped; then the posterior is not changed. One way to resolve the label-switching problem is to insert known or artificial constraints to ensure the latent states are identified. In the casino example below, the distribution of one of the two latent states is known, and this information resolves label switching. If nothing is known about the latent states, then elements of $\mathbf{A}$, $\mathbf{B}$ or $\mathbf{\Sigma}_u$ can be fixed. A common constraint is that $\mathbf{A}$ is lower-triangular with positive diagonal elements. Another challenge is selecting the number of latent states; model-fit criteria such as DIC, WAIC or cross validation can be used.

Hidden Markov models: HMMs [174] have a similar structure and interpretation as DLMs except for the key difference that the latent process $X_t \in \{1, ..., q\}$ is discete with only q levels. This adds the interpretation that the latent state represents different modes of behaviour. For example, in animal movement modeling where $\mathbf{Y}_t$ is the speed and direction of an animal at time t, the latent state may have $q = 3$ levels: $X_t = 1$ indicates the animal is resting, $X_t = 2$ that they are hunting and $X_t = 3$ that they are in transit. Clearly the distribution of the response will differ based on the current state. The objectives of an HMM analysis include identifying the latent states and understanding the temporal process in each state.

The latent state X_t is assumed to be a Markov process; that is, given all previous states $X_1, ..., X_{t-1}$, X_t depends only on X_{t-1}. Under this assumption, its distribution is completely determined by the $q \times q$ transition probability matrix with elements

$$P_{ij} = \text{Prob}(X_t = j | X_{t-1} = i).$$

Given the previous state is $X_{t-1} = i$, the PMF of the current state is given by the q probabilities $\mathbf{P}_i = (P_{i1}, ..., P_{iq})$. The conjugate prior for a vector of probabilities is a Dirichlet prior (Section 2.1) $\mathbf{P}_i \sim \text{Dirichlet}(a_{i1}, ..., a_{iq})$. Alternatively, if there are external covariates that are thought to affect the transition probabilities, then $\mathbf{P}_i$ can be modeled using a multinomial logistic regression model (Section 4.3.2).

The parameters in the response distribution are determined by the current state. If the data have support on the real line, then the simplest model is

$$\mathbf{Y}_t | X_t = k \overset{indep}{\sim} \text{Normal}(\boldsymbol{\mu}_k, \mathbf{\Sigma})$$

where $\boldsymbol{\mu}_k = (\mu_{k1}, ..., \mu_{kp})^T$ and $\mathbf{\Sigma}$ is the $p \times p$ covariance matrix. This model can be extended in many ways, including a state-specific covariance, covariates in the mean with state-specific regression coefficients, dependence in the response over time (called a regime-switching model), or a generalized linear model with state-specific coefficients for non-Gaussian data. If the number of states is small, then it is reasonable to use uninformative priors for the state-specific parameters; they can also be treated as random effects if the number of latent states is large.

Dishonest casino example: The "dishonest casino" data are from the `R` package `aphid` [164]. The data consist of $n = 300$ rolls of a die. The assumption is that there are $q = 2$ latent states: one is a fair die with equal probability on all six outcomes, and the other is a rigged die with unequal (and unknown) probabilities. The dealer switches in the rigged die at times during the game, and the goal is to compute the posterior probability that the rigged die is being used at a given time. The data are plotted in Figure 7.3 (top). The sample proportions of the six outcomes are 0.13, 0.14, 0.16, 0.12, 0.14 and 0.30, so there is a substantially higher proportion of sixes than expected under a fair die.

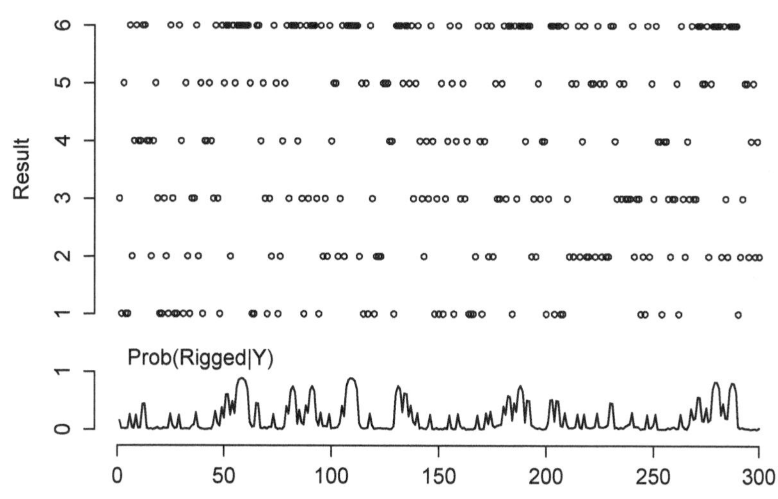

FIGURE 7.3
Hidden Markov Model for the "dishonest casino". Plot of the results (1-6) of each roll of the die and (bottom) the posterior probability of being in the rigged state. The results are plotted as solid circles for roll number with at least 0.5 probability of being in the rigged state.

Let $Y_t \in \{1, ..., 6\}$ be the result of roll number $t \in \{1, ..., n\}$. The HMM is defined through the latent state at time t that takes values $X_t = 1$ if the fair die is in play and $X_t = 2$ if the rigged die is in play. The distribution of Y_t is $\text{Prob}(Y_t = y|X_t = 1) = \frac{1}{6}$ under the fair die and $\text{Prob}(Y_t = y|X_t = 2) = p_y$ under the rigged die. The distribution of the latent states is determined by the transition probability matrix with $P_{ij} = \text{Prob}(X_t = j|X_{t-1} = i)$. Uninformative Dirichlet priors are given to the rigged die probabilities $(p_1, ..., p_6) \sim \text{Dirichlet}(1/6, ..., 1/6)$ and the rows of the transition probability matrix $(P_{11}, P_{12}) \sim \text{Dirichlet}(1/2, 1/2)$ and $(P_{21}, P_{22}) \sim \text{Dirichlet}(1/2, 1/2)$. Note that since there are only two states and one (fair die) has known data distribution, the model does not suffer from the label-switching issue discussed above.

Under the rigged state, the posterior mean (standard deviation) of the probability of rolling a six is 0.80 (0.13) and less than 0.10 for the other five outcomes, so the rigged die appears to be weighted towards a six. The posterior probabilities of being in the rigged state at time t, $\text{Prob}(X_t = 2|Y_1, ..., Y_n)$, are plotted in Figure 7.3 (bottom). There are some intervals, e.g., around roll number 60, with a high proportion of sixes and thus a high likelihood of the rigged die being in play. The posterior mean (standard deviation) of the elements of the transition probability matrix are

$$\begin{pmatrix} 0.89\ (0.05) & 0.11\ (0.05) \\ 0.44\ (0.18) & 0.56\ (0.18) \end{pmatrix}.$$

Therefore, given the current state is fair, the probability of transitioning to the rigged die in the next roll is estimated to be 0.11, and given the current state is rigged, the probability of transitioning back to a fair die is 0.44. In summary, the analysis reveals that there is a second die weighted toward sixes that is used for shorter intervals in the game.

Listing 7.1

JAGS model statement for the HMM analysis of the dishonest casino data (p2 is a vector with elements 1/2; p6 is a vector with elements 1/6).

```
# Distribution of the data given the latent state
 for(i in 1:n){
   Y[i] ~ dcat(die_probs[state[i],1:6])
 }

# Prior for the latent state
 state[1]  ~ dcat(p2[1:2])
 for(i in 2:n){
   state[i]  ~ dcat(trans_probs[state[i-1],1:2])
 }

 #Prior distributions
 die_probs[1,1:6] <- p6[1:6]          # Probs for fair die
 die_probs[2,1:6]  ~ ddirch(p6[1:6]) # Probs for loaded die

 trans_probs[1,1:2] ~ ddirch(p2[1:2]) # Probs from fair state
 trans_probs[2,1:2] ~ ddirch(p2[1:2]) # Probs from rigged state
```

7.1.3 Multilevel statistical models

In many applications, physical models of the system of interest are unavailable and are replaced with purely statistical models built on simple relationships, such as linear trends and correlations. Hierarchical modeling is also useful for building rich statistical models. The general idea of building a hierarchical model through the data, process and prior layers remains valid, but with a statistical instead of physical view of the process layer. For example, a purely statistical model to replace the Reed-Frost process in the infectious disease analysis above is $I_{t+1}x|S_t, I_t \sim \text{Binomial}(S_t, q_t)$ with $\text{logit}(q_t) = \beta_0 + \beta_1 I_t$. This process layer lacks the physical interpretation of the Reed-Frost model, but is a valid statistical model that may perform well in practice.

A general idea for building a multilevel model is to split the data into homogeneous groups and then pool information across the groups via the prior distribution. The random slopes model for the jaw bone density data (plotted in Figure 4.8) discussed in Section 4.4 is an example of a hierarchical model built this way. The model is

$$\begin{aligned} \text{Data layer}: &\quad Y_{ij}|\boldsymbol{\beta}_i \sim \text{Normal}(\mathbf{X}_j\boldsymbol{\beta}_i, \sigma^2) \\ \text{Process layer}: &\quad \boldsymbol{\beta}_i|\boldsymbol{\mu}, \boldsymbol{\Sigma} \sim \text{Normal}(\boldsymbol{\mu}, \boldsymbol{\Sigma}) \\ \text{Prior layer}: &\quad \boldsymbol{\mu} \sim \text{Normal}(\mathbf{0}, c^2\mathbf{I}_2), \boldsymbol{\Sigma} \sim \text{InvWishart}(\nu, \boldsymbol{\Omega}) \end{aligned}$$

for $i = 1, ..., n = 20$ and $j = 1, ..., m = 4$. Each of the 20 patients gets their own simple linear regression model, and these regressions are combined in the process layer that specifies the distribution of the regression coefficients across individuals in the patient population.

This model is visualized as a DAG in Figure 7.4. The DAG shows how information moves through the hierarchical model. For example, how does the data from patient 1 help us predict the bone density for patient 2's next visit? To traverse the DAG from $\mathbf{Y}_1$ to $\mathbf{Y}_2$ requires going through the population parameters $\boldsymbol{\mu}$ and $\boldsymbol{\Sigma}$. That is, $\mathbf{Y}_1$ informs the model about $\boldsymbol{\beta}_1$ which shapes the random effects distribution that enters the model for $\boldsymbol{\beta}_2$ and thus $\mathbf{Y}_2$. If we only had data from patient 2, then we would likely resort to an uninformative prior for $\boldsymbol{\beta}_2$ and with only a few observations for patient 2 the posterior would be unstable.

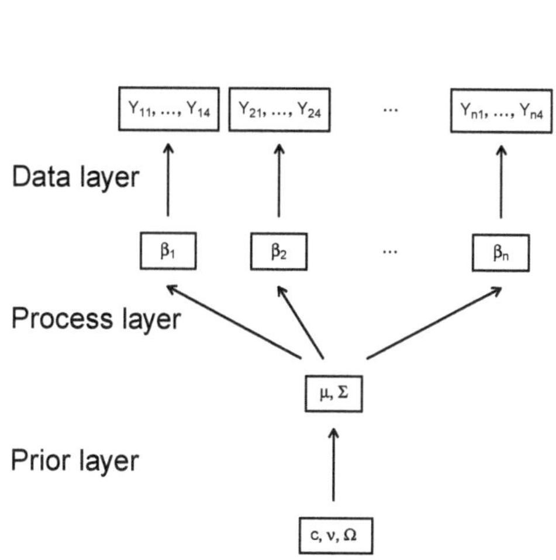

FIGURE 7.4
Directed acyclic graph. Visual representation of the random slopes model $Y_{ij}|\boldsymbol{\beta}_i \sim \text{Normal}(\mathbf{X}_j\boldsymbol{\beta}_i, \sigma^2)$ for $i = 1, ..., n$ and $j = 1, ..., 4$ with random effect distribution $\boldsymbol{\beta}_i|\boldsymbol{\mu}, \boldsymbol{\Sigma} \sim \text{Normal}(\boldsymbol{\mu}, \boldsymbol{\Sigma})$ and priors $\boldsymbol{\mu} \sim \text{Normal}(\mathbf{0}, c^2\mathbf{I}_2)$ and $\boldsymbol{\Sigma} \sim \text{InvWishart}(\nu, \boldsymbol{\Omega})$.

However, the hierarchical model allows us to borrow strength across patients to stabilize the results.

MCMC is a natural choice for fitting hierarchical models; just as hierarchical models build complexity by layering simple conditional distributions, MCMC samples from the complex posterior distribution by sequentially updating parameters from simple full conditional distributions. In fact, displaying the hierarchical model as a DAG not only helps us understand the model but it also aids in coding the MCMC sampler because the full conditional distribution for a parameter depends only on terms with an arrow to or from the parameter's node in the DAG. From Figure 7.4 it is clear that the full conditional distribution for $\boldsymbol{\beta}_2$ depends only on the data-layer terms for $Y_{21}, ..., Y_{2m}|\boldsymbol{\beta}_2$ and the process-layer term for $\boldsymbol{\beta}_2|\boldsymbol{\mu}, \boldsymbol{\Sigma}$. If we view the model only through these terms then it is immediately clear that the full conditional distribution of $\boldsymbol{\beta}_2$ is exactly the full conditional distribution of the regression coefficients in standard Bayesian linear regression (Section 4.2).

When to stop adding layers? The hierarchical models in this section have three levels: data, process and prior. However, it is likely the values that define the prior layer will not be known exactly, and it is therefore tempting to add a fourth (and a fifth, etc.) layer to explain this uncertainty. A general rule of thumb is to stop adding layers when there is no replication to estimate parameters. For example, referring to the DAG in Figure 7.4, it is reasonable to add a layer to estimate the random effect mean $\boldsymbol{\mu}$ and covariance $\boldsymbol{\Sigma}$ because there are repeated random effects $\boldsymbol{\beta}_1, ..., \boldsymbol{\beta}_n$ that can be leveraged to estimate these parameters. However, adding an additional layer to estimate prior mean $\boldsymbol{\Omega}$ of the random effect covariance $\boldsymbol{\Sigma}$ would not be reasonable because there is only one covariance $\boldsymbol{\Sigma}$ in the model, and even if we knew $\boldsymbol{\Sigma}$ exactly, we would not be able to estimate its distribution from a single sample.

7.1.4 Hierarchical models for missing and censored data

Missing data and censoring are complicating factors in many analyses. A Bayesian analysis seamlessly handles these irregularities by treating the missing or censored variables in the same way as unknown parameters. As with parameters, this requires assigning prior distributions for the missing values. For missing or censored responses, their distribution is the same as the observed responses and thus no additional assumptions are needed, except that the data are missing/censored at random. Handling missing or censored covariate values, e.g., the covariates in a linear regression, does require a new prior distribution for the missing values because distributional assumptions are not required for covariates in standard analyses. Assuming reasonable distributions can be identified and modeling assumptions are met, a Bayesian analysis with missing data proceeds as usual and has the benefit of accounting for uncertainty in the missing values when summarizing the posterior distribution of the parameters or predictions of interest.

Missing data: The easiest resolution to the missing-data problem is to simply discard observations with missing data and proceed with a complete-case linear regression. However, this can dramatically reduce the size of the dataset and discard useful information in the partially-observed responses. Another simple approach is to impute the missing values using a regression with the other variables as covariates. A drawback of single imputation is that it does account for uncertainty in the missing observations, and thus the resulting posterior inference is questionable. Multiple imputation techniques are also available [134] (and are often motivated using Bayesian ideas).

A Bayesian analysis using a hierarchical model is a natural way to handle missing data. The Bayesian approach handles missing data in the same way as unknown parameters; we represent our uncertainty about them by treating them as random variables in a hierarchical Bayesian model. As with an unknown parameter, posterior inference on an unknown missing covariate requires assigning it a prior distribution. As an example, consider the multiple linear regression model in Section 4.2,

$$Y_i|\mathbf{X}_i \sim \text{Normal}(\mathbf{X}_i\boldsymbol{\beta}, \sigma^2) \text{ where } \mathbf{X}_i \sim \text{Normal}(\boldsymbol{\mu}, \boldsymbol{\Sigma}) \tag{7.4}$$

where $\boldsymbol{\mu}$ is the mean vector of length p and $\boldsymbol{\Sigma}$ is the $p \times p$ covariance matrix of the covariates. In the absence of subject-matter prior information, the hyperparameters $\boldsymbol{\mu}$ and $\boldsymbol{\Sigma}$ are given priors and estimated as part of the Bayesian analysis, resembling a random-effects model. In this way, the complete cases inform the model about the distribution of the covariates across observations (via $\boldsymbol{\mu}$ and $\boldsymbol{\Sigma}$), and this information is used to impute the missing values.

Of course, this approach requires a reasonable model for the covariate distribution. This is challenging when p is large and/or the covariates are non-Gaussian. For example, if the covariates are a mix of continuous and binary variables, sufficiently capturing their joint distribution is difficult. A simple approach is to model the p covariates with independent priors. This is inefficient but at least supplies a reasonable approximation in many situations. More sophisticated modeling, such as the methods outlined in Chapter 8, may improve this aspect of the analysis. A crucial (and often unverifiable) assumption is that there are no systematic biases in the missing data; i.e., they are missing completely at random. If the probability that X_{ij} is missing depends on Y_i, then assuming the X_{ij} are missing at random would likely result in a biased estimator for β_j.

Assuming that the covariate model is correct, a hierarchical Bayesian analysis appropriately accounts for uncertainty in the missing values. A Gibbs sampler would update each parameter ($\boldsymbol{\beta}$, σ, etc.) and then cycle through the missing observations (Y_i, X_{ij}) and update them from their full conditional distributions, treating them exactly the same as the model parameters. Therefore, each sample of the regression coefficient $\boldsymbol{\beta}$ is updated using a complete set of covariates, but the imputed covariates vary across iterations follow-

Listing 7.2
JAGS model statement for the missing data analysis of the marathon data.

```
# Likelihood
 for(i in 1:n){
   Y[i] ~ dnorm(alpha + inprod(X[i,],beta[]),taue)
 }

# Missing-data model
 for(i in 1:n){
   X[i,1] ~ dnorm(0,tau1)
   for(j in 2:p){
     X[i,j] ~ dnorm(rho*X[i,j-1],tau2)
   }
 }

# Priors
 alpha ~ dnorm(0,0.01)
 for(j in 1:p){
   beta[j] ~ dnorm(0,0.01)
 }
 taue  ~  dgamma(0.1, 0.1)
 tau1  ~  dgamma(0.1, 0.1)
 tau2  ~  dgamma(0.1, 0.1)
 rho   ~  dnorm(0, 0.01)
```

ing their posterior distribution. The missing values are thus effectively treated as nuisance parameters, and the analysis produces both the posterior predictive distribution of the missing values but more importantly the posterior distribution of the regression coefficients marginally over the uncertainty in the missing observations, as desired.

Marathon example with missing data: In this analysis of the 2016 Boston Marathon data from Section 2.1.7, we build a linear regression model to predict the speed (minutes per mile) of the final mile (mile 26) of the top female runners as a function of their speeds in the first 25 miles. Let Y_i be the speed of mile 26 for runner $i = 1, ..., n = 149$ and X_{ij} be the speed for runner i in mile $j = 1, ..., p = 25$. Approximately 8% of the speed measurements are missing: the number of missing observations ranges from 0 to 19 across runners, and the percentage of missing observations ranges from 0% to 55% (miles 6 and 7) across miles (Figure 7.5a). Discarding observations with at least one missing X_{ij} would reduce the sample size from 149 runners to 58. Given that most of the missing values are for miles 6 and 7 and these covariates are likely not important predictors of the response, it would be wasteful to discard all of these observations from the analysis.

In Section 2.1.7, we modeled the covariance of the covariates (Σ) using an inverse Wishart model. This model did not assume any structure among the miles, but the posterior mean of the covariance matrix revealed that subsequent miles are highly correlated. Therefore, here we model the standardized (to have mean zero and variance one) covariates using the first-order autoregressive time series model $X_{i1} \sim \text{Normal}(0, \sigma_1^2)$ and $X_{ij+1}|X_{ij} \sim \text{Normal}(\rho X_{ij}, \sigma_2^2)$ for $j \geq 1$. `JAGS` code for this model is given in Listing 7.2. All hyperparameters have uninformative priors.

Figure 7.5b plots the observed covariates (dots) and the posterior distribution of the missing covariates (boxplots) for two representative runners. Because of the time series model for the missing covariates, the posterior distributions of the missing X_{ij} are close to the speeds of the adjacent miles for both runners. The posterior distributions of the

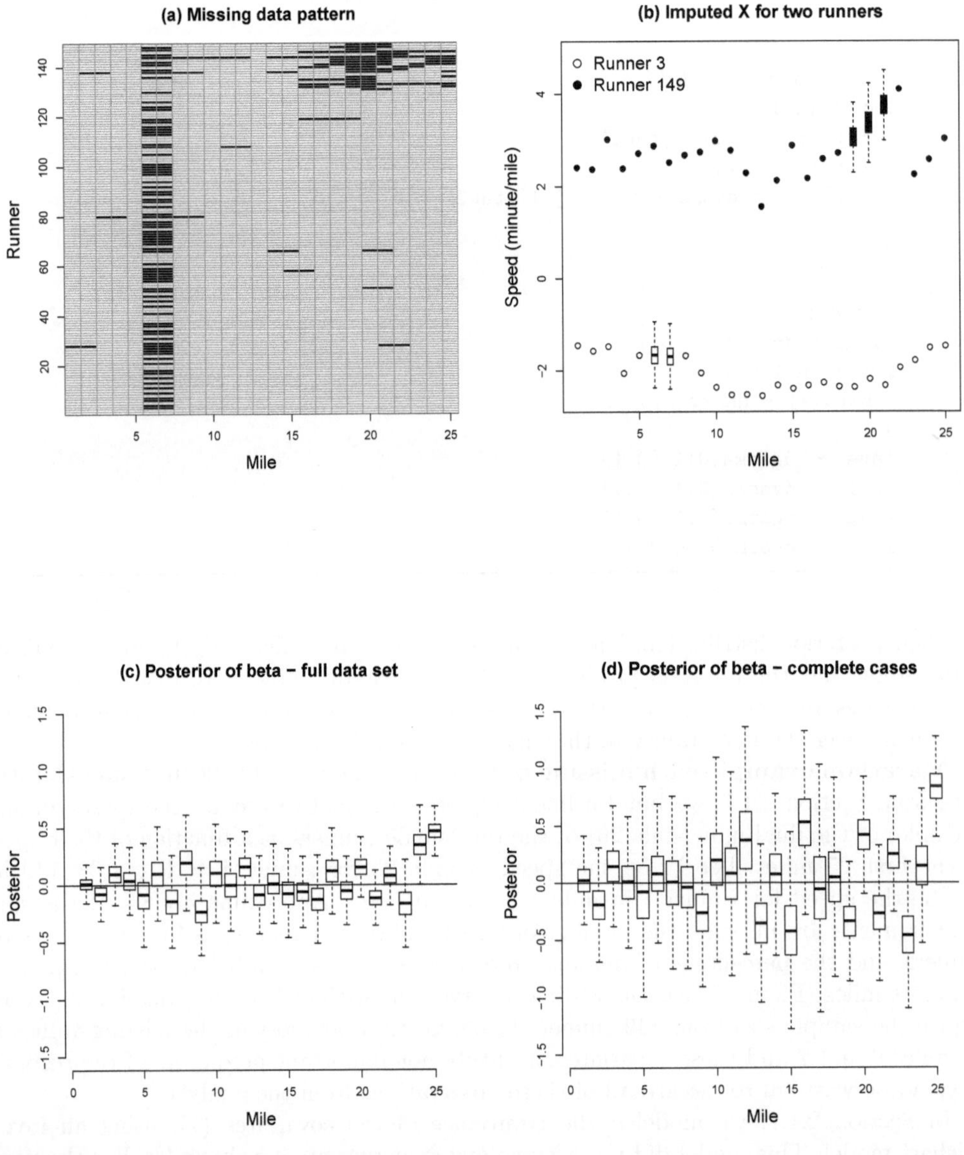

FIGURE 7.5
Missing data analysis of the 2016 Boston Marathon data. Panel (a) shows the missing (black) and non-missing (white) X_{ij} by runner (i) and mile (j). Panel (b) plots the observed (points) standardized covariates (X_{ij}) and the posterior distributions (boxplots) of the missing covariates for two runners. Panels (c) and (d) plot the posterior distribution of each regression coefficient, β_j, for the missing-data model and complete-case analysis, respectively.

covariates, β_j, are plotted in Figure 7.5c. Only miles 24 and 25 appear to be useful predictors of the speed in the final mile. The posterior variances in this missing-data analysis are much smaller than the posterior variances in the complete-case analysis with $n = 58$ in Figure 7.5d, illustrating the benefit of the missing data model.

Covariate measurement error: Error-in-variables models, also called measurement-error models [66], are used to handle covariates that are not missing, but are measured with error. For example, say controlled experiments show that an environmental contaminant measured in a person's blood has error standard deviation τ. Then a regression of a person's health status, measured by Y_i, onto their level of exposure to the contaminant, X_i, would suffer from covariate measurement error.

If the variance of the covariate measurement error is known and a distribution of the error can be assumed (which are strong assumptions), then a Bayesian hierarchical model can address this error. A joint model is

$$Y_i|\boldsymbol{\beta}, X_i^* \sim \text{Normal}(\beta_0 + \beta_1 X_i^*, \sigma^2) \text{ and } X_i|X_i^* \sim \text{Normal}(X_i^*, \tau^2),$$

where X_i^* is the true value of the covariate and has prior $X_i^* \sim \text{Normal}(\mu, \sigma_X^2)$. In MCMC, the true value handled like missing data, i.e., is imputed for each observation at each iteration, and all other parameters ($\boldsymbol{\beta}$ and σ^2) are updated as in regular linear regression.

Censored data: A censored observation is known partially but not exactly. Typically a censored observation is known to have occurred in an interval, say $Y_i \in \mathcal{I}_i = (L_i, U_i)$. Special cases of interval censoring include: if $\mathcal{I}_i = (-\infty, U_i)$, then it is left-censored; if $\mathcal{I}_i = (L_i, \infty)$, then it is right-censored; if $\mathcal{I}_i = (-\infty, \infty)$, then it is missing; and if $L_i = U_i$, then it is uncensored. The canonical example of censored data is survival analysis (e.g., [82]), where Y_i represents the time from patient i entering the study to an event occurring (e.g., developing a disease). Right-censoring results from patients being lost to follow-up or events occurring only after trial completion. In this case, a patient who experienced an event during the trial is not censored and $L_i = U_i = Y_i$, and a patient known to have survived until L_i but who was then lost to follow-up is right-censored with $U_i = \infty$.

A convenient model for right-censored data is $Y_i|\lambda \overset{iid}{\sim} \text{Exponential}(\lambda)$ with conjugate prior $\lambda \sim \text{Gamma}(a, b)$. The contribution of observation i to the likelihood is the PDF $\lambda \exp(-\lambda Y_i)$ if it is uncensored and the survival time $\text{Prob}(Y_i > L_i|\lambda) = \exp(-\lambda_i L_i)$ if it is right censored at L_i. If we denote $\delta_i = 0$ and $T_i = L_i$ if observation i is censored and $\delta_i = 1$ and $T_i = Y_i$ otherwise, then

$$\lambda|\delta_1, ..., \delta_n, T_1, ..., T_n \sim \text{Gamma}\left(\sum_{i=1}^n \delta_i + a, \sum_{i=1}^n T_i + b\right).$$

Under this model inference does not require MCMC sampling. However, the exponential model is usually too restrictive, and adding covariates, e.g., $\log(\lambda_i) = \mathbf{X}_i\boldsymbol{\beta}$ breaks conjugacy. Therefore, more flexible models are needed.

An attractive model to include convariates is the accelerated failure time model [25] that regresses the log time to event, $Z_i = \log(Y_i)$, onto covariates, $Z_i = \mathbf{X}_i\boldsymbol{\beta} + \varepsilon_i$. Under this model, if covariate j increases by one, then the median (or any other quantile) time until event in multiplied by $\exp(\beta_j)$. If there is no censoring and $\varepsilon_i \overset{iid}{\sim} \text{Normal}(0, \sigma^2)$, then this reduces to standard multiple regression as in Section 4.2. Further, selecting conditionally conjugate priors for $\boldsymbol{\beta}$ and σ^2 leads to Gibbs sampling.

However, if Z_i is censored to the interval $[L_i, U_i]$, then conjugacy is broken. Assuming normality for ε_i, Z_i follows a truncated normal distribution $Z_i \sim \text{TruncatedNormal}(\mu_i, \sigma^2, L_i, U_i)$ with $\mu_i = \mathbf{X}_i\boldsymbol{\beta}$ and density function

$$\frac{1}{\sigma\sqrt{2\pi}} \exp\left[-\frac{(Z_i - \mu_i)^2}{2\sigma^2}\right] \frac{1}{\Phi[(U_i - \mu_i)/\sigma] - \Phi[(L_i - \mu_i)/\sigma]} \tag{7.5}$$

where Φ is the standard normal CDF. Since $\boldsymbol{\beta}$ is included (via μ_i) inside the normal CDF, its posterior is not conjugate.

Conjugacy is restored by treating the censored value Z_i using Bayesian imputation inside MCMC (see Listing 7.3 for a simple example). For each iteration, the censored value Z_i is drawn from the truncated normal distribution $Z_i \sim \text{TruncatedNormal}(\mu_i, \sigma^2, L_i, U_i)$ which has support $[L_i, U_i]$ and thus retains the censoring information. After imputation, the responses are complete, and updates of the regression coefficients and variance can be performed using the same Gibbs sampling methods as in usually multiple linear regression. As with missing data, new samples of the censored observations are generated for each MCMC iteration, and so this Bayesian imputation approach is equivalent to fitting the model with the likelihood function composed of terms such as (7.5). Also, Bayesian multiple imputation ensures that uncertainty about the censored values is incorporated in the posterior distribution of the regression parameters. This method of imputation is very general, and can thus be applied to the flexible regression methods in Chapter 8 to relax assumptions about normality and linearity.

7.2 Case studies

The remainder of this chapter is formatted as a sequence of case studies in hierarchical modeling. The three case studies each pose different challenges:

1. **Species distribution mapping via data fusion**: Combining information from multiple data streams while accounting for their bias and uncertainty
2. **Tyrannosaurid growth curves**: Pooling information across subpopulations (species) and quantifying uncertainty in non-linear models with a small number of observations
3. **Climate reconstruction using proxy data**: Accounting for bias and uncertainty in proxy data to reconstruct the temperature distribution in the recent past

In these analyses we demonstrate the flexibility of hierarchical modeling, and also illustrate complete Bayesian analyses including model and prior specification, model comparisons, and presentation of the results.

7.2.1 Species distribution mapping via data fusion

The data for this case study come from [119]. The objective is to map the spatial distribution of a small song bird, the brown-headed nuthatch (BHNU; *Sitta pusilla*), in the southeastern US. There are two data sources, each with different strengths. The first data source is the Breeding Birds Survey (BBS). BBS is a network of hundreds of routes surveyed by thousands of volunteers that has been active since 1966 [138]. Data are collected systematically by trained volunteers, and sites are visited annually to monitor for changes. However, even with this immense sampling effort, there are spatial and temporal gaps in BBS coverage. An emerging line of research is to supplement systematic survey data with massive citizen science data such as the Cornell Lab of Ornithology's eBird database [150], which consists of millions of data points from thousands of citizen scientists each year. These data are not collected by trained birders, but have far greater spatial and temporal coverage.

Listing 7.3
Sketch of an Gibbs sampler with censoring.

```
# The model is Z[i] ~ Normal(mu,sigma^2).
# Observations with delta[i]=1 are observed
# Observations with delta[i]=0 are censored to (L[i],U[i])
# The improper prior for mu is pi(mu) = 1
# Prior for sigma^2 ~ InvGamma(a,b)

# Lung cancer survival data from the survival package

 Z     <- log(lung$time) # Final visit time
 n     <- length(Z)
 delta <- lung$status-1  # status is 1 for censored, 2 for died
 L     <- Z              # Data are right censored on (L[i],U[i])
 U     <- rep(Inf,n)

# Initial values

 mu      <- mean(Z[delta==1])
 s2      <- var(Z[delta==1])
 keep_mu <- rep(0,10000)          # Store the samples
 a       <- b <- 0.1              # Prior for s2

# Gibbs sampling
 library(truncnorm)

 for(s in 1:10000){

   # Impute censored values
    for(i in 1:n){if(delta[i]==0){
       Z[i] <- rtruncnorm(1,L[i],U[i],mu,sqrt(s2))
    }}

   # Update parameters using the ''complete'' data
    mu <- rnorm(1,mean(Z),sqrt(s2/n))
    s2 <- 1/rgamma(1,a+n/2,b+sum((Z-mu)^2)/2)

   # Store the results
    keep_mu[s] <- mu
}
```

For this analysis, the southeast US is partitioned into $n = 741$, 0.25×0.25 degree lat/lon cells, and we analyze data from 2012. Let N_{1i} be the number of BBS sampling occasions in cell i and $Y_{1i} \in \{0, 1, ..., N_{1i}\}$ be the number of occasions with a sighting of the BHNU. Many cells do not have a BBS route and thus $N_{1i} = Y_{i1} = 0$. Similarly, for cell i define N_{2i} as the number of hours logged by eBird citizen scientists and Y_{2i} as the number of BHNU eBird sightings. Figure 7.6 maps the data. The BBS sampling effort is fairly uniform, whereas the eBird effort is more concentrated in populated areas. Both maps show more BHNU sightings in Alabama, Georgia and the Carolinas.

The true process of interest in cell i is the abundance $\lambda_i \geq 0$, defined as the expected number of birds present in the region during one unit of surveying effort. For a cell that is not inhabited by the BHNU $\lambda_i = 0$. The data layer relates the process to the observed data, and requires careful consideration of the strengths of the two data sources. First we make the assumption that the BBS and eBird datasets are independent given λ_i. This seems reasonable as most eBird users are not following BBS updates. The BBS data are gathered by expert birders, and thus we assume that there are no false positives or negatives (although more flexible models are available and likely preferable; see [119] for example). If the number of birds present during a survey is distributed as Poisson(λ_i), then the probability that there is at least one bird present is $1 - \exp(-\lambda_i)$, and so we model the BBS data as $Y_{i1}|\lambda_i \sim \text{Binomial}[N_{1i}, 1 - \exp(-\lambda_i)]$.

For the eBird data, we do allow for false positives and false negatives. The mean is $N_{2i}\tilde{\lambda}_i$ with rate $\tilde{\lambda}_i = \theta_1\lambda_i + \theta_2$, where $\theta_1 > 0$ controls the difference between observation rates by BBS observers and eBird observers and $\theta_2 > 0$ is the eBird false positive rate, so that if the cell is truly uninhabited and $\lambda_i = 0$, then $\text{E}(Y_{2i}) = N_{2i}\theta_2$. To allow for over-dispersion we fit the model

$$Y_{2i}|\lambda_i, \boldsymbol{\theta} \sim \text{NegBinomial}(q_i, m) \tag{7.6}$$

with probability $q_i = m/(\tilde{\lambda}_i + m)$ and size m. Combining these two contributions to the likelihood, the data layer is

$$\begin{aligned}
\text{Data layer} : Y_{1i}|\lambda_i, \theta_1, \theta_2 &\sim \text{Binomial}[N_{1i}, 1 - \exp(-\lambda_i)] \\
Y_{2i}|\lambda_i, \theta_1, \theta_2 &\sim \text{NegBinomial}(q_i, m).
\end{aligned}$$

The latent process of interest is the abundance in each cell, λ_i. It is difficult to model all the λ_i without prior knowledge because some cells do not have BBS data. In the absence of other prior knowledge, we may simply assume that nearby cells have similar abundance. This is a reasonable assumption if the underlying factors that drive abundance (climate, habitat, etc.) vary spatially and allows us to pool information locally to estimate abundance. Many spatial models can be used for model abundance [53], but here we use spline regression as in Section 8.3. The log abundance (we use a log transformation to ensure λ_i is non-negative) is a smooth function of the grid cell's spatial location, $\mathbf{s}_i = (s_{1i}, s_{2i})$. Since this is a two-dimensional function, we use spline basis expansions in both the longitude and latitude,

$$\text{Process layer} : \log(\lambda_i) = \sum_{j=1}^{J}\sum_{k=1}^{K} B_j(s_{i1})D_k(s_{i2})\beta_{jk} \tag{7.7}$$

where B_j are B-spline basis functions of longitude, D_k are B-spline basis functions of latitude, and $\beta_{jk} \stackrel{iid}{\sim} \text{Normal}(\beta_0, \sigma^2)$ (here we use different notation for the basis function in the latitude and longitude directions because they have a different number of basis functions and thus a different form). Some of the products $B_j(s_{i1})D_k(s_{i2})$ are near zero for all $\mathbf{s}_i$ in the domain and are discarded. Since the spatial domain spans a wider range of longitudes than latitudes, we take $K = 2L$ and $J = L$ for a total of $p = 2L^2$ terms of the form

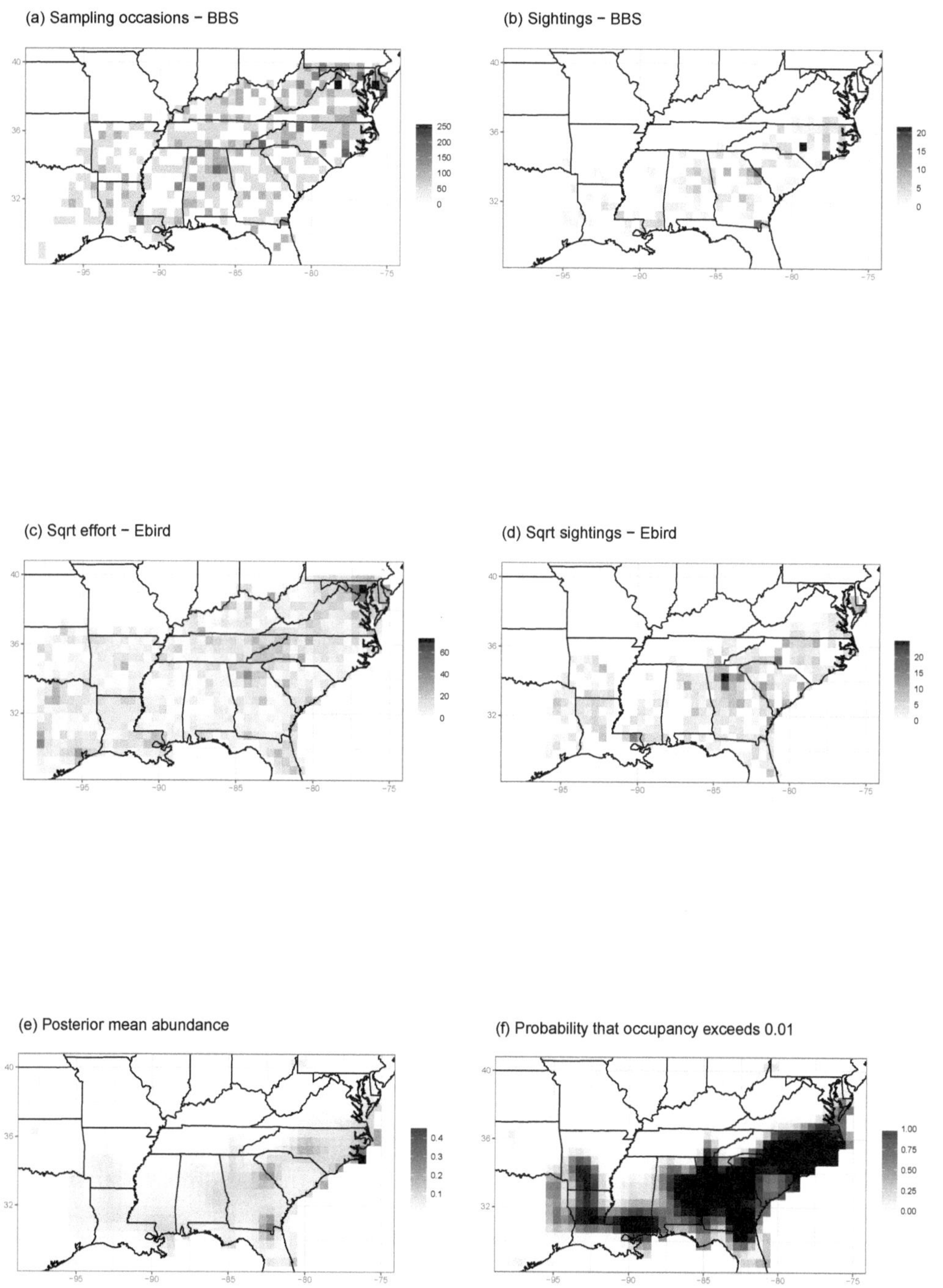

FIGURE 7.6
2012 Brown-headed nuthatch data. Panels (a) and (b) plot the number of BBS sampling occasions (N_{1i}) and number of BBS sightings (Y_{1i}); Panels (c) and (d) plot the square root of Ebird effort ($\sqrt{N_{2i}}$) and the square root of the number of Ebird sightings ($\sqrt{Y_{2i}}$). Panel (e) plots the posterior mean abundance, λ_i, and Panel (f) plots the posterior probability that the occupancy probability exceeds 0.01, i.e., $\text{Prob}[1 - \exp(-\lambda_i) > 0.01|\mathbf{Y}]$.

Listing 7.4
Spatial data fusion model for BHNU abundance.

```
# Data layer
  for(i in 1:n){
    Y1[i]   ~ dbin(phi[i],N1[i]) # BBS
    phi[i] <- 1-exp(-lam[i])

    Y2[i]   ~ dnegbin(q[i],m)      # eBird
    q[i]   <- m/(m+N2[i]*(theta1*lam[i]+theta2))
  }

# Process layer
  for(j in 1:p){beta[j]~dnorm(beta0,tau)}
  for(i in 1:n){
    log(lam[i])  <- inprod(X[i,],beta[])
  }

# Prior layer
  theta1 ~ dgamma(0.1,0.1)
  theta2 ~ dgamma(0.1,0.1)
  m      ~ dgamma(0.1,0.1)
  tau    ~ dgamma(0.1,0.1)
  beta0  ~ dnorm(0,1)
```

$X_l = B_j(s_{i1})D_k(s_{i2})$ and select L using DIC. To complete the Bayesian hierarchical model, we specify uninformative priors

$$\text{Prior layer} : \theta_1, \theta_2, m, \sigma^{-2} \sim \text{Gamma}(0.1, 0.1) \text{ and } \beta_0 \sim \text{Normal}(0, 100). \quad (7.8)$$

JAGS code to implement this model is given in Listing 7.4.

Two chains are run, each with a burn-in of 10,000 iterations and 50,000 post-burn-in samples. The samples are thinned by 5 leaving 20,000 samples to approximate the posterior. The DIC (p_D) is 3107 (30) for $L = 4$, 3056 (58) for $L = 6$, 3015 (89) for $L = 8$, 2999 (127) for $L = 10$, 3014 (177) for $L = 12$ and 3009 (209) for $L = 14$, and so we proceed with $L = 10$. With $L = 10$, the effective sample size is greater than 1,000 for all β_{jk}, indicating the sampler has mixed well and sufficiently explored the posterior.

Table 7.1 presents the posterior distributions of the hyperparameters. Of note, the eBird false positive rate, θ_2, is estimated to be near zero, and thus the eBird data appears to be a reliable source of information. The posterior mean of λ_i and the posterior probability that cells are occupied (i.e., at least one individual is present) are mapped in Figures 7.6e and 7.6f, respectively. As expected, the estimated abundance is the largest in Georgia and the Carolinas, but the occupancy probability is also high farther west in Louisiana and Arkansas. The occupancy probabilities would be lower in these western states if the eBird data were excluded.

7.2.2 Tyrannosaurid growth curves

We analyze the data from 20 fossils to estimate the growth curves of four tyrannosaurid species: Albertosaurus, Daspletosaurus, Gorgosaurus and Tyrannosaurus. The data are taken from Table 1 of [48] and plotted in Figure 7.7. The objective is to establish the growth curve, i.e., expected body mass by age, for each species. The data exhibit non-

TABLE 7.1
Posteriors for the BHNU analysis. Posterior median and 95% intervals mean for the final fit with $L = 10$.

	Median	95% Interval
Scaling factor, θ_1	11.5	(9.2, 14.4)
False positive rate, θ_2	0.00	(0.00, 0.00)
Over-dispersion parameter, m	0.45	(0.37, 0.55)
Mean abundance parameter, β_0	−5.81	(−6.69, −4.86)
Spline standard deviation, σ	5.58	(4.52, 7.03)

linear relationships between age and mass, and there are commonalities between species. We therefore pursue a non-linear hierarchical model.

The original analysis of these data used non-linear least squares (fitted curves shown in the left panel of Figure 7.7). Quantifying uncertainty in this fit is challenging. The sampling distribution of the estimator does not have a closed form due to the non-linear mean structure, and with only a handful of observations to estimate roughly the same number of parameters, large-sample normal approximations are not valid, and resampling techniques such as the bootstrap may have insufficient data to approximate the sampling distribution. As shown below, a Bayesian analysis powered by MCMC fully quantifies posterior uncertainty.

Let Y_{ij} and X_{ij} be the body mass and age, respectively, of sample i from species $j = 1, ..., 4$. We model the data as

$$Y_{ij} = f_j(X_{ij})\epsilon_{ij}, \tag{7.9}$$

where f_j is the true growth curve for species j and $\epsilon_{ij} > 0$ is multiplicative error with mean one. We use multiplicative error rather than additive error because variation in the population likely increases with mass/age. Assuming the errors are log-normal with $\log(\epsilon_{ij}) \sim \text{Normal}(-\sigma_j^2/2, \sigma_j^2)$, then $\text{E}(\epsilon_{ij}) = 1$ as required, and the model becomes

$$\log(Y_{ij}) \sim \text{Normal}\left(\log[f_j(X_{ij})] - \sigma_j^2/2, \sigma_j^2\right), \tag{7.10}$$

with $\text{E}(Y_{ij}) = f_j(X_{ij})$, where σ_j^2 controls the error variance for species j.

The data in Figure 7.7 (left) clearly exhibit nonlinearity. However, after taking a log transformation of both mass and age, their relationship becomes fairly linear (Figure 7.7, right). Therefore, one model we consider is the log-linear model

$$\log[f_j(X)] = a_j + b_j \log(X) \tag{7.11}$$

where a_j and b_j are the intercept and slope, respectively, for species j. On the original scale, the corresponding growth curve is $f_j(X) = \exp(a_j)X^{b_j}$. If as expected b_j is positive, then the growth curve increases indefinitely, which may not be realistic. Therefore, we compare the log-linear model with the logistic growth curve

$$f_j(X) = a_j + b_j \frac{\exp[d_j(x - c_j)]}{1 + \exp[d_j(x - c_j)]}, \tag{7.12}$$

where $x = \log(X)$. This model has four parameters:

1. a_j is the expected mass at age 0.
2. b_j is the expected lifetime gain in mass.

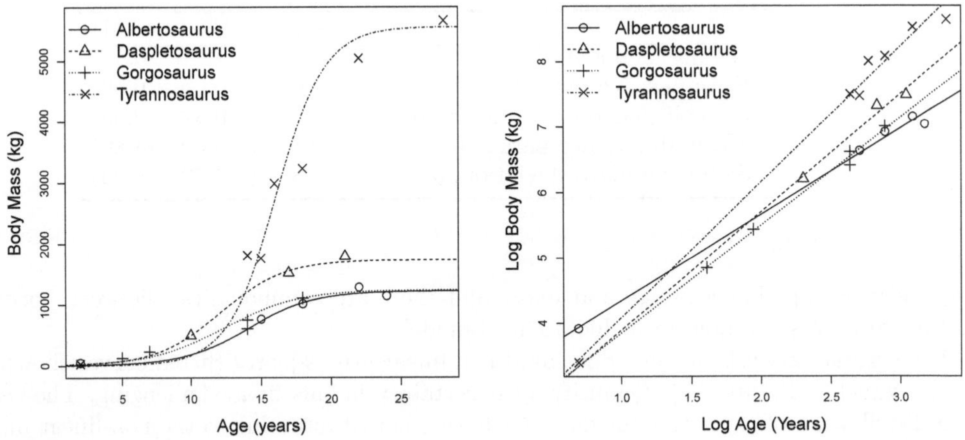

FIGURE 7.7
Tyrannosaurid growth curve data. The left panel gives scatter plots of the estimated age and body mass (kg) of 20 samples of four tyrannosaurid species; the right panel plots the same data after a log transformation of both variables. The curves plotted in the left panel are the fitted logistic curves from [48], and the lines in the right panel are least squares fits.

3. $\log(c_j)$ is the age at which the species reaches half its expected gain.
4. $d_j > 0$ determines the rate of increase with age.

The form of the curve is increasing (assuming $b_j > 0$) and plateaus at $a_j + b_j$ rather than continuing to increase with age. This is the same function fit by [48], except that we transform to log age.

In addition to comparing these two forms of growth curves, we also compare two priors. The first prior ("unpooled") fits each species separately using uninformative priors. For the log-linear model the priors are $a_j, b_j \sim \text{Normal}(0, 10)$ and $\sigma_j^2 \sim \text{InvGamma}(0.1, 0.1)$, and for the logistic model the priors are $\log(a_j), \log(b_j), c_j, \log(d_j) \sim \text{Normal}(0, 10)$ and $\sigma_j^2 \sim \text{InvGamma}(0.1, 0.1)$. The normal prior is replaced with the log-normal prior for a_j, b_j and d_j to ensure these parameters are positive, and thus $f_j(X)$ is positive and increasing for all X. The second prior ("pooled") is a Bayesian hierarchical model that borrows information across the four species. In the pooled analysis we assume the variance is the same for all species, $\sigma_j^2 = \sigma^2$, and has uninformative prior $\sigma^2 \sim \text{InvGamma}(0.1, 0.1)$. For the log-linear model priors for the intercepts are $a_j \sim \text{Normal}(\mu_a, \sigma_a^2)$, where $\mu_a \sim \text{Normal}(0, 10)$ and $\sigma_a^2 \sim \text{InvGamma}(0.1, 0.1)$. The same hierarchical model is applied to the $\log(a_j)$, $\log(b_j)$, c_j and $\log(d_j)$ in the logistic model. The JAGS code for this model is given in Listing 7.5.

This hierarchical model treats the parameters across the four species as random effects, and learning about the random effects distribution (i.e., μ_a and σ_a^2) stabilizes the posterior by providing additional information via the priors. It is debatable whether these parameters are truly random effects, i.e., whether there is an infinite distribution of exchangeable species from which these four were randomly selected for the study. However, analyzing the data

Listing 7.5
JAGS code for hierarchical growth curve modeling.

```
# n is the total number of observations for all species
# x[i] is the log age of individual i
# y[i] is the log mass of individual i
# sp[i] is the species number (1, 2, 3, or 4) of individual i

# Data layer
for(i in 1:n){
  y[i]     ~ dnorm(muY[i],taue)
  muY[i]  <- log(a[sp[i]] + b[sp[i]]/(1+exp(-part[i])) - 0.5/taue
  part[i] <- (x[i]-c[sp[i]])/d[sp[i]]
}

# Process layer
for(j in 1:N){
  a[j]  <- exp(alpha[j,1])
  b[j]  <- exp(alpha[j,2])
  c[j]  <- alpha[j,3]
  d[j]  <- exp(alpha[j,4])

  for(k in 1:4){alpha[j,k] ~ dnorm(mu[k],tau[k])}
}

# Prior layer
for(k in 1:4){
  mu[k]  ~ dnorm(0,0.1)
  tau[k] ~ dgamma(0.1,0.1)
}
taue ~ dgamma(0.1,0.1)
```

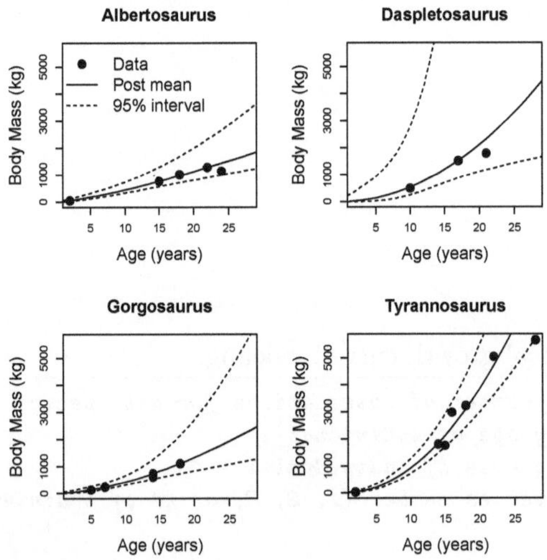

FIGURE 7.8
Fitted log-linear growth curves – unpooled. Observations (points) versus the posterior mean (solid lines) and 95% intervals (dashed lines) of the tyrannosaurid growth curves for the unpooled log-linear model.

from these four species using a random effects model clearly improves the results (as shown below) by pooling information across species to reduce uncertainty.

We fit the model with log-linear and logistic growth curves, each separate by species (unpooled) and using a hierarchical model (pooled). DIC (p_D) for the four fits are: 29 (25) for log-linear unpooled, −3 (9) for log-linear pooled, 64 (41) for logistic unpooled and −2 (12) for logistic pooled. The pooled models reduce model complexity (as measured by p_D), and this leads to smaller (better) DIC. DIC for the log-linear and logistic growth curves are similar.

Figures 7.8–7.10 plot the posterior mean and pointwise 95% credible interval for f_j for each model and each species (the interval estimates are for f_j and not Y_{ij}, so they should not include 95% of the observations). The posterior means of the four methods are fairly similar, and all fit the data well. The main difference between the fits is that by borrowing information across species, the pooled analyses have narrower credible sets. Visually, the log-linear fits in Figure 7.10 appear to sufficiently model the growth curves. However, given that the logistic curve fits nearly as well and possesses the intuitive property of plateauing at an advanced age, this model is arguably preferable when considering the entire life course.

7.2.3 Climate reconstruction

This analysis uses the data of [39] and a simplified version of the statistical approach of [27] to reconstruct past annual precipitation in Brisbane, AU. The analysis is challenging because it combines several data types in a hierarchical model, and each data type has a high proportion of missing values. The rainfall index (RFI) is measured (in mm) from 1889-2017 and plotted in Figure 7.12. Proxy data, e.g., ice cores and tree rings, dating as far back as 1612 are used to reconstruct RFI prior to 1889. Let Y_t be the observed RFI

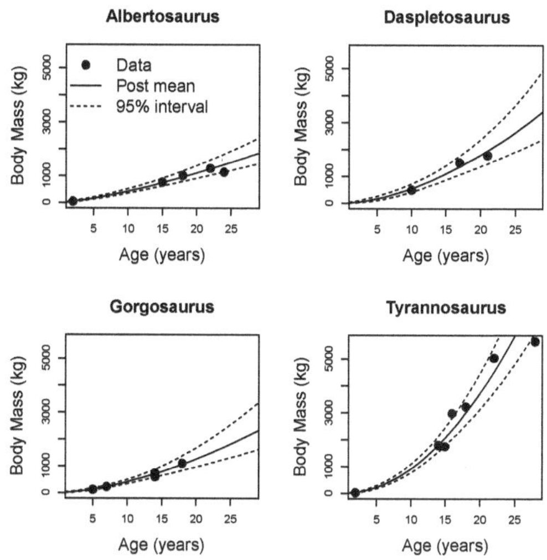

FIGURE 7.9
Fitted log-linear growth curves – pooled. Observations (points) versus the posterior mean (solid lines) and 95% intervals (dashed lines) of the tyrannosaurid growth curves for the pooled (hierarchical) log-linear model.

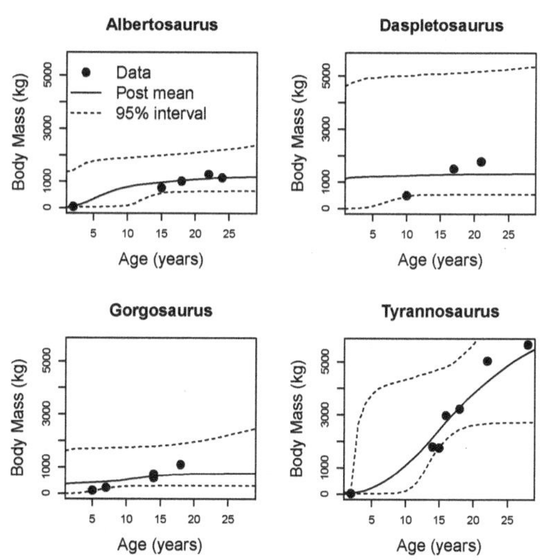

FIGURE 7.10
Fitted log-linear growth curves – logistic, unpooled. Observations (points) versus the posterior mean (solid lines) and 95% intervals (dashed lines) of the tyrannosaurid growth curves for the unpooled logistic model.

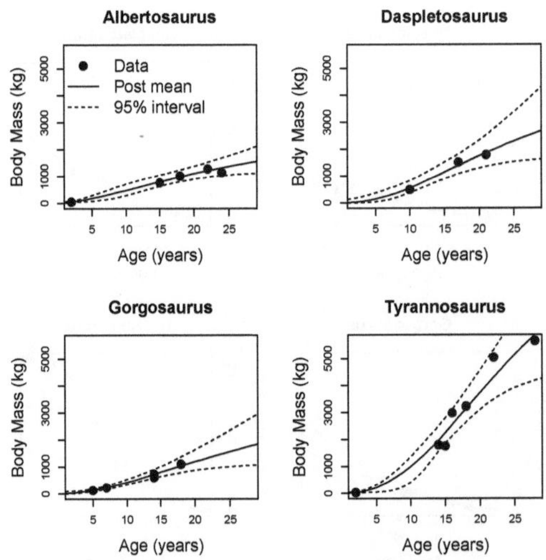

FIGURE 7.11
Fitted log-linear growth curves – logistic, pooled. Observations (points) versus the posterior mean (solid lines) and 95% intervals (dashed lines) of the tyrannosaurid growth curves for the pooled (hierarchical) logistic model.

in year $t + 1611$ (so Y_1 corresponds to year 1612). The autoregressive time-series model is $Y_1 \sim \text{Normal}(\mu_0, \sigma_0^2)$ and

$$Y_t|Y_{t-1} \sim \text{Normal}(\mu_0 + \phi_0(Y_{t-1} - \mu_0), (1 - \phi_0^2)\sigma_0^2),$$

where μ_0 is the mean, σ_0^2 is the variance and ϕ_0 controls the temporal dependence. The form of the mean and variance are such that the marginal (over the Y_j for $j \neq t$) mean and variances are $\text{E}(Y_t) = \mu_0$ and $\text{Var}(Y_t) = \sigma_0^2$ for all t.

Unfortunately, precipitation is measured only for the past 129 years. To impute these missing observations, we use several proxy variables shown in Figure 7.13. The proxy variables also have a high proportion of missing values; however, there is some overlap between the years with proxies and RFI measurements, and the correlation between RFI and the proxies is as high as 0.40 depending on the proxy. Therefore, the relationships between precipitation and the proxies can be exploited to impute RFI back to 1612. A hierarchical Bayesian model is a natural way to handle this task because of the large number of parameters and missing data, and thus uncertainty, that can be propagated through to the final analysis of precipitation across time.

Denote proxy variable j in year $t+1611$ as X_{tj}. One option is to use these proxy variables as covariates in the model for the observed precipitation. However, a hierarchical model that better matches physical reality is to have precipitation be a covariate in the mean of the proxy variables because in reality precipitation is thought to be a driver of these proxy variables. The proxy variables are related to precipitation using a linear relationship

$$\text{E}(X_{jt}) = \mu_{jt} = \alpha_j + \beta_j Y_t$$

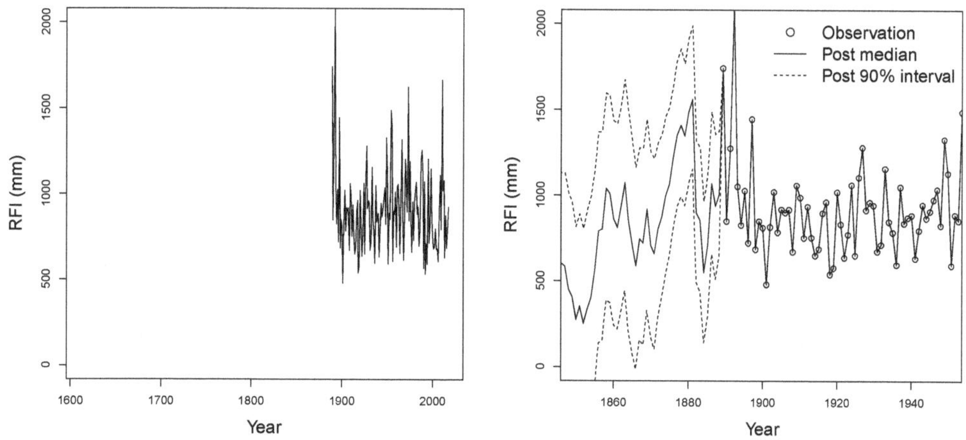

FIGURE 7.12
Climate reconstruction rainfall index data. The left panel shows all values of the rainfall index (RFI), and the right panel shows the data and posterior median and 90% interval for years 1850-1950.

where the intercept α_j and slope β_j account for additive and multiplicative bias. To account for temporal dependence, we use another autoregressive model with $X_{j1} \sim \text{Normal}(\mu_{j1}, \sigma_j^2)$ and

$$X_{jt}|X_{jt-1} \sim \text{Normal}(\mu_{jt} + \phi_j(X_{jt-1} - \mu_{jt-1}), (1 - \phi_j^2)\sigma_j^2).$$

JAGS code is given in Listing 7.6.

Since the proxies are on very different scales, before the analysis they are standardized separately to have mean zero and variance one (see Figure 7.13). The Bayesian model is completed with uninformative priors (see Listing 7.6). Both Y_t and X_{jt} have missing values, which is handled naturally using Bayesian multiple imputation as described in Section 7.1.4. Therefore, the resulting posterior distribution of Y_t for years with missing data accounts for uncertainty in the parameters as well as missing values in the RFI and proxy data.

Because of the large number of missing values, we run long MCMC chains. We fit two MCMC chains in JAGS, each with burn-in of 20,000 iterations followed by 100,000 iterations thinned by 10. The effective sample size is between 845 and 2626 across parameters and imputed RFI values. MCMC sampling takes around 7 minutes on an ordinary PC.

The strength of the relationship between proxy j and RFI is controlled by β_j. The posterior median (standard deviation) of β_j for the six proxy variables are 0.00077 (0.00026), 0.00135 (0.00034), −0.00008 (0.00023), 0.00089 (0.00030), 0.00100 (0.00030), 0.00156 (0.00033). Therefore, all but the third proxy have a 90% interval that excludes zero. All posterior medians are small, but this is because the proxy data are standardized while RFI remains unstandardized and ranges from 0 to 2000. The right panel of Figure 7.12 shows the fitted value of Y_t for a subset of the years in the study. As expected, the variability increases moving back in time, but even in the mid-1800's there is considerable variation in the estimated RFI across years due to the fitted relationship with the proxy data.

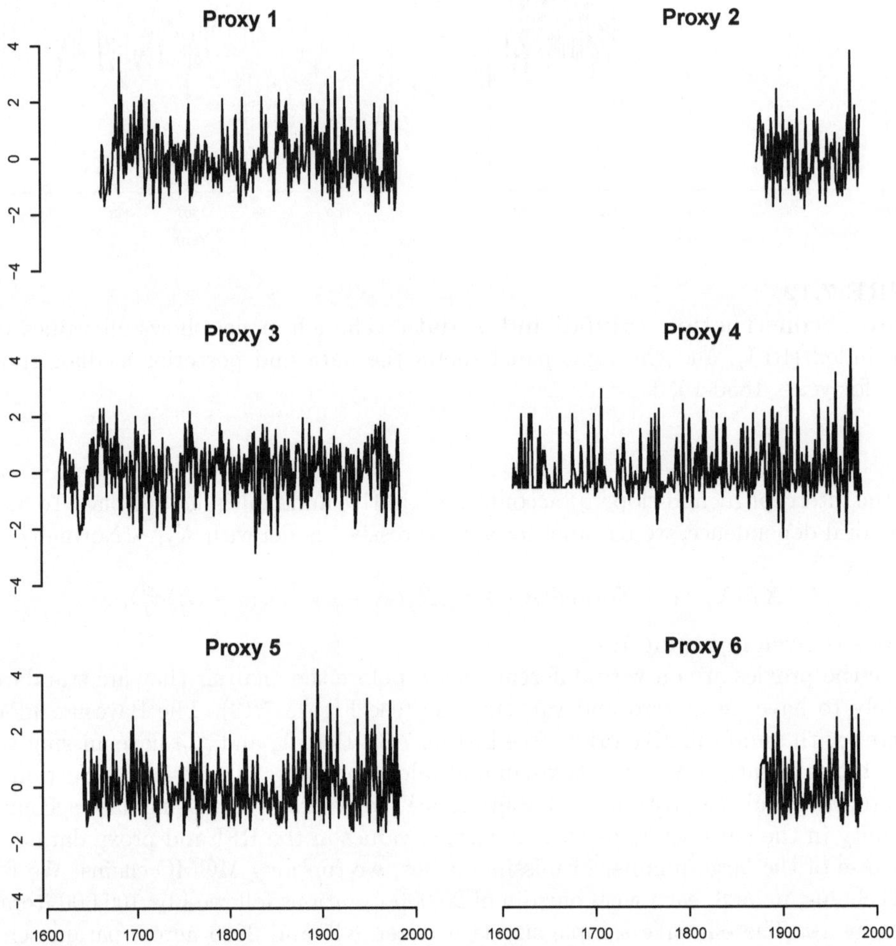

FIGURE 7.13
Climate reconstruction proxy data. The six proxy variables are plotted after standardization.

Listing 7.6
JAGS model statement for the Brisbane climate reconstruction.

```
# RFI observation model
Y[1] ~ dnorm(mu0,tau0)
for(t in 2:n){
  Y[t] ~ dnorm(mu0+phi0*(Y[t-1]-mu0),tauAR0)
}

phi0    ~ dunif(0,1)
mu0     ~ dnorm(0,0.001)
tau0    ~ dgamma(0.1, 0.1)
tauAR0 <- tau0/(1-phi0*phi0)

# Proxy model
for(j in 1:6){
  mu[1,j] <- Alpha[j] + Beta[j]*Y[1]
  X[1,j]   ~ dnorm(mu[1,j],tau[j])
  for(t in 2:n){
    mu[t,j] <- Alpha[j] + Beta[j]*Y[t]
    X[t,j]   ~ dnorm(mu[t,j] + phi[j]*(X[t-1,j]-mu[t-1,j]),
                     tauAR[j])
  }
  Alpha[j]  ~ dnorm(0, 0.1)
  Beta[j]   ~ dnorm(0, 0.1)
  phi[j]    ~ dunif(0,1)
  tau[j]    ~ dgamma(0.1, 0.1)
  tauAR[j] <- tau[j]/(1-phi[j]*phi[j])
}
```

TABLE 7.2
Posterior distributions of the average RFI in Brisbane by time period. Posterior mean, standard deviation and 90% intervals of the mean RFI $\bar{Y}_k$ by time period and the posterior probability that each period's mean RFI is higher than the RFI of the present period (2000-2017).

Period	Mean	SD	90% interval	Prob > present
1612-1649	809	213	(455, 1129)	0.42
1650-1699	685	151	(399, 888)	0.09
1700-1749	731	149	(460, 946)	0.17
1750-1799	533	173	(203, 746)	0.00
1800-1849	600	164	(305, 806)	0.02
1850-1899	923	95	(759, 1074)	0.78
1900-1949	860	0	(860, 860)	
1950-1999	938	0	(938, 938)	
2000-2017	851	0	(851, 851)	

Year-to-year variation in RFI is high, so studying changes over time is more reliable using average RFI over time periods. Below we study average RFI over periods 1612-1649, 1650-1699, ... ,1950-1999 and 2000-2017. Letting P_k be the indices of the n_k years that are included in period k, the average precipitation in period k is $\bar{Y}_k = \sum_{t \in P_k} Y_t / n_k$. The posterior of $\bar{Y}_k$ is approximated using MCMC by computing $\bar{Y}_k$ for each MCMC iteration using the imputed values, which produces posterior draws of $\bar{Y}_k$.

Table 7.2 summarizes the posterior distribution of $\bar{Y}_k$ for each period, and gives the posterior probability that each $\bar{Y}_k$ is larger than $\bar{Y}_k$ for 2000-2017. For recent periods, there are no missing data, and thus $\bar{Y}_k$ has no uncertainty, whereas for distant periods $\bar{Y}_k$ is based solely on dependence with the proxy data and thus has large uncertainty. The present period, 2000-2017, is drier than the previous period, 1950-1999. However, there is considerable variation across periods in the posterior mean, and due to missing data the posterior standard deviations are also large. There are even two historical periods spanning 1750-1850 with probability near one of being drier than the present period.

7.3 Exercises

1. Since full conditional distributions are used in many MCMC algorithms, it is tempting to specify the model via its conditional distributions. For example, consider the conditional distributions:

$$Y|X \sim \text{Normal}(aX, 1) \quad \text{and} \quad X|Y \sim \text{Normal}(bY, 1).$$

   (a) Select values of a and b so that these full conditional distributions are incompatible; i.e., there is no valid joint distribution for X and Y that gives these full conditional distributions. Argue, but do not formally prove, your assertion that the full conditional distributions are incompatible.
   (b) Explain why building a model that produces a valid DAG will always lead to a valid joint distribution.

2. Draw a DAG for the hierarchical model in Section 7.2.3.

3. In this problem we will conduct a meta analysis, i.e., an analysis that combines the results of several studies. The data are from the `rmeta` package in `R`:

```
> library(rmeta)
> data(cochrane)
> cochrane
          name ev.trt n.trt ev.ctrl n.ctrl
1     Auckland     36   532      60    538
2        Block      1    69       5     61
3        Doran      4    81      11     63
4        Gamsu     14   131      20    137
5     Morrison      3    67       7     59
6 Papageorgiou      1    71       7     75
7      Tauesch      8    56      10     71
```

   The data are from seven randomized trials that evaluate the effect of corticosteroid therapy on neonatal death. For trial $i \in \{1, ..., 7\}$ denote Y_{i0} as the number of events in the N_{i0} control-group patients and Y_{i1} as the number of events in the N_{i1} treatment-group patients.

   (a) Fit the model $Y_{ij}|\theta_j \overset{indep}{\sim} \text{Binomial}(N_{ij}, \theta_j)$ with $\theta_0, \theta_1 \sim \text{Uniform}(0, 1)$. Can we conclude that the treatment reduces the event rate?
   (b) Fit the model $Y_{ij}|\theta_{ij} \overset{indep}{\sim} \text{Binomial}(N_{ij}, \theta_{ij})$ with $\text{logit}(\theta_{ij}) = \alpha_{ij}$ and $\boldsymbol{\alpha}_i = (\alpha_{i0}, \alpha_{i1})^T \overset{iid}{\sim} \text{Normal}(\boldsymbol{\mu}, \boldsymbol{\Sigma})$, $\boldsymbol{\mu} \sim \text{Normal}(0, 10^2 I_2)$, and $\boldsymbol{\Sigma} \sim \text{InvWishart}(3, I_2)$. Summarize the evidence that the treatment reduces the death rate.
   (c) Discuss the advantages and disadvantages of both models.
   (d) Which model is preferred for these data?

4. Download the marathon data of Section 6.4 from the course webpage. Let Y_{ij} be the speed of runner i in mile j. Fit the hierarchical model $Y_{i1} \sim \text{Normal}(\mu_i, \sigma_0^2)$ and

$$Y_{ij}|Y_{ij-1} \sim \text{Normal}(\mu_i + \rho_i(Y_{ij-1} - \mu_i), \sigma_i^2),$$

   where $\mu_i \overset{iid}{\sim} \text{Normal}(\theta_1, \theta_2)$, $\rho_i \overset{iid}{\sim} \text{Normal}(\theta_3, \theta_4)$, and $\sigma_i^2 \overset{iid}{\sim} \text{InvGamma}(\theta_5, \theta_6)$.

(a) Draw a DAG for this model, and give an interpretation for each parameter in the model.
(b) Select uninformative prior distributions for $\theta_1, ..., \theta_6$.
(c) Fit the model in JAGS using three chains each with 25,000 iterations, and thoroughly assess MCMC convergence for the θ_j.
(d) Are the data informative about the θ_j? That is, are the posterior distributions more concentrated than the prior distributions?
(e) In light of (c) and (d), are there any simplifications you might consider, and if so, how would you compare the full and simplified models?

5. Download the `lung` data from the survival package in R. Let Y_i be the survival time (days) for subject $i \in \{1, ..., n\}$. We will make the unrealistic assumption that the survival times follow an exponential distribution, $Y_i|\lambda \sim \text{Exponential}(\lambda)$. The study runs from time 0 to time T, with patient i joining the study at time $T - T_i$ so that they are part of the study for T_i days. The analysis is complicated by censoring; i.e., some patients have not experienced their event at the conclusion of the study. An event is observed if $Y_i < T_i$ and censored if $Y_i > T_i$, with $\delta_i = I(Y_i > T_i)$ denoting the binary indicator that the event is censored. For uncensored patients with $\delta_i = 0$, the contribution to the likelihood function is the exponential PDF, $\lambda \exp(-\lambda y_i)$; for censored patients with $\delta_i = 1$, the contribution to the likelihood function is the probability that they survive beyond T_i days, $\exp(-\lambda T_i)$.

   (a) Write the likelihood function in terms of the parameter λ and the data $(Y_1, T_1, \delta_1), ..., (Y_n, T_n, \delta_n)$.
   (b) Derive a conjugate prior distribution for λ, and propose hyperparameters that lead to an uninformative prior distribution.
   (c) Using the full dataset, summarize the posterior distribution of λ under a few conjugate priors (i.e., different hyperparameters). Are the results sensitive to the prior?
   (d) Using an uninformative conjugate prior, analyze the data separately for men and women, and test whether the survival distribution varies by sex.

6. Download the Gambia data

```
> library(geoR)
> data(gambia)
> ?gambia
```

   The data consist of 2,035 children that live in 65 villages. For village $v \in \{1, ..., 65\}$, denote n_v as the number of children in the sample, Y_v as the number of children that tested positive for malaria, and p_v as the true probability of testing positive for malaria. We use the spatial model $\alpha_v = \text{logit}(p_v)$, where $\boldsymbol{\alpha} = (\alpha_1, ..., \alpha_{65})^T$ follows a multivariate normal distribution with mean $\text{E}(\alpha_v) = \mu$, variance $\text{Var}(\alpha_v) = \sigma^2$ and correlation $\text{Cor}(\alpha_u, \alpha_v) = \exp(-d_{uv}/\rho)$, where d_{uv} is the distance between villages u and v. For priors assume $\mu \sim \text{Normal}(0, 10^2)$, $\sigma^2 \sim \text{InvGamma}(0.1, 0.1)$, and $\rho \sim \text{Uniform}(0, d^*)$ where d^* is the maximum distance between villages.

   (a) Specify the data layer, process layer and prior layer for this hierarchical model.
   (b) Fit the model using JAGS and assess convergence.

(c) Summarize the data and results using five maps: the sample size n_v, the sample proportion Y_v/n_v, the posterior means of the p_v, the posterior standard deviations of the p_v, and the posterior probabilities that p_v exceeds 0.5.

8

Machine learning

CONTENTS

A classic Bayesian analysis strives for a fairly simple statistical model with a handful of interpretable parameters. This is ideal for testing specific hypotheses or understanding the main drivers of the process of interest. This approach requires careful model building and checking to ensure that the chosen model fits the data well and collaboration with domain experts to ensure the model is scientifically plausible. Due to this need for supervision, it is difficult to deploy at scale.

As a concrete example, consider the linear regression model in Section 4.2. The multiple linear regression model for response Y_i and covariates $X_{i1}, ..., X_{ip}$ is $Y_i = \sum_{j=1}^{p} X_{ij}\beta_j + \varepsilon_i$, where the errors are $\varepsilon_i \stackrel{iid}{\sim} \text{Normal}(0, \sigma^2)$. Fitting this model requires first selecting the response variable and the p covariates to include and then verifying four key assumptions:

(1) Linearity: The mean of $Y_i|\mathbf{X}_i$ is linear in $X_{i1}, ..., X_{ip}$.

(2) Equal variance: The residual variance (σ^2) is the same for all i.

(3) Normality: The errors ε_i are Gaussian.

(4) Independence: The errors ε_i are independent.

In real analyses, most if not all of these assumptions will be violated to some extent. Minor violations will not invalidate statistical inference, but glaring model misspecifications should be addressed.

Machine learning (ML) is a branch of statistics that attempts to automate this process by introducing robust statistical procedures that are less dependent on expert knowledge and make fewer statistical assumptions that require empirical verification. In this chapter, we discuss ML approaches to generalizing the Bayesian linear regression model. The first step in fitting a linear model is to select the covariates to be included. A classical analysis proceeds by fitting models sequentially and comparing their fits in combination with graphical checks. ML alternatives presented in Section 8.1 gather a large collection of covariates and fit a single Bayesian model that simultaneously selects important covariates and estimates their effects. This is generalized in Section 8.2 to identify the relationships between a large number of variables using graphical models. The linearity assumption is relaxed in Section 8.3 with a collection of methods that specify a prior for the mean response that is potentially a non-linear functions of the covariates. Moving from the mean to the residuals, Section 8.4 introduces models for the residual distribution that are non-Gaussian and depend on the covariates.

These ML methods are closely related to Bayesian nonparametrics [62]. The classic model is called a parametric model because it assumes the only unknowns are a small number of parameters, and all other aspects of the model including linearity, Gaussian residuals, etc. are given. To avoid making invalid assumptions, a Bayesian nonparametric analysis specifies priors on functions (i.e., infinitely-many parameters) rather than the few parameters that determine the functions. For example, rather than assuming the residuals follow a Gaussian distribution with unknown variance σ^2, a nonparametric analysis specifies a model for the residual distribution function that allows the distribution to be any valid continuous distribution function, encompassing left-skewed, right-skewed or even multi-modal distributions.

Combined with large datasets, Bayesian ML/nonparametric models often give better fidelity to the data and more accurate out-of-sample predictions than parametric models. However, a consequence of selecting a complicated approach is that the results are less interpretable. In the extreme, the statistical model becomes essentially a "black box" that produces accurate predictions but does not enhance scientific understanding of the relationships between variables. To close this gap, Section 8.5 provides several techniques to extract scientific meaning from fitted ML models.

Armed with these Bayesian ML methods, Section 8.6 demonstrates several applications ranging from causal inference to equation learning. In most of these applications, the ML methods are not used merely as black-box prediction algorithms, but rather they are incorporated in analyses to uncover information about biological or physical systems such as determining the causal effect of a treatment or the most important drivers of extreme hurricane winds.

Bayesian ML/nonparametric methods offer several advantages over optimization-based ML. These are largely the same advantages discussed in previous chapters for parametric models. The most important advantage is uncertainty quantification. Frequentist uncertainty quantification (p-values, confidence intervals) often relies on asymptotic arguments with the sample size far exceeding the number of parameters, which does not apply to nonparametric methods with infinitely-many parameters. Bayesian uncertainty quantification methods (posterior probabilities, credible intervals) maintain their validity even in nonparametric problems. Another advantage is the potential for incorporating prior information. ML methods are flexible but require large datasets for training for their potential to be realized. Prior distributions that center ML models on simple parametric models, or incorporate information from related datasets (see transfer learning methods in Section 8.6.4) can supplement ML methods when training data is lacking. One limitation of Bayesian ML methods is computation. Therefore, in this section we largely move away from JAGS in favor of tailored ML packages.

8.1 High-dimensional linear regression

In classic linear regression, the sample size n is much larger than the number of covariates p. In this case, the likelihood overwhelms the prior, and Bayesian results mirror those from least squares (Section 4.2). However, many modern applications use high-throughput data and have p on the order of n or even much larger than n. As an example, consider the data from [94] that can be downloaded from the course webpage. In this study, $n = 60$ mice (31 female) were sampled, and the physiological phenotype stearoyl-CoA desaturase 1 (SCD1) is taken as the response to be regressed onto the expression levels of 22,575 genes. Following [19] and [172], we use only the $p = 1,000$ genes with highest pairwise correlation with the response as predictors in the model. Even after this simplification, this leaves a high-dimensional problem with $p > n$.

The regression coefficients cannot all be estimated in high-dimensional problems without strong assumptions. In a Bayesian analysis, assumptions are encoded in the prior distribution. A common assumption is sparsity, which presumes that many of the regression coefficients are equal to or at least close to zero, and thus only a small subset of the covariates are needed to predict the response. In Section 4.2.2.3 the assumption of sparsity was induced by the Bayesian LASSO prior. This prior is a continuous PDF with a peak at the origin. While this prior gives probability zero to the regression coefficients being exactly zero, the peak shrinks parameter estimates towards zero. The Stochastic Search Variable Selection (SSVS) prior in Section 6.2 goes further by adding discrete point-mass distribution so that the regression coefficients have positive probability of being exactly zero. In this chapter, we more thoroughly discuss continuous and discrete shrinkage priors for high-dimensional regression.

Continuous shrinkage priors: The multiple linear regression model is

$$Y_i|\boldsymbol{\beta}, \sigma^2 \sim \text{Normal}\left(\sum_{j=1}^{p} X_{ij}\beta_j, \sigma^2\right).$$

Denote the prior for the regression coefficients as $\beta_j \sim \pi(\beta)$ for $j = 2, ..., p$. Continuous shrinkage priors assume that π is a continuous function of β, but with a large peak at zero to encourage sparsity and heavy tails to allow signals to emerge. One example is the double exponential prior in Figure 8.1, which leads to the posterior mode being the LASSO solution (4.15).

The double exponential prior is just one example of a shrinkage prior with peak at zero and heavy tails. The horseshoe prior [30] is

$$\beta_j|\tau_j, \sigma^2 \sim \text{Normal}(0, \sigma^2\tau_0^2\tau_j^2) \quad \text{and} \quad \tau_j \stackrel{iid}{\sim} \text{HC}(1) \tag{8.1}$$

where τ_0^2 is global variance common to all regression coefficients and τ_j is a local prior standard deviation specific to β_j. The local standard deviations are given half-Cauchy priors (i.e., Student's t-distribution with location 0, scale 1 and one degree of freedom, restricted to being positive). The prior derives its name from the special case with $Y_i|\beta_i, \sigma^2 \sim \text{Normal}(\beta_i, \sigma^2)$ for $i \in \{1, ..., p\}$ and $\tau_0 = 1$. In this case, the posterior mean given τ_i is $\text{E}(\beta|Y_i) = \{1 - \kappa(\tau_i)\}Y_i$ for $\kappa(\tau_i) = 1/(1 + \tau_i^2)$. Shrinkage is controlled by $\kappa(\tau_i)$, with $\kappa(\tau_i) = 0$ giving $\text{E}(\beta|Y_i) = Y_i$ and no shrinkage and $\kappa(\tau_i) = 1$ giving $\text{E}(\beta|Y_i) = 0$ and complete shrinkage. Under the half-Cauchy prior for τ_i, $\kappa(\tau_i) \sim \text{Beta}(1/2, /1/2)$ which is shaped like a horseshoe with mass near zero and one, allowing for no shrinkage and complete shrinkage.

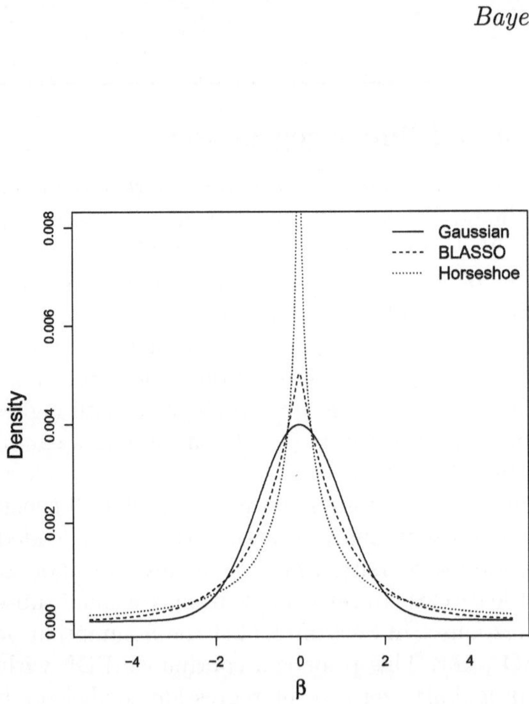

FIGURE 8.1
Comparison of the Gaussian and shrinkage priors. Plotted are the standard normal, double exponential/BLASSO and horseshoe PDFs, all set to have the same interquartile range.

The inclusion of local shrinkage parameters allows the global-local skrinkage prior the flexibility to shrink null coefficients towards zero by having small variance while the true signals to have uninformative priors with large variance. The marginal distribution of β_j over τ_j has a complicated expression,

$$\pi_j(\beta|\sigma^2, \tau_0) = \int_0^\infty \phi(\beta, 0, \sigma^2\tau_0^2\tau_j^2)h(\tau_j)d\tau_j$$

where ϕ is the Gaussian PDF and h is the half-Cauchy PDF. However, Figure 8.1 shows that the marginal horseshoe prior gives more (actually infinite) mass at zero and in the tails than the Bayesian LASSO prior.

The Dirichlet–Laplace prior [18] gives even more shrinkage towards zero by supplementing the Bayesian LASSO with local shrinkage parameters and a Dirichlet prior on the shrinkage parameters. The R2D2 prior [172] is another global-local shrinkage prior for the regression coefficients that is constructed so that the model's coefficient of determination (i.e., R-squared) has a beta prior as in (4.18). The R2D2 prior has the most mass at zero and heaviest tail among these priors.

To illustrate the use of continuous shrinkage priors for high-dimensional regression, we analyze the genetic data described above using the horseshoe prior as in Listing 8.1. Recall that if $\tau|b \sim \text{Gamma}(1/2, b)$ and $b \sim \text{Gamma}(1/2, 1)$, then $\sigma = 1/\sqrt{\tau} \sim \text{HC}(1)$. Therefore, conveniently, with the parameterization in Listing 8.1 the MCMC can be entirely Gibbs sampling. We compare the results using the horseshoe prior with Gaussian and BLASSO priors. The posterior distribution of selected β_j are plotted in Figure 8.2. It is difficult to plot the posterior for all $p = 1,000$ regression coefficients, so only the 50 with smallest (left) and

Listing 8.1
JAGS code for regression with horseshoe priors.

```
for(i in 1:n){
  Y[i]    ~ dnorm(mu[i],taue)
  mu[i] <- alpha + inprod(X[i,],beta[])
}
for(j in 1:p){ # Horseshoe priors
  beta[j]  ~ dnorm(0,prec[j])
  prec[j] <- taue*tau0*tau[j]
  tau[j]   ~ dgamma(1/2,b[j]); b[j]    ~ dgamma(1/2, 1)
}
alpha   ~ dnorm(0, 0.01)
taue    ~ dgamma(1/2,be); be    ~ dgamma(1/2, 1)
tau0    ~ dgamma(1/2,b0); b0    ~ dgamma(1/2, 1)
```

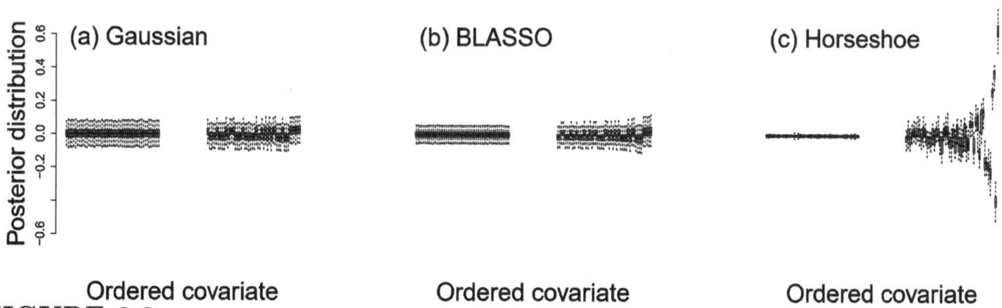

FIGURE 8.2
Comparison of shrinkage priors for the genetic data. For each methods, the posterior distribution (boxplot) is plotted for the 50 regression coefficients β_j with smallest (left) and largest (right) absolute posterior mean.

largest (right) absolute posterior mean are shown in Figure 8.2. The Gaussian and BLASSO methods result in posterior distribution centered near zero and similar posterior variances for all β_j; therefore these methods are unable to isolate important predictors and separate them for null predictors. In contrast, the horseshoe prior results in many coefficients with posterior mean and variance near zero, effectively removing these covariates from the model. Despite this aggressive shrinkage for some of the parameters, the effects of other covariates have large posterior mean and variance and credible sets that exclude zero. In this example, the global-local shrinkage prior results in a more satisfactory sparse solution.

Stochastic search variable selection: While continuous shrinkage priors provide excellent estimation and prediction, they do not exactly set regression coefficients to zero, and technically all covariates are included in the model with probability one, even if their effects are negligible. Therefore, selecting a single reduced model requires some post-processing ([19, 71]). On the other hand, SSVS does set coefficients exactly to zero which allows for intuitive posterior summaries such as the most likely model or the probability that a covariate is included in the model.

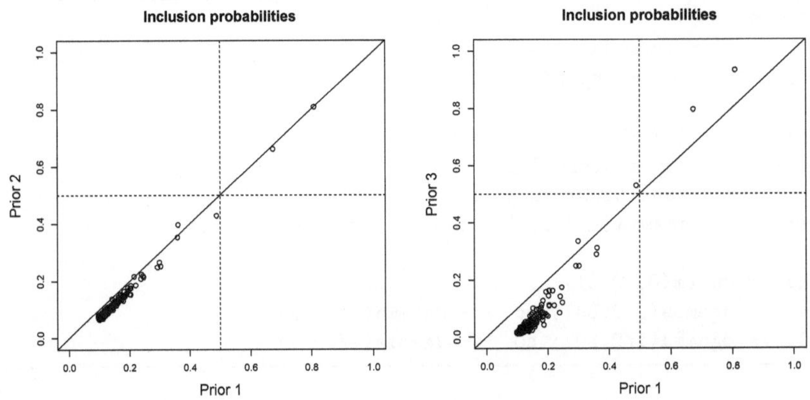

FIGURE 8.3
High-dimensional regression example. Marginal inclusion probabilities ($MIP_j = \text{Prob}(\beta_j \neq 0|\mathbf{Y})$) under three priors: (1) $q \sim \text{Beta}(1,1)$ and $\tau^2 \sim \text{InvGamma}(0.1, 0.1)$, (2) $q \sim \text{Beta}(1,2)$ and $\tau^2 \sim \text{InvGamma}(0.1, 0.1)$, and (3) $q \sim \text{Beta}(1,1)$ and $\tau^2 \sim \text{InvGamma}(0.5, 0.5)$.

The linear regression model with SSVS prior is

$$\begin{aligned} Y_i &\sim \text{Normal}\left(\sum_{j=1}^{p} X_{ij}\beta_j, \sigma^2\right) \\ \beta_j &= \gamma_j\delta_j \text{ for } j > 1 \\ \gamma_j &\sim \text{Bernoulli}(q) \\ \delta_j &\sim \text{Normal}(0, \sigma^2\tau^2). \end{aligned} \tag{8.2}$$

Because p is large, it is possible to learn about the hyperparameters q and τ^2. We select priors $q \sim \text{Beta}(1,1)$ and $\sigma^2, \tau^2 \sim \text{InvGamma}(0.1, 0.1)$ (Prior 1) and present the marginal inclusion probabilities $\text{Prob}(\beta_j \neq 0|Y)$ as a measure of variable importance. In high-dimensional problems, sensitivity to the prior is especially concerning. Therefore, we also refit with priors $q \sim \text{Beta}(1,2)$ (Prior 2) and $\tau^2 \sim \text{InvGamma}(0.5, 0.5)$ (Prior 3). The models are fit using Gibbs sampling with 200,000 iterations, a burn-in of 20,000 iterations discarded and the remaining samples thinned by 20. This MCMC code requires 1–2 hours on an ordinary PC.

Applied to the genetic data above, only two genes have inclusion probability greater than 0.5 (Figure 8.3). The ordering of the marginal inclusion probabilities is fairly robust to the prior, but the absolute value of the marginal inclusion probabilities varies with the prior. The posterior median (95% interval) of the number of variables included in the model $p_{in} = \sum_{j=2}^{p} \gamma_j$ are 33 (12, 535) for Prior 1, 28 (12, 392) for Prior 2 and 20 (11, 64) for Prior 3. In this analysis, the results are more sensitive to the prior for the variance than the inclusion probability.

8.2 Graphical models

In contrast with linear regression where one variable is designated as the response variable of interest and the others are deemed covariates, in a graphical model, the objective is to uncover relationships between p variables without prespecified roles. In this sense, graphical models are less structured and thus truer to the machine learning paradigm than linear regression.

Let $\mathbf{Y} = (Y_1, ..., Y_p)^T$ be the vector of outcomes; e.g., $\mathbf{Y}$ may include the activity levels of p brain regions, and the objective is to uncover functional relationships between the regions. If the interest is primarily in region j, then we would perform a linear regression with outcome Y_j and $p-1$ covariates Y_k for $k \neq j$,

$$Y_j|Y_k \text{ for } k \neq j \sim \text{Normal}\left(\sum_{k\neq j} Y_k\beta_{jk}, \sigma_j^2\right).$$

However, in a graphical analysis, the goal is to uncover relationships among all the variables. This could be accomplished with p regressions, one with each variable as the outcome and the other as the responses. However, this might lead to incoherent results, that is, since the regressions are fit separately, there may not be a valid joint distribution for all p variables that gives the fitted conditional distributions.

The simplest graphical model is the multivariate normal model $\mathbf{Y} \sim \text{Normal}(\mathbf{0}, \mathbf{\Sigma})$. Since the goal is to study relationships between the p variables, we focus on the $p \times p$ covariance matrix $\mathbf{\Sigma}$ and assume the data have been centered to have mean zero. This coherent joint model gives interpretable conditional distributions in terms of the precision matrix $\mathbf{\Omega} = \mathbf{\Sigma}^{-1}$ with (j, k) element Ω_{jk},

$$Y_j|Y_k \text{ for } k \neq j \sim \text{Normal}\left(\sum_{k\neq j} Y_k b_{jk}, \tau_j^2\right),$$

where $b_{jk} = -\Omega_{jk}/\Omega_{jj}$ and $\tau_j^2 = 1/\Omega_{jj}$. The conditional mean is linear in the other Y_k with b_{jk} determining the effect of Y_k on Y_j after accounting for the effects of the other $p-2$ variables. Therefore, fitting a multivariate normal model to $\mathbf{Y}$ is effectively equivalent to fitting p linear regressions but with restrictions that the models are coherent.

The interpretation of the elements of $\mathbf{\Sigma}$ and $\mathbf{\Omega}$ in the Gaussian graphical model are subtly different. If the (j, k) element of $\mathbf{\Sigma}$ is zero, then Y_j and Y_k are marginally independent; i.e., they are independent when ignoring the effects of the other variables. In contrast, if the (j, k) element of $\mathbf{\Omega}$ is zero, then Y_j and Y_k are conditionally independent; i.e., they are independent after removing the effects of the other $p-2$ variables. Formally, Y_j and Y_k are said to be conditionally (on the other $p-2$ variables) independent if and only if $\Omega_{jk} = 0$, so testing whether elements of $\mathbf{\Omega}$ are zero gives a test of conditional independence.

Under some conditions, conditional dependence can be interpreted as causal relationships (see Section 8.6.1) rather than simply correlation. To see this, say a particular brain activity is caused by region 1 sending a signal to region 2 that is then passed to region 3 then region 4. This could be encoded in the model $Y_1 \sim \text{Normal}(0, 1)$ and $Y_j|Y_{j-1} \sim \text{Normal}(Y_{j-1}, 1)$ for $j \in \{2, 3, 4\}$, giving covariance and precision matrices

$$\mathbf{\Sigma} = \begin{bmatrix} 1 & 1 & 1 & 1 \\ 1 & 2 & 2 & 2 \\ 1 & 2 & 3 & 3 \\ 1 & 2 & 3 & 4 \end{bmatrix} \text{ and } \mathbf{\Omega} = \begin{bmatrix} 2 & -1 & 0 & 0 \\ -1 & 2 & -1 & 0 \\ 0 & -1 & 2 & -1 \\ 0 & 0 & -1 & 1 \end{bmatrix}.$$

Since there are no zeros in Σ, all four regions are correlated. However, the pattern of zeros in $\mathbf{\Omega}$ reflects the causal pathway through the four regions. Therefore, viewing the variables as nodes that are connected by edges if the corresponding elements of $\mathbf{\Omega}$ are non-zero produces a graph of the relationship between the variables, hence the term "graphical model".

While the Gaussian graphical model provides a more comprehensive understanding of the relationship between variables than a linear regression model, this comes at the expense of having to estimate the $p(p-1)/2$ elements in the covariance matrix as opposed to the p linear regression coefficients. Therefore, a prior distribution that encodes sparsity with many elements of $\mathbf{\Omega}$ shrunk towards zero to avoid over-fitting is even more critical here than in linear regression. The conjugate prior for $\mathbf{\Omega}$ is the Wishart prior (Section 2.1.7); however, this does not encode sparsity. The ingenious prior of [161] for $\mathbf{\Omega}$ implies a Bayesian LASSO prior on the off-diagonal elements of $\mathbf{\Omega}$ and thus encourages sparsity and permits Gibbs sampling updates for the elements of $\mathbf{\Omega}$.

8.3 Nonparametric regression

Linear regression is one of the most successful tools in all of statistics owing to its stability, ease of implementation and convenient interpretation. However, when the linear assumption is grossly violated, the linear model can suffer from poor prediction and inferential properties. In this chapter, we introduce a rich collection of models for the mean of the response distribution. The model is assumed to be

$$Y_i = g(\mathbf{X}_i) + \varepsilon_i. \tag{8.3}$$

As before, we use Gaussian errors $\varepsilon_i \overset{iid}{\sim} \text{Normal}(0, \sigma^2)$. However, the mean response, denoted as $\text{E}(Y|\mathbf{X}) = g(\mathbf{X}) = g(X_1, ..., X_p)$, is now allowed to be an arbitrarily complex function of the covariates including non-linear effects and interactions between covariates. We describe the models for g for an arbitrary covariate vector $\mathbf{X} = (X_1, ..., X_p)$ but continue to use $\mathbf{X}_i = (X_{i1}, ..., X_{ip})$ to describe a specific covariate vector for a specific observation.

Of course, the linear model remains appealing for its simplicity and transparency, and thus it is important to consider simple modifications to the linear model before launching into a more sophisticated analysis. For example, if a linear model is found to be unsatisfactory, an alternative could be the second-order model

$$g(\mathbf{X}) = \sum_{j=1}^{p} X_j\beta_j + \sum_{j=1}^{p} X_j^2\alpha_j + \sum_{j<k} X_jX_k\gamma_{jk} \tag{8.4}$$

that includes linear (with effects β_j), non-linear/quadratic (with effects α_j) and interaction (with effects γ_{jk}) terms.

For small p, the second-order model is tidy and stable. However, for even a moderately-sized problem with $p = 10$, the full second-order model has 65 parameters, and interpretation with quadratic and interaction effects is not straightforward. It is possible to trim the model by adding only the non-linear and interaction terms that are motivated by exploratory analysis or scientific insights, but this is somewhat subjective. It is also possible to employ statistical model building tools to prune the model, such as stochastic search variable selection (Section 8.1), but this is now of comparable sophistication to the non-parametric regression models that are the focus of this chapter.

A fully nonparametric model allows for any continuous function g. That is, the target changes from a finite number of parameters (e.g., $\boldsymbol{\beta}$) in a parametric analysis to a function (e.g., a curve or surface g) in a nonparametric analysis. A model this flexible requires infinitely many parameters. We will primarily focus on semiparametric models that specify the mean function in terms of a finite number of parameters in a way that increasing the number of parameters can approximate any function g. For example, if there is only $p = 1$ covariate (X), we could fit a Jth order polynomial function

$$g(X) = \sum_{j=0}^{J} X^j \beta_j. \tag{8.5}$$

This model has $J+1$ parameters, and by increasing J the polynomial function can approximate any continuous function g.

We explore many Bayesian semiparametric/nonparametric regression models, including regression splines [36], Gaussian process regression [126], Bayesian adaptive regression trees [33] and neural networks [114]. While generalizations to non-Gaussian data are available (using similar tools to generalized linear mixed model in Section 4.5), we restrict our attention to the Gaussian case in (8.3). In this case, twice the negative log likelihood function is an irrelevant constant plus

$$\frac{1}{\sigma^2}\sum_{i=1}^{n}\{Y_i - g(\mathbf{X}_i)\}^2.$$

Therefore, point-estimators, particularly the maximum a posteriori estimator, have similarity with optimization-based machine learning algorithms that use squared-error loss as the objective function. However, as with connection between LASSO and Bayesian LASSO regression in Section 4.2, the interpretation and uncertainty quantification of the Bayesian methods differs from optimization-based approaches.

Hurricane wind forecasting application: We illustrate and compare nonparametric regression methods using the data in [6, 35]. The data consist of forecasted and observed tropical cyclone intensity, defined as the maximum 1-min average 10-m wind speed (knots), for 120 storms in the Atlantic and Pacific Oceans from 2014 to 2017. Two baseline methods are available for comparison. The first is the Hurricane Weather Research and Forecasting (HWRF) prediction. HWRF is a physics-based numerical model that takes current conditions as inputs and predicts the path and intensity of the storm. The second baseline is the forecast from the National Hurricane Center (NHC). Both predictions are made for lead times (hours) $l \in \{3, 6, ..., 48\}$, i.e., the prediction made on one day for the same time the following day is the 24-hour lead-time prediction. Excluding missing observations, across storm and lead time there are 37,207 observations. Figure 8.4 plots the mean absolute error $\sum_{i=1}^{n} |Y_i - \hat{Y}_i|/n$ for each baseline method and lead time; as expected the error increases with lead time.

In addition to the forecasted intensity, the HWRF model has many other state variables that describe the storm including the spatial location, sea-surface temperature, humidity, speed of the storm, current intensity, etc. [35] used both the HWRF forecast and these state variables as covariates (see Table 1 in [35]) in a machine learning model to refine the HWRF prediction, and showed that this improved forecasting precision, especially for rapid intensification events.

In this chapter, we revisit these data and model storm intensity (Y) based on $p = 18$ state variables ($\mathbf{X}$). The covariates are all standardized. In all analyses, models are trained separately by lead time using data from 2014-2016, and evaluated using data from 2017. As an initial analysis, including these state variables in a Bayesian linear regression improves predictions compared to the HWRF model alone (Figure 8.4) for all forecast lags. The

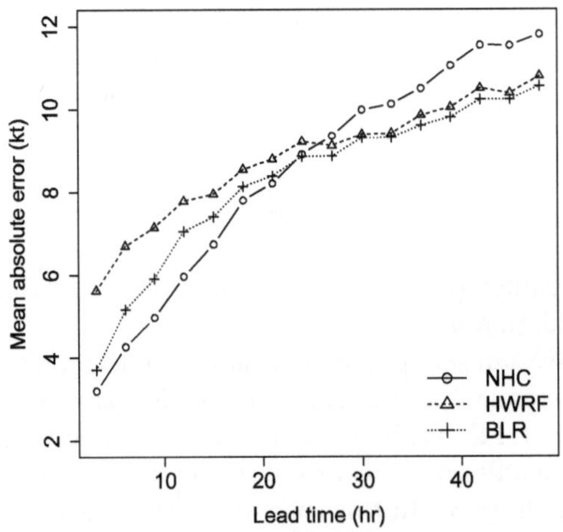

FIGURE 8.4
Hurricane intensity forecast error. Mean absolute prediction error by lead time for the Hurricane Weather Research and Forecasting (HWRF) model, the National Hurricane Center (NHC) and Bayesian linear regression (BLR) using HWRF state variables.

objectives in this chapter are to use nonparametric regression methods to reduce mean absolute error and access which covariates are driving the predictive model.

8.3.1 Spline regression and additive models

As an introduction, consider the `mcycle` data from the `MASS R` package plotted in Figure 8.5a. The predictor, X_i, is the time since a motorcycle makes impact (scaled to the unit interval) and the response, Y_i, is the acceleration of a monitor on the head of a crash test dummy. Clearly, a linear model will not fit these data well as the mean is flat for the first quarter of the experiment, then dramatically dips, rebounds, and then plateaus until the end of the experiment.

In spline regression, we construct non-linear functions of the original covariates and use these constructed covariates as the predictors in multiple linear regression. We begin by assuming a single covariate, i.e., $p = 1$. From this single covariate, we construct J covariates $B_1(X), ..., B_J(X)$ to capture non-linearity. In polynomial regression above, $B_j(X) = X^j$, but there are many other choices. For example, Figure 8.5b plots $J = 10$ B-spline basis functions. These functions are appealing because they are smooth and local, i.e., non-zero only for some values of X.

After constructing new covariates, the model is simply multiple linear regression model (Section 4.2),

$$Y_i|\boldsymbol{\beta} \overset{indep}{\sim} \text{Normal}\left\{g(X_i), \sigma^2\right\} \text{ and } g(X_i) = \beta_0 + \sum_{j=1}^{J} B_j(X_i)\beta_j.$$

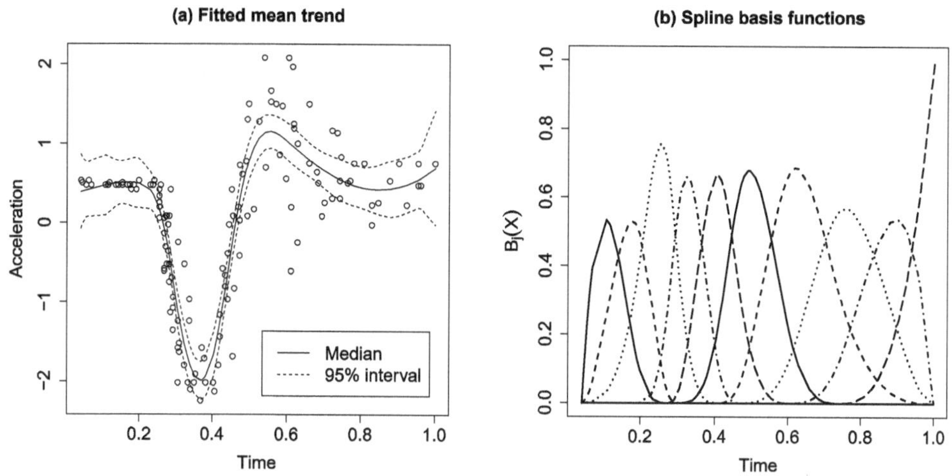

FIGURE 8.5
Nonparametric regression for the motorcycle data. Panel (a) plots the time since impact (scaled to be between 0 and 1) and the acceleration (g) along with the posterior median and 95% interval for the mean function from the homoskedastic fit; Panel (b) shows the $J = 10$ spline basis functions $B_j(X)$.

Note that each basis function in Figure 8.5b has $B_j(0) = 0$, and so an intercept (β_0) is required.

Selecting the number of basis function, J, is a critical step in fitting a spline model. By increasing J, any smooth mean function can be approximated as a linear combination of the B-spline basis functions, but large J risks over-fitting. One approach to selecting J is to assume uninformative priors, $\beta_j \stackrel{iid}{\sim} \text{Normal}(0, \sigma^2\tau^2)$ with τ^2 a large fixed constant, and then compare several J using model selection criteria such as cross-validation, DIC or WAIC. Another approach is to set J to be large, e.g., $J = n/2$, and allow the prior distribution for the β_j to prevent over-fitting via $\tau \sim \text{HC}(1)$. Especially when J is large, over-fitting can be guarded against by specifying a prior distribution that smooths across the splines coefficients,

$$\beta_j|\beta_{j-1} \sim \text{Normal}(\phi\beta_{j-1}, \sigma^2\tau^2)$$

where $\phi \in (0, 1)$ and τ^2 are both given prior distributions to allow the data to determine the amount of prior shrinkage.

Motorcycle example: To fit the mean curve to the data plotted in Figure 8.5a, we use $J = 10$ B-spline basis functions and priors $\beta_j \sim \text{Normal}(0, \tau^2\sigma^2)$ and $\sigma^2, \tau^2 \sim \text{InvGamma}(0.1, 0.1)$. The model is fit using MCMC with the code in Listing 4.2. In this code, the basis functions have been computed using the `bs` package in `R` and passed to `JAGS` as $X_{ij} = B_j(X_i)$. For each iteration, we compute $g(X_i)$ for all $i = 1, ..., n$ as a function of that iteration's posterior sample of $\boldsymbol{\beta}$. This produces the entire posterior distribution of the mean function g for all n sample points (or any other X we desire), and Figure 8.5a plots the posterior median and 95% interval of g at each sample point. The fitted model accurately captures the main trend including the valley around $X = 0.4$.

Monotonic regression: Nonparametric regression is used when the true form of the mean function is unknown. However, it can also incorporate partial knowledge. For example, if X is the dose of drug and Y is the response, then we might expect the dose-response curve $\text{E}(Y|X) = g(X)$ to be an increasing function of X. Monotonicity is easily accommodated using Bayesian splines regression. Using a B-spline basis expansion for g, the mean curve is increasing so that $g(X + \epsilon) > g(X)$ for all X and $\epsilon > 0$ if (though not only if) the regression coefficients are increasing, $\beta_1 < ... < \beta_J$. This restriction is straightforward to implement by reparameterizing the coefficients as $\beta_j = \beta_{j-1} + \delta_j$ for $j > 1$ and specifying a prior distribution for the δ_j with positive support, e.g., a gamma or truncated normal distribution. A Bayesian analysis including uncertainty quantification proceeds as usual, although depending on the prior distribution for δ_j a Metropolis-Hastings step may be required. Other shape-restricted regression models are available [105] to model curves that are decreasing, convex, unimodal, etc.

Generalized additive models: There are many ways to generalize spline regression to the multivariate case with $p > 1$ covariates. The most straightforward to implement and interpret is the generalized additive model (GAM) [166]. The main-effects-only GAM model represents the mean as the sum of p nonlinear functions,

$$g(\mathbf{X}) = \beta_0 + \sum_{j=1}^{p} g_j(X_j) \tag{8.6}$$

where g_j is a non-linear function the jth covariate. The non-linear function is interpreted analogously to linear regression coefficients; with all other covariates fixed, the mean response increases following $g_j(X_j)$ as X_j increases.

Each curve can be modeled using a spline basis expansion,

$$g_j(X_j) = \sum_{k=1}^{J} B_{jk}(X_j)\beta_{jk} \tag{8.7}$$

where $B_{j1}, ..., B_{jJ}$ are spline basis functions and $\beta_{jk} \overset{indep}{\sim} \text{Normal}(0, \sigma^2\tau_j^2)$ are the basis coefficients. Inserting (8.7) into (8.6) gives a standard Bayesian linear regression with constructed covariates $B_{jk}(X_j)$ and regression coefficients β_{jk}. The covariate-specific variance components, τ_j^2, allow for a differential amount of smoothing by covariate and can be given inverse gamma or half-Cauchy prior distributions. If p is large, Bayesian variable selection methods can be applied to simplify the model by shrinking irrelevant covarates. For example, [162] use a continuous shrinkage priors with differential shrinkage by covariate.

The main-effects-only GAM model includes non-linear effects, but not interactions. Following (8.4), the GAM model can be extended to include interactions

$$g(\mathbf{X}) = \beta_0 + \sum_{j=1}^{p} g_j(X_j) + \sum_{j<l} g_{jl}(X_j, X_l) \tag{8.8}$$

where g_{jl} is the interaction function for covariates j and l. The interaction function is a surface, rather than a curve, because it returns a value for any combination of the two covariates. The interaction surfaces can be modeled as outer products of spline basis functions

$$g_{jl}(X_j, X_l) = \sum_{k=1}^{J}\sum_{h=1}^{J} B_{jk}(X_j)B_{lh}(X_l)\beta_{jlkh}$$

Listing 8.2

GAM model in JAGS. `B` are the spline basis functions and `term[l]=j` indicates that column `l` of `B` corresponds to covariate `j`.

```
for(i in 1:n){
  Y[i]  ~ dnorm(g[i],taue)
  g[i] <- beta0 + inprod(B[i,1:q],beta[1:q])
}
for(l in 1:q){
  beta[l] ~ dnorm(0,taue*taub[term[l]])
}
for(j in 1:p){
 # Half-Cauchy priors
 taub[j]  ~ dgamma(0.5,b[j]); b[j] ~ dgamma(0.5,1)
}
taue ~ dgamma(0.5,be); be ~ dgamma(0.5,1)
beta0 ~ dnorm(0,0.01)
```

with prior distribution $\beta_{jlkh} \sim \text{Normal}(0, \sigma^2\tau_{jl}^2)$. As the number of basis functions increase, any smooth two-dimensional surface can be approximated using this representation. Extensions to three-way interactions and higher are possible, but are computationally demanding.

With the basis functions serving as covariates, both the main-effects or interactions models can be fit using standard MCMC algorithms including Gibbs updates. Posterior uncertainty about the curves g_j naturally follows from MCMC by computing the g_j at covariate values of interest at each MCMC iteration, and then summarizing the samples as in Figure 8.5. However, computation can be challenging for even moderate p. First, the number of parameter can be large, with roughly pJ parameters for the main effects and on the order of p^2J^2 for interactions. The computational burden can be eased by reducing the number of basis functions for the interaction terms or including only a subset of covariates for the interaction terms. Another complication is separating the effect of different covariates. If the basis functions are fixed at zero for some value of the covariates (see Figure 8.5b) then technically the curves can be identified. However, concurvity, a functional generalization of collinearity, makes estimation difficult. For example, if J is large than either g_1 or g_2 can effectively interpolate the observations. Therefore, even if the correlation between X_1 and X_2 is moderate, $g_1(X_1)$ and $g_2(X_2)$ compete to explain the same signal in the data and so their posterior distributions can be highly correlated.

Analysis of the hurricane intensity data: We fit the first-order GAM model in (8.6)-(8.7) with $J = \{3, 5\}$. Listing 8.2 gives the `JAGS` code to implement this model, but for faster computation we use the `R` package `mgcv` [165] with Metropolis sampling. The results are similar for different number of basis functions, so we focus on $J = 5$. Including the state variables reduces MAE compared to using HWRF directly for most prediction lead times, with the largest reductions for lead times 3-12 hours. For lead times greater 36 hours, the GAM model actually has smaller MAE than HWRF or NHC.

To understand the covariate effects, Figure 8.7 plots the mean and 90% intervals for $g_j(X)$ for three covariates and two lead times. The three covariates are: sea surface temperature (SST), the hurricane intensity forecast from HWRF (HWRF) and the intensity of the storm at the time of the forecast (VMAX_T0). SST does not appear to be an important predictor since its credible interval includes zero for all X. The effect of the current intensity is negative for large intensities. That is, if the storm is currently intense, the GAM model reduces the predicted intensity resembling a regression towards the mean. Finally, as might be expected, the effect for the HWRF prediction is an increasing, nearly-linear function.

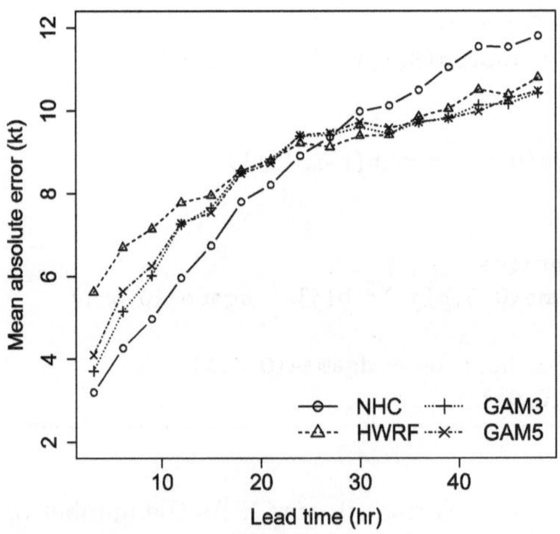

FIGURE 8.6
Hurricane intensity forecast error using a GAM. Mean absolute prediction error by lead time for the Hurricane Weather Research and Forecasting (HWRF) model, the National Hurricane Center (NHC) and Bayesian generalized additive models (GAM) using 3 and 5 basis functions per covariate.

8.3.2 Gaussian process regression

The spline regression model $Y_i|\beta \overset{indep}{\sim} \text{Normal}\{g(\mathbf{X}_i), \sigma^2\}$ with $g(\mathbf{X}_i) = \sum_{j=1}^{J} B_j(\mathbf{X}_i)\beta_j$ and $\beta_j \overset{iid}{\sim} \text{Normal}(0, \tau^2)$ resembles the random effects model with the β_j serving as random effects. As discussed in Section 4.4, the random effects induce dependence between observations. Averaging over the prior distribution of the β_j with $p = 1$ covariate,

$$\text{Cov}(Y_i, Y_k) = \tau^2 \sum_{j=1}^{J} B_j(X_i)B_j(X_k). \tag{8.9}$$

Therefore, spline regression can be viewed as a model for the dependence between observations with different covariate values, and the choice of basis functions determines the form of the covariance. As an example, if $B_j(X) = \exp\{-(X - v_j)^2\}$ for a sequence of fixed values v_j covering the range of X, then the sum in (8.9) is

$$\text{Cov}(Y_i, Y_k) \propto \sum_{j=1}^{J} \exp\left\{-(X_i - v_j)^2\right\} \exp\left\{-(X_k - v_j)^2\right\},$$

which for large J converges to

$$\begin{aligned} \text{Cov}(Y_i, Y_k) &\propto \int \exp\left\{-(X_i - v)^2\right\} \exp\left\{-(X_k - v)^2\right\} dv \\ &\propto \exp\left\{-\eta(X_i - X_k)^2\right\} \end{aligned}$$

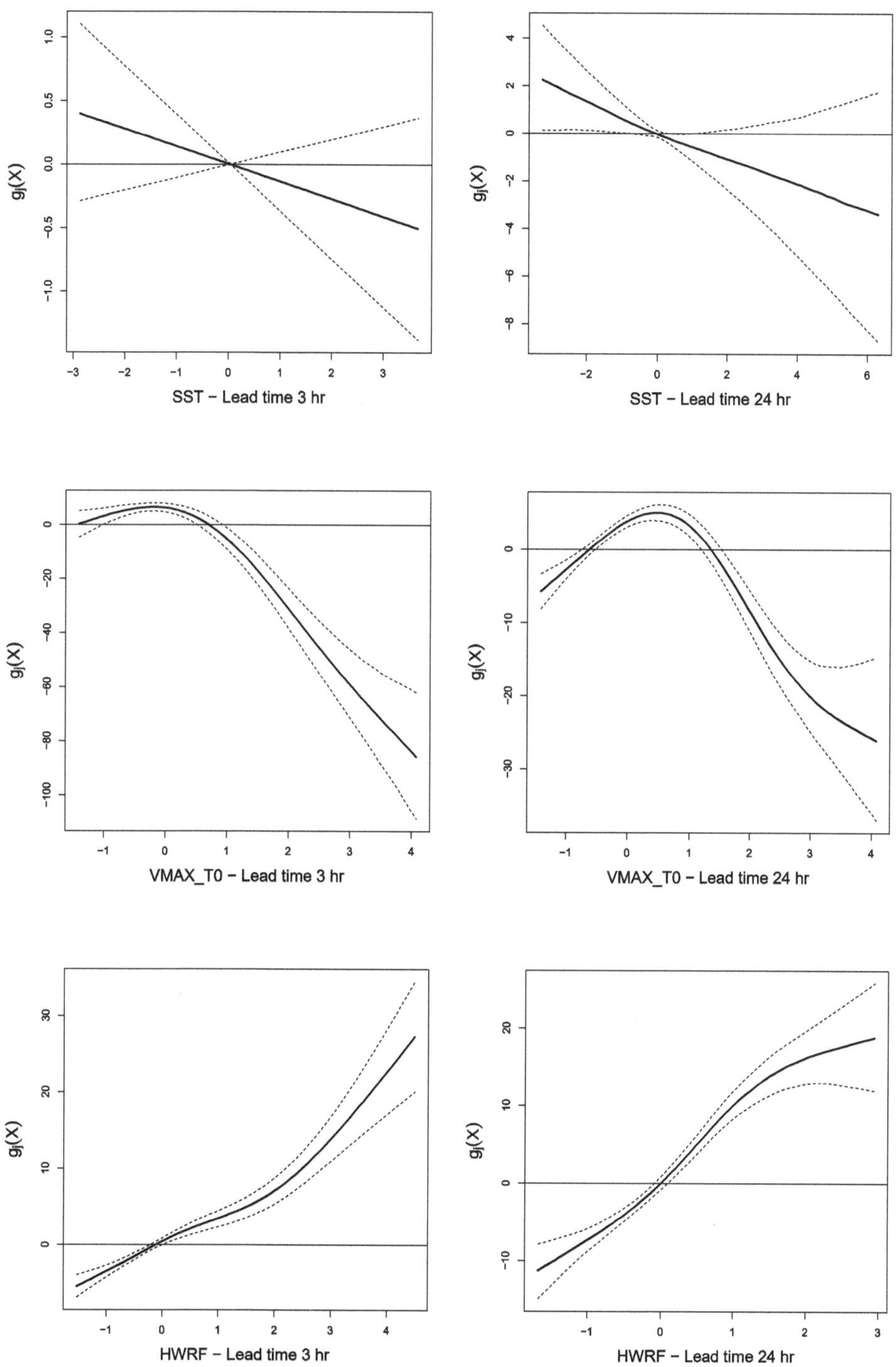

FIGURE 8.7
Hurricane intensity forecast main effects. Posterior mean (solid) and 90% intervals (dashed) for the main effects $g_j(X_j)$ for sea surface temperature (SST), current storm intensity (VMAX) and the Hurricane Weather Research and Forecasting (HWRF) intensity for lead times 3 and 24 hours. All covariates are standardized.

for some constant η. The correlation between two observations thus decays with the distance between their covariate values.

Gaussian process (GP) regression [126] views g as a stochastic process over covariate space. That is, it is not approximated using a finite number of basis functions, but rather it can be viewed as the process that arises in the limit as the number of basis function tends to infinity for a specific sequence of basis functions. This is accomplished by modeling its covariance directly as in Section 4.6. The function g is modeled as a GP, defined as a process such that for any N covariate values $\mathbf{X}_1, ..., \mathbf{X}_N$ the random vector $\mathbf{g} = \{g(\mathbf{X}_1), ..., g(\mathbf{X}_N)\}^T$ follows a multivariate normal distribution. The definition that a GP applies to any set of locations is key because in practice we will define a joint model for the n observed data points and an unknown set of $N - n$ prediction points.

The Gaussian process is defined by the mean and covariance functions used to determine the mean vector and covariance matrix of $\mathbf{g}$. The simplest mean function is $\text{E}\{g(\mathbf{X}_i)\} = \mu$ for all $\mathbf{X}_i$, although it can also be a linear function of the covariates. Similar to the spatial and temporal covariance functions in Section 4.6, the covariance functions is defined by distance between the covariate vectors of two observations, $d_{ik} = \{\sum_{j=1}^p (X_{ij} - X_{kj})^2\}^{1/2}$. Since this distance weights all p covariates equally it is said to be isotropic; an anisotropic extension is $d_{ik} = \{\sum_{j=1}^p w_j (X_{ij} - X_{kj})^2\}^{1/2}$ where $w_j > 0$ are weights and can be given prior distributions to allow the data to determine the best weighting scheme. The simplest covariance function is the Gaussian (squared exponential) covariance function

$$\text{Cov}\{g(\mathbf{X}_i), g(\mathbf{X}_k)\} = \tau^2 \exp\left\{-(d_{ik}/\phi)^2\right\},$$

where τ^2 is the variance (since $\tau^2 \exp\{-(d/\phi)^2\} = \tau^2$ if $d = 0$) and $\phi > 0$ determines the decay of the correlation as a function of the distance between $\mathbf{X}_i$ and $\mathbf{X}_k$. Other covariance functions are available including the Matérn function [38].

The mean and covariance functions define the joint distribution of $\mathbf{g}$. Assuming the constant mean and isotropic Gaussian covariance function (as we do throughout), the marginal distribution of $g(\mathbf{X}_i)$ is Normal(μ, τ^2) for all i and the joint distribution is

$$\mathbf{g}|\boldsymbol{\mu}, \boldsymbol{\Sigma} \sim \text{Normal}(\boldsymbol{\mu}, \boldsymbol{\Sigma}) \tag{8.10}$$

where $\boldsymbol{\mu} = (\mu, ..., \mu)^T$ is the mean vector of length n and $\boldsymbol{\Sigma}$ is the $n \times n$ covariance matrix with (i, k) element equal to $\tau^2 \exp\{-(d_{ik}/\phi)^2\}$.

The Bayesian model is completed with priors for hyperparameters. The mean and variance parameters can be given prior distributions as in linear mixed models, e.g., $\mu \sim \text{Normal}(0, c^2)$ for large scale parameter c and $\sigma, \tau \sim \text{HC}(1)$ or $\sigma^2, \tau^2 \sim \text{InvGamma}(\epsilon, \epsilon)$. The prior for the spatial range parameter ϕ depends on the scale of the d_{ik}. That is, if the units of distance change, so should the prior for ϕ. One way to account for this is the prior $\phi \sim \text{Uniform}(0, d_{max})$ where $d_{max} = \max\{d_{ik}\}$ is the maximum of the observed distances between observations.

Given the likelihood $Y_i|\mathbf{g} \stackrel{indep}{\sim} \text{Normal}\{g(\mathbf{X}_i), \sigma^2\}$ and prior in (8.10), MCMC can be used to sample from the posterior distribution of $\mathbf{g}$ and the parameters $\boldsymbol{\theta} = (\mu, \sigma^2, \tau^2, \phi)$. Assuming conjugate priors, the updates for $\mathbf{g}$, μ, σ^2 and τ^2 are simple Gibbs updates; however, since ϕ appears inside the covariance matrix there is no conjugate prior and a Metropolis step is required. While the algorithm used to draw samples is familiar, MCMC is notoriously slow for GP regression even with moderate n (say $n > 1,000$) because each step requires manipulating the covariance matrix $\boldsymbol{\Sigma}$. This bottleneck can be alleviated by fixing ϕ (e.g., using maximum likelihood methods) in advance so that all matrix manipulation can be done outside of the MCMC algorithm. Even after this concession, GP regression is slow and requires large memory allocation for large n (say $n > 10,000$).

In some cases, the objective of the analysis is to estimate the mean $\mathbf{g}$ and MCMC can return samples of $\mathbf{g}$ for this purpose. However, if the goal of the analysis is to estimate the parameters $\boldsymbol{\theta}$ to make predictions, then $\mathbf{g}$ can be marginalized over to give the distribution of the observed data $\mathbf{Y} = (Y_1, ..., Y_n)^T$ given $\boldsymbol{\theta}$ as the likelihood function. The marginal distribution of $\mathbf{Y}$ is

$$\mathbf{Y}|\boldsymbol{\theta} \sim \text{Normal}(\boldsymbol{\mu}, \boldsymbol{\Sigma}_Y) \tag{8.11}$$

where $\boldsymbol{\Sigma}_Y = \text{Cov}(\mathbf{Y}|\boldsymbol{\theta}) = \sigma^2 \mathbf{I}_n + \boldsymbol{\Sigma}$. MCMC then proceeds using (8.11) as the likelihood function and gives the same posterior for $\boldsymbol{\theta}$ but often converges faster than the algorithm that includes $\mathbf{g}$ since there are fewer parameters to estimate.

One of the main advantages of GP regression is provably optimal predictions. Denote Y_0 as the prediction for an observation with covariate $\mathbf{X}_0$. Then the mean and covariance functions can be used to define the $n+1$ dimensional multivariate normal distribution for Y_0 and $\mathbf{Y}$, and the properties of the multivariate normal distribution give the prediction distribution

$$Y_0|\mathbf{Y}, \boldsymbol{\theta} \sim \text{Normal}\left\{\mu + \boldsymbol{\Sigma}_0^T \boldsymbol{\Sigma}_Y^{-1}(\mathbf{Y} - \boldsymbol{\mu}), \sigma^2 + \tau^2 - \boldsymbol{\Sigma}_0^T \boldsymbol{\Sigma}_Y^{-1} \boldsymbol{\Sigma}_0\right\}, \tag{8.12}$$

where $\boldsymbol{\Sigma}_0$ is the vector of length n with i^{th} element equal $\text{Cov}(Y_0, Y_i) = \tau^2 \exp\{-(d_{i0}/\phi)^2\}$. A similar expression gives the posterior distribution for only the mean at $\mathbf{X}_0$,

$$g(\mathbf{X}_0)|\mathbf{Y}, \boldsymbol{\theta} \sim \text{Normal}\left\{\mu + \boldsymbol{\Sigma}_0^T \boldsymbol{\Sigma}_Y^{-1}(\mathbf{Y} - \boldsymbol{\mu}), \tau^2 - \boldsymbol{\Sigma}_0^T \boldsymbol{\Sigma}_Y^{-1} \boldsymbol{\Sigma}_0\right\}, \tag{8.13}$$

which is identical to (8.12) except the error variance σ^2 is excluded from the prediction variance. The mean of the prediction distribution in (8.12) is known as the Kriging prediction, and is the best linear unbiased predictor assuming the mean and covariance functions are correct. Of course, in practice the mean and covariance functions will be unknown and so there is no guarantee the predictor is optimal, but this result at least gives a path forward: using exploratory analysis or model comparisons we try to identify a reasonable covariance function and then use Kriging for theoretically-justified prediction.

Figure 8.8 plots a synthetic dataset with $p = 1$ covariate and $n = 25$ observations along with GP predictions and prediction intervals (connected by lines) for a sequence of $\mathbf{X}_0$ covering the range of the observed $\mathbf{X}_i$. Here the parameters are set to $\mu = 10$, $\sigma^2 = 0.2$, $\tau^2 = 2$ and $\phi = 0.05$. The predicted values effectively smooth over the observed responses, and the prediction intervals are wider where there is less data. In a real data analysis, the parameters $\boldsymbol{\theta}$ would be unknown and rather than make prediction for one fixed value of $\boldsymbol{\theta}$ we would make a prediction for each MCMC sample of $\boldsymbol{\theta}$, thereby averaging over parametric uncertainty and giving the posterior predictive distribution $p(Y_0|\mathbf{Y})$.

8.3.3 Bayesian additive regression trees

The GAM approach of Section 8.3.1 represents the regression function g as a linear combination of simple spline basis functions. However, to capture non-linearity and interaction effects requires a huge number of basis functions which is prohibitive for large datasets. Bayesian adaptive regression trees (BART) [33] also represents g as the sum of simple functions, but with BART these simple functions are trees with parameters that adapt to the data so that fewer terms are required to give a sufficient approximation.

The fundamental unit of BART is a regression tree. Trees are relatively simple structures that can capture non-linear relationships and interaction effects. The covariate space is partitioned into intervals ("terminal nodes") and each interval has a different mean response.

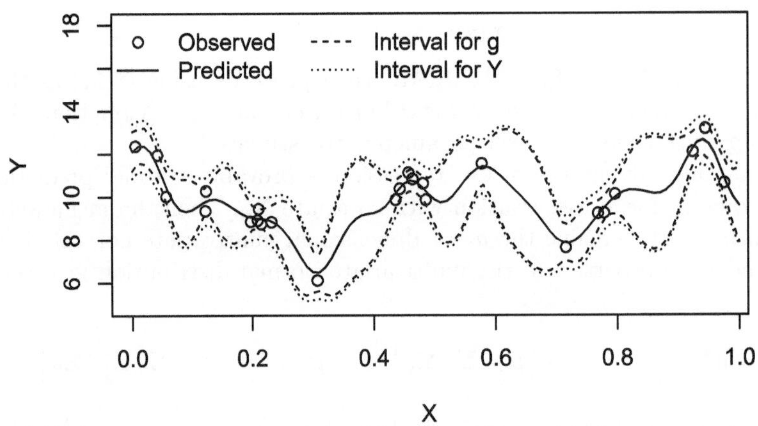

FIGURE 8.8
Illustration of Gaussian process regression. The data are show as points and the predicted values and 95% intervals for the true mean function g and test set observations Y are plotted as lines for a sequence of X covering (0,1).

For example, the tree on the left of Figure 8.9 is defined mathematically as

$$\text{tree}(\mathbf{X}; T, \boldsymbol{\mu}) = \begin{cases} \mu_1 & x_3 < c_1 \text{ and } x_7 < c_2 \\ \mu_2 & x_3 < c_1 \text{ and } x_7 > c_2 \\ \mu_3 & x_3 > c_1. \end{cases} \tag{8.14}$$

The tree is a function of the tree structure, T, that includes the number of branches ($m = 3$) and the variables (x_3 and x_7) and cutpoints (c_1 and c_2) used to define each branch, and the values at each of the m terminal nodes, $\boldsymbol{\mu} = (\mu_1, ..., \mu_m)$. Although the notation to precisely define T is complicated, the ideas is that the function is defined sequentially through simple "splits" determined by a covariate and a cutpoint, such as $x_3 > c_1$. Each split results in two "branches", and when splits are defined sequentially the branches form the tree. When a covariate vector $\mathbf{X}$ is evaluated, based on the splits it falls in one of the terminal nodes where it is a assigned a value from $\boldsymbol{\mu}$. To fit the data well, both T and $\boldsymbol{\mu}$ must be learned from the data.

Unlike GP regression or even GAMs, the tree structure is appealing because the entire mean function can be written in a simple form and visualized as in Figure 8.9. However, it is unlikely that a single tree provides a good fit for complex data, i.e., the tree above takes only $m = 3$ values which is too simple for practical application. However, it has been shown that taking the average over several trees ("weak learners") gives a rich modeling framework. The most common approach for aggregating trees in the machine learning literature is random forest regression; below we describe BART.

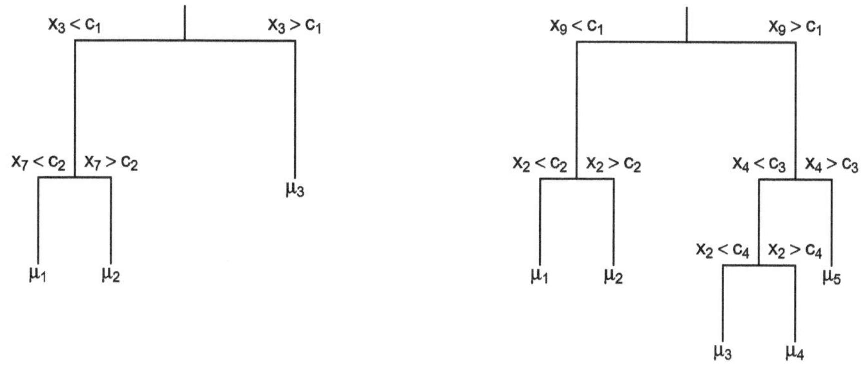

FIGURE 8.9
Regression trees. Two hypothetical regression trees to be used in BART.

The BART model for the mean response function aggregates over J trees

$$g(\mathbf{X}) = \sum_{j=1}^{J} \text{tree}(\mathbf{X}; T_j, \boldsymbol{\mu}_j), \tag{8.15}$$

where T_j is the structure of the jth tree and $\boldsymbol{\mu}_j$ is the collection of mean values for each of the m_j terminal nodes in T_j. BART is a fully-Bayesian model and thus requires prior distributions for T_j and $\boldsymbol{\mu}_j$. The prior distributions are independent across the J trees. The prior for the number of branches is defined sequentially; as the tree grows, the probability that each branch splits into two is $\alpha(1+d)^{-\beta}$ where d is the depth of the tree and α and β are hyperparameters. The covariate that defines each split has Uniform($\{1, ..., p\}$) prior and the cutpoint c_j has uniform prior over the observed covariate values. Finally, the mean parameters $\boldsymbol{\mu}_j$ have Gaussian priors with mean zero and variance chosen based on the scale of the data.

BART is fit using MCMC with T_j and $\boldsymbol{\mu}_j$ as the unknown parameters. An advantage of using MCMC is that posterior draws can be used to obtain posterior credible intervals for $g(\mathbf{X})$ and prediction intervals that account for both the error variance σ^2 and uncertainty in $g(\mathbf{X})$. BART also performs variable selection in a similar fashion as stochastic search variable selection (Section 6.2). A variable is excluded from the model if it is not included in any of the splits for any of the trees, and MCMC returns an approximation of the posterior probability of this event. An alternative is the proportion of the trees that include the variable, and MCMC also returns the posterior distribution of this variable importance measure.

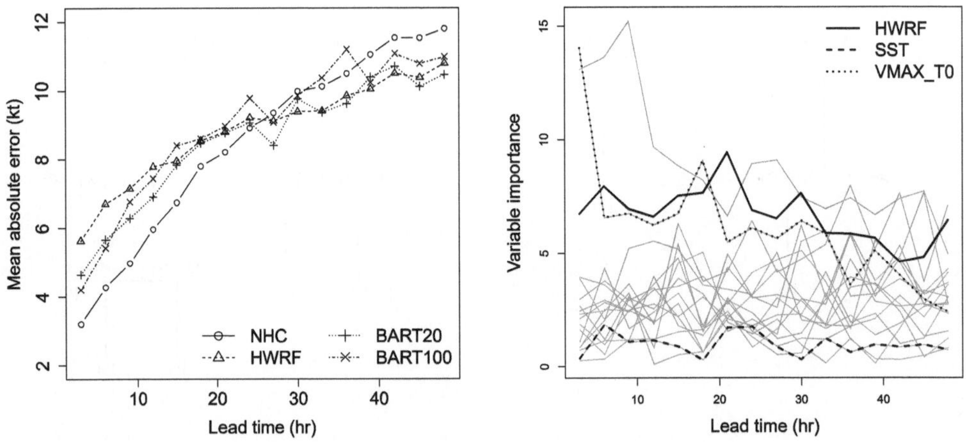

FIGURE 8.10
Hurricane intensity forecasting with BART. Mean absolute prediction error by lead time for the Hurricane Weather Research and Forecasting (HWRF) model, the National Hurricane Center (NHC) and BART using 20 or 100 trees (left) and variable importance for all covariates by lead time (gray) with emphasis of HWRF, sea surface temperature (SST) and intensity (VMAX) at the time of the forecast.

Analysis of the hurricane intensity data: We fit the BART model to the hurricane forecasting problem using the `wbart` function in the `BART` package [144]. We use the default settings except for the number of trees is set to either $J = 20$ or $J = 100$. The MAE by lead time in Figure 8.10 (left) is similar to GAM, with larger reductions compared to HWRF for early lead times. The number of trees does not have a large impact on the results, so for simplicity we use $J = 20$ to summarize covariate effects.

Figure 8.10 (right) plots the posterior mean of the number of trees that include each covariate. The three covariates evaluated in the GAM analysis are highlighted. SST is unimportant for all lead times. As expected, the HWRF prediction is one of the most influential covariates for all lead times. Surprisingly, for short-term forecasts with 3-hour lead time, the current intensity is actually a stronger predictor than the HWRF forecast according to this measure of variable importance.

8.3.4 Deep learning

Deep learning has revolutionized ML with its ability to process large datasets and analyze and generate complex data structures such as images and text [65]. In this section, we will introduce the basic concepts of deep learning through the nonparametric regression problem in (8.3). The common deep learning model for this task is a feed-forward neural network (FFNN) (also known as multi-layer perceptron). After introducing the FFNN, we focus on

Bayesian implementation and its advantages [114] including proper uncertainty quantification for prediction and variable importance. We then very briefly discuss extensions to more complex data structures such as text and images.

One way to approach DL is as a generalization of GAMs in Section 8.3.1. In the main effects GAM model in (8.6) is $g(\mathbf{X}) = \beta_0 + \sum_{j=1}^{p} g_j(X_j)$. The functions g_j are interpretable as the effect of X_j with all other covariates held fixed. While interpretable, this makes the tenuous assumption that the mean response is additive in the covariates and breaks down with large p because estimating p non-linear functions requires many parameters.

A simple way to avoid the curse of dimensionality is the single-index model [29]. As in linear regression, the covariates are reduced to a single index, $b + \sum_{j=1}^{p} X_j W_j$, that enters the mean through a non-linear function A,

$$g(\mathbf{X}) = \beta_0 + A\left(b + \sum_{j=1}^{p} X_j W_j\right) \tag{8.16}$$

where A can be modeled using spline basis functions as in a GAM. This is efficient because adding an additional covariate requires adding only one additional element to $\mathbf{W} = (W_1, ..., W_p)$. Obviously, if A is the identity function, then the model reduces to Bayesian linear regression with regression coefficients W_j.

Despite its efficiency, the single-index model has limitations. The first is identifibility caused by needing to estimate both A and $\mathbf{W}$. For example, doubling $\mathbf{W}$ and replacing $A(x)$ with $A(x/2)$ gives the function $g(\mathbf{X})$. This is usually resolved by restricting the shape and scale of A (via restrictions on the spline basis function coefficients as discussed in Section 8.3.1). More critically, this model is not nonparametric because not all mean functions can be expressed in terms of a single index.

A multiple-index model generalizes the single-index model to have L indices each with a different set of regression coefficients

$$Z_l(\mathbf{X}) = A_l\left(b_{1l} + \sum_{j=1}^{p} X_j W_{1lj}\right).$$

Since it is difficult to estimate all L non-linear functions, they can all be taken to be the same, $A_l = A$ for all l. Further, to avoid complications in estimating A with (restricted) spline regression, it can be taken as a simple known function, e.g., the Rectified Linear Unit (ReLU) function $A(x) = (x)_+$, i.e., $A(x) = x$ if x is positive and $A(x) = 0$ otherwise (Figure 8.11). The mean is then linear in the indices to allow them to have differential importance,

$$g(\mathbf{X}) = b_2 + \sum_{l=1}^{L} Z_l(\mathbf{X}) W_{2l} \tag{8.17}$$

$$Z_l(\mathbf{X}) = A\left(b_{1l} + \sum_{j=1}^{p} X_j W_{1lj}\right) \quad \text{for} \quad l \in \{1, ..., L\} \tag{8.18}$$

This is in fact a single-layer feed-forward neural network.

An interpretation of this model is that the original covariates X_j are transformed to Z_l before entering the linear model. This interpretation resembles the arguments for spline

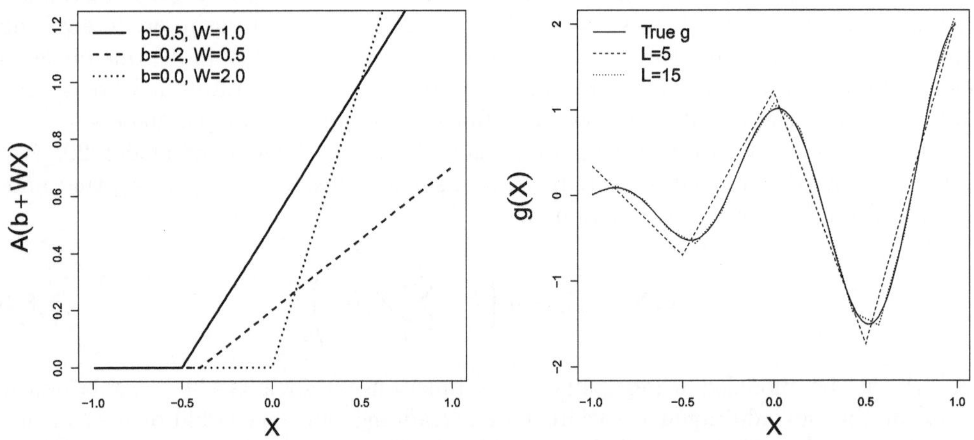

FIGURE 8.11
Deep learning activation functions. ReLU activation function $A(b+WX) = (b+WX)_+$ for several b and W (left) and approximation to the true curve $g(X) = (1+X)\cos(2\pi X)$ using a linear combination of L neurons.

basis functions $B_j(X)$ in (8.6) and (8.7). A key difference is that in the FFNN the Bayesian model simultaneously learns the transformed variables (via the b_{1l} and W_{1lj}) and their effect (via b_2 and W_{2l}), making the FFNN more efficient for high-dimensional problems.

Terminology is often a stumbling block in connecting neural network and statistical models. In statistical linear regression, the parameters b_2 and W_{2l} in (8.17) can be referred to as the intercept and slopes, respectively. However, when non-linear activation functions are introduced as in (8.18), this linear terminology is misleading, and so the terms (in both layers) b are referred to as biases and the terms W are referred to as weights. Also, in a statistical model A and L might be referred to as the basis function and the number of basis functions, but in deep learning they are called the activation function and number of neurons. This stems from a neuroscience analogy. If we take the activation function to be the ReLU function then we can think of $Z_l(\mathbf{X})$ as the output of a neuron that fires in response to a particular stimulus, and the information from this stimulus in the data as measured $b_{1l} + \sum_{j=1}^{p} X_j W_{1lj}$. If the information is negative, the neuron does not fire and $A_l(\mathbf{X}) = 0$, but once the information exceeds zero, the neuron fires with strength proportional to the information.

According to the universal approximation theorem [81], under general conditions the model in (8.17)-(8.18) can approximate any smooth function $g(\mathbf{X})$ for large enough L, and is thus a nonparametric model. This is illustrated by Figure 8.11, which plots the true mean function $g(X) = (1+X)\cos(2\pi X)$ by single input X (so $p=1$). This smooth curve is approximated by a deep FFNN with L neurons and ReLU activation function. For illustration, the input layer weights are set to $W_{lj} = 1$, and biases are a grid of L values spanning $[-1, 1]$, so $-1 = b_1 < ... < b_L = 1$, and the output layer parameters b_0 and W_{0l} are set to least squares estimates; of course, in reality all of these parameters would be estimated using MCMC. The fitted curves are piecewise linear functions, and it is clear that increasing

L and selecting the weights and biases appropriate any smooth curve can be approximation by a curve with form (8.17)-(8.18). Further, for large L and assuming the weights W_{0l} are Gaussian, arguments similar to those below (8.9) suggest that g approximates a Gaussian process with covariance function that depends on the activation function [95].

While a single-layer FFNN in (8.17)-(8.18) is nonparametric in that it possesses the universal approximation property, it has been found that empirical performance can be improved by adding more layers, making it a *deep* learning model. This is a familiar concept in Bayesian methods where layers are often added to hierarchical models to add flexibility (Sections 2.2 and 7.1). Adding another layer gives

$$g(\mathbf{X}) = b_3 + \sum_{k=1}^{L_2} Z_{2k}(\mathbf{X})W_{3k} \tag{8.19}$$

$$Z_{2k}(\mathbf{X}) = A\left(b_{2k} + \sum_{l=1}^{L_1} Z_{1l}(\mathbf{X})W_{2kl}\right) \quad \text{for} \quad k \in \{1, ..., L_2\} \tag{8.20}$$

$$Z_{1l}(\mathbf{X}) = A\left(b_{1l} + \sum_{j=1}^{p} X_j W_{1jl}\right) \quad \text{for} \quad l \in \{1, ..., L_1\}. \tag{8.21}$$

In this model, (8.21) is called the input layer because it is the initial transformation of the inputs, $\mathbf{X}$. The final layer (8.19) is called the output layer because it gives the final output of the network. The middle layer (8.20) is called the hidden layer because it does not directly involve either the covariates or response. Additional hidden layers can be added in the obvious way.

As with any Bayesian analysis, the model parameters must be given prior distributions (for an extensive review, see [50]). Selecting informative prior distributions for the weights and biases is challenging because these parameters not interpretable. For example, in (8.19), the weights W_{3k} resemble linear regression effects of the covariates Z_{2k}, but the Z_{2k} are actually obscure combinations of the original p covariates. Further, the parameters are invariant to label switching; e.g., if $Z_{21}(\mathbf{X})$ and W_{31} are exchanged with $Z_{22}(\mathbf{X})$ and W_{32}, the function $g(\mathbf{X})$ is unchanged.

Given that the individual parameters are not of interest, but rather represent a distribution of weights, it is natural to model them (and similarly for the biases) as random effects (Section 4.4): $W_{3k} \stackrel{iid}{\sim} \text{Normal}(0, \tau_3^2)$, $W_{2kl} \stackrel{iid}{\sim} \text{Normal}(0, \tau_2^2)$ and $W_{1jl} \stackrel{iid}{\sim} \text{Normal}(0, \tau_1^2)$. Although the individual weights are not interpretable, it is still possible to perform Bayesian variable selection because covariate j is removed from the model if $W_{1j1} = 0$ for all l. This motivates a grouped discrete or continuous shrinkage priors such as the horseshoe prior [63] with $W_{1jl} \stackrel{iid}{\sim} \text{Normal}(0, \tau_{1j}^2)$ and $\tau_{1j} \sim \text{HC}(1)$. Discrete priors can be constructed similarly, with τ_{1j} having a prior with positive mass at $\tau_{1j} = 0$. Variable selection priors can also be used to select the number of neurons per layer, since if $W_{3k} = 0$ for some k, then the effective number of neurons in the output layer is reduced by one. However, using cross-validation to select the number of layers and neurons is more common.

Computation is also challenging for neural networks due to the large number of parameters and label-switching. Fortunately, the gradient of the posterior with respect to the parameters is straight-forward to compute. This enables Hamiltonian Monte Carlo (Section 9.4) for moderately-sized datasets. For large datasets, more advanced methods are required such as stochastic gradient MCMC and variational Bayesian methods (discussed briefly in Section 3.4.4).

Despite these computational challenges, the Bayesian approach has advantages over frequentist approach in terms of uncertainty quantification. Prediction uncertainty follows

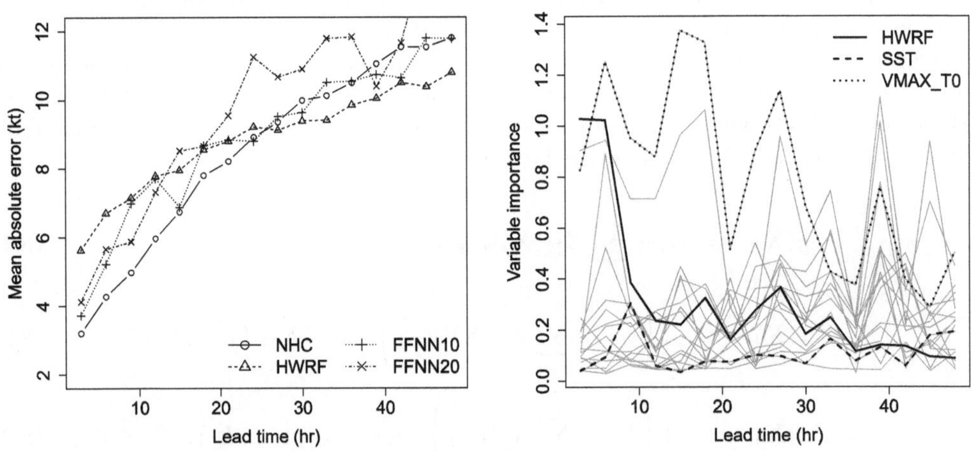

FIGURE 8.12
Hurricane intensity forecasting with FFNN. Mean absolute prediction error by lead time for the Hurricane Weather Research and Forecasting (HWRF) model, the National Hurricane Center (NHC) and FFNN using 10 or 20 neurons (left) and variable importance for all covariates by lead time (gray) with emphasis on HWRF, sea surface temperature (SST) and intensity (VMAX) at the time of the forecast.

trivially from MCMC output, as each iteration produces a different posterior sample of g and draw of the response given g, and thus the posterior predictive distribution accounts for both epistemic and aleatoric uncertainty. Also, with the full posterior distribution of g, it is possible to quantify the relative importance of the each covariate and make careful probabilistic statements about which variables are the most important (see Section 8.5). It is also possible to incorporate information from other fitted networks [1], although this is more complicated than for parametric models.

Our discussion has been limited to FFNN for the classic nonparametric regression case with a scale response variable and p scalar predictors. As we have seen, there are connections between the FFNN and GAM and GP regressions, and so there is no clear choice between these methods in practice. However, deep learning uniquely excels in settings where the data are not scalars and have more complex structure such as images or paragraphs of text. In these cases, different layers of the network are able to extract and combine features of the data in a computationally efficient manner. These neural networks are sophisticated with many layers and ingenious ways to process information, but the basic approach to applying Bayesian methods to analyze them is similar to the FFNN, and so we leave this as future reading.

Analysis of the hurricane intensity data: The Bayesian neural network model is fitted using one hidden layer as in (8.17)-(8.18) using the ReLU activation function and the horseshoe prior for all standard deviation parameters. We compare $L = 10$ and $L = 20$ neurons in the hidden layer. MCMC uses Gibbs sampling for the variance and output-layer parameters, and HMC (20 leapfrog steps and step-size tuned to give 80% acceptance rate) for the input-layer parameters b_l and W_{lj} for $l \in \{1, ..., L\}$. The results are shown in Figure 8.12.

As with the other semiparametric regression methods, the feed-forward neural network reduces MAE compared to HWRF for shorter lead times with mixed results for longer lead times. Figure 8.12 (right) plots the posterior mean of σ_w for each covariate and lead time as a crude measure of variable importance. As with other analyses, sea surface temperature is less important than the HWRF prediction and current storm intensity; see Section 8.5 for a more formal analysis of variable importance for this neural network model.

8.4 Models for the residual distribution

The focus of nonparametric regression is a flexible model for the mean response. If the objective is merely a point prediction, then an accurate estimate of the mean function is likely sufficient. However, if the objective is to perform statistical inference on parameters or provide valid prediction intervals, then a model for the residual distribution is required. Checking that the assumed residual distribution fits well is part of any standard analysis. If a plot of the residuals from the estimated mean function are clearly non-Gaussian, then a transformation of the response could be performed or a non-Gaussian error distribution (e.g., Student's t-distribution) could be fit. In the spirit of machine learning, this chapter discusses two approaches to automating this process. In Section 8.4.1, we extend the nonparametric repression approach from the mean function to the Gaussian variance function. Section 8.4.2 dispenses with the Gaussian, or any other parametric error distribution, in favor of modeling the error distribution semiparametrically.

8.4.1 Heteroskedastic models

The standard linear regression analysis in Section 4.2 assumes a homoskedastic variance $\text{Var}(\varepsilon_i) = \sigma^2$ for all i. A more flexible heteroskedastic model allows the covariates to affect both the mean and the variance. A natural approach is to build a linear model for the variance as a function of the covariates so that $\text{Var}(\varepsilon_i) = \sigma^2(\mathbf{X}_i)$. Since the variance is positive, we must transform the linear predictor to be positive before linking to the variance. For example,

$$\sigma^2(\mathbf{X}_i) = \exp\left(\sum_{j=1}^{p} X_{ij}\alpha_j\right). \tag{8.22}$$

The parameter α_j determines the effect of the covariate j on the variance and must be estimated. `JAGS` code for this model is in Listing 8.3. The variance model does not need to be a log-linear function of covariates, and any of the nonparametric regression models for the mean can be applied to the variance. For example, in the example below we use spline regression for the variance.

Motorcycle example: The variance of the observations about the mean trend in Figure 8.5a clearly depends on X, with small variance at the beginning of the experiment and large variance in the middle. To capture this heteroskedasticity in a flexible way, we model the log variance using the same J B-spline basis functions used for the mean,

$$g(X) = \beta_0 + \sum_{j=1}^{p} B_j(X)\beta_j \text{ and } \log[\sigma^2(X)] = \alpha_0 + \sum_{j=1}^{p} B_j(X)\alpha_j, \tag{8.23}$$

where $\beta_j \sim \text{Normal}(0, \sigma_b^2)$ and $\alpha_j \sim \text{Normal}(0, \sigma_a^2)$. The hyperparameters have uninformative priors $\sigma_a^2, \sigma_b^2 \sim \text{InvGamma}(0.1, 0.1)$.

Listing 8.3
JAGS model statement for heteroskedastic Gaussian regression.

```
for(i in 1:n){
  Y[i]     ~ dnorm(mu[i],prec[i])
  mu[i]    <- inprod(x[i,],beta[])
  prec[i] <- 1/sig2[i]
  sig2[i] <- exp(inprod(x[i,],alpha[]))
}
for(j in 1:p){beta[j] ~ dnorm(0,taub)}
for(j in 1:p){alpha[j] ~ dnorm(0,taua)}
taub ~ dgamma(0.1,0.1)
taua ~ dgamma(0.1,0.1)
```

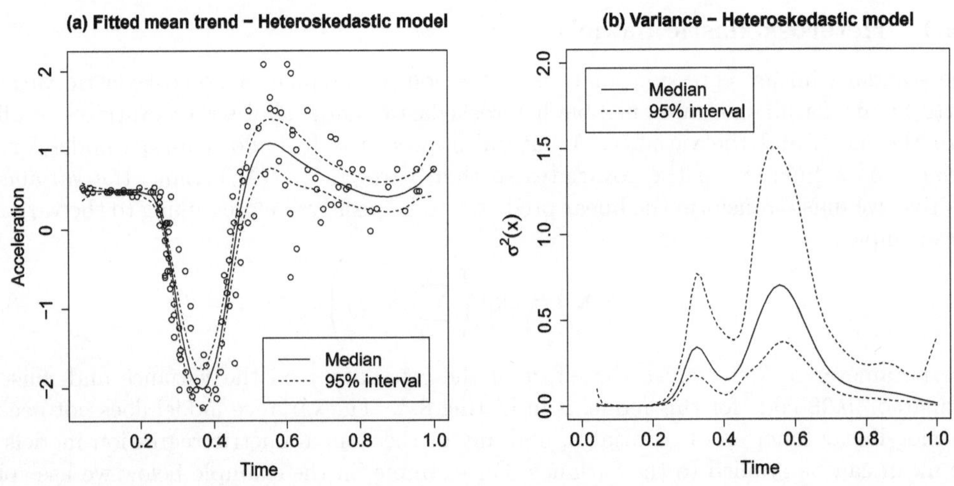

FIGURE 8.13
Nonparametric regression for the motorcycle data. Panel (a) plots the time since impact (scaled to be between 0 and 1) and the acceleration (g) along with the posterior median and 95% interval for the mean function from the heteroskedastic fit; Panel (d) shows the posterior median and 95% intervals for the variance functions from the heteroskedastic model.

The pointwise 95% intervals of $\sigma^2(X)$ in Figure 8.13b suggest that the variance is indeed small at the beginning of the experiment and increases with X. Comparing the posterior distributions of the mean trend $g(X)$ for the homoskedastic (Figure 8.5a) and heteroskedastic (Figure 8.13a) models, the posterior means are similar but the heteroskedastic model produces more realistic 95% intervals with widths that vary according to the pattern of the error variance. Therefore, it appears that properly quantifying uncertainty about the mean trend requires a realistic model for the error variance.

8.4.2 Bayesian density estimation

Most Bayesian regression models assume Gaussian errors, but more flexible methods are easily constructed. For example, to accommodate heavy tails, the errors could be modeled using a Student's t- or double-exponential (Laplace) distribution. To further allow for asymmetry, generalizations such as the skew-t and asymmetric Laplace distributions would be used. Listing 8.4a provides code for regression with Student's t-distribution errors.

In most analyses, a suitable parametric distribution can be found. However, this process is subjective and difficult to automate. Just as a mixture of conjugate priors (Section 2.1.8) can be used to approximate virtually any prior distribution, the mixture-of-normals distribution can be used to approximate virtually any residual distribution. The mixture-of-normals density for ε is

$$f(\varepsilon) = \sum_{k=1}^{K} \pi_k \phi(\varepsilon; \theta_k, \sigma^2), \tag{8.24}$$

where K is the number of mixture components, $\pi_k \in (0, 1)$ is the probability on mixture component k, and $\phi(\varepsilon; \theta, \tau^2)$ is the Gaussian PDF with mean θ and variance τ^2. This model is equivalent to the clustering model where

$$Y_i | g_i \sim \text{Normal}\left(\sum_{j=1}^{p} X_{ij}\beta_j + \theta_{g_i}, \sigma^2\right) \tag{8.25}$$

and $g_i \in \{1, ..., K\}$ is the cluster label for observation i with $\text{Prob}(g_i = k) = \pi_k$. By letting the number of mixture components increase to infinity, any smooth distribution can be approximated, and by selecting priors that are Gaussian, inverse gamma, and Dirichlet priors for θ_k, σ^2, τ^2 and $(\pi_1, ..., \pi_K)$, respectively, all full conditional distributions for all parameters are conjugate permitting Gibbs sampling (Listing 8.4b).

For a fixed number of mixture components (K), the mixture-of-normals model is a semiparametric estimator of the density $f(\varepsilon)$. There is a rich literature on nonparametric Bayesian density estimation [62]. The most common model is the Dirichlet process mixture model that has infinitely many mixture components and a particular model for the mixture probabilities.

Density regression: The most flexible model for a residual distribution combines the heteroskedastic and mixture-of-normal models, giving Bayesian density regression (e.g., [46]). The idea is to develop a framework so that the distribution can deviate arbitrarily from normality and across the covariates. For instance, the distribution might be left-skewed for one range of the covariates but right-skewed for another. If this model can be fit, then it provides a complete understanding of the covariates' effects on the response distribution.

Listing 8.4
Model statement for Gaussian regression with non-normal errors.

```
# (a) Regression with Student's t-distribution errors
for(i in 1:n){
  Y[i]   ~ dt(mu[i],tau,df)
  mu[i] <- inprod(X[i,],beta[])
}
for(j in 1:p){beta[j] ~ dnorm(0,taub)}
tau ~ dgamma(0.1,0.1)
df  ~ dgamma(0.1,0.1)


# (b) Regression with mixture-of-normals errors
for(i in 1:n){
  Y[i]   ~ dnorm(mu[i]+theta[g[i]],tau1)
  mu[i] <- inprod(X[i,],beta[])
  g[i]   ~ dcat(pi[])
}
for(k in 1:K){theta[k] ~ dnorm(0,tau2)}
for(j in 1:p){beta[j] ~ dnorm(0,tau3)}
tau1     ~ dgamma(0.1,0.1)
tau2     ~ dgamma(0.1,0.1)
tau3     ~ dgamma(0.1,0.1)
pi[1:K] ~ ddirch(alpha[1:K])
```

The mixture-of-normals model in (8.24) has three types of parameters: mixture probabilities (π), means (θ) and variances (σ^2). Any or all of these parameters could theoretically depend on the covariates, but conceptually the easiest approach is to allow the covariates to affect the mixture means

$$f(\varepsilon|\mathbf{X}) = \sum_{k=1}^{K} \pi_k \phi\left\{\varepsilon; \theta_k(\mathbf{X}), \sigma^2\right\}. \tag{8.26}$$

The mean functions θ_k can be modeled using linear regression, $\theta_k = \mathbf{X}\boldsymbol{\alpha}_k$, with different regression coefficients ($\boldsymbol{\alpha}_k$) in each mixture component or using nonparametric regression methods such as spline regression. If the mean is sufficiently modeled in density regression, then the overall mean $g(\mathbf{X})$ can be removed.

To illustrate the flexibility of this model, Figure 8.14 plots the distribution for $p = 1$ covariate and $K = 3$ mixture components with $\pi_1 = 0.5$, $\pi_2 = 0.3$, $\pi_3 = 0.2$, $\sigma = 0.5$ and the mean functions $\theta_k(X)$ given in the left panel. Even with this fairly simple model, the distribution varies from symmetric to left- and right-skewed over X.

Assuming a moderately-complex specification for the θ_k, the model can be fit using code similar to Listing 8.4b. However, care must be taken to monitor convergence due potential label-switching. That is, the model with $\theta_1(\mathbf{X})$ and $\theta_2(\mathbf{X})$ relabeled as $\theta_2(\mathbf{X})$ and $\theta_1(\mathbf{X})$ gives an identical fit, and so without restrictions, the individual parameters will not converge to unique values. Instead of monitoring convergence of individual parameters, it is better to monitor convergence of the distribution $f(\varepsilon|\mathbf{X})$ for a few $\mathbf{X}$ because this quantity is invariant to label switching.

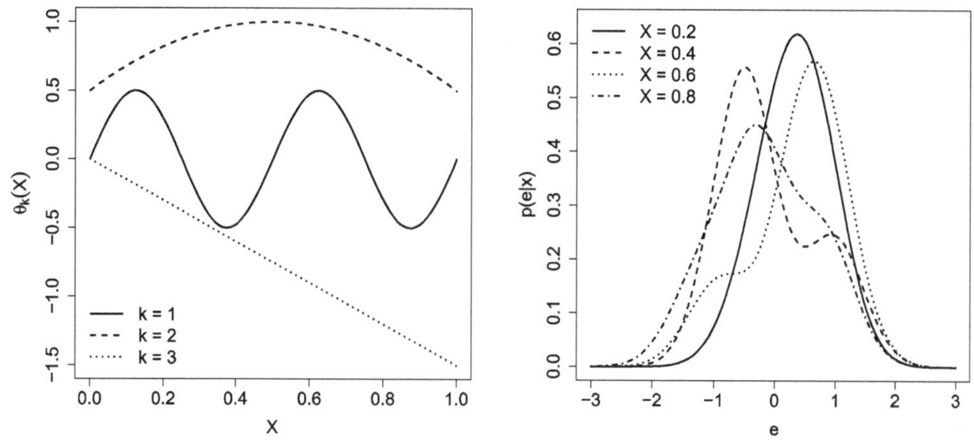

FIGURE 8.14
Illustration of Bayesian density regression. Panel (a) plots the mean functions $\theta_k(X)$; Panel (b) shows the distribution of $\varepsilon|X$.

8.5 Interpretable machine learning

Machine learning models are designed to be flexible and when trained with large datasets often provide more accurate predictions than classical statistical methods. However, in many cases, accurate predictions alone are insufficient, and understanding of factors that that drive the predictive model are required in order to understand and trust the prediction algorithm. More fundamentally, understanding the relationships between variables is how science advances. Balancing predictive power with understanding the fitted model is the objective of interpretable machine learning (e.g., [108]).

One approach to interpretable machine learning is to select a transparent model such as linear or logistic regression. However, if the transparent model fits poorly, the resulting analysis is questionable. Another approach is to fit a robust machine learning model and post-process the fitted model to elucidate the covariate effects. This is the topic of the chapter. The Bayesian approach is natural for this task. MCMC produces samples from the model parameters, and these parameters can be transformed to any quantity that summarizes the fit of the model and automatically gives the posterior distribution of this quantity. Viewed this way, interpetable machine learning is simply the analyst using the fitted Bayesian model to creatively examine relationships between variables.

Before discussing general methods for interpreting machine learning models, we first examine the trade-off between fitting a simple but incorrect linear regression model versus a complicated but robust machine learning model. Consider the synthetic data in Figure 8.15 (left). The data are generated with a cubic relationship between the single covariate and the response, but is only slightly non-linear and perhaps close enough to linear that a simple linear regression model is an appealing choice. Fitting this model results in a one-number summary of the covariate effect (the slope) and a simple test of statistical significance

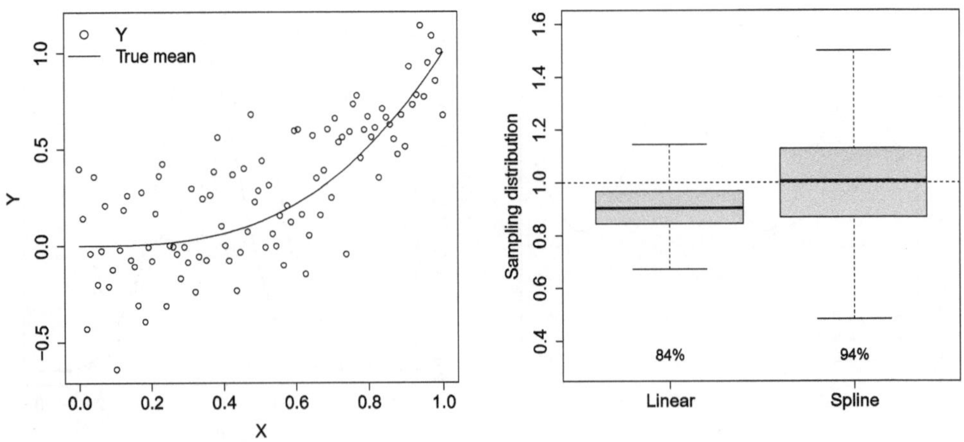

FIGURE 8.15
Simulation study comparing linear and non-linear regression. The left panel plots a synthetic dataset generated with true mean equal to $g(X) = X^3$; the right panel shows the sampling distribution of the posterior mean of $\Delta = g(1) - g(0)$ for a Bayesian linear model and Bayesian spline regression model with $J = 5$ basis function (coverage of 95% intervals is given below the boxplots).

(e.g., a Bayes factor). Alternatively we could fit a spline regression model which would likely give a better fit, but require tuning (selecting the number of basis functions), and results in a model with many parameters and a curve, rather than a scalar, to describe the covariate effect. However, with some post-processing of the posterior, we can still describe the covariate effect with a single value, such as the slope of the mean curve evaluated at the two endpoints, $\Delta = [g(1) - g(0)]/[1 - 0]$. In a Bayesian analysis, MCMC produces the full posterior of this interpretable quantity without having to assume the true model is linear. So, which approach to choose, fit a linear model or fit a non-linear model and summarize the posterior using interpretable quantities?

We conduct a simulation study to answer this question. For each of 1000 simulated datasets, we generate $n = 100$ observations from the true model $X_i \stackrel{iid}{\sim} \text{Uniform}(0, 1)$ and $Y_i|X_i \sim \text{Normal}(X_i^3, 0.25^2)$. We then fit each dataset twice using the default settings in the BLR package [41], once using a linear and once using a semiparameteric spline model:

- Linear: $Y_i|\boldsymbol{\beta} \sim \text{Normal}(\beta_0 + X_i\beta_1, \sigma^2)$
- Spline: $Y_i|\boldsymbol{\beta} \sim \text{Normal}(\beta_0 + \sum_{j=1}^{5} B_j(X_i)\beta_j, \sigma^2)$

where B_j are B-spline basis functions. For the linear model $\Delta = \beta_1$ and for the spline model $\Delta = \beta_5$ because for $X = 1$ all basis functions are zero except the last one which is $B_5(1) = 1$ (Figure 8.5b).

The linear model is clearly incorrect, and the spline model includes the cubic mean function as a special case but also includes many irrelevant terms. For each simulated dataset, we approximate the posterior mean and 95% interval of the parameter of interest, Δ, and Figure 8.15 reports the sampling distribution of the posterior mean and coverage probability of the credible intervals. The incorrectly specified linear model exhibits negative bias and undercoverage. On the other hand, the spline regression model is unbiased with valid coverage, but at the expense of increased variability. We elaborate on this bias/variance trade-off in Chapter 9, but for now note that fitting the incorrect linear model has poor statistical properties.

As with selecting the prior distribution, there is no single best way to summarize the fit of a machine learning model. One general approach is to use properties of the fitted model to summarize covariate effects, such as the proportion of the trees that include each covariate in BART. While these model-specific summaries are valuable, here we focus on model-agnostic summaries that view the mean function $g(\mathbf{X})$ as a black box and can thus be applied to virtually any nonparametric regression model.

Accumulated local effects and variable importance: The accumulated local effects (ALE) method [8] summarizes the effect of a covariate using the partial derivatives of the mean function $g(\mathbf{X})$. Let the partial derivative of $g(\mathbf{X})$ with respect to the jth covariate equal

$$d_j(\mathbf{X}) = \frac{\partial}{\partial X_j} g(\mathbf{X}).$$

Under the linear model, $g(\mathbf{X}) = \sum_{j=1}^{p} X_j\beta_j$, $d_j(\mathbf{X}) = \beta_j$ reduces to the usual summary of a covariate effect in linear regression.

If $g(\mathbf{X})$ is non-linear or has interactions, then $d_j(\mathbf{X})$ is interpreted as the local (in covariate space) linear effect of X_j. That is, with all other variables fixed *at* $\mathbf{X}$, $d_j(\mathbf{X})$ is the effect of slightly increasing X_j. Since the local effect of X_j depends on the value of all other covariates, $d_j(\mathbf{X})$ is a complex and uninterpretable function. To reduce the complexity of this function, the local effect (LE) simply average of these local linear effects across the distribution of the other covariates,

$$LE_j(x) = \mathrm{E}\{d_j(\mathbf{X})|X_j = x\}$$

where the expectation is with respect to the distribution of $\mathbf{X}$ given $X_j = x$. A plot of $LE_j(x)$ versus x summarizes the importance of X_j; if $LE_j(x) \approx 0$ for all x then X_j on average does not affect the mean response. A limitation of this summary measure is that if X_j has a positive effect for some $\mathbf{X}$ and a negative effect for others, then these effects could offset and $LE_j(x) \approx 0$ even though X_j impacts the mean response.

To plot the covariate effects on the original scale rather than the derivative scale, the local effects are integrated (accumulated) giving the accumulated local effect (ALE)

$$\widetilde{ALE}_j(x) = \int_{l_j}^{x} LE_j(x')dx',$$

where the support of X_j is $[l_j, u_j]$. Since the ALE is a function of only g's derivative, it is invariant to an additive constant. Therefore, the ALE is centered around zero by subjecting the mean given the final, centered, version

$$ALE_j(x) = \widetilde{ALE}_j(x) - \mathrm{E}\left\{\widetilde{ALE}_j(X_j)\right\} \tag{8.27}$$

where the expectation is with respect to the distribution of $\mathbf{X}$. A plot of X_j versus $ALE_j(X_j)$ summarizes the main effect of X_j on the mean response; if $d_j(x) \approx 0$ and $ALE_j(x)$ is flat, then this indicates that covariate j is, averaging over covariate space, null. The ALE summarizes the main effect of covariate j on the mean response, similar to the main effect in the GAM model in Section 8.3.1. A conceptual distinction is that the ALE plot does not build a model assuming an additive structure; rather it summarizes the mean function fitted from a flexible model using the idea of a main effect.

ALE can also be computed for categorical predictors and interactions. Quantifying interaction effects using ALE is very similar to the main effects. The local effect of the interaction between X_j and X_k is

$$d_{jk}(X_j, X_k) = \frac{\partial^2}{\partial X_j \partial X_k} g(\mathbf{X}).$$

This local effect is then converted to the $ALE_{jk}(x_j, x_k)$ by taking its expected value with respect to $\mathbf{X}$ given $X_j = x_j$ and $X_k = x_k$, then using bivariate integration and centering. The ALE can then be viewed as a surface over X_j and X_k. Higher-order interaction terms can be computed following similar steps, but are difficult to compute and interpret. For categorical variables, derivatives are replaced with the difference between $g(\mathbf{X})$ evaluated for different categories.

While the ALEs are useful measures of covariate effects, they are curves or surfaces, and it useful to further distill this information down to a single number to rank variables in terms of their overall importance. Intuitively, if $ALE_j(x)$ is variable across x but $ALE_k(x)$ is flat, then X_j is a more important predictor than X_k. This intuition leads to using the standard deviation of $ALE_j(x)$ as a measure of variable importance [68],

$$VI_j = \text{SD}\{ALE_j(X_j)\} \tag{8.28}$$

where the standard deviation is computed with respect to the distribution of X_j. A similar formula applies to interaction effects, but for categorical predictors the standard deviation is replaced by one fourth of the range across categories.

Computing the ALE and VI measures require differentiation and integration, which is challenging for black-box prediction methods and large datasets. Fortunately, [8] derive an efficient algorithm to approximate the ALE and VI, and the methods are available in the `R` package `ALEplot` [7]. While the details of this approximation are beyond the scope of this chapter, a key idea is that a Bayesian implementation can quantify posterior uncertainty about ALE_j and VI_j. The mean function g, and thus the derived quantities ALE_j and VI_j, are functions of unknown parameters (e.g., the basis function coefficients in GAM). Therefore, posterior uncertainty about ALE_j and VI_j can be quantified using MCMC output. As with any summary of the parameters, if ALE_j and VI_j are computed at several MCMC iterations, then these samples approximate their posterior distributions. These samples can then be used to give credible bands for each ALE_j and credible intervals and posterior rankings of the VI_j.

Analysis of the hurricane intensity data: Figure 8.16 shows the variable importance and ALE measures for the neural network with $M = 10$ neurons for three-hour lead time as described in Section 8.3.4. Perhaps due to the horseshoe prior, a few covariates dominate the variable importance. In fact, the posterior probability that HWFI is the most important variable is 0.97 (i.e., the proportion of the MCMC samples for which its VI is the largest). The posterior uncertainty provided by the Bayesian analysis shows that there is definitive separation between variable importances.

The centered ALE plots in Figure 8.16 show the effects of three predictors. Although conceptually different, the ALE plots resemble the GAM main effect plots in Figure 8.7 (left). SST is not important, HWRF has an increasing effect, and the current intensity has a more quadratic effect.

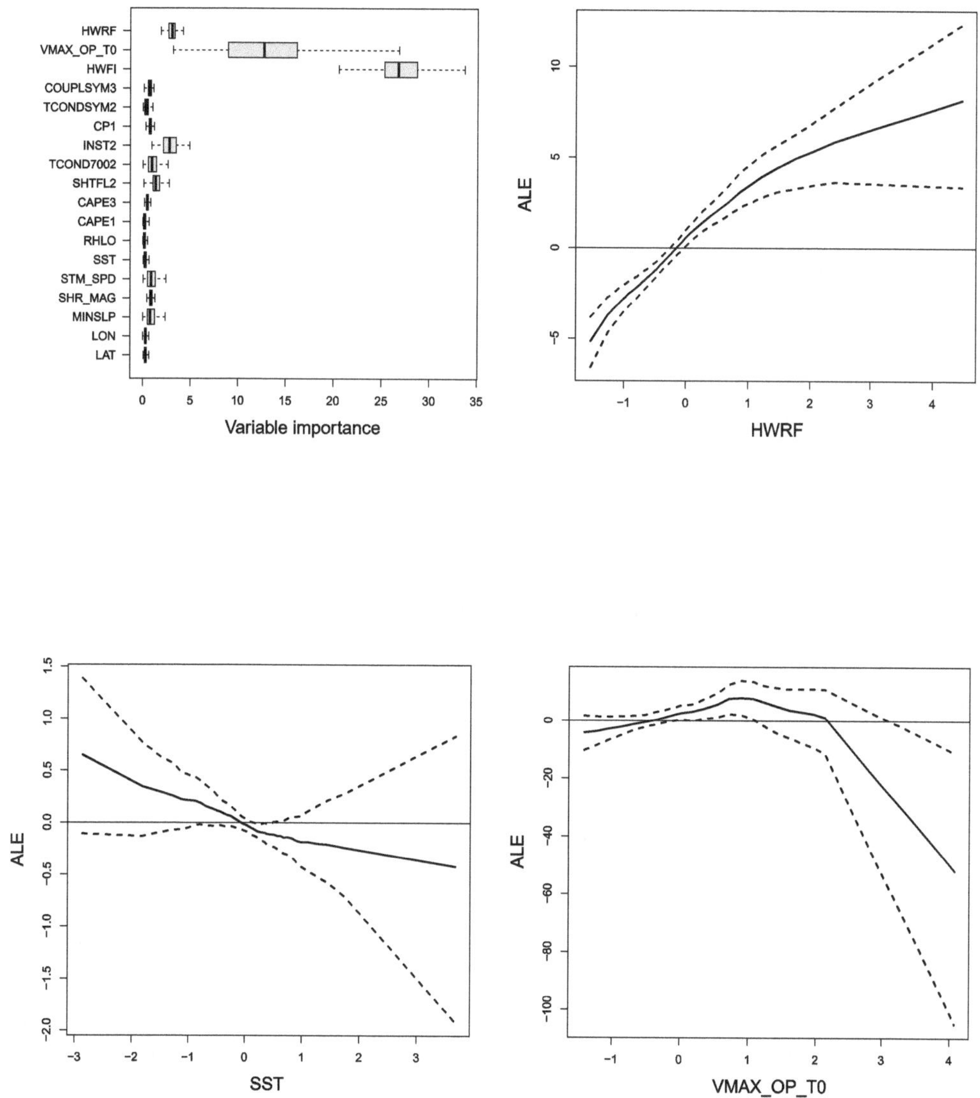

FIGURE 8.16

Covariate effects in the neural network model of hurricane intensity. The top left panel plots the posterior of the variable importance measures VI_j, and the remaining panels show the posterior of $ALE_j(X)$ for the HWRF prediction, sea surface temperature and current intensity (VMAX_T0), all for the three-hour lead time.

8.6 Applications of Bayesian machine learning

The core machine-learning/nonparametric methods introduced above are designed to make predictions and quantify prediction uncertainty. However, these powerful methods have been creatively used for many other statistical tasks. Below we give a flavor of these applications and emphasise the strengths of the Bayesian approach in terms of quantifying uncertainty and ease of interpretation.

8.6.1 Causal inference

The ultimate goal of most scientific inquiry is to understand causal relationships between variables. Only a causal understanding leads to a reliable intervention strategy. The gold-standard approach for establishing causality is a controlled experiment that often relies on randomization. In a carefully-designed experiment, methods such as the Bayesian two-sample t-test (Section 4.1.2) are sufficient, and there is little need for advanced statistical modeling. However, due to ethical or financial considerations, not all questions can be answered with controlled experiments, and in these cases observational data is used in lieu of an experiment. This introduces many sources of potential bias and can never result in a definitive causal conclusion. However, methods from causal inference can be used to reduce bias and clearly state the assumptions needed to interpret a relationship as causal. This does require advanced statistical analysis, often incorporating machine learning methods. In this role, machine learning is not used as a black-box prediction tool; on the contrary, it is a key component of the foundational task of understanding causal mechanisms.

In this chapter, we largely follow [96] to give an overview of causal inference via the Rubin causal framework [132] and Bayesian machine learning. This framework begins with potential outcomes. Each unit (e.g., patient) is associated with a different potential outcome under two treatments that we refer to as treatment and control. For unit $i = 1, ..., n$ with covariates $\mathbf{X}_i$, let $Y_i(0)$ be the potential outcome if the unit is given the control and $Y_i(1)$ be the potential outcome if the unit is given the treatment. If both of these potential outcomes were observed, then the average of $Y_i(1) - Y_i(0)$ over the n units would be the causal effect estimate,

$$\Delta = \sum_{i=1}^{n} \{Y_i(1) - Y_i(0)\}/n.$$

However, we only observe one of the two potential outcomes and the challenge is estimating the causal effect using the available data, denoted as outcome Y_i and treatment indicator $A_i \in \{0, 1\}$, where $A_i = 0$ indicates control and $A_i = 1$ indicates treatment.

There are several quantities of interest in a causal analysis. Ideally, we would estimate the treatment for individual units through the individual treatment effect $ITE_i = Y_i(1) - Y_i(0)$ because this could guide interventions. This is generally too difficult because only one potential outcome is observed, so instead we might average over units with the same covariate values giving the conditional (on the covariates) average treatment effect

$$CATE(\mathbf{x}) = \text{E}\{Y_i(1) - Y_i(0)|\mathbf{X}_i = \mathbf{x}\} = g_1(\mathbf{x}) - g_0(\mathbf{x})$$

where $g_a(\mathbf{x})$ is the expected value of $Y_i(a)$ given $\mathbf{X}_i = \mathbf{x}$. Similar to nonparametric regression, g_0 and g_1 are complex functions that might be difficult to interpret. A one-number summary of the overall treatment effect is the population average treatment effect

$$PATE = \text{E}\{CATE(\mathbf{X})\},$$

where the expectation is with respect to the distribution of the covariates $\mathbf{X}$ and can be approximated as the mixed average treatment effect $MATE = \sum_{i=1}^{n} CATE(\mathbf{X}_i)/n$.

To connect potential outcome with standard Bayesian methods, say $m = n/2$ observations are assigned to the treatment group and the rest to the control group. These data could be analyzed using the Bayesian two-sample t-test in Section 4.1.2. The posterior mean estimator in (4.8) is the difference between group means, $\bar{Y_1} - \bar{Y_0}$ for group means $\bar{Y_1} = \sum_{i=1}^{n} A_i Y_i/m$ and $\bar{Y_0} = \sum_{i=1}^{n}(1 - A_i)Y_i/m$. To connect this estimator back to the potential outcomes, we assume that the observed treatment assignment dictates the potential outcome that is observed; i.e., if $A_i = 0$, then $Y_i = Y_i(0)$, and if $A_i = 1$, then $Y_i = Y_i(1)$. Then (recalling $1/m = 2/n$)

$$\bar{Y_1} - \bar{Y_0} = \frac{2}{n}\sum_{i=1}^{n} A_i Y_i(1) + (1 - A_i)Y_i(0).$$

Further assumptions are needed to relate this estimator to the causal effect.

The most robust way to estimate the casual effect is a randomized trial. Say the m units given the treatment are a complete random sample from $\{1, ..., n\}$ so that $\text{E}(A_i) = 1/2$ for all i. Then the simple difference in sample means is an unbiased estimator of the causal effect because the expected value (where the treatment allocation $\mathbf{A} = (A_1, ..., A_n)$ is the random variable) of the difference is

$$\text{E}(\bar{Y_1} - \bar{Y_0}) = \Delta = \sum_{i=1}^{n}\{Y_i(1) - Y_i(0)\}/n.$$

The difference in means estimator does not use the covariates, $\mathbf{X}_i$. The precision of the estimator can be increased by incorporating covariate information, but randomization breaks the dependence between covariates and treatment assignment, so simply treating the covariates as unobserved error is sufficient to obtain an unbiased estimator.

Treatments are not randomized in observational studies, and this can lead to systematic bias. Consider the extreme scenario that units are given treatment $A_i = 1$ if $Y_i(1) > Y_i(0)$ and $A_i = 0$ otherwise. This obviously induces positive bias in the estimation of the treatment effect. While this is a pathological case, similar factors may be lurking in a study where A_i is an indicator that subject i has a particular diet and Y_i is a measure of their health. In an observational study, the subject chooses their own diet, and they may have experimented with different diets and settled on the one they found personally optimal, so that A_i is more likely to be one if $Y_i(1) > Y_i(0)$. The goal of causal inference is to mitigate biases in observational studies to the extent possible.

Mitigating bias requires some information about its source. The most common method for dealing with bias is to assume it is explained by measured confounding variables in $\mathbf{X}_i$. A variable is a confounding variable if is correlated with both the treatment variable and the response. In this approach, a causal inference can only be achieved if all confounding variables are included in the model, and thus careful consideration of the sampling design and biasing mechanism is crucial. Therefore, the first and most important step in a causal inference is to understand the problem and select appropriate confounding variables. Given that the appropriate confounding variables have been identified, they can be included in the statistical model as covariates in a linear regression or machine learning models, as described below.

To illustrate the effects of confounding variables, the two datasets in Figure 8.17 are generated from the models:

- Linear: $Y_i(a) \sim \text{Normal}(a + X_i, 0.1^2)$

- Non-linear: $Y_i(a) \sim \text{Normal}(0.1a + aX_i^2 + X_i^2, 0.1^2)$.

In both cases, 100 observations have $A_i = 0$, and 100 have $A_i = 1$, and the single covariate is generated as $X_i|A_i \sim \text{Beta}(A_i + 0.5, 1.5 - A_i)$. The observed response is set to the potential outcome determined by A_i, so $Y_i = Y_i(A_i)$. The true CATE is $CATE(x) = 1$ for the linear model and $CATE(x) = 0.1 + x^2$ for the non-linear model.

The treatment is dependent on the covariate, and thus ignoring the covariate leads to bias. For the linear model, the mean response is 0.22 for observations with $A_i = 0$ and 1.72 for observations with $A_i = 1$; therefore a crude estimate of the treatment effect ignoring the covariate is $1.72 - 0.22 = 1.50$, which is much larger than the true value of one. This discrepancy arises because the mean of X_i is smaller for $A_i = 0$ than $A_i = 1$, and X_i is correlated with the response (top left of Figure 8.17).

To link the observed data and the potential outcomes and establish the role of confounding variables requires assumptions about the data-generating model. Common assumptions include:

(A1) **Stable unit treatment value assumption (SUTVA)**: For each unit, the observed response equals the potential outcome dictated by the observed treatment assignment, $Y_i = Y_i(A_i)$.

(A2) **No unmeasured confounders**: Given $\mathbf{X}_i$, the treatment A_i is independent of the potential outcomes $Y_i(0)$ and $Y_i(1)$.

(A3) **Covariate overlap**: The propensity score $e(\mathbf{x}) = \text{Prob}(A_i = 1|\mathbf{X}_i = \mathbf{x})$ is bounded away from 0 and 1 for all i, $0 < e(\mathbf{X}_i) < 1$.

Assumption A1 seems trivial, but it is key to relating the potential outcomes to the observed data. A common violation is interference/spillover, which occurs if the response for unit i is affected by the treatment status of other units, such as another unit that is close to unit i in space. Assumption A2 states that the observed covariates are sufficient to explain all non-random sampling of the treatment variable, so that if the covariates are incorporated correctly the results can resemble the results of a randomized experiment. Assumption A3 is required for estimation of the CATE; e.g., if $e(x) = 0$, then all observations with $\mathbf{X}_i = \mathbf{x}$ have $A_i = 0$, so $g_1(\mathbf{x})$ cannot be estimated without extrapolation. Generally, Assumptions A1 and A2 are not verifiable with observed data, but provide a basis to evaluate the validity of causal claims; Assumption A3 can and should be checked using the observed data.

A special case of Assumption A2 that we adopt throughout is

$$Y_i(a)|\mathbf{X}_i \stackrel{indep}{\sim} \text{Normal}\left\{g_a(\mathbf{X}_i), \sigma^2\right\},$$

so that confounding variables determine the mean but not the variance or other aspects of the potential outcome distribution. With Assumption A1 and this special case of Assumption A2, the model for the observed data is

$$Y_i|A_i, \mathbf{X}_i \stackrel{indep}{\sim} \text{Normal}\left\{(1 - A_i)g_0(\mathbf{X}_i) + A_i g_1(\mathbf{X}_i), \sigma^2\right\}. \tag{8.29}$$

Assumption A3 implies that both $g_0(\mathbf{X}_i)$ and $g_1(\mathbf{X}_i)$ can be estimated because both $A_i = 0$ and $A_i = 1$ are possible for all observed $\mathbf{X}_i$. Therefore, under these assumptions, if we fit a Bayesian model to estimate the parameters in (8.29), these estimates relate directly to the treatment effects CATE and MATE defined by the potential outcomes. An advantage of the Bayesian approach is that summaries of posterior uncertainty about causal effects are directly available from MCMC sampling. Specifically, the JAGS code in Listing 8.5 returns estimates of the full posterior distribution of CATE and MATE.

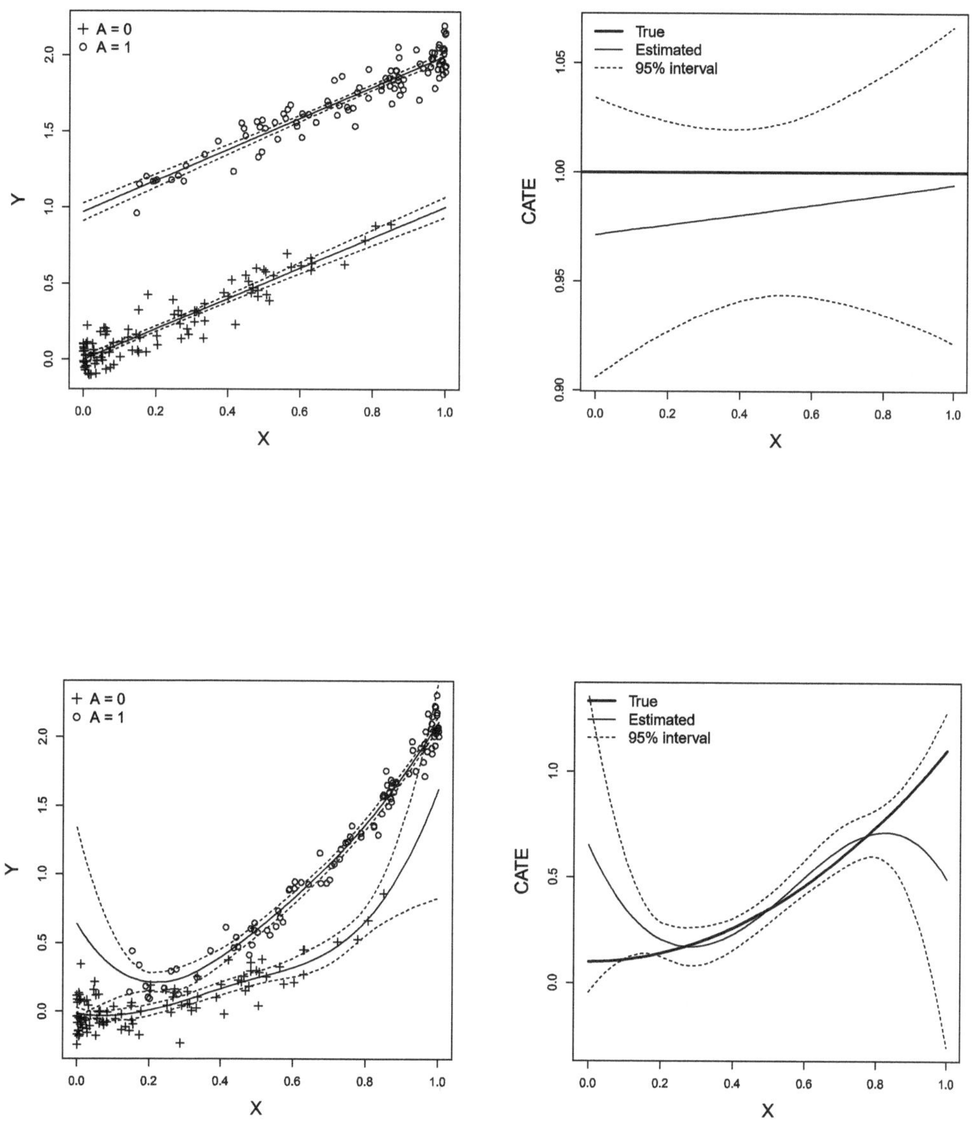

FIGURE 8.17

Illustration of confounding variables in causal inference. The top row plots data and fitted values for a linear model and the true and estimated conditional average treatment effect (CATE); the bottom row repeats this for a non-linear example. This figure is adopted from Figure 1 of [96].

Listing 8.5
JAGS model for causal inference. B includes an intercept and spline basis functions of the covariate X_i.

```
# mn[i] = mn0[i] if A[i]=0 and mn1[i] if A[i]=1
for(i in 1:n){
   Y[i]      ~ dnorm(mn[i],tau)
   mn[i]    <- (1-A[i])*mn0[i] + A[i]*mn1[i]
   mn0[i]   <- inprod(B[i,1:5],beta[1:5])
   mn1[i]   <- inprod(B[i,1:5],beta[6:10])
   CATE[i] <- mn1[i] - mn0[i]
}
for(j in 1:10){beta[j] ~ dnorm(0,0.01)}
tau    ~ dgamma(0.1,0.1)
MATE <- mean(CATE[1:n])
```

Returning to the data in Figure 8.17 (top left), we could estimate the causal effect using a Bayesian analysis of (8.29). Visual inspection of the data suggests that linear models are sufficient for both treatment groups, so $g_a(x) = \alpha_a + \gamma_a x$ for $a = 0, 1$. Rearranging terms gives a linear model that includes A_i, X_i and their interaction, i.e., with $\mathrm{E}(Y_i|\boldsymbol{\beta}) = \beta_1 + \beta_2 A_i + \beta_3 X_i + \beta_4 A_i X_i$. If the model assumptions are correct, this removes the effect of X_i and isolates the true treatment effect (top right of Figure 8.17). Averaging over X_i, the true MATE is one, and the posterior median and 95% interval are 0.98 and (0.94,1.02), so after adjusting for the confounder variable, the Bayesian linear regression returns a good estimate of the causal effect.

Turning to the non-linear example in Figure 8.17 (lower left), fitting a linear model is clearly inappropriate for these data. Any of the nonparameteric regression models in Section 8.3 could be used to model the mean functions g_0 and g_1. [75] and [72] recommend BART. Listing 8.5 gives JAGS code for a spline regression with five basis functions for each treatment group. By using a flexible machine learning model, the covariate effects are properly accounted for, and the CATE (lower right of Figure 8.17) is estimated fairly well. The true MATE is $\sum_{i=1}^{n}(0.1 + X_i^2)/n = 0.45$, and the posterior mean (95% interval) of the MATE is 0.47 (0.28, 0.64), so the method provides an accurate estimate of the causal effect.

In this example, the covariate overlap assumption is questionable because X_i is never less than 0.1 for $A_i = 1$ or greater than 0.9 for $A_i = 0$. However, in the linear example, because a simple linear model is assumed (and in this simulated example justified) the variability is not too large for extrapolation, and the variance of the CATE is only slightly larger near the boundary of X_i's support. In contrast, for the nonlinear example the violation of the overlap assumptions leads to huge varability in the CATE near the boundaries which propagates to large posterior variance for the MATE.

Of course, the discussion of Bayesian casual inference merely scratches the surface of this rich and rapidly-evolving area. Some other notable approaches are to include the propensity score in the analysis, either by first fitting the propensity score model using logistic regression and then using it as a covariate, or using a joint Bayesian analysis of the propensity score and the response [173]. Another current thread is accounting for dependence between observations, such as spatiotemporal dependence [128].

8.6.2 Precision medicine

Assuming a causal relationship is established, either through a randomized experiment or a careful observational study, then the results can be used for decision making. If a study establishes that the MATE of a treatment versus control is statistically significant, then one might recommend using this treatment. Rather than a single decision applying to the entire population, precision medicine [93] devises targeted recommendations (medication, dose, frequency, etc.) for specific patients or subgroups of patients defined by covariates, e.g., treatment if the patient is over 65 years old and control otherwise.

The target for precision medicine is a decision rule that recommends an action based on the covariates. Let $D(\mathbf{X})$ be the decision rule for a patient with covariate $\mathbf{X}$. The rule can be either deterministic or stochastic. If there are two treatments, and the rule is deterministic, then $D(\mathbf{X}) = 1$ assigns the patient to treatment, and $D(\mathbf{X}) = 0$ assigns the patient to control. If the rule is stochastic, then $D(\mathbf{X})$ could be the probability of assigning the patient to the treatment. The decision rule differs from the distribution of the treatments assigned in the training data, $A_1, ..., A_n$. In causal inference, it is assumed that the covariate effects on treatment assignment are captured in the propensity score, $\text{Prob}(A_i = 1|\mathbf{X}_i) = e(\mathbf{X}_i)$. In precision medicine, we use these training data to estimate the causal effect, and then assign future treatment following the optimal decision rule, i.e., $\text{Prob}(A_{n+1} = 1|\mathbf{X}_{n+1}) = D(\mathbf{X}_{n+1})$.

The treatment effect for a patient with covariates $\mathbf{X}$ is summarized by the conditional average treatment effect, $CATE(\mathbf{X})$, introduced in Section 8.6.1. Assuming a large value of the response variable is desirable, then all things being equal (cost, side effects, etc.), the treatment is beneficial compared to control for a patient with covariates $\mathbf{X}$ if $CATE(\mathbf{X}) > 0$ and not beneficial if $CATE(\mathbf{x}) \leq 0$. In the hypothetical nonlinear example in Figure 8.17 (bottom right), the posterior mean of $CATE(\mathbf{X})$ is positive for all $\mathbf{X}$, but the treatment appears to be more beneficial for patients with $\mathbf{X}$ between 0.5 and 0.8. This Bayesian analysis powered by MCMC returns not only an estimate of the CATE, but also the full posterior distribution including the posterior probability that $CATE(\mathbf{X})$ is positive, denoted $pCATE^+(\mathbf{X})$.

The decision rule that assigns the treatment most likely to have the optimal CATE is $D_O(\mathbf{X}) = 1$ if $pCATE^+(\mathbf{X}) > 0.5$ and $D_O(\mathbf{X}) = 0$ otherwise. This rule is optimal in the sense that it maximizes the probability that patients are given the optimal treatment. Thompson sampling (Section 5.3) is a stochastic version of this rule and assigns treatment 1 with probability $pCATE^+(\mathbf{X})$. With a single covariate, the CATE (and thus these decision rules) can be visualized as in Figure 8.17. However, if the treatment depends on multiple variables and is fit using a machine learning model such as BART, then it is difficult to understand and implement such a complex treatment rule.

Interpretable machine learning can be used explain the operating characteristics of a complicated decision rule, or to develop a new and simpler rule that approximates the complicated rule. Since the decision rule $D(\mathbf{X})$ is a function of the covariates just like the nonlinear mean function $g(\mathbf{X})$ in Section 8.3, methods such as the accumulated local effect in Figure 8.15 can be used to understand the effects of individual covariates on the decision rule and rank covariates in terms of importance. An alternative to interpreting a complex rule is to devise a simpler and more interpretable rule. For example, a regression tree (Section 8.3.3) might approximate the complicated decision rule D_O, e.g.,

$$D_O(\mathbf{X}) \approx D_T(\mathbf{X}) = \begin{cases} 1 & x_3 < c_1 \text{ and } x_7 < c_2 \\ 0 & \text{Otherwise} \end{cases} \quad (8.30)$$

where the thresholds c_j and variables to include in the rule (x_2 and x_7) are selected to give the best approximation. One way to fit the tree is to compute $D_O(\mathbf{X})$ for the n training observations and then regress $D_O(\mathbf{X}_1), ..., D_O(\mathbf{X}_n)$ onto $\mathbf{X}_1, ..., \mathbf{X}_n$ using the `rpart` package [153] in `R`. The decision rule can also be approximated with logistic regression to give a linear treatment rule.

Markov decision process: Thus far we have considered a simple case where each patient in the training data produces a single data point and there is a single decision to be made for each potential patient. Optimal decision making is more challenging when data are collected and decisions are made longitudinally. This occurs routinely in medical settings where a patient visits a doctor periodically to decide whether to switch medications or change the dose of their current medication.

To formalize the problem, say training patient $i \in \{1, ..., n\}$ enters the study with covariates $\mathbf{X}_i$ that do not evolve over time. They return for regular visits, and at visit number t their health response is Y_{it}, and they are assigned treatment A_{it}. Treatment assignment for the training data could be observational and influenced by the covariates or the patient's health status and past treatments. The patient's history going back m time points is denoted as $\mathcal{H}_{it} = \{X_i, Y_{it-m}, ..., Y_{it}, , A_{it-m}, ..., A_{it-1}\}$.

The decision problem is to devise a rule to assign treatment given $\mathcal{H}_{it}$. This is an example of a Markov decision process [152], where decisions are made based on available data. A Bayesian analysis begins with fitting a model for future responses given the history $\mathcal{H}_{it}$ and the current treatment A_{it} using the training data. The model could be

$$Y_{it+1}|\mathcal{H}_{it}, A_{it} \stackrel{indep}{\sim} \text{Normal}\left\{g(\mathcal{H}_{it}, A_{it}), \sigma^2\right\}, \tag{8.31}$$

where g is modeled using a machine learning model such as the causal inference model with BART in Section 8.6.1. This model can be fit using standard machine learning techniques. Because the distribution of Y_{it+1} depends on Y_{it}, this accounts for longitudinal dependence. Assuming m is sufficiently large, this model can even capture treatment lag effects where the treatment at time t has no immediate effect but does impact future health responses. Further complicating the problem is that future performance depends on the allocation of future treatments, which may depend on yet-to-be-observed changes in health status.

Policy search is a general strategy for optimizing a treatment assignment rule that maximizes performance across the population [116]. Although not necessary, an interpretable approach is a parametric policy that depends only on a few tuning parameters. For example, $D(\mathcal{H}_{it}) = 1$ with probability $\eta(\mathcal{H}_{it})$ and

$$\text{logit}\{\eta(\mathcal{H}_{it})\} = \alpha_0 + \sum_{j=1}^{P} H_{itj}\alpha_j$$

where H_{itj} are the P elements in $\mathcal{H}_{it}$ and $\boldsymbol{\alpha} = (\alpha_0, ..., \alpha_P)$ are the parameters that determine the policy. We assume that all treatments, both at the current time and at all future time steps, are assigned following this rule. This assumption makes it possible to compute the expected health response in the future and for unobserved patients.

With the assumed parametric model established, policy search boils down to finding the value of $\boldsymbol{\alpha}$ that optimizes future health status. Defining an "optimal" policy requires defining a metric of success, i.e., the reward function. This is of course case- and user-specific, but we give an example below. Say the data consist of n patients, each measured at times $\{1, 2, ..., T\}$. Then we might want to select the policy $\boldsymbol{\alpha}$ that maximizes the average value over the next L time steps

$$R(\boldsymbol{\alpha}) = \text{E}\left(\frac{1}{nL}\sum_{i=1}^{n}\sum_{t=T+1}^{T+L} Y_{it}\right)$$

where the expectation is with respect to future realizations of Y_{it} from the posterior predictive distribution with treatment assigned according to the policy parameters $\boldsymbol{\alpha}$. That is, we envision realizations of the (8.31) with A_{it} assigned following the policy determined by $\boldsymbol{\alpha}$.

The reward function $R(\boldsymbol{\alpha})$ generally cannot be computed analytically, but, as we have seen repeatedly (Section 1.5), sampling from the posterior predictive distribution is straightforward using MCMC, so a Monte Carlo approximation of $R(\boldsymbol{\alpha})$ can be computed for any $\boldsymbol{\alpha}$. The optimal policy then searches for the value of $\boldsymbol{\alpha}$ that maximizes the reward function. However, since the reward function is approximated using Monte Carlo (stochastic) sampling for each candidate $\boldsymbol{\alpha}$, methods from stochastic optimization [73] must be used to find the optimal $\boldsymbol{\alpha}$.

8.6.3 Quantile regression

Multiple linear regression models the effect of covariates on the mean response. While this leads to beautiful mathematical results that form the basis of our understanding of statistical methods, restricting the analysis to the mean response could miss important trends including covariate effects on the spread of the distribution or the likelihood of extreme values. The heteroskedastic model (Section 8.4.1) extends mean regression to allow covariates to affect the variance, but quantile regression goes even further. Quantile regression allows for a potentially different relationship between the covariates and each quantile of the response distribution. For example, a covariate can have a positive effect on the median (the 0.5 quantile) but a negative effect on the spread (the interquartile range is the difference between the 0.75 and 0.25 quantiles) and extreme values (the 0.99 quantile). Therefore, quantile regression offers a systematic and comprehensive way to study covariate effects.

For $\tau \in (0,1)$, the τ quantile given covariates $\mathbf{X}$ is defined as the value $q(\tau;\mathbf{X})$, so that

$$\text{Prob}\{Y < q(\tau;\mathbf{X})\} = \tau.$$

For instance, setting $\tau = 0.5$ gives the median of Y given $\mathbf{X}$. Linear quantile regression assumes $q(\tau;\mathbf{X}) = \mathbf{X}\boldsymbol{\beta}_\tau$, so that $\boldsymbol{\beta}_\tau$ gives the effects of the covariates on the τ quantile, which is allowed to vary by quantile level. A typical analysis estimates the regression coefficients separately for several quantile levels, e.g., $\tau \in \{0.1, 0.2, ..., 0.9\}$ to study how the covariate varies by quantile level.

Classical linear quantile regression (e.g., [92]) is framed as an optimization problem. The estimator for quantile level $\tau \in (0,1)$ is the minimizer of the check loss

$$\widehat{\boldsymbol{\beta}}_\tau = \underset{\boldsymbol{\beta}}{\text{argmin}} \sum_{i=1}^{n} \rho_\tau(Y_i - \mathbf{X}_i\boldsymbol{\beta}) \tag{8.32}$$

where $\rho_\tau(x) = \tau x$ if $x \geq 0$ and $\rho_\tau(x) = (\tau - 1)x$ if $x < 0$. An informative special case is median regression with $\tau = 0.5$, which simplifies to

$$\widehat{\boldsymbol{\beta}}_{0.5} = \underset{\boldsymbol{\beta}}{\text{argmin}} \sum_{i=1}^{n} |Y_i - \mathbf{X}_i\boldsymbol{\beta}|. \tag{8.33}$$

Analogous to the robustness of the median compared to the mean as an estimator of the center of a distribution, quantile regression can be shown to be more robust to outliers than mean regression.

This idea can be extended to the Bayesian setting [170] by assuming the observations follow the asymmetric Laplace distribution (see Appendix A.1). The model is

$$Y_i|\boldsymbol{\beta}_\tau \overset{indep}{\sim} \text{ASL}(\mathbf{X}_i\boldsymbol{\beta}_\tau, \sigma, \tau), \tag{8.34}$$

where $\boldsymbol{\beta}_\tau$ and $\sigma > 0$ are parameters to be estimated and τ is fixed. This distribution is a convenient choice for Bayesian quantile regression because both the mode and τ-quantile level are $\mathbf{X}_i\boldsymbol{\beta}_\tau$. The connection with classical quantile regression is that assuming a flat prior, the maximum a posteriori estimator of $\boldsymbol{\beta}_\tau$ is exactly (8.32). However, unlike the classical estimator, this Bayesian approach specifies a full likelihood, and therefore it is important to check the fit of the model before trusting the results.

Applying the model in (8.34) for several values of τ can give the posterior distribution of several $\boldsymbol{\beta}_\tau$ and can shed light on the covariate effects at different quantile levels. However, it is unlikely that the asymmetric Laplace distribution fits well for all quantile levels (it is right skewed for $\tau < 0.5$ and left skewed for $\tau > 0.5$), and a separate analysis by τ does not preserve quantile ordering that $\mathbf{X}\boldsymbol{\beta}_{\tau_1} < \mathbf{X}\boldsymbol{\beta}_{\tau_2}$ if $\tau_1 < \tau_2$. Another option is to fit the quantile function simultaneously for all quantile levels with $\mathbf{X}\boldsymbol{\beta}_\tau$ and $\boldsymbol{\beta}_\tau$ modeled using splines or a Gaussian process that assures quantile ordering ([127, 169]). However, this leads to computational challenges, and assuming all quantiles are linear in the covariates is likely too restrictive.

Rather than assume a linear quantile regression model, the Semi-Parametric Quantile Regression (SPQR) approach of [167] fits a density regression model similar to (8.26) that makes no assumptions about the structure of the quantiles, and then uses ideas from interpretable machine learning (Figure 8.15) to estimate quantile effects. This approach gives a flexible model that separates model fitting from interpretation and automatically preserves quantile ordering. This approach is appealing because it avoids strong assumptions about the data-generating models and provides a comprehensive study of covariate effects across quantile levels with full Bayesian uncertainty quantification.

The density regression model is

$$f(y|\mathbf{X}) = \sum_{k=1}^{K} \pi_k(\mathbf{X}) B_k(y) \tag{8.35}$$

where the the mixture weights π_k are modeled using a neural network and the B_k are m-spline basis functions. The simplest network is a one-layer FFNN for the probabilities $\pi_k(\mathbf{X})$,

$$\pi_k(\mathbf{X}) = \frac{\exp\{b_{2k} + \sum_{l=1}^{L} W_{2kl} Z_l(\mathbf{X})\}}{\sum_{h=1}^{K} \exp\{b_{2h} + \sum_{l=1}^{L} W_{2hl} Z_l(\mathbf{X})\}}$$

with ReLU activation function $Z_l(\mathbf{X}) = (b_{1l} + \sum_{j=1}^{p} W_{1lj} X_j)_+$. This formulation ensures that the probabilities are positive and sum to one, $\sum_{k=1}^{K} \pi_k(\mathbf{X}) = 1$, and since the m-spline basis functions are proper density functions, $f(y|\mathbf{X})$ is a proper density for any value of the weights and biases. Letting the number of basis functions K and the width of the neural network L increase, this density function can approximate any continuous density function of $Y|\mathbf{X}$. This gives an arbitrarily flexible model from the quantile function $q(\tau; \mathbf{X})$, which can be computed from (8.35).

The effects of coviarates on the quantile function are measured using the accumulated local effects and variable importance metrics introduced in Figure 8.15. For a given quantile level τ, these measures are computed for $q(\tau; \mathbf{X})$ (i.e., $g(\mathbf{X}) = q(\tau; \mathbf{X})$ is taken as the function to be summarized). This is repeated for several quantile levels of interest, giving estimates with posterior uncertainty of each covariate's effect on each quantile level.

Semi-Parametric Quantile Regression example: We illustrate SPQR with a small simulated dataset. The $n = 1000$ observations plotted in Figure 8.18 are generated from the model $X_1, X_2 \overset{iid}{\sim} \text{Uniform}(-1, 1)$ and $Y|X_1, X_2 \sim \text{Beta}\{2, \exp(3X_1^2)\}$. The first covariate affects the entire response distribution with a more prominent effect on the upper than lower tail. The second covariate is irrelevant.

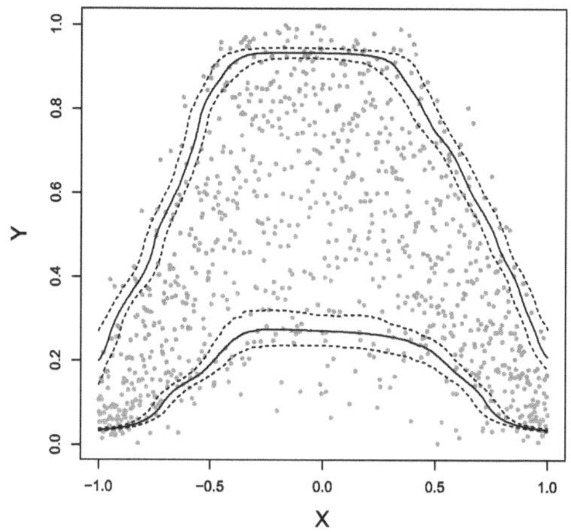

FIGURE 8.18
Semi-parametric quantile regression. Simulated data (points) and posterior mean (solid lines) and 90% interval (dashed lines) of the conditional 0.1 (bottom) and 0.9 (top) quantile functions.

We fit the SPQR model using Hamiltonian Monte Carlo with $K = 20$ basis functions and $L = 25$ neurons and priors $b_{jk}, W_{jkl}|\tau \stackrel{iid}{\sim} \text{Normal}(0, \tau^2)$ and $\tau \sim \text{HC}(1)$. For each MCMC draw of the model parameters b_{jk} and W_{jkl}, we numerically compute the quantile function $q(\tau; \mathbf{X})$, giving draws of the posterior for the quantile function to be used to summarize covariate effects.

We first fit the data with only the first covariate. Figure 8.18 plots the fitted $\tau = 0.1$ and $\tau = 0.9$ quantile functions by X_1. The semi-parametric model fits the data well. For both quantile levels, the quantile function increases for X_1 near zero and decreases in the extremes.

In a second analysis, we include both covariates and study their effects using ALE and VI measures by quantile levels. Figure 8.19 confirms that X_1 has a significant effect on the response because the credible bands for its quantile function exclude zero for some values of X_1. This plot also confirms that the effect is larger for the 0.9 than 0.1 quantiles. In contrast, X_2 has ALE near zero for both quantile levels and is thus deemed insignificant. Table 8.1 gives the posterior distribution of the variable importance measure by quantile level. As in the ALE plots, it is clear that X_1 is more important than X_2 and that X_1 has a larger impact on high than low quantiles.

8.6.4 Transfer learning

Machine learning via nonparametric regression provides greater flexibility than traditional parametric modeling, but often requires large datasets for adequate training. Transfer learning (TL) is a possible remedy. In TL, the information from a related analysis (or analyses) with ample data (called the source) are leveraged to improve the analysis of interest

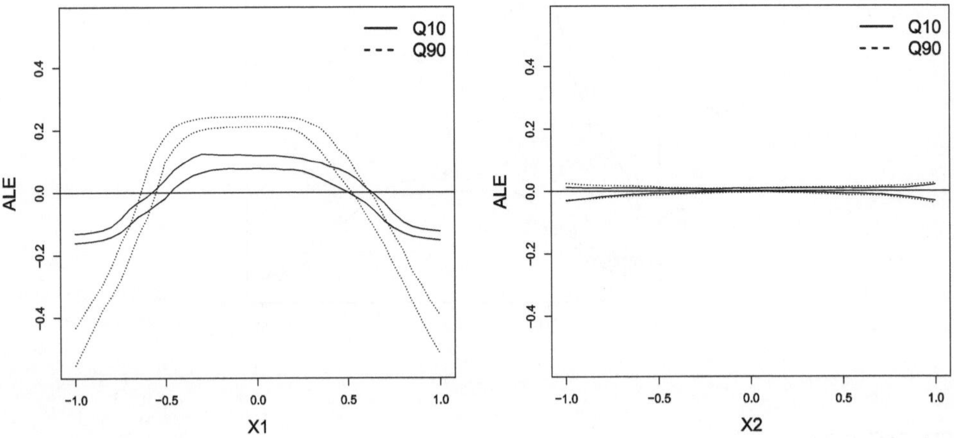

FIGURE 8.19
SPQR ALE plots for density regression. 90% credible intervals for the accumulated local effects for the 0.1 and 0.9 quantile level, for the two covariates in the simulated SPQR example.

TABLE 8.1
Variable importance by quantile level. Posterior median and 95% intervals of variable importance from SPQR by covariate and quantile level (τ) for the simulated example.

Quantile level	X_1	X_2
0.1	1.4 (0.9,1.9)	0.5 (0.3, 1.3)
0.2	1.8 (1.0,2.5)	0.6 (0.4, 1.5)
0.3	1.7 (1.2,2.8)	0.6 (0.4, 1.5)
0.4	1.9 (1.3,2.8)	0.6 (0.4, 1.6)
0.5	2.0 (1.5,2.9)	0.7 (0.4, 1.6)
0.6	2.1 (1.6,3.0)	0.7 (0.4, 1.5)
0.7	2.2 (1.3,3.0)	0.7 (0.4, 1.6)
0.8	2.4 (1.3,3.3)	0.7 (0.4, 1.6)
0.9	2.7 (1.0,3.6)	0.8 (0.4, 1.6)

with sparse data (called the target). Canonical examples include using data from a large metropolitan hospital to supplement data from a small rural hospital, or using classifiers trained on images from ImageNet [43] to assist in tumor classification for a rare type of cancer. The Bayesian framework is natural for transfer learning. Informative priors are an obvious way to incorporate the results from related analyses, and hierarchical models for multiple datasets are a means of borrowing information. In this section, we briefly define TL notation and terminology and discuss some methods for TL with Bayesian deep learning.

Although TL can be used for other tasks, for simplicity we restrict our discussion to regression with a single source dataset. We adopt the notation of [149], who define a regression of $\mathbf{X}$ onto $\mathbf{Y}$ by its domain and task: the domain, $\mathcal{D} = \{\mathcal{X}, P\}$, consists of the support $\mathcal{X}$ so that $\mathbf{X} \in \mathcal{X}$ and the distribution of $\mathbf{X}$, P; the task, $\mathcal{T} = \{\mathcal{Y}, g\}$, consists of the support of $\mathbf{Y}$ so that $\mathbf{Y} \in \mathcal{Y}$, and g is the true regression function $g(\mathbf{X}) = \text{E}(\mathbf{Y}|\mathbf{X})$. TL borrows information from the source domain and task, $\mathcal{D}_1 = \{\mathcal{X}_1, P_1\}$ and $\mathcal{T}_1 = \{\mathcal{Y}_1, g_1\}$, to the target domain and task, $\mathcal{D}_0 = \{\mathcal{X}_0, P_0\}$ and $\mathcal{T}_0 = \{\mathcal{Y}_0, g_0\}$.

The simplest example is linear regression with the same domains but different regression functions, $g_j(\mathbf{X}) = \mathbf{X}\boldsymbol{\beta}_j$, with $\boldsymbol{\beta}_1$ and $\boldsymbol{\beta}_0$ the true regression coefficients in the source and target distributions. An obvious two-stage approach is to use the source data to estimate $\boldsymbol{\beta}_1$ and then use this estimate as the prior mean for β_0; [2] show that adopting a continuous shrinkage prior centered on the source-task estimator can lead to improvements over using only the target data.

TL is more complex when the domain and/or task differ between the source and target. Returning to the hospital example, say the metropolitan and rural hospitals have the same p covariates, so $\mathcal{X}_0 = \mathcal{X}_1 = \mathcal{R}^p$, but the metropolitan hospital has a continuous measure of the health outcome, so $\mathcal{Y}_1 = \mathcal{R}$, and the task is linear regression $g_1(\mathbf{X}) = \mathbf{X}\boldsymbol{\beta}_1$, but the rural hospital has a binary response, so $\mathcal{Y}_0 = \{0, 1\}$, and the task is logistic regression $\text{logit}\{g_0(\mathbf{X})\} = \mathbf{X}\boldsymbol{\beta}_0$. How to combine these two disparate data sources? There is not a single path forward, and the analyst must use subject-matter knowledge and model-building techniques to arrive at a suitable approach. One possibility is a stochastic search variable selection model (as in Section 8.1) that analyzes both datasets simultaneously with the same binary inclusion indicators but different non-zero coefficients for the source and target models. TL is even more challenging when the two hospitals have different covariates and thus $\mathcal{X}_0 \neq \mathcal{X}_1$. Unless there is some common structure between the two domains, this may be beyond the scope of TL.

Deep learning is one of the most common applications of TL because these models have many parameters and thus require large training data. Consider the feed-forward neural network in (8.19) with the same domains, but potentially different parameters (weights and biases), for the source and target. A two-stage approach might first compute a point estimate, say MAP estimates (see (1.39)) for the parameters using the source data, denoted $\widetilde{W}$ (and similar for the biases), and then use these estimates as priors $W_{3k} \stackrel{iid}{\sim} \text{Normal}(\widetilde{W}_{3k}, \tau_3^2)$, $W_{2kl} \stackrel{iid}{\sim} \text{Normal}(\widetilde{W}_{2kl}, \tau_2^2)$, and $W_{1jl} \stackrel{iid}{\sim} \text{Normal}(\widetilde{W}_{1jl}, \tau_1^2)$. This reduces the problem to specifying the variances τ_l^2 or their priors. One approach is to freeze certain layers by setting some of the $\tau_l^2 = 0$. For example, fitting $\tau_1^2 = \tau_2^2 = 0$ sets $W_{1jl} = \widetilde{W}_{1jl}$ and $W_{2kl} = \widetilde{W}_{2kl}$, leaving only the output-layer parameters W_{3k} to be fit using the target data. This assumption that the latent structure of the regression model is shared across tasks reduces the TL problem to the linear regression model above.

Freezing layers is a drastic simplification, but it is unlikely that the latent structure is identical in the source and target, and so it may be preferable to specify priors for the variances. The prior should have some mass near zero to retain the information from the source data, but it is challenging to determine an appropriate prior for deep learning because of the large number of uninterpretable parameters. [1] propose a penalized complexity prior (Section 2.4.5) that treats the source model as the base model and places an exponential prior on the KL divergence between the target and source models. However, this is computationally intensive, and so it may be preferable to simply select the prior based on cross-validation or some other measure of fit.

Transfer learning is an emerging field with novel methods developing for the challenging cases of transferring information between regressions with different domains and tasks. In this chapter we have only scratched the surface. We refer to [149] for a comprehensive review.

8.6.5 Equation learning

Thus far we have used ML methods to estimate statistical quantities including the mean function in a regression model. In this context, ML models are often described as black boxes that provide excellent predictions but do not contribute to scientific understanding of the relationship between variables. However, ML methods are increasingly being used in science and engineering to study, and even uncover, the mathematical equations that govern physical and biological systems, thus bringing new scientific insights.

A classic example is an engineer designing a wind turbine to maximize efficiency. The mathematical model (i.e., differential equations based on physical laws) might depend on some independent variables, $\mathbf{X}$, such as the height of the turbine and length of the blades, and some unknown parameters, $\boldsymbol{\theta}$, such as average wind speed in the region or the strength of one of the turbine's components. Given these values, the efficiency of the turbine is a deterministic function $g(\mathbf{X}, \boldsymbol{\theta})$ that follows from a set of trusted differential equations. The engineer may use this model to predict performance in a new setting, estimate the parameters $\boldsymbol{\theta}$ that cannot be directly measured, or find the design values $\mathbf{X}$ that optimize performance.

In some cases, data is collected to learn about unknown parameters $\boldsymbol{\theta}$ or to evaluate and correct for an imperfect model g. A Bayesian analysis is appealing for this task because uncertainty in the parameters or the true physical model can be propagated to the final prediction, but standard MCMC methods are prohibitive because it would require thousands of evaluations of g. Say n observations are collected with $(\mathbf{X}_i, Y_i)$. A model for these data is [90]

$$Y_i = g(\mathbf{X}_i, \boldsymbol{\theta}) + d(\mathbf{X}_i) + \epsilon_i \tag{8.36}$$

where g is the mathematical model output, d is the discrepancy/bias between the model and the data and $\epsilon_i \stackrel{iid}{\sim} \text{Normal}(0, \sigma^2)$ is measurement error. The unknowns to be learned from the data are the model parameters $\boldsymbol{\theta}$, the discrepancy function d and the error variance σ^2. Prior distributions for $\boldsymbol{\theta}$ and σ^2 are ideally based on scientific knowledge, but can also be uninformative. The discrepancy term d is a function and this is modeled using nonparametric regression, e.g., a Gaussian process prior.

If the model g were easy to compute, then Bayesian inference could proceed as usual with, say, MCMC to estimate the unknown parameters. If a Metropolis update is used for $\boldsymbol{\theta}$, then each iteration would propose a candidate for $\boldsymbol{\theta}$ and then evaluate the model $g(\mathbf{X}_i, \boldsymbol{\theta})$ for all n observations to compute the acceptance probability. However, if each evaluation of g takes hours, then this algorithm is prohibitively expensive.

An alternative is to approximate g using machine learning. This is often called a statistical emulator or surrogate model of a mathematical model [67]. To train this model, the analyst would select m training cases $(\widetilde{\mathbf{X}}_1, \widetilde{\boldsymbol{\theta}}_1), ..., (\widetilde{\mathbf{X}}_m, \widetilde{\boldsymbol{\theta}}_m)$ and evaluate the model at these m values, $Z_j = g(\widetilde{\mathbf{X}}_j, \widetilde{\boldsymbol{\theta}}_j)$. The statistical model is

$$Z_j = \widetilde{g}(\widetilde{\mathbf{X}}_j, \widetilde{\boldsymbol{\theta}}_j) + \varepsilon_j \tag{8.37}$$

where $\widetilde{g}$ is an approximation for g and ε_j is approximation error. Since it is presumed that the mathematical model is a complex function of its inputs, a nonparametric regression model such as Gaussian process regression is selected for the function $\widetilde{g}$. The idea is that it may take a long time to produce the m training observations and fit the machine learning model to estimate $\widetilde{g}$, but once this work is completed, it is easy to evaluate $\widetilde{g}$ for a new set of inputs. An advantage of a fully Bayesian analysis involving emulators is that uncertainty about the emulator's accuracy can be measured and propagated to the final output.

The emulator can be used for Bayesian optimization. That is, if the objective is to find the input $\mathbf{X}$ that maximizes the function g, that can be done quickly by instead maximizing the surrogate function, i.e.,

$$\mathbf{X}_{opt} = \underset{\mathbf{X}}{\arg\max}\ g(\mathbf{X}, \boldsymbol{\theta}) \approx \underset{\mathbf{X}}{\arg\max}\ \widetilde{g}(\mathbf{X}, \boldsymbol{\theta}). \tag{8.38}$$

While evaluating $\widetilde{g}$ is fast, optimization is complicated by the fact that $\widetilde{g}$ is a statistical prediction and thus is a random function, requiring advanced methods from stochastic optimization [73].

The emulator can also be included in the model for the observations to learn about unknown parameter and model error. The model in (8.36) then becomes

$$Y_i = \widetilde{g}(\mathbf{X}_i, \boldsymbol{\theta}) + d(\mathbf{X}_i) + \epsilon_i. \tag{8.39}$$

With the fast statistical prediction $\widetilde{g}$ replacing the cumbersome mathematical model solution g, standard Bayesian computational tools such as MCMC can be used to estimate the parameters $\boldsymbol{\theta}$ and discrepancy function d.

Equation learning: Methods that emulate the g are useful for improving computational efficiency, and modeling the discrepancy term d can shed some light on model performance. However, an emerging area of research [20, 139, 24] is to replace specifying g based on physical laws and instead learn the underlying equations that dictate g. This method then replaces mathematical modeling based on physical or biological laws with machine learning based on training data and statistical analysis.

To understand this approach, consider the problem of mathematically modeling for the spread of an infectious disease briefly discussed in Section 7.1. A standard model in epidemiology is the SIR model, where the number of susceptible, infected and recovered individuals evolve over time through the equations

$$\frac{dS(t)}{dt} = -\beta S(t)I(t),\ \frac{dI(t)}{dt} = \beta S(t)I(t) - \gamma I(t),\ \text{and}\ \frac{dR(t)}{dt} = \gamma I(t).$$

The model has two parameters, $\boldsymbol{\theta} = (\beta, \gamma)$. Say the output of interest is $g(t; \boldsymbol{\theta}) = I(t)$; then a reasonable analysis is to use the observed number of infected individuals at time t, $Y(t)$, to fit the model $Y(t) = g(t, \boldsymbol{\theta}) + d(t) + \varepsilon(t)$ and learn the parameters $\boldsymbol{\theta}$ while accounting for model error d and measurement error ε.

If the mathematical model (8.40) is approximately correct, then learning the parameters $\boldsymbol{\theta}$ and making forecasts from the model is an effective approach, even for scenarios beyond the range of data. However, if these equations are incorrect, which is almost certain for a process as complicated as the spread of an infectious disease, then no amount of data can correct for this model misspecification. An alternative is to model the derivatives using ML, e.g.,

$$\frac{dS(t)}{dt} = g_S\{S(t), I(t), R(t)\} \tag{8.40}$$

where g_S, and similarly g_I and g_R for derivatives of the infected and recovered classes, is modeled using a nonparametric regression (Section 8.3) such as GAMs or deep learning

[101, 109, 157, 158, 102, 111]. This approach has the advantages of being able to learn equations from the data without strict assumptions. However, as with any statistical model, prediction outside the range of the training data is questionable.

Another option is symbolic regression, where the goal is to uncover the parametric form of the equations along with the unknown parameters. Symbolic regression assumes the differential equations can be constructed using basic operations of the state variables, such as polynomials, products and trigonometric functions. A simple model that would be effective for the SIR model is to assume the second-order polynomial model

$$\begin{aligned} g_S\{S(t), I(t), R(t)\} &= b_0 + S(t)b_1 + I(t)b_2 + R(t)b_3 && (8.41)\\ &+ S(t)^2b_4 + I(t)^2b_5 + R(t)^2b_6 && (8.42)\\ &+ S(t)I(t)b_7 + S(t)R(t)b_8 + I(t)R(t)b_9. && (8.43)\end{aligned}$$

This simple model includes the true process with $b_7 = -\beta$ and $b_j = 0$ for all $j \neq 7$, so Bayesian variable selection methods such as stochastic search variable selection (Section 6.2) may be able to select the true model. Tree-based approaches similar to Bayesian adaptive regression trees (Section 8.3.3) can also be used [84]. Of course, this approach relies on including the necessary basic operations to be considered, which cannot be guaranteed in a real application.

If observations are available at times $t_1, ..., t_n$, then the fitted model could be

$$\frac{S(t_{i+1}) - S(t_i)}{t_{i+1} - t_i} = g_S\{S(t_i), I(t_i), R(t_i)\} + \varepsilon_i \quad (8.44)$$

and similar for the other two classes. The prior for $g_S(S, I, R)$ could be centered on the parametric model, e.g., a Gaussian process with mean $-\beta SI$, and standard deviation with mass near zero such as a Cauchy prior and correlation that decays with distance in (S, I, R). Further modeling could add biological constraints, such as $S(t) + I(t) + R(t)$ being fixed at the population size and $g_S(S, I, R) = 0$ if $I = 0$ since with no infected individuals there can be no new cases and thus no change in the number of susceptible individuals. The modeling opportunities are endless, making this a promising area of future research.

8.7 Exercises

1. Say $Y|\beta \sim \text{Normal}(\beta, \sigma^2)$ and $\beta \sim \text{Normal}(0, \tau^2\sigma^2)$.
   (a) Prove that the posterior mean of β is $E(\beta|Y) = \{1 - \kappa(\tau)\}Y$ for $\kappa(\tau) = 1/(1 + \tau^2)$.
   (b) Discuss how small and large τ lead to different degrees of shrinkage.
   (c) Prove that if $\tau \sim \text{HC}(1)$ (as in the horseshoe prior), then $\kappa(\tau) \sim \text{Beta}(1/2, 1/2)$.
   (d) Argue that $\kappa(\tau) \sim \text{Beta}(1/2, 1/2)$ is a reasonable shrinkage prior.
2. In Section 6.2, we used stochastic search variable selection (SSVS) to analyze childhood malaria data.
   (a) Fit the model in (6.5) using horseshoe priors for the regression coefficients, and verify MCMC convergence.
   (b) Compare the results using horseshoe regression with the SSVS results in Table 6.2.

   (c) In general, what are some pros and cons of SSVS versus horseshoe regression?

3. Download the `airquality` dataset in `R`. Fit the model

$$\text{ozone}_i \sim \text{Normal}\{f(\text{solar.R}_i), \sigma^2\}$$

   where the function f is approximated using spline basis functions.

   (a) Use model selection tools to select the number of spline basis functions.
   (b) Make a scatterplot of solar radiation (x-axis) versus ozone (y-axis) and overlay the fitted curve f with a 95% interval for f for each value of solar radiation. Visually, does the estimated curve fit the data well? Can you conclude that there is a statistically significant relationship between these two variables?
   (c) Do 95% of the observations fall in the 95% credible interval in part (b)? Is this a concern? If not, why? If so, how would you address this concern?
   (d) Looking at the data, does it appear that f is monotonically increasing? How would you modify your analysis to incorporate the restriction that f is a strictly increasing function of solar radiation?

4. Using the `airquality` data from the previous question with response ozone and solar radiation, wind speed and temperature as the covariates, fit the heteroskedastic linear model

$$Y_i|\boldsymbol{\beta}, \boldsymbol{\alpha} \sim \text{Normal}\left\{\sum_{j=1}^{p} X_{ij}\beta_j, \exp\left(\sum_{j=1}^{p} X_{ij}\alpha_j\right)\right\}.$$

   (a) Confirm that the MCMC chains have converged.
   (b) Summarize the covariate effects on the mean and variance.
   (c) Does the heteroskedastic fit better than the usual model with constant variance?

5. Using the `airquality` data, fit the GAM model

$$\text{ozone}_i = f_1(\text{solor.R}_i) + f_2(\text{wind.speed}_i) + f_3(\text{temp}_i) + \varepsilon_i$$

   where $\varepsilon_i \stackrel{iid}{\sim} \text{Normal}(0, \sigma^2)$ and f_1, f_2 and f_3 are each modeled using B-spline basis expansions with five knots. Which are important predictors of ozone, and which appear to have non-linear effects?

6. Given a strength and a weakness for each of the following approaches:

   (a) Generalized additive modeling
   (b) Gaussian process regression
   (c) Deep learning
   (d) Bayesian additive regression trees
   (e) Quantile regression

7. Fit a semiparametric model to the `mcycle` data from the `MASS` package with the time since a motorcycle makes impact as the only predictor and acceleration as the outcome using the `wbart` function in the `BART` package.

   (a) Describe the role of the tuning parameters `ntree`, `power` and `k`.

(b) Fit the model and plot the fitted values for several values of these tuning parameters. Are the results sensitive to these tuning parameters? If so, which ones?

(c) In practice, how would you select these tuning parameters in a computationally efficient way?

8. Download the daily bike sharing data from the book webpage. The response variable is the daily count (`cnt`) of bike rentals (the first three values are 985, 801 and 1349), which we model as Gaussian.

   (a) Describe and justify how you process the other variables to be used as predictors.

   (b) Regress `cnt` onto the predictors from (a), and determine the most important predictors from your model.

   (c) Use five-fold cross-validation to compare BART and multiple linear regression in terms of mean squared prediction error and coverage of 95% prediction intervals.

9. Simulate 100 observations from (8.29) with response Y, treatment A and potential confounders X using the following `R` code:

```
set.seed(919)
X  <- matrix(runif(500),100,5)
A  <- rbinom(100,1,X[,1])
g0 <- 2 + cos(10*X[,3])
g1 <- 1 + 2*X[,2] + cos(10*X[,3])
Y  <- rnorm(100,(1-A)*g0+A*g1,1)
```

   (a) Assuming this model is known, what is the optimal decision rule? That is, for an observation with a given value of `X`, which value (0 or 1) of `A` would maximize the fitted mean response? Express the optimal rule as an equation, e.g., $g_1 > g_0$ if $X_2 > X_3$.

   (b) Fit a Bayesian machine learning model to these data, and summarize the fitted model.

   (c) Using the fitted model, for each of the 100 observations in the training data, which value of the treatment maximizes the mean response? For what proportion of the observations does this match the optimal treatment rule from (a)?

   (d) Give the main assumptions needed to have a casual interpretation of these results.

10. Load the `black_politicians` dataset from the `causaldata` package. To simplify the problem, we will analyze only data with `treat_out` equal to one. The data consist of the results of emails sent to legislators from black voters from outside the legislator's district. The response variable is the binary indicator of whether the legislator responded to the email, `responded`. The treatment variable is whether the legislator is black, `leg_black`, and the remaining 11 variables are used as potential confounders.

   (a) Perform a Bayesian logistic regression that includes only the treatment variable as a predictor. Summarize the treatment effect. What are some potential sources of bias in this analysis?

(b) Perform a Bayesian logistic regression that includes both the treatment variable and the confounders (as linear terms), and summarize the treatment effect. Are the results affected by including the potential confounders?

(c) Perform a semiparametric Bayesian logistic regression using the function `lbart` in the package `BART`. Compute the posterior distribution of the mixed average treatment effect (MATE), and summarize the results.

11. Use the `R` package `covid19.analytics` to generate data from the SIR model similar to (8.40) for Uruguay as

```
data <- covid19.data("ts-confirmed")
SIR  <- generate.SIR.model(data,"Uruguay",
          tot.population=3500000)
```

Assume that you observe noisy (with multiplicative gamma errors) versions of the number of susceptible and infected,

```
set.seed(1)
S     <- SIR$model$S*rgamma(90,100,100)
I     <- SIR$model$I*rgamma(90,100,100)
```

Fit a Bayesian equation learning model that regresses $dI(t)/dt$ onto $S(t)$ and $I(t)$, and summarize the disease dynamics you learn from the data.

9

Statistical properties of Bayesian methods

CONTENTS

In this chapter we briefly discuss some of the most important concepts of statistical theory as they relate to Bayesian methods. This book is primarily dedicated to the practical application of Bayesian statistical methods, and thus this chapter can be skipped on first read. However, even an applied statistician should be familiar with statistical theory at least at a high level to plan and defend their work.

As an example, consider the problem of estimating a normal mean, θ. An estimator is a function that takes the data as input and returns an estimate of the parameter of interest. An estimator of the parameter θ is denoted $\hat{\theta}(\mathbf{Y})$. A natural estimator of a normal mean is the sample mean, $\hat{\theta}(\mathbf{Y}) = \sum_{i=1}^{n} Y_i/n$. However, there are other estimators including the sample median or the trimmed mean. In Section 2.1.3 we discussed Bayesian estimators such as the posterior mean $\hat{\theta}(\mathbf{Y}) = \mathrm{E}(\theta|\mathbf{Y}) = \sum_{i=1}^{n} Y_i/(n+m)$. How does one justify that the Bayesian estimator is a better estimator than the sample mean? In what sense is an estimator optimal? Questions such as these motivate the development of statistical theory.

In addition to investigating the properties of specific estimators and determining the settings for which they are preferred, much of statistical theory is dedicated to developing general procedures for deriving estimators with good properties. General estimation procedures include maximum likelihood estimation, the method of moments, estimating equations, and of course, Bayesian methods. Comparisons of these statistical frameworks are often made using frequentist criteria. As we will see, even if the objective if to generate a procedure with good frequentist properties, Bayesian methods often perform well.

In Section 9.1 we focus on Bayesian methods and optimally summarizing the posterior distribution. In Sections 9.2 and 9.3 we study the frequentist properties of Bayesian methods using mathematical and computational tools, respectively.

9.1 Decision theory

One of the primary goals of collecting data is to make decisions about the scientific problem. The decision can be devising the "best" estimate of a population level summary or the parameter θ based on the observed data $D = (y_1, y_2, \ldots, y_n)$. It can also be of interest to test a hypothesis about the parameter θ; for example if $\theta = (\theta_1, \theta_2)$, we may be interested to test the null hypothesis $H_0 : \theta_1 \leq \theta_2$ based on the observed data D. All of these cases can be considered within the general notion of *Decision Theory Framework* that consists of a loss function $L(\theta, d)$ which measures the price one would pay if θ is the true (but unobserved) value of the parameter, and we take a decision d based on the data D.

Point Estimation: Suppose we would like to compare two estimators $T_1(D)$ and $T_2(D)$, which are some functions of the data D, to estimate a real-valued parameter θ, how would we compare their performances? Consider a loss function (known as squared error loss) given by $L(\theta, T) = (T - \theta)^2$. Clearly, we can't directly compare the values $L_j = L(\theta, T_j(D)$ for $j = 1, 2$ for two obvious reasons: (i) L_j is random as it depends on the observed data D; and (ii) L_j's depend on θ, which is not known in the first place. This is exactly where the solutions obtained by frequentist and Bayesian methods differ and has been the subject of controversy over many decades if not centuries!

Frequentist solutions to resolve the two issues have been the following: (i) as L_j's are random, average out the loss function over all possible data that could have been observed, and consider the risk function given by $R(\theta, T) = E_\theta[L(\theta, T(D))]$, where the expectation E_θ is taken with respect to the sampling distribution of D given θ, viz., $f(D|\theta)$. This clearly takes care of the first problem, and there's no controversy about this step. The difference (between frequentist and Bayesian methods) arises on how the second problem (ii) is resolved. Clearly, $R(\theta, T_j)$'s of the two estimators T_1 and T_2 are still not comparable directly as these risks still depend on θ.

The Minimax approach: One popular approach in frequentist theory (known as the *minimax method*) has been to find an estimator that would minimize the maximum risk, $R_{\max}(T) = \sup_\theta R(\theta, T)$. E.g., if we have two competing estimators T_1 and T_2, we can choose the one with smaller $R_{\max}(T_j)$ for $j = 1, 2$. In theory, the minimax approach may sound reasonable, but in practice it can be often infeasible, e.g., $R_{\max}(T)$ may not be finite: if $Y_i \sim N(\theta, \theta^2)$ and we use $T = \bar{Y}$, then $R(\theta, T) = \theta^2/n$ and $R_{\max}(T) = \infty$ for any fixed sample size n. Moreover, there might be possibilities like $R(\theta, T_1) < R(\theta, T_2)$ for all $\theta \neq \theta_0$ but $R_{\max}(T_1) = R(\theta_0, T_1) = R(\theta_0, T_2) = R_{\max}(T_2)$, thus making T_1 and T_2 not comparable, and yet T_1 may have significantly smaller risk than that of T_2, except perhaps at the point $\theta = \theta_0$, at which the maximum risk is attained by both estimators. These types of problems are not uncommon in practice and led frequentist methods to consider alternative solutions.

Minimum Variance Unbiased Estimators: The primary idea is to work with a smaller class of estimators by restricting the class of possible estimators to the class of so-called *unbiased estimators* satisfying $E_\theta[T(D)] = \theta$ for all values of θ; it tries to find the estimator with minimum variance among the class of all unbiased estimators, which leads to so-called minimum variance unbiased estimators (MVUE). The problem with this approach is often that for simple problems, there might not exist any unbiased estimator for a fixed sample size n. E.g., if $Y_i \sim Poisson(\theta)$ and we want to estimate the population standard deviation $\sqrt{\theta}$, there is no (finite sample) unbiased estimator of $\sqrt{\theta}$ and hence no MVUE!

Thus, both minimax and MVUE approaches are not practically feasible for a broader class of problems (especially when θ itself is of higher dimensions). The way out of these theoretical problems has been to consider asymptotic methods as $n \to \infty$ and find the "best" estimator that has the smallest asymptotic risk, which leads to the popular *maximum likelihood estimator* defined as $\hat{\theta}_{MLE} = \arg\max_\theta L(\theta; D)$, has been shown to have the least possible asymptotic risk.

However, a finite sample solution to the point estimation has remained illusive and unsatisfactory from a frequentist point of view, and so the final alternative is to consider a

Bayesian approach which eliminates the problem of non-comparable risk functions $R(\theta, T)$ by considering the average risk $r(T) = E_\pi[R(\theta, T)]$ where the expectation E_π is taken with respect to a prior distribution $\pi(\theta)$, which can be viewed as a weighting function.

Bayes risk: The Bayes risk of an estimator $T = T(D)$ corresponding to a loss function $L(\theta, T)$ is defined as

$$r(T) = E[L(\theta, T(Y))] = \int\int L(\theta, T(y)) f(y|\theta)\pi(\theta) dy\, d\theta \tag{9.1}$$

where the expectation is taken with respect to the joint distribution of the data and parameter, (Y, θ) given by $f(y|\theta)\pi(\theta)$. Clearly, the Bayes risk for two estimators (using the same prior $\pi(\theta)$) are directly comparable, and the Bayes estimator is the one that minimizes the Bayes risk among all estimators $T = T(D)$. For many standard loss functions, the Bayes estimators can be derived explicitly. E.g., for the squared error loss $L(\theta, T) = ||\theta - T||^2$ which measures the squared Euclidean distance between θ and T, the Bayes estimator is given by $\hat{\theta}_B = E[\theta|Y]$, the posterior mean of θ given the data Y.

The main point of controversy lies in the choice of the prior density function $\pi(\theta)$ as the Bayes estimators do depend on such a (subjective) choice. However, both frequentist and Bayes estimators also depend on the (subjective) choice of the sampling density $f(y|\theta)$ and also the loss function $L(\theta, T)$. It can be argued that there's no way out of such subjective choices in making a statistical inference or any other scientific method, and the completely objective inference is more of a myth than a real objective!

We provide a simple example of estimating a proportion θ based on the observations $Y = (y_1, \ldots, y_n)$ and derive the frequentist and Bayes estimators.

Twin Birth Example: Suppose we would like to predict the chance of a twin birth for the next day given that no twin birth was observed over the last 30 days in given city (e.g., in Raleigh, NC). What is the best estimate of $\theta = \Pr[\text{Twin birth}]$ given the data $D = (y_1 = 0, y_2 = 0, \ldots, y_n = 0)$, where y_i is a binary variable taking a value of 1 if there's a twin birth on day i and the value of 0 otherwise.

For this example, as the childbirths can be regarded as independent events, we may assume $Y_i|\theta \overset{iid}{\sim} Ber(\theta)$, for $i = 1, \ldots, n$, where $Ber(\theta)$ denotes the Bernoulli distribution with parameter $\theta \in [0, 1]$. Thus, in this case, $f(y|\theta) = \theta^y(1-\theta)^{1-y}$ for $y \in \{0, 1\}$, and the likelihood function is given by $L(\theta; D) = \theta^S(1-\theta)^{n-S}$, where $S = \sum_{i=1}^n y_i$ denotes total number of twin births in a sample of n observations.

Notice that for the twin birth example, we have $S = 0$, and hence it follows that the "best" frequentist estimate (MVUE or MLE) of θ is given by $\hat{\theta} = S/n = 0$. The difficulty lies with the fact that the estimated standard error of the estimator is given by $\hat{\theta}(1-\hat{\theta})/n = 0$. So, we are almost absolutely certain that the chance of a twin birth for next day is zero!! But is this really reasonable?

How about using the available background information based on historical data? A simple "Google search" will reveal that the chance of a twin birth is only about 1 in 30, and could we not possibly use such 'prior' information? E.g., we could use the prior $\theta \sim Beta(1, 29)$ as a prior distribution, which assures $E[\theta] = 1/30$, and the posterior distribution is $\theta|S \sim Beta(1 + S, 30 + n - S)$, and the Bayes estimator is $\hat{\theta} = 1/60 \approx 0.017$, which is small but positive.

Before studying the frequentist properties of Bayesian estimators, we discuss how to optimally summarize the posterior distribution. An appealing feature of a Bayesian analysis is that it produces the full posterior distribution of each parameter. However, in many cases a point estimate (i.e., a single-number estimator) is desired. Common one-number summaries of the posterior used as point estimators are the posterior mean, median or mode. Rather than arbitrarily choosing one of these posterior summaries as the estimator, Bayesian decision theory provides a formal way to select the optimal Bayesian point estimator.

Defining the optimal point estimator requires defining the measure of estimation accuracy that is to be optimized. If θ is the true value of the parameter and $\hat{\theta}(\mathbf{Y})$ is the estimator, then the loss function is defined as $l(\theta, \hat{\theta}(\mathbf{Y}))$. For example, we might choose squared error loss, $l(\theta, \hat{\theta}(\mathbf{Y})) = [\theta - \hat{\theta}(\mathbf{Y})]^2$, or absolute loss $l(\theta, \hat{\theta}(\mathbf{Y})) = |\theta - \hat{\theta}(\mathbf{Y})|$.

The loss function depends on the true value of the parameter, and therefore cannot be evaluated in a real data analysis. From the Bayesian perspective, given the data, uncertainty about the fixed but unknown parameter θ is quantified by its posterior distribution. In this view, θ, and thus $l(\theta, \hat{\theta}(\mathbf{Y}))$, are random variables. The Bayesian risk is the expected (with respect to the posterior distribution of θ) loss, $R(\hat{\theta}(\mathbf{Y})) = \mathrm{E}[l(\theta, \hat{\theta}(\mathbf{Y}))|\mathbf{Y}]$. The *Bayes' rule* is the estimator $\hat{\theta}(\mathbf{Y})$ that minimizes Bayesian risk.

For example, assume squared error loss and define the posterior mean as $\bar{\theta} = \mathrm{E}(\theta|\mathbf{Y})$. The Bayesian risk is

$$\begin{aligned}
R(\hat{\theta}(\mathbf{Y})) &= \mathrm{E}[l(\theta, \hat{\theta}(\mathbf{Y}))|\mathbf{Y}] \\
&= \mathrm{E}\{[\theta - \hat{\theta}(\mathbf{Y})]^2|\mathbf{Y}\} \\
&= \mathrm{E}\{[(\theta - \bar{\theta}) - (\hat{\theta}(\mathbf{Y}) - \bar{\theta})]^2|\mathbf{Y}\} \\
&= \mathrm{E}\{(\theta - \bar{\theta})^2|\mathbf{Y}\} - 2\mathrm{E}\{(\theta - \bar{\theta})(\hat{\theta}(\mathbf{Y}) - \bar{\theta})|\mathbf{Y}\} + \mathrm{E}\{[\hat{\theta}(\mathbf{Y}) - \bar{\theta}]^2|\mathbf{Y}\} \\
&= \mathrm{Var}(\theta|\mathbf{Y}) - 2(\hat{\theta}(\mathbf{Y}) - \bar{\theta})\mathrm{E}(\theta - \bar{\theta}|\mathbf{Y}) + [\hat{\theta}(\mathbf{Y}) - \bar{\theta}]^2 \\
&= \mathrm{Var}(\theta|\mathbf{Y}) + [\hat{\theta}(\mathbf{Y}) - \bar{\theta}]^2.
\end{aligned}$$

In the last equation, $\mathrm{E}(\theta - \bar{\theta}|\mathbf{Y}) = \mathrm{E}(\theta) - \bar{\theta} = 0$ by definition. The estimator $\hat{\theta}(\mathbf{Y})$ does not affect the posterior variance $\mathrm{Var}(\theta|\mathbf{Y})$, and so under squared error loss, the posterior mean $\hat{\theta}(\mathbf{Y}) = \bar{\theta}$ minimizes Bayesian risk and is the Bayes rule estimator. This justifies the use of the posterior mean to summarize the posterior in cases where squared error loss is reasonable.

Different loss functions give different Bayes rule estimators. Absolute loss gives the posterior median as the Bayes rule estimator. More generally, if over-estimation and under-estimation are weighted differently, a useful loss function is the asymmetric check-loss function

$$l(\theta, \hat{\theta}(\mathbf{Y})) = \begin{cases} (1-\tau)(\hat{\theta}(\mathbf{Y}) - \theta) & \text{if } \hat{\theta}(\mathbf{Y}) > \theta \\ \tau(\theta - \hat{\theta}(\mathbf{Y})) & \text{if } \hat{\theta}(\mathbf{Y}) < \theta \end{cases} \tag{9.2}$$

for $\tau \in (0, 1)$. In this loss function, if τ is close to zero, then over-estimation (top row) is penalized more than under-estimation (bottom row). The Bayes rule for this loss function is the τth posterior quantile. If θ is a discrete random variable, then the MAP estimator $\hat{\theta}(\mathbf{Y}) = \arg\max_\theta f(\theta|\mathbf{Y})$ is the Bayes rule under zero-one loss $l(\theta, \hat{\theta}(\mathbf{Y})) = I[\theta \neq \hat{\theta}(\mathbf{Y})]$.

Decision theory can also formalize Bayesian hypothesis testing. Let $H \in \{0, 1\}$ denote the true state, with $H = 0$ if the null hypothesis is true and $H = 1$ if the alternative hypothesis is true. A Bayesian analysis produces posterior probabilities $\mathrm{Prob}(H = h|\mathbf{Y})$ for $h = \{0, 1\}$ (see Section 5.1). The decision to be made is which hypothesis to select. Let $d(\mathbf{Y}) = 0$ if we select the null hypothesis and $d(\mathbf{Y}) = 1$ if we select the alternative hypothesis.

To determine the Bayes rule for $d(\mathbf{Y})$ we must specify the loss incurred if we select the wrong hypothesis. If we assign a Type I error the loss λ_1 and a Type II error the loss λ_2, the loss function can be written

$$l(H, d(\mathbf{Y})) = \begin{cases} 0 & \text{if } H = d(\mathbf{Y}) \\ \lambda_1 & \text{if } H = 0 \text{ and } d(\mathbf{Y}) = 1 \\ \lambda_2 & \text{if } H = 1 \text{ and } d(\mathbf{Y}) = 0. \end{cases} \tag{9.3}$$

The Bayesian risk is

$$R(d(\mathbf{Y})) = \begin{cases} \lambda_1 \text{Prob}(H = 0|\mathbf{Y}) & \text{if } d(\mathbf{Y}) = 1 \\ \lambda_2 \text{Prob}(H = 1|\mathbf{Y}) & \text{if } d(\mathbf{Y}) = 0. \end{cases} \tag{9.4}$$

Therefore, the Bayes rule is to reject the null hypothesis and conclude that the alternative is true; i.e., select $d(\mathbf{Y}) = 1$, if the posterior probability of the alternative hypothesis exceeds

$$\text{Prob}(H = 1|\mathbf{Y}) > \frac{\lambda_1}{\lambda_1 + \lambda_2}. \tag{9.5}$$

Note that unlike classical hypothesis testing, if we swap the roles of the hypotheses, the decision rule remains the same. Often the loss of a Type 1 error is assumed to be larger than a Type II error, e.g., $\lambda_1 = 10\lambda_2$, so that the null hypothesis is rejected only if the posterior probability of the alternative hypothesis is near one.

These decision-theory results hold for prediction as well. For example, if predictions are evaluated using squared-error prediction loss, then the mean of the posterior predictive distribution is the Bayes rule. Also, if the response is binary, then the Bayes rule for classification is to predict $Y^{pred} = 1$ if the posterior predictive probability $\text{Prob}(Y^{pred} = 1|\mathbf{Y}) > \lambda_1/(\lambda_1 + \lambda_2)$, where λ_1 is the loss for a false positive and λ_2 is the loss of a false negative. In this case, it might be reasonable to set $\lambda_1 = \lambda_2$ and predict $Y^{pred} = 1$ if $\text{Prob}(Y^{pred} = 1|\mathbf{Y}) > 0.5$.

9.2 Frequentist properties

Comparison and evaluation of statistical procedures proposed for general use are commonly evaluated using frequentist criteria. A frequentist evaluation of an estimator focuses on its sampling distribution, i.e., the distribution of $\hat{\theta}(\mathbf{Y})$ over repeated samples of the data $\mathbf{Y}$. The distribution of the data of course depends on the true value of θ, denoted in this discussion of frequentist properties as θ_0. For example, if $Y_i \stackrel{iid}{\sim} \text{Normal}(\theta_0, \sigma^2)$, then the sample mean $\hat{\theta}(\mathbf{Y}) = \sum_{i=1}^n Y_i/n$ is an estimator of θ, and its sampling distribution is $\hat{\theta}(\mathbf{Y}) \sim \text{Normal}(\theta_0, \sigma^2/n)$.

There are many ways to compare the sampling distributions of two estimators. For example, we could study the bias

$$\text{Bias}[\hat{\theta}(\mathbf{Y})] = E[\hat{\theta}(\mathbf{Y})] - \theta_0. \tag{9.6}$$

If the bias is positive, this means that on average the estimator over-estimates the parameter and vice versa. With all else equal, an unbiased estimator with bias equal to zero is preferred. However, the bias only evaluates the center of the sampling distribution and ignores its spread. Therefore we also study the variance of the sampling distribution $\text{Var}[\hat{\theta}(\mathbf{Y})]$. For two estimators that are unbiased, we prefer low variance because this means the sampling

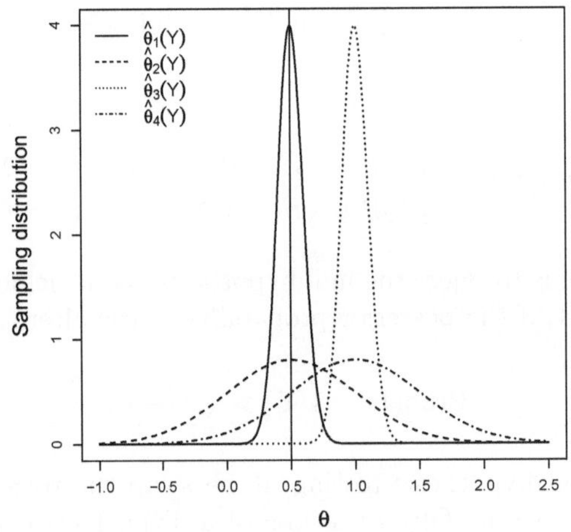

FIGURE 9.1
Hypothetical sampling distributions. Of the four hypothetical sampling distributions, the first two are unbiased and the last two are biased, and the first and third have small variance while the second and fourth have large variance. The true value θ_0 is denoted with the vertical line at $\theta = 0.5$. The mean squared errors of the four estimators are 0.01, 0.25, 0.26 and 0.50, respectively.

distribution is concentrated around the true parameter; on the other hand, a small variance can be undesirable for biased estimators (Figure 9.1). The most common method of comparison is mean squared error

$$MSE[\hat{\theta}(\mathbf{Y})] = \text{E}[\hat{\theta}(\mathbf{Y}) - \theta_0]^2 = \text{Bias}[\hat{\theta}(\mathbf{Y})]^2 + \text{Var}[\hat{\theta}(\mathbf{Y})], \tag{9.7}$$

which combines bias and variance into a single measure.

Evaluating estimators using bias, variance and MSE is complicated by the fact that these summaries depend on the true parameter value, θ_0, and the sample size, n. For example, it may be that an estimator performs well if θ_0 is close to zero and the sample is small, but performs poorly in other settings. Therefore it is important to evaluate the procedure over a range of settings. To deal with the sample size, we often consider asymptotic arguments with $n \to \infty$. A method is asymptotically unbiased if its bias converges to zero as the sample size goes to infinity. An estimator is said to be consistent if it converges in probability to the true value, i.e., the probability of the estimator being within ϵ (for any ϵ) of the true value increases to one as the sample size increases to infinity. An asymptotically unbiased estimator is consistent if its variance decreases to zero with the sample size.

Frequentist evaluation of statistical methods extends to interval estimation and testing. For Bayesian credible intervals we evaluate their frequentist coverage probability, i.e., the probability that they include the true value when applied repeatedly to many datasets. Similarly, for a Bayesian testing procedure we compute the probability of making Type I and Type II errors when the test is applied to many random datasets.

TABLE 9.1
Bias-variance tradeoff for a normal-mean analysis. Assuming $Y_i, ..., Y_n \stackrel{iid}{\sim}$ Normal(θ, σ^2), this table gives the bias, variance and mean squared error (MSE) of the estimators $\hat{\theta}_1(\mathbf{Y}) = \bar{Y}$ and $\hat{\theta}_2(\mathbf{Y}) = c\bar{Y}$, where $c = n/(n+m) \in [0,1]$ and θ_0 is the true value.

Estimator	Bias	Variance	MSE
$\hat{\theta}_1(\mathbf{Y})$	0	$\frac{\sigma^2}{n}$	$\frac{\sigma^2}{n}$
$\hat{\theta}_2(\mathbf{Y})$	$(c-1)\theta_0$	$\frac{c^2\sigma^2}{n}$	$(c-1)^2\theta_0^2 + \frac{c^2\sigma^2}{n}$

9.2.1 Bias-variance tradeoff

For difficult problems such as analyses with many parameters or regression with correlated predictors, adding prior information can stabilize the statistical analysis by reducing the variance of the estimator. On the other hand, if the prior information is erroneous, then this can lead to bias. The balance between these two consequences of incorporating prior information can be formalized by studying the bias/variance trade-off. Recall that the mean squared error is the sum of the variance and the squared bias,

$$\text{MSE}[\hat{\theta}(\mathbf{Y})] = \text{Bias}[\hat{\theta}(\mathbf{Y})]^2 + \text{Var}[\hat{\theta}(\mathbf{Y})]. \tag{9.8}$$

Therefore it is possible for a biased estimator to have smaller MSE than an unbiased estimator if the reduction in variance is large enough to offset the squared bias.

The optimal estimator as given by the bias-variance tradeoff often depends on the true value of the parameter and the sample size. As a concrete example, consider the simple case with $Y_1, ..., Y_n \stackrel{iid}{\sim}$ Normal(θ, σ^2). For mathematical simplicity we assume σ is fixed. We compare two estimators:

(1) $\hat{\theta}_1(\mathbf{Y}) = \frac{1}{n}\sum_{i=1}^n Y_i = \bar{Y}$

(2) $\hat{\theta}_2(\mathbf{Y}) = \frac{1}{n+m}\sum_{i=1}^n Y_i = c\bar{Y}$

where $c = n/(n+m) \in [0,1]$. The first estimator is the usual sample mean (i.e., the MLE or posterior mean under Jeffreys prior) and the second is the posterior mean assuming prior $\theta \sim$ Normal$(0, \sigma^2/m)$.

Table 9.1 gives the bias, variance and MSE of the two estimators. The sample mean is unbiased but always has larger variance than the Bayesian estimator. The relative MSE is

$$RMSE = \frac{MSE[\hat{\theta}_2(\mathbf{Y})]}{MSE[\hat{\theta}_1(\mathbf{Y})]} = \frac{nm^2(\theta_0/\sigma)^2 + n^2}{(n+m)^2}. \tag{9.9}$$

The Bayesian procedure is preferred when this ratio is less than one. When the prior mean is $\theta_0 = 0$, then the ratio is less than one for all n and m. That is, when the prior mean is the true value, the Bayesian estimator is preferred over the sample mean. The Bayesian estimator can be preferred even if the prior mean is not exactly the true value. As long as the true value is close to zero,

$$|\theta_0| < \sigma\sqrt{\frac{1}{n} + \frac{2}{m}} = B(n, m), \tag{9.10}$$

the Bayesian estimator is preferred. The bound $B(n, m)$ shrinks to zero as the sample size increases and so in this case the advantage of the Bayesian approach specifically applies for small datasets. As $n \to \infty$ or $m \to 0$ the RMSE converges to one, and thus for large sample sizes or uninformative priors, the two estimators perform similarly.

9.2.2 Asymptotics

In Section 3.1.3 we discussed the Bayesian central limit theorem that states that under general conditions the posterior converges to a normal distribution as the sample size increases. It can also be shown that under these conditions, and assuming the prior includes the true value, that the sampling distribution of the posterior mean $\hat{\boldsymbol{\theta}}_B = \text{E}(\boldsymbol{\theta}|\mathbf{Y})$ converges (as $n \to \infty$) to the sampling distribution of the maximum likelihood estimator

$$\hat{\boldsymbol{\theta}}_B \sim \text{Normal}\left(\boldsymbol{\theta}_0, \hat{\boldsymbol{\Sigma}}_{MLE}\right) \tag{9.11}$$

where $\boldsymbol{\theta}_0$ is the true value and $\hat{\boldsymbol{\Sigma}}_{MLE}$ is defined in Section 3.1.3. Therefore, the posterior mean is asymptotically unbiased. Further, it follows from classic results for maximum likelihood estimators that its variance decreases to zero as the sample size increases, and thus the posterior mean is a consistent estimator for $\boldsymbol{\theta}$ under essentially the same conditions as the maximum likelihood estimator.

An asymptotic property that is uniquely Bayesian is posterior consistency. For a given dataset $\mathbf{Y}$, the posterior probability that θ is within ϵ of the true value is

$$\text{Prob}(||\theta - \theta_0|| < \epsilon|\mathbf{Y}). \tag{9.12}$$

A Bayesian procedure is said to possess posterior consistency if this probability is assured to converge to one as the sample size increases (Figure 9.2). Appendix A.3 provides a proof that the posterior distribution is consistent for parameters with discrete support under very general conditions on the likelihood and prior, and posterior consistency has been established for most finite-dimensional problems and many Bayesian nonparametric methods. For a more thorough discussion of posterior consistency see [62].

Both the Bayesian CLT and the posterior consistency result in Appendix A.3 hold for any prior as long as the prior does not change with the sample size and has positive mass/density around the true value. *This confirms the argument that for large datasets, any reasonable prior distribution should lead to the same statistical inference and that the resulting posterior will converge to the true value.*

9.3 Simulation studies

For simple models such as the normal mean example in Section 9.2.1 it is possible to derive the frequentist properties of an estimator using algebra. However, for more complicated models, a purely mathematical study is impossible. Especially in these complicated settings, it is important to understand the operating characteristics of a statistical procedure, and a simulation study is a general way to carry out this evaluation.

Just as MCMC is used to approximate complicated posterior distributions, simulation studies use Monte Carlo sampling to approximate complicated sampling distributions. To approximate, say, the mean squared error of an estimator, we generate S independent and identically distributed datasets given the true parameter value θ_0 and sample size n, and

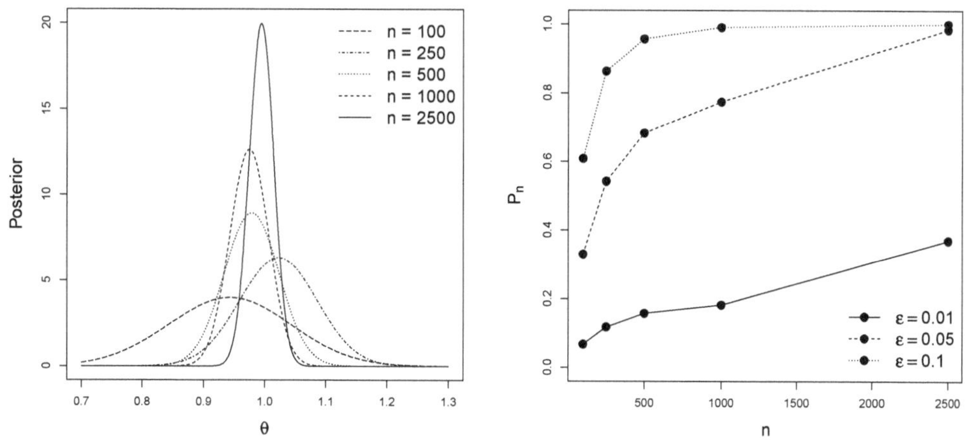

FIGURE 9.2
Illustration of posterior consistency. The data are generated as $Y_i, ..., Y_N \overset{iid}{\sim}$ Normal$(\theta_0, 1)$ with $N = 2,500$ and $\theta_0 = 1$, and we fit the Bayesian model with flat (improper) prior for the mean and variance fixed at one using the first n observations $\mathbf{Y}_n = (Y_1, ..., Y_n)$. The left panel plots the posterior $f(\theta|\mathbf{Y}_n)$ by n, and the right panel plots the corresponding posterior probability $P_n = \text{Prob}(|\theta - \theta_0| < \epsilon|\mathbf{Y}_n)$.

then compute the estimator for each simulated dataset. If we denote the S simulated datasets as $\mathbf{Y}_1, ..., \mathbf{Y}_S$, then $\hat{\theta}(\mathbf{Y}_1), ..., \hat{\theta}(\mathbf{Y}_S)$ are S draws from the sampling distribution of the estimator $\hat{\theta}(\mathbf{Y})$. The mean squared error is then approximated as

$$\text{MSE}[\hat{\theta}(\mathbf{Y})] \approx \frac{1}{S}\sum_{s=1}^{S}[\hat{\theta}(\mathbf{Y}_s) - \theta_0]^2. \tag{9.13}$$

Other summaries of the sampling distribution such as bias and coverage are computed similarly. Of course, this is only a Monte Carlo estimate of the true MSE, and a different simulation experiment will give a different estimate. Therefore, the approximation should be accompanied by a standard error, $s_{MSE}/\sqrt{S}$, where s_{MSE} is the sample standard deviation of the S squared errors, $[\hat{\theta}(\mathbf{Y}_1) - \theta_0]^2, ..., [\hat{\theta}(\mathbf{Y}_S) - \theta_0]^2$.

A typical simulation study will generate data from a few values of θ_0 and n to understand when the method performs well and when it does not. Note that if $\hat{\theta}(\mathbf{Y})$ is a Bayesian estimator, say the posterior mean, then computing each $\hat{\theta}(\mathbf{Y}_s)$ may require MCMC and thus Bayesian simulation experiments may need many applications of MCMC and can be time consuming (running the S chains in parallel is obviously helpful). For a thorough description of simulation studies, see [21] (Chapter 9).

Bayesian LASSO simulation study: As an example, we conduct a simulation study to compare ordinary least squares (OLS) regression with Bayesian LASSO regression (BLR; Section 4.2). The sampling distribution of the OLS estimator is known (multivariate Student's t-distribution) but the sampling distribution of the posterior mean under the BLR model is quite complicated and difficult to study without simulation. We generate $X_{ij} \overset{iid}{\sim}$ Normal$(0, 1)$ and $Y_i|\mathbf{X}_i \sim$ Normal$(\sum_{j=1}^{p} X_{ij}\beta_{j0}, \sigma_0^2)$. The data are generated with

Listing 9.1
R simulation study code.

```
# Set up the simulation
library(BLR)
n       <- 25      # Sample size
p_null <- 15      # Number of null covariates
p_act  <- 5       # Number of active covariates
nsims  <- 100     # Number of simulated datasets
sigma  <- 1       # True value of sigma
beta   <- c(rep(0,p_null),rep(1,p_act)) # True beta

# Define matrices to store the results
p     <- p_null + p_act
EST1 <- VAR1 <- matrix(0,nsims,p)
EST2 <- VAR2 <- matrix(0,nsims,p)

# Start the simulation
for(sim in 1:nsims){
 set.seed(sim*1234)
  # Generate a dataset
   X <- matrix(rnorm(n*p),n,p)
   Y <- X%*%beta+rnorm(n,0,sigma)

  # Fit ordinary least squares
   ols          <- summary(lm(Y~X))$coef[-1,]
   EST1[sim,] <- ols[,1]
   VAR1[sim,] <- ols[,2]^2

  # Fit the Bayesian LASSO
   blr          <- BLR(y=Y,XL=X)
   EST2[sim,] <- blr$bL
   VAR2[sim,] <- blr$SD.bL^2
  }

# Compute the results
 E     <- sweep(EST1,2,beta,"-")
 MSE <- mean(E^2)
 BIAS <- mean(E)
 VAR  <- mean(VAR1)
 COV  <- mean(abs(E/sqrt(VAR1))<2)

 E     <- sweep(EST2,2,beta,"-")
 MSE  <- c(MSE,mean(E^2))
 BIAS <- c(BIAS,mean(E))
 VAR  <- c(VAR,mean(VAR2))
 COV  <- c(COV,mean(abs(E/sqrt(VAR2))<2))

 out <- cbind(BIAS,VAR,MSE,COV)
```

TABLE 9.2

LASSO simulation study results. The simulation study compares ordinary least squares ("OLS") with Bayesian LASSO regression ("BLR") for estimating regression coefficients in terms of bias, variance, mean squared error ("MSE") and coverage of 95% intervals (all metrics are averaged over covariates and datasets). The simulations vary based on the sample size (n), the number of null covariates (p_0) and the number of active covariates (p_1). All values are multiplied by 100.

			Bias		Variance		MSE		Coverage	
n	p_0	p_1	OLS	BLR	OLS	BLR	OLS	BLR	OLS	BLR
40	20	0	−1.65	−0.08	5.59	0.19	5.40	0.03	94.7	100.0
	15	5	0.63	−3.14	5.38	3.47	5.71	3.45	93.8	96.0
	0	20	−0.88	−11.71	5.59	7.09	5.40	9.47	93.7	91.6
100	20	0	−0.43	−0.04	1.28	0.09	1.17	0.02	95.8	100.0
	15	5	0.44	−0.56	1.22	1.02	1.27	0.98	94.5	95.5
	0	20	0.11	−1.27	1.33	1.36	1.22	1.26	96.0	95.6

true values $\sigma_0 = 1$, and the first p_0 elements of $\boldsymbol{\beta}_0$ equal to zero, and the final p_1 elements equal to one, so that $p = p_0 + p_1 = 20$. We generate $S = 100$ datasets each from six combinations of n, p_0 and p_1 using the `R` code in Listing 9.1. Each dataset is analyzed using least squares (the `lm` function in `R`) and Bayesian LASSO (the `BLR` function in `R` with default values). For dataset $s = 1, ..., S$, let $\hat{\beta}_{js}$ be the estimate of β_j (either the least squares solution for OLS or the posterior mean for BLR) and v_{js} its estimated variance (either the squared standard error for OLS or the posterior variance for BLR). Methods are compared using

$$\begin{aligned}
\text{Bias} &= \frac{1}{Sp}\sum_{s=1}^{S}\sum_{j=1}^{p}(\hat{\beta}_{js} - \beta_{j0}) \\
\text{Variance} &= \frac{1}{Sp}\sum_{s=1}^{S}\sum_{j=1}^{p} v_{js} \\
\text{MSE} &= \frac{1}{Sp}\sum_{s=1}^{S}\sum_{j=1}^{p}(\hat{\beta}_{js} - \beta_{j0})^2 \\
\text{Coverage} &= \frac{1}{Sp}\sum_{s=1}^{S}\sum_{j=1}^{p} I(|\hat{\beta}_{js} - \beta_{j0}| < 2\sqrt{v_{js}}).
\end{aligned}$$

For coverage, we approximate the posterior distribution as Gaussian to avoid saving all posterior samples. Results are averaged across covariates, and standard errors are omitted to make the presentation concise. It may also be interesting to study these coefficients separately, in particular, to evaluate performance separately for null and active covariates.

Table 9.2 shows the results. For the small sample size ($n = 40$), the Bayesian method gives a large reduction in variance and thus mean squared error in the first two cases with more null covariates than active covariates. In these cases the prior information that many of the coefficients are near zero is valid and improves the stability of the algorithm. However, when this prior information is wrong and all covariates are active, as in case three, the BLR method has larger bias and MSE than OLS, and the empirical coverage of the Bayesian credible sets dips below the nominal level. For the large samples size cases ($n = 100$), these same trends are apparent but, the differences between methods are smaller, as expected.

Variational Bayesian simulation study: This simulation study examines the frequentist properties and a (naive) variational Bayesian (VB) analysis of a linear model. The data-generating model is the same multiple linear regression as the Bayesian LASSO simulation. For this simulation, we set $n = 1000$, $p = 500$ and $\sigma_0^2 = 1$. The covariates are drawn from a multivariate normal distribution with mean zero, variance one and $\text{Cor}(X_{ij}, X_{ik}) = 0.9^{|j-k|}$ to give strong collinearity. The model is not sparse with true parameters drawn (differently for each dataset) $\beta_{j0} \sim \text{Normal}(0, 0.05^2)$. We generate and analyze 100 datasets from this model.

The simulated data are analyzed using the conjugate prior $\beta_j \stackrel{iid}{\sim} \text{Normal}(0, \tau^2)$ with $\tau^2 = 10$ and error variance fixed at the true value. With this model, the posterior has a closed form,

$$\boldsymbol{\beta}|\mathbf{Y} \sim \text{Normal}\left\{(\mathbf{R} + g\mathbf{I}_p)^{-1}\mathbf{R}\hat{\beta}, \sigma^2(\mathbf{R} + g\mathbf{I})^{-1}\right\},$$

where $g = \sigma^2/\tau^2$, $\mathbf{R} = \mathbf{X}^T\mathbf{X}$ has diagonal elements $R_j = \sum_{i=1}^n X_{ij}^2$ and $\hat{\beta} = (\mathbf{X}^T\mathbf{X})^{-1}\mathbf{X}^T\mathbf{Y}$. Of course, we do not need a VB approximation in this case, but we implement a naive VB approximation to illustrate how one might evaluate the properties of an approximation in more challenging settings. If we select the multivariate normal posterior distribution for $\boldsymbol{\beta}$ with unknown mean and covariance matrix, then the VB posterior would be the exact posterior. However, we use a mean-field approximation

$$\beta_j|\mathbf{Y} \stackrel{indep}{\sim} \text{Normal}(m_j, u_j).$$

It can be shown that the posterior mean estimators, m_j, are the exact posterior mean, but the posterior variances are $u_j = \sigma^2/(R_j + g)$, which are generally smaller than the exact posterior variance.

Since the point estimates are the same for both the exact and VB posteriors, we compare methods using only the average variance and coverage (defined as in Table 9.2). The average posterior variance is 0.019 for the exact posterior compared to 0.001 for the VB approximation. This leads to poor coverage, 39%, for the VB method compared to 95% for the exact posterior. Therefore, in this case with strong collinearity, the mean-field approximation does not give reliable frequentist uncertainty quantification and thus a better approximation to the posterior (i.e., a richer parametric model for the posterior) should be considered.

9.4 Exercises

1. Assume $Y|\mu \sim \text{Normal}(\mu, 2)$ and $\mu \sim \text{Normal}(0, 2)$ (i.e., $n = 1$). The objective is to test the null hypothesis $H_0 : \mu \le 0$ versus the alternative hypothesis that $H_1 : \mu > 0$. We will reject H_0 if $\text{Prob}(H_1|Y) > c$.
   (a) Compute the posterior of μ.
   (b) What is the optimal value of c if Type I and Type II errors have the same costs?
   (c) What is the optimal value of c if a Type I error costs ten times more than a Type II error?
   (d) Compute the Type I error rate of the test as a function of c.
   (e) How would you pick c to control Type I error at 0.05?

2. Given data from a small pilot study, your current posterior probability that the new drug your company has developed is more effective than the current treatment is $\theta \in [0,1]$. Your company is considering to run a large clinical trial to confirm that your drug is indeed preferred. If you run the trial it will cost \$X. If in fact your drug is better, then the probability that you will confirm this in the trial is 80%; if in fact your drug is not better there is still a 5% chance the trial will conclude it is better. If the trial suggests your drug is preferred, you will make $\$cX$. For which values of θ and c would you initiate the trial?

3. Assume $Y|\theta \sim \text{Binomial}(n, \theta)$. Consider two estimators of θ: the sample proportion $\hat{\theta}_1(Y) = Y/n$ and the posterior mean under the Jeffreys prior $\hat{\theta}_2(Y) = (Y + 1/2)/(n + 1)$.

   (a) Compute the bias, variance and MSE for each estimator, and comment on the bias-variance tradeoff.

   (b) In terms of n and the true value θ_0, when is $\hat{\theta}_2(Y)$ preferred?

4. Assume $Y_1, ..., Y_n|\sigma^2 \overset{iid}{\sim} \text{Normal}(0, \sigma^2)$ and $\sigma^2 \sim \text{InvGamma}(a, b)$.

   (a) Give an expression for two estimators of σ^2: the posterior mean and the posterior mode.

   (b) Plot the posterior density functions, and then give the two estimators for a sample with $n = 10$ and $\sum_{i=1}^{n} Y_i^2 = 200$ and $a = b = 0.001$.

   (c) Argue that both estimators are consistent (i.e., their MSE goes to zero as n increases) for any a and b.

5. Assume $Y|\theta \sim \text{Binomial}(n, \theta)$ and $\theta \sim \text{Beta}(1/2, 1/2)$. Use a simulation study to compute the empirical coverage of the equal-tailed 95% credible set for $n \in \{1, 5, 10, 25\}$ and true value $\theta_0 \in \{0.05, 0.10, ..., 0.50\}$. Comment on the frequentist properties of the Bayesian credible set.

6. A paper reports the results of a logistic regression analysis using maximum likelihood estimation. They estimate β_1 to be $\hat{\beta}_1 = 1.8$, with the 95% confidence interval (0.42, 3.17). Further, the p-value for the test that $H_0 : \beta_1 = 0$ versus $H_1 : \beta_1 \neq 0$ is 0.01. Conduct a Bayesian analysis using only this information and asymptotic arguments.

7. You are designing a study to estimate the success probability of a new marketing strategy. When the data have been collected, you will analyze them using the model $Y|\theta \sim \text{Binomial}(n, \theta)$ and $\theta \sim \text{Uniform}(0, 1)$. Before collecting any data, you suspect (based on past studies) that $\theta \approx 0.4$. Which value of n should be used to ensure the posterior standard deviation will be approximately 0.01?

8. Suppose $Y_1, ..., Y_n \overset{iid}{\sim} f(y|\eta)$ where f is the canonical exponential family

$$f(y|\eta) \propto \exp\{y\eta - \psi(\eta)\}$$

   for some known function ψ. We are interested in estimating the parameter $\theta = E(Y_i|\eta) = d\psi(\eta)/d\eta$.

   (a) Show that the Gaussian density with variance fixed at one and the Poisson density can be written in this form.

   (b) Obtain the maximum likelihood estimator of θ. Is it unbiased for θ?

(c) Find a class of conjugate priors for η.

(d) Obtain the Bayes estimator of θ using a conjugate prior and squared error loss.

(e) Is there any (proper) conjugate prior for which the Bayes estimator that you obtained in (d) is unbiased? Justify your answer.

Appendices

A.1: Probability distributions

Univariate discrete

In the following plots the probability mass functions for several combinations of parameters are denoted with points; lines connect the points for visualization, but the probability is non-zero only at the points.

Beta-binomial

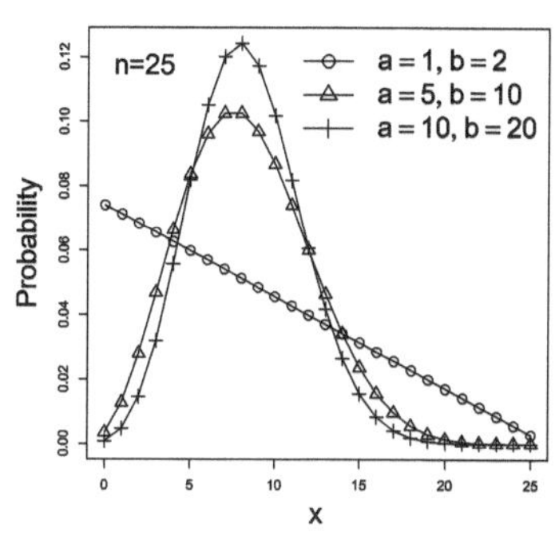

Notation: $X \sim \text{BetaBinomial}(n, a, b)$
Support: $X \in \{0, 1, ..., n\}$
Parameters: $n \in \{1, 2, ...\}$, $a, b > 0$
PMF: $\frac{\Gamma(n+1)\Gamma(x+a)\Gamma(n-x+b)\Gamma(a+b)}{\Gamma(x+1)\Gamma(n-x+1)\Gamma(n+a+b)\Gamma(a)\Gamma(b)}$
Mean: $na/(a+b)$
Variance: $nab(a+b+n)/[(a+b)^2(a+b+1)]$
Notes: If $X|\theta \sim \text{Binomial}(n, \theta)$ and $\theta \sim \text{Beta}(a, b)$, then $X \sim \text{BetaBinomial}(n, a, b)$. If $a = b = 1$ then $X \sim \text{DiscreteUniform}(0, n)$.

Binomial

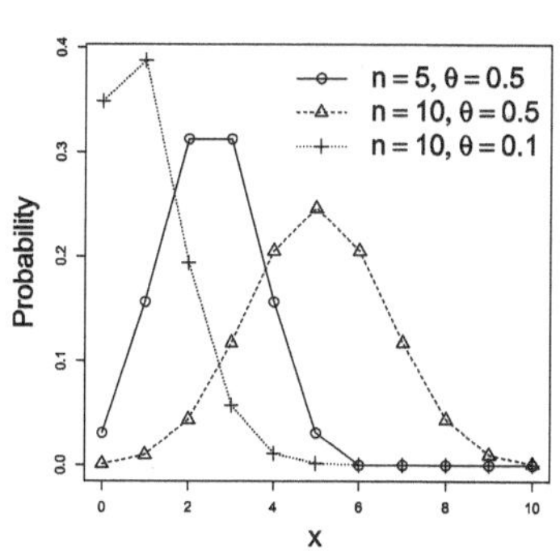

Notation: $X \sim \text{Binomial}(n, \theta)$
Support: $X \in \{0, 1, ..., n\}$
Parameters: $n \in \{1, 2, ...\}$, $\theta \in [0, 1]$
PMF: $\binom{n}{x}\theta^x(1-\theta)^{n-x}$
Mean: $n\theta$
Variance: $n\theta(1-\theta)$
Notes: If X is the number of successes in n independent trials each with success probability θ, then $X \sim \text{Binomial}(n, \theta)$; if $n = 1$ then $X \sim \text{Bernoulli}(\theta)$.

Discrete uniform

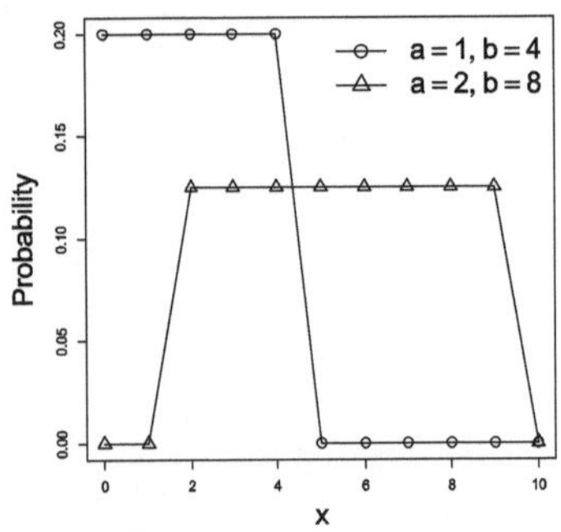

Notation: $X \sim \text{DiscreteUniform}(a, b)$
Support: $X \in \{a, a+1, ..., b\}$
Parameters: $a, b \in \{..., -1, 0, 1, ...\}$ with $a < b$
PMF: $1/(b - a + 1)$
Mean: $(a + b)/2$
Variance: $[(b - a + 1)^2 - 1]/12$
Notes: The discrete uniform can be applied to any finite set. For example, we could say that X is distributed uniformly over the set $\{1/10, 2/10, ..., 10/10\}$.

Negative Binomial

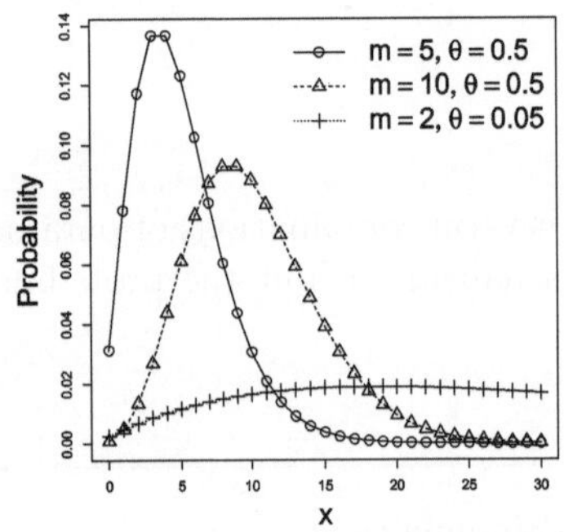

Notation: $X \sim \text{NegBinomial}(\theta, m)$
Support: $X \in \{0, 1, 2, ...\}$
Parameters: $m > 0$, $\theta \in [0, 1]$
PMF: $\binom{x+m-1}{x}\theta^m(1-\theta)^x$
Mean: $m(1 - \theta)/\theta$
Variance: $m(1 - \theta)/\theta^2$
Notes: In a sequence of independent trials each with success probability θ, if X is the number of failures that occur before the mth success (assuming m is an integer), then $X \sim \text{NegBinomial}(\theta, m)$; if $m = 1$ then $X \sim \text{Geometric}(\theta)$. (The distribution can also be defined with m as the number of failures, but we use the `JAGS` parameterization.)

Poisson

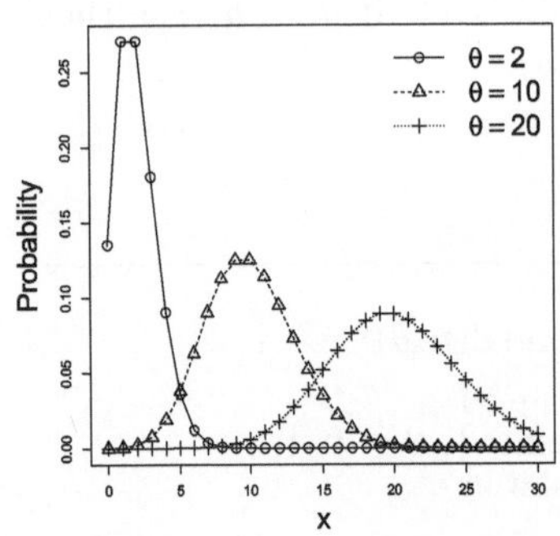

Notation: $X \sim \text{Poisson}(\theta)$
Support: $X \in \{0, 1, 2, ...\}$
Parameters: $\theta > 0$
PMF: $\frac{\theta^x \exp(-\theta)}{x!}$
Mean: θ
Variance: θ
Notes: If events occur independently and uniformly over time (space) with the expected number of events in a given time interval (region) equal to θ, then the number of events that occur in the interval (region) follows a Poisson(θ) distribution.

Multivariate discrete

Multinomial

Notation: $\mathbf{X} = (X_1, ..., X_p) \sim \text{Multinomial}(n, \boldsymbol{\theta})$
Support: $X_j \in \{0, 1, ..., n\}$ with $\sum_{j=1}^{p} X_j = n$
Parameters: $\boldsymbol{\theta} = (\theta_1, ..., \theta_p)$ with $\theta_j \in [0, 1]$ and $\sum_{j=1}^{p} \theta_j = 1$
PMF: $\frac{n!}{\prod_{j=1}^{p} x_j!} \prod_{j=1}^{p} \theta_j^{x_j}$
Mean: $\text{E}(X_j) = n\theta_j$
Variance: $\text{Var}(X_j) = n\theta_j(1 - \theta_j)$
Covariance: $\text{Cov}(X_j, X_k) = -n\theta_j\theta_k$
Marginal distributions: $X_j \sim \text{Binomial}(n, \theta_j)$
Notes: If n independent trials each have p possible outcomes with the probability of outcome j being θ_j and X_j is the number of the trials that result in outcome j, then $\mathbf{X} = (X_1, ..., X_p) \sim \text{Multinomial}(n, \boldsymbol{\theta})$.

Univariate continuous

Asymmetric Laplace distribution

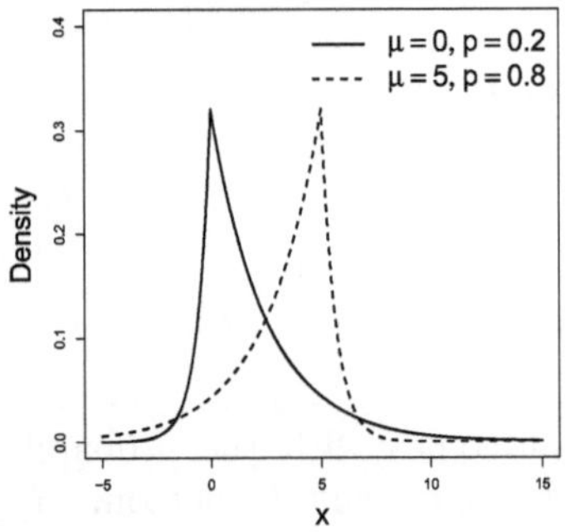

Notation: $X \sim \text{ASL}(\mu, \sigma, p)$
Support: $X \in (-\infty, \infty)$
Parameters: location $\mu \in (-\infty, \infty)$, scale $\sigma > 0$, shape $p \in (0, 1)$
PDF: $\frac{2p(1-p)}{\sigma} \exp\left(-2\frac{\rho_p(x-\mu)}{\sigma}\right)$
where $\rho_p(x) = px$ if $x \geq 0$ and $\rho_p(x) = (p-1)x$ if $x < 0$
Mode: μ
Quantile: the p quantile is μ
Notes: if $p = 0.5$, then $X \sim \text{DE}(\mu, \sigma)$.

Beta

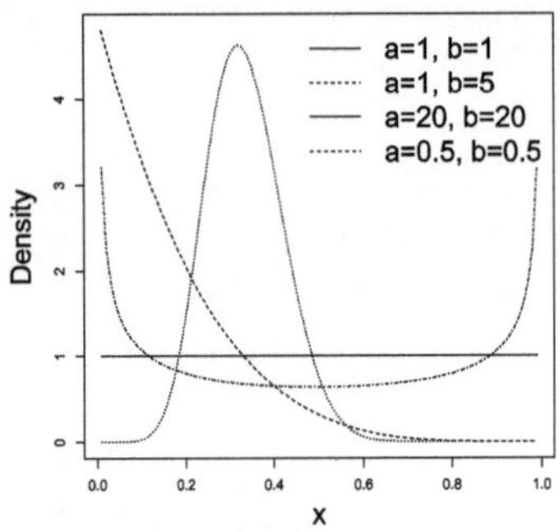

Notation: $X \sim \text{Beta}(a, b)$
Support: $X \in [0, 1]$
Parameters: $a > 0$, $b > 0$
PDF: $\frac{\Gamma(a+b)}{\Gamma(a)\Gamma(b)} x^{a-1}(1-x)^{b-1}$
Mean: $\frac{a}{a+b}$
Variance: $\frac{ab}{(a+b)^2(a+b+1)}$
Notes: If $a = b = 1$ then $X \sim \text{Uniform}(0, 1)$; $1 - X \sim \text{Beta}(b, a)$; $X/(1-X) \sim \text{BP}(a, b)$.

Beta prime

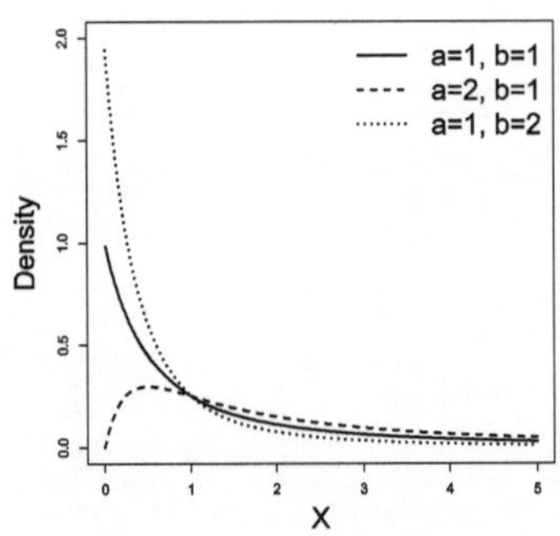

Notation: $X \sim \text{BP}(a, b)$
Support: $X \in [0, \infty]$
Parameters: shapes $a, b > 0$
PDF: $\frac{\Gamma(a+b)}{\Gamma(a)\Gamma(b)} x^{a-1}(1+x)^{-a-b}$
Mean: $a/(b-1)$ for $b > 1$
Variance: $\frac{a(a+b-1)}{(b-2)(b-1)^2}$ for $b > 2$
Notes: $X/(1+X) \sim \text{Beta}(a, b)$; $dX^{1/c} \sim \text{GPB}(a, b, c, d)$.

Gamma

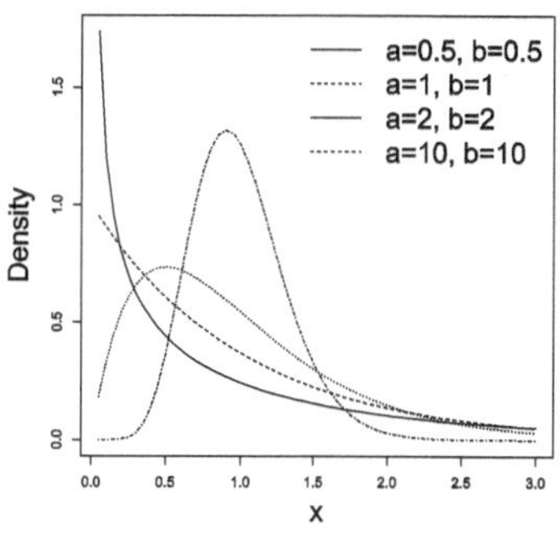

Notation: $X \sim \text{Gamma}(a, b)$
Support: $X \in [0, \infty]$
Parameters: shape $a > 0$, scale $b > 0$
PDF: $\frac{b^a}{\Gamma(a)} x^{a-1} \exp(-bx)$
Mean: a/b
Variance: a/b^2
Notes: $cX \sim \text{Gamma}(a, b/c)$; if $a = 1$ then $X \sim \text{Exponential}(b)$; if $a = \nu/2$ and $b = 1/2$ then $X \sim \text{Chi-squared}(\nu)$; $1/X \sim \text{InvGamma}(a, b)$.

Generalized beta prime

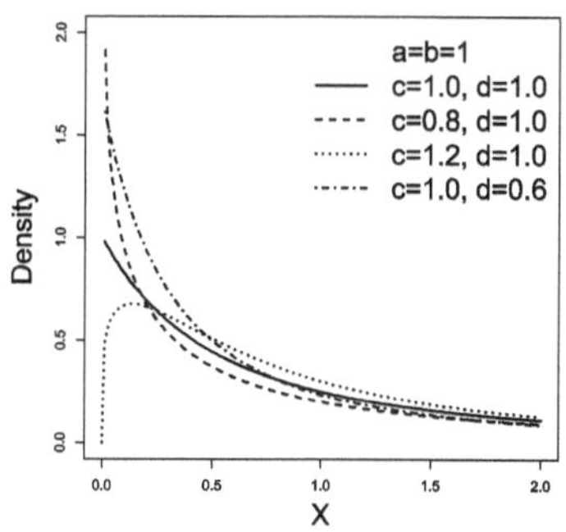

Notation: $X \sim \text{GBP}(a, b, c, d)$
Support: $X \in (0, \infty)$
Parameters: shapes $a, b, c > 0$ and scale $d > 0$
PDF: $\frac{c}{d} \frac{\Gamma(a+b)}{\Gamma(a)\Gamma(b)} \left(\frac{x}{d}\right)^{ac-1} \left\{1 + \left(\frac{x}{d}\right)^c\right\}^{-a-b}$
Mean: $\frac{d(a+1/c)\Gamma(b-1/c)}{\Gamma(a)\Gamma(b)}$ if $bc > 1$
Notes: If $a = b = 1/2$ and $c = 1$ then $\sqrt{X} \sim \text{HC}(\sqrt{d})$.

Half Cauchy

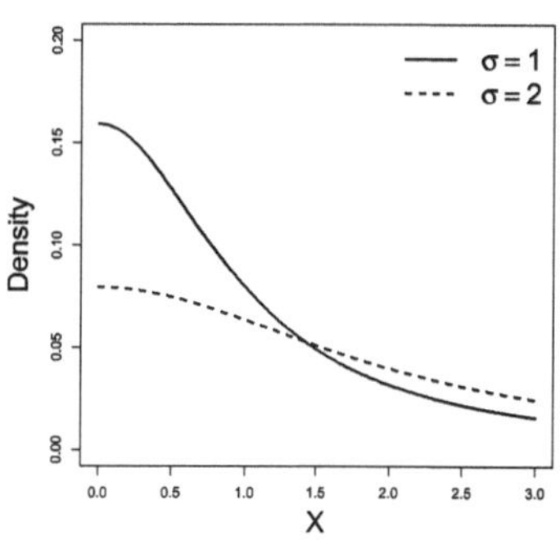

Notation: $X \sim \text{HC}(\sigma)$
Support: $X \in (0, \infty)$
Parameters: scale $\sigma > 0$
PDF: $\frac{2}{\pi\sigma}(1 + x^2/\sigma^2)^{-1}$
Mode: 0
Notes: $cX \sim \text{HC}(c\sigma)$; if $Y \sim t_1(0, \sigma)$, then $|Y| \sim \text{HC}(\sigma)$.

Inverse gamma

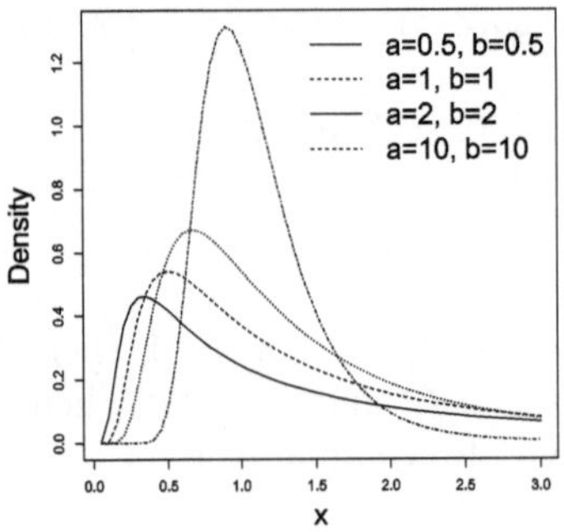

Notation: $X \sim \text{InvGamma}(a, b)$
Support: $X \in [0, \infty]$
Parameters: shape $a > 0$, scale $b > 0$
PDF: $\frac{b^a}{\Gamma(a)} x^{-a-1} \exp(-b/x)$
Mean: $\frac{b}{a-1}$ (if $a > 1$)
Variance: $\frac{b^2}{(a-1)^2(a-2)}$ (if $a > 2$)
Notes: $cX \sim \text{InvGamma}(a, cb)$; $1/X \sim \text{Gamma}(a, b)$.

Laplace/Double exponential

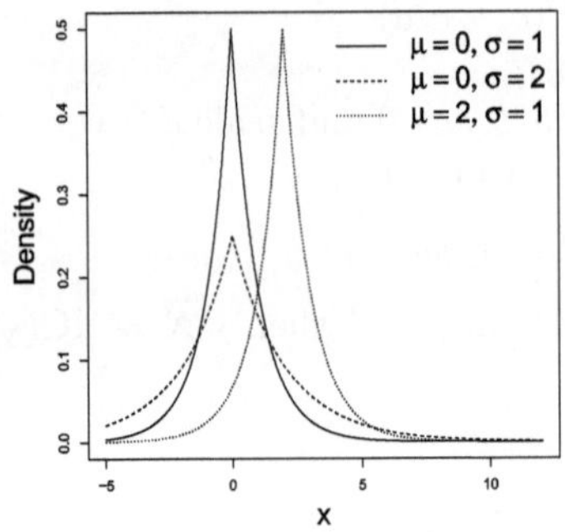

Notation: $X \sim \text{DE}(\mu, \sigma)$
Support: $X \in (-\infty, \infty)$
Parameters: location $\mu \in (-\infty, \infty)$, scale $\sigma > 0$
PDF: $\frac{1}{2\sigma} \exp\left(-\frac{|x-\mu|}{\sigma}\right)$
Mean: μ
Variance: $2\sigma^2$
Notes: $c + dX \sim \text{DE}(c + d\mu, d\sigma)$.

Logistic

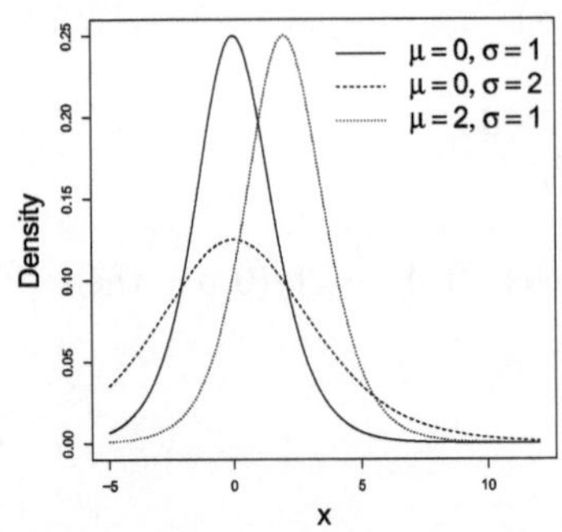

Notation: $X \sim \text{Logistic}(\mu, \sigma)$
Support: $X \in (-\infty, \infty)$
Parameters: location $\mu \in (-\infty, \infty)$, scale $\sigma > 0$
PDF: $\frac{1}{\sigma} \frac{\exp[-(x-\mu)/\sigma]}{\{1+\exp[-(x-\mu)/\sigma]\}^2}$
Mean: μ
Variance: $\pi^2\sigma^2/3$
Notes: $c + dX \sim \text{Logistic}(c + d\mu, d\sigma)$; if $U \sim \text{Uniform}(0, 1)$, then $\mu + \sigma\text{logit}(U) \sim \text{Logistic}(\mu, \sigma)$.

Normal/Gaussian

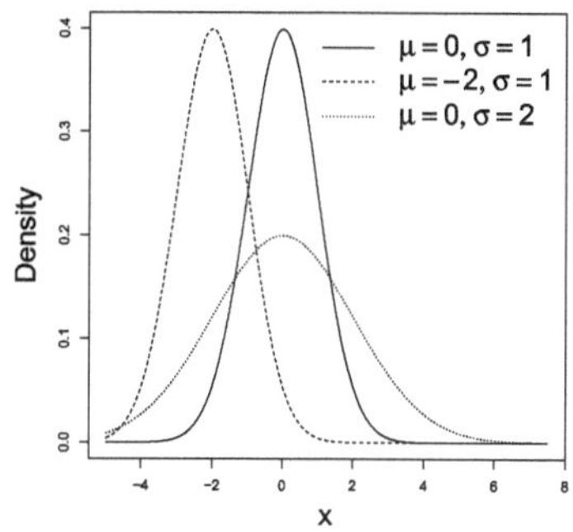

Notation: $X \sim \text{Normal}(\mu, \sigma^2)$
Support: $X \in (-\infty, \infty)$
Parameters: location $\mu \in (-\infty, \infty)$, scale $\sigma > 0$
PDF: $\frac{1}{\sqrt{2\pi}\sigma} \exp\left[-\frac{(x-\mu)^2}{2\sigma^2}\right]$
Mean: μ
Variance: σ^2
Notes: $c+dX \sim \text{Normal}(c+d\mu, d^2\sigma^2)$; if $\mu = 0$ and $\sigma^2 = 1$ then X follows the standard normal distribution.

Student's t

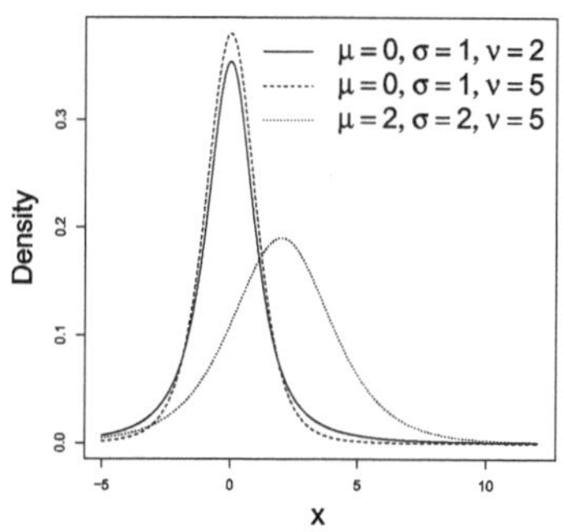

Notation: $X \sim t_\nu(\mu, \sigma^2)$
Support: $X \in (-\infty, \infty)$
Parameters: location $\mu \in (-\infty, \infty)$, scale $\sigma > 0$, degrees of freedom $\nu > 0$
PDF: $\frac{\Gamma(\frac{\nu+1}{2})}{\Gamma(\nu/2)\sqrt{\nu\pi}\sigma}\left[1 + \frac{(x-\mu)^2}{\nu\sigma^2}\right]^{-(\nu+1)/2}$
Mean: μ (if $\nu > 1$)
Variance: $\sigma^2\frac{\nu}{\nu-2}$ (if $\nu > 2$)
Notes: $c + dX \sim t_\nu(c + d\mu, d^2\sigma^2)$; if $\mu = 0$ and $\sigma^2 = 1$, then X follows the standard t distribution; if $Z \sim \text{Normal}(0, 1)$ independent of $W \sim \text{Gamma}(\nu/2, 1/2)$, then $\mu + \sigma Z/\sqrt{W/\nu} \sim t_\nu(\mu, \sigma^2)$; if $\nu = 1$, then X follows the Cauchy distribution; X is approximately $\text{Normal}(\mu, \sigma^2)$ for large ν.

Uniform

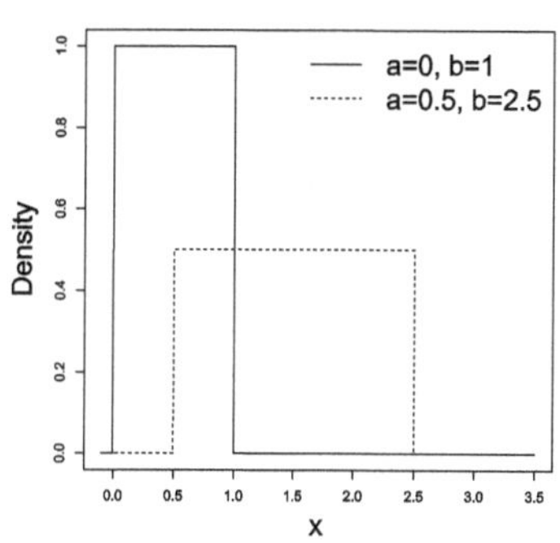

Notation: $X \sim \text{Uniform}(a, b)$
Support: $X \in [a, b]$
Parameters: $-\infty < a < b < \infty$
PDF: $\frac{1}{b-a}$
Mean: $(a + b)/2$
Variance: $(b - a)^2/12$
Notes: If $X_1, X_2 \overset{iid}{\sim} \text{Uniform}(0, 1)$, then $\sqrt{-2\log(X_1)}\cos(2\pi X_2) \sim \text{Normal}(0, 1)$; if $X \sim \text{Uniform}(0, 1)$ and F is a continuous CDF, then $F^{-1}(X)$ has CDF F.

Multivariate continuous

Dirichlet

Notation: $\mathbf{X} = (X_1, ..., X_p) \sim \text{Dirichlet}(\boldsymbol{\theta})$
Support: $X_j \in [0,1]$ with $\sum_{j=1}^p X_j = 1$
Parameters: $\boldsymbol{\theta} = (\theta_1, ..., \theta_p)$ with $\theta_j > 0$
PDF: $\frac{\Gamma(\sum_{j=1}^p \theta_j)}{\prod_{j=1}^p \Gamma(\theta_j)} \prod_{j=1}^p x_j^{\theta_j - 1}$
Mean: $\text{E}(X_j) = \theta_j / (\sum_{k=1}^p \theta_k)$
Variance: $\text{Var}(X_j) = \frac{\theta_j(\sum_{k \neq j} \theta_k)}{(\sum_{k=1}^p \theta_k)^2 (1 + \sum_{k=1}^p \theta_k)}$
Covariance: $\text{Cov}(X_j, X_k) = \frac{-\theta_j \theta_k}{(\sum_{k=1}^p \theta_k)^2 (1 + \sum_{k=1}^p \theta_k)}$
Marginal distributions: $X_j \sim \text{Beta}(\theta_j, \sum_{k \neq j} \theta_k)$
Notes: If $W_j \stackrel{indep}{\sim} \text{Gamma}(\theta_j, b)$ and $X_j = W_j / (\sum_{k=1}^p W_k)$, then $\mathbf{X} = (X_1, ..., X_p) \sim \text{Dirichlet}(\boldsymbol{\theta})$.

Inverse Wishart

Notation: $\mathbf{X} \sim \text{InvWishart}(\nu, \boldsymbol{\Omega})$
Support: $\mathbf{X} = \{X_{jk}\}$ is a $p \times p$ symmetric positive definite matrix
Parameters: degrees of freedom $\nu > p - 1$ and $p \times p$ symmetric positive definite matrix $\boldsymbol{\Omega} = \{\Omega_{jk}\}$
PDF: $\frac{|\boldsymbol{\Omega}|^{\nu/2}}{2^{p\nu/2}\Gamma_p(\nu/2)} |\mathbf{X}|^{-(\nu+p+1)/2} \exp[-\text{trace}(\boldsymbol{\Omega}\mathbf{X}^{-1})/2]$
Mean: $\text{E}(X_{jk}) = \frac{1}{\nu - p - 1} \Omega_{jk}$ (for $\nu > p + 1$)
Variance: $\text{Var}(X_{jk}) = \frac{(\nu - p + 1)\Omega_{kk}^2 + (\nu - p - 1)\Omega_{jj}\Omega_{kk}}{(\nu - p)(\nu - p - 1)^2(\nu - p - 3)}$ (for $\nu > p + 3$)
Marginal distributions: $X_{jj} \sim \text{InvGamma}((\nu - p + 1)/2, \Omega_{jj}/2)$
Notes: If $\mathbf{Y} \sim \text{Wishart}(\nu, \boldsymbol{\Omega}^{-1})$ then $\mathbf{Y}^{-1} \sim \text{InvWishart}(\nu, \boldsymbol{\Omega})$; if $\nu = p + 1$ and $\boldsymbol{\Omega}$ is a diagonal matrix then the correlation $X_{jk}/\sqrt{X_{jj}X_{kk}} \sim \text{Uniform}(-1, 1)$.

Multivariate normal

Notation: $\mathbf{X} = (X_1, ..., X_p)^T \sim \text{Normal}(\boldsymbol{\mu}, \boldsymbol{\Sigma})$
Support: $X_j \in (-\infty, \infty)$
Parameters: mean vector $\boldsymbol{\mu} = (\mu_1, ..., \mu_p)$ with $\mu_j \in (-\infty, \infty)$ and $p \times p$ positive definite covariance matrix $\boldsymbol{\Sigma}$
PDF: $(2\pi)^{-p/2} |\boldsymbol{\Sigma}|^{-1/2} \exp[-\frac{1}{2}(\mathbf{X} - \boldsymbol{\mu})^T \boldsymbol{\Sigma}^{-1} (\mathbf{X} - \boldsymbol{\mu})]$
Mean: $\text{E}(X_j) = \mu_j$
Variance: $\text{Var}(X_j) = \sigma_j^2$ where σ_j^2 is the (j, j) element of $\boldsymbol{\Sigma}$
Covariance: $\text{Cov}(X_j, X_k) = \sigma_{jk}$ where σ_{jk} is the (j, k) element of $\boldsymbol{\Sigma}$
Marginal distributions: $X_j \sim \text{Normal}(\mu_j, \sigma_j^2)$
Notes: For q-vector $\mathbf{a}$ and $q \times p$ matrix $\mathbf{b}$, $\mathbf{a} + \mathbf{b}\mathbf{X} \sim \text{Normal}(\mathbf{a} + \mathbf{b}\mu, \mathbf{b}\boldsymbol{\Sigma}\mathbf{b}^T)$.

Multivariate Student's t

Notation: $\mathbf{X} = (X_1, ..., X_p)^T \sim \mathrm{t}_\nu(\boldsymbol{\mu}, \boldsymbol{\Sigma})$
Support: $X_j \in (-\infty, \infty)$
Parameters: location $\boldsymbol{\mu} = (\mu_1, ..., \mu_p)$ with $\mu_j \in (-\infty, \infty)$, $p \times p$ positive definite matrix $\boldsymbol{\Sigma}$ and degrees of freedom $\nu > 0$
PDF: $\frac{\Gamma(\nu/2+p/2)}{\Gamma(\nu/2)(\nu\pi)^{p/2}}|\boldsymbol{\Sigma}|^{-1/2}\left[1 + \frac{1}{\nu}(\mathbf{X} - \boldsymbol{\mu})^T\boldsymbol{\Sigma}^{-1}(\mathbf{X} - \boldsymbol{\mu})\right]^{-(\nu+p)/2}$
Mean: $\mathrm{E}(X_j) = \mu_j$ (if $\nu > 1$)
Variance: $\mathrm{Var}(X_j) = \frac{\nu}{\nu-2}\sigma_j^2$ where σ_j^2 is the (j, j) element of $\boldsymbol{\Sigma}$ (if $\nu > 2$)
Covariance: $\mathrm{Cov}(X_j, X_k) = \frac{\nu}{\nu-2}\sigma_{jk}$ where σ_{jk} is the (j, k) element of $\boldsymbol{\Sigma}$ (if $\nu > 2$)
Marginal distributions: $X_j \sim \mathrm{t}_\nu(\mu_j, \sigma_j^2)$
Notes: For q-vector $\mathbf{a}$ and $q \times p$ matrix $\mathbf{b}$, $\mathbf{a} + \mathbf{bX} \sim \mathrm{t}_\nu(\mathbf{a} + \mathbf{b}\mu, \mathbf{b}\boldsymbol{\Sigma}\mathbf{b}^T)$; $\mathbf{X}$ is approximately Normal$(\boldsymbol{\mu}, \boldsymbol{\Sigma})$ for large ν; if $\mathbf{X}|W \sim$ Normal$(\boldsymbol{\mu}, \boldsymbol{\Sigma}/W)$ and $W \sim$ Gamma$(\nu/2, \nu/2)$, then $\mathbf{X} \sim \mathrm{t}_\nu(\boldsymbol{\mu}, \boldsymbol{\Sigma})$.

Wishart

Notation: $\mathbf{X} \sim$ Wishart$(\nu, \boldsymbol{\Omega})$
Support: $\mathbf{X} = \{X_{jk}\}$ is a $p \times p$ symmetric positive definite matrix
Parameters: degrees of freedom $\nu > p - 1$ and $p \times p$ symmetric positive definite matrix $\boldsymbol{\Omega} = \{\Omega_{jk}\}$
PDF: $\frac{1}{2^{p\nu/2}|\boldsymbol{\Omega}|^{\nu/2}\Gamma_p(\nu/2)}|\mathbf{X}|^{(\nu-p-1)/2}\exp[-\mathrm{trace}(\boldsymbol{\Omega}^{-1}\mathbf{X})/2]$ where Γ_p is the multivariate gamma function
Mean: $\mathrm{E}(X_{jk}) = \nu\Omega_{jk}$
Variance: $\mathrm{Var}(X_{jk}) = \nu(\Omega_{jk}^2 + \Omega_{jj}\Omega_{kk})$
Marginal distributions: $X_{jj} \sim$ Gamma$(\nu/2, \Omega_{jj}/2)$
Notes: If ν is an integer and $\mathbf{Z}_1, ..., \mathbf{Z}_\nu \stackrel{iid}{\sim}$ Normal$(\mathbf{0}, \boldsymbol{\Omega})$, then $\sum_{i=1}^{\nu}\mathbf{Z}_i\mathbf{Z}_i^T \sim$ Wishart$(\nu, \boldsymbol{\Omega})$.

A.2: List of conjugacy pairs

Below is a partial list of conjugacy pairs. In these derivations, all parameters not assigned a prior are assumed to be fixed.

1. **Binomial proportion**

   Likelihood: $Y|\theta \sim \text{Binomial}(n, \theta)$
   Prior: $\theta \sim \text{Beta}(a, b)$
   Posterior: $\theta|Y \sim \text{Beta}(a + Y, b + n - Y)$

2. **Negative-binomial proportion**

   Likelihood: $Y|\theta \sim \text{NegBinomial}(\theta, m)$
   Prior: $\theta \sim \text{Beta}(a, b)$
   Posterior: $\theta|Y \sim \text{Beta}(a + m, b + Y)$

3. **Multinomial probabilities**

   Likelihood: $\mathbf{Y} = (Y_1, ..., Y_p)|\boldsymbol{\theta} \sim \text{Multinomial}(n, \boldsymbol{\theta})$
   Prior: $\boldsymbol{\theta} \sim \text{Dirichlet}(\boldsymbol{\alpha})$ with $\boldsymbol{\alpha} = (\alpha_1, ..., \alpha_p)$
   Posterior: $\boldsymbol{\theta}|\mathbf{Y} \sim \text{Dirichlet}(\boldsymbol{\alpha} + \mathbf{Y})$

4. **Poisson rate**

   Likelihood: $Y_1, ..., Y_n|\lambda \overset{indep}{\sim} \text{Poisson}(N_i\lambda)$ with N_i fixed
   Prior: $\lambda \sim \text{Gamma}(a, b)$
   Posterior: $\lambda|\mathbf{Y} \sim \text{Gamma}(a + \sum_{i=1}^n Y_i, b + \sum_{i=1}^n N_i)$

5. **Mean of a normal distribution**

   Likelihood: $Y_1, ..., Y_n|\mu \overset{iid}{\sim} \text{Normal}(\mu, \sigma^2)$
   Prior: $\mu \sim \text{Normal}(\theta, \sigma^2/m)$
   Posterior: $\mu|\mathbf{Y} \sim \text{Normal}\left(\frac{n\bar{Y}+m\theta}{n+m}, \frac{\sigma^2}{n+m}\right)$ for $\bar{Y} = \sum_{i=1}^n Y_i/n$

6. **Variance of a normal distribution**

   Likelihood: $Y_1, ..., Y_n|\sigma^2 \overset{indep}{\sim} \text{Normal}(\mu_i, \sigma^2)$
   Prior: $\sigma^2 \sim \text{InvGamma}(a, b)$
   Posterior: $\sigma^2|\mathbf{Y} \sim \text{InvGamma}(a + n/2, b + \sum_{i=1}^n (Y_i - \mu_i)^2/2)$

7. **Precision of a normal distribution**

   Likelihood: $Y_1, ..., Y_n|\tau^2 \overset{indep}{\sim} \text{Normal}(\mu_i, 1/\tau^2)$
   Prior: $\tau^2 \sim \text{Gamma}(a, b)$
   Posterior: $\tau^2|\mathbf{Y} \sim \text{Gamma}(a + n/2, b + \sum_{i=1}^n (Y_i - \mu_i)^2/2)$

8. **Mean vector of a multivariate normal distribution**

   Likelihood: $\mathbf{Y}_1, ..., \mathbf{Y}_n|\boldsymbol{\mu} \overset{indep}{\sim} \text{Normal}(\mathbf{X}_i\boldsymbol{\mu}, \Sigma_i)$
   Prior: $\mu \sim \text{Normal}(\boldsymbol{\theta}, \boldsymbol{\Omega})$
   Posterior: $\mu|\mathbf{Y} \sim \text{Normal}(\mathbf{VM}, \mathbf{V})$ with $\mathbf{V} = (\sum_{i=1}^n \mathbf{X}_i^T \boldsymbol{\Sigma}_i^{-1} \mathbf{X}_i + \boldsymbol{\Omega}^{-1})^{-1}$ and $\mathbf{M} = \sum_{i=1}^n \mathbf{X}_i^T \Sigma_i^{-1} \mathbf{Y}_i + \boldsymbol{\Omega}^{-1}\boldsymbol{\theta}$
   Special case: If $\mathbf{X}_i = \mathbf{I}$ and $\boldsymbol{\Sigma}_i = \boldsymbol{\Sigma}$ for all i, then $\mathbf{V} = (n\boldsymbol{\Sigma}^{-1} + \boldsymbol{\Omega}^{-1})^{-1}$ and $\mathbf{M} = n\boldsymbol{\Sigma}^{-1}\bar{\mathbf{Y}} + \boldsymbol{\Omega}^{-1}\boldsymbol{\theta}$

9. **Covariance matrix of a multivariate normal distribution**

   Likelihood: $\mathbf{Y}_1, ..., \mathbf{Y}_n|\boldsymbol{\Sigma} \overset{indep}{\sim} \text{Normal}(\boldsymbol{\mu}_i, \boldsymbol{\Sigma})$
   Prior: $\boldsymbol{\Sigma} \sim \text{InvWishart}(\nu, \mathbf{R})$
   Posterior: $\boldsymbol{\Sigma}|\mathbf{Y} \sim \text{InvWishart}\,(n + \nu, \mathbf{S} + \mathbf{R})$, where $\mathbf{S} = \sum_{i=1}^n (\mathbf{Y}_i - \boldsymbol{\mu}_i)(\mathbf{Y}_i - \boldsymbol{\mu}_i)^T$

10. **Precision matrix of a multivariate normal distribution**

    Likelihood: $\mathbf{Y}_1, ..., \mathbf{Y}_n|\mathbf{\Omega} \stackrel{indep}{\sim} \text{Normal}(\boldsymbol{\mu}_i, \mathbf{\Omega}^{-1})$
    Prior: $\mathbf{\Omega} \sim \text{Wishart}(\nu, \mathbf{R})$
    Posterior: $\mathbf{\Omega}|\mathbf{Y} \sim \text{Wishart}\left(n + \nu, \left[\mathbf{S} + \mathbf{R}^{-1}\right]^{-1}\right)$, where $\mathbf{S} = \sum_{i=1}^{n}(\mathbf{Y}_i - \boldsymbol{\mu}_i)(\mathbf{Y}_i - \boldsymbol{\mu}_i)^T$

11. **Scale parameter of a gamma distribution**

    Likelihood: $Y_1, ..., Y_n|\mu \stackrel{iid}{\sim} \text{Gamma}(a_i, w_i b)$
    Prior: $b \sim \text{Gamma}(u, v)$
    Posterior: $b|\mathbf{Y} \sim \text{Gamma}(\sum_{i=1}^{n} a_i + u, \sum_{i=1}^{n} w_i Y_i + v)$

12. **Arbitrary parameter with discrete prior**

    Likelihood: $\mathbf{Y}_1, ..., \mathbf{Y}_n|\boldsymbol{\theta} \stackrel{indep}{\sim} f_i(\mathbf{Y}_i|\boldsymbol{\theta})$
    Prior: $\text{Prob}(\boldsymbol{\theta} = \boldsymbol{\theta}_k) = \pi_k$ for $\theta \in \{\boldsymbol{\theta}_1, ..., \boldsymbol{\theta}_m\}$
    Posterior: $\text{Prob}(\boldsymbol{\theta} = \boldsymbol{\theta}_k|\mathbf{Y}) = L_k/[\sum_{j=1}^{m} L_j]$ where $L_k = \pi_k \prod_{i=1}^{n} f_i(\mathbf{Y}_i|\boldsymbol{\theta}_k)$

A.3: Derivations

Normal-normal model for a mean

Say $Y_i|\mu \overset{iid}{\sim} \text{Normal}(\mu,\sigma^2)$ for $i = 1,...,n$ with σ^2 known and prior $\mu \sim \text{Normal}(\theta,\sigma^2/m)$. Since the $Y_1,...,Y_n$ are independent, the likelihood factors as

$$f(\mathbf{Y}|\mu) = \prod_{i=1}^{n} f(Y_i|\mu) = \prod_{i=1}^{n} \frac{1}{\sqrt{2\pi}\sigma} \exp\left[-\frac{(Y_i-\mu)^2}{2\sigma^2}\right].$$

Discarding constants that do not depend on μ and expressing the product of exponentials as the exponential of the sum, the likelihood is

$$f(\mathbf{Y}|\mu) \propto \exp\left[-\sum_{i=1}^{n}\frac{(Y_i-\mu)^2}{2\sigma^2}\right] \propto \exp\left[-\frac{1}{2}\left(-2\frac{n\bar{Y}}{\sigma^2}\mu + \frac{n}{\sigma^2}\mu^2\right)\right]$$

where $\bar{Y} = \sum_{i=1}^{n} Y_i/n$. The last equality comes from multiplying the quadratic terms, collecting them as a function of their power of μ, and discarding terms without a μ. Similarly, the prior can be written

$$\pi(\mu) \propto \exp\left[-\frac{m(\mu-\theta)^2}{2\sigma^2}\right] \propto \exp\left[-\frac{1}{2}\left(-2\frac{m\theta}{\sigma^2}\mu + \frac{m}{\sigma^2}\mu^2\right)\right].$$

Because both the likelihood and prior are quadratic in μ, they can be combined as

$$\begin{aligned} p(\mu|\mathbf{Y}) &\propto f(\mathbf{Y}|\mu)\pi(\mu) \\ &\propto \exp\left[-\frac{1}{2}\left(-2\frac{n\bar{Y}+m\theta}{\sigma^2}\mu + \frac{n+m}{\sigma^2}\mu^2\right)\right] \\ &\propto \exp\left[-\frac{1}{2}\left(-2M\mu + \frac{1}{V}\mu^2\right)\right], \end{aligned}$$

where $M = (n\bar{Y}+m\theta)/\sigma^2$ and $V = \sigma^2/(n+m)$. The exponent of the posterior is quadratic in μ, and we have seen that a Gaussian PDF is quadratic in the exponent. Therefore, we rearrange the terms in the posterior to reveal its Gaussian PDF form. Completing the square in the exponent (and discarding and/or adding terms that do not depend on μ) gives

$$p(\mu|\mathbf{Y}) \propto \exp\left[-\frac{1}{2}\left(-2M\mu + \frac{1}{V}\mu^2\right)\right] \propto \exp\left[-\frac{(\mu-VM)^2}{2V}\right].$$

Therefore, the posterior is $\mu|\mathbf{Y} \sim \text{N}(VM, V)$. Plugging in the above expressions for M and V gives

$$\mu|\mathbf{Y} \sim \text{N}\left(w\bar{Y} + (1-w)\theta, \frac{\sigma^2}{n+m}\right)$$

where $w = n/(n+m)$.

Normal-normal model for a mean vector

The model is

$$\mathbf{Y}|\boldsymbol{\beta} \sim \text{Normal}(\mathbf{X}\boldsymbol{\beta}, \boldsymbol{\Sigma}) \text{ and } \boldsymbol{\beta} \sim \text{Normal}(\boldsymbol{\mu}, \boldsymbol{\Omega}).$$

As with the normal-normal model in Section 9.4, we proceed by expressing the exponential of the posterior as a quadratic form in $\boldsymbol{\beta}$ and then comparing this expression to a multivariate normal to determine the posterior. Using precision matrices $\mathbf{U} = \boldsymbol{\Sigma}^{-1}$ and $\mathbf{V} = \boldsymbol{\Omega}^{-1}$, the posterior is

$$\begin{aligned} p(\boldsymbol{\beta}|\mathbf{Y}) &\propto f(\mathbf{Y}|\boldsymbol{\beta})\pi(\boldsymbol{\beta}) \\ &\propto \exp\left[-\frac{1}{2}(\mathbf{Y} - \mathbf{X}\boldsymbol{\beta})^T\mathbf{U}(\mathbf{Y} - \mathbf{X}\boldsymbol{\beta})^T\right]\left[-\frac{1}{2}(\boldsymbol{\beta} - \boldsymbol{\mu})^T\mathbf{V}(\boldsymbol{\beta} - \boldsymbol{\mu})^T\right] \\ &\propto \exp\left\{-\frac{1}{2}\left[-2(\mathbf{Y}^T\mathbf{U}\mathbf{X} + \boldsymbol{\mu}^T\mathbf{V})\boldsymbol{\beta} + \boldsymbol{\beta}(\mathbf{X}^T\mathbf{U}\mathbf{X} + \mathbf{V})\boldsymbol{\beta}\right]\right\} \\ &\propto \exp\left[-\frac{1}{2}\left(-2\mathbf{W}^T\boldsymbol{\beta} + \boldsymbol{\beta}^T\mathbf{P}\boldsymbol{\beta}\right)\right] \end{aligned}$$

where $\mathbf{W} = \mathbf{X}^T U\mathbf{Y} + \mathbf{V}\boldsymbol{\mu}$ and $\mathbf{P} = \mathbf{X}^T\mathbf{U}\mathbf{X} + \mathbf{V}$. If $\boldsymbol{\beta}|\mathbf{Y} \sim \text{Normal}(\mathbf{M}, \mathbf{S})$ for some mean vector $\mathbf{M}$ and covariance matrix $\mathbf{S}$, then its PDF can be written

$$\begin{aligned} p(\boldsymbol{\beta}|\mathbf{Y}) &\propto \exp\left[-\frac{1}{2}(\boldsymbol{\beta} - \mathbf{M})^T\mathbf{S}^{-1}(\boldsymbol{\beta} - \mathbf{M})\right] \\ &\propto \exp\left[-\frac{1}{2}\left(-2\mathbf{M}^T\mathbf{S}^{-1}\boldsymbol{\beta} + \boldsymbol{\beta}^T\mathbf{S}^{-1}\boldsymbol{\beta}\right)\right]. \end{aligned}$$

To reconcile these two expressions of the posterior, we must have posterior covariance $\mathbf{S} = \mathbf{P}^{-1}$ and $\mathbf{S}^{-1}\mathbf{M} = \mathbf{W}$ and thus $\mathbf{M} = \mathbf{S}\mathbf{W} = \mathbf{P}^{-1}\mathbf{W}$. Inserting the expressions for $\mathbf{W}$ and $\mathbf{P}$ and replacing precision matrices with covariance matrices gives the posterior

$$\boldsymbol{\beta}|\mathbf{Y} \sim \text{Normal}\left[\boldsymbol{\Sigma}_\beta(\mathbf{X}^T\boldsymbol{\Sigma}^{-1}\mathbf{Y} + \boldsymbol{\Omega}^{-1}\boldsymbol{\mu}), \boldsymbol{\Sigma}_\beta\right],$$

where $\boldsymbol{\Sigma}_\beta = (\mathbf{X}'\boldsymbol{\Sigma}^{-1}\mathbf{X} + \boldsymbol{\Omega}^{-1})^{-1}$

Normal-inverse Wishart model for a covariance matrix

The model for the p-vectors $\mathbf{Y}_1, ..., \mathbf{Y}_n$ given the $p \times p$ covariance matrix $\boldsymbol{\Sigma}$ is

$$\mathbf{Y}_i \stackrel{indep}{\sim} \text{Normal}(\mu_i, \boldsymbol{\Sigma}) \text{ and } \boldsymbol{\Sigma} \sim \text{InvWishart}_p(\nu, \mathbf{R}).$$

Using the facts that for arbitrary matrices $\mathbf{A}$, $\mathbf{B}$ and $\mathbf{C}$, Trace($\mathbf{A}$+$\mathbf{B}$) = Trace($\mathbf{A}$)+Trace($\mathbf{B}$) and Trace($\mathbf{ABC}$) = Trace($\mathbf{BCA}$), the likelihood can be written

$$\begin{aligned}
f(\mathbf{Y}|\boldsymbol{\Sigma}) &\propto \prod_{i=1}^{n} |\boldsymbol{\Sigma}|^{-1/2} \exp\left[-\frac{1}{2}(\mathbf{Y}_i - \boldsymbol{\mu}_i)^T \boldsymbol{\Sigma}^{-1}(\mathbf{Y}_i - \boldsymbol{\mu}_i)\right] \\
&\propto |\boldsymbol{\Sigma}|^{-n/2} \exp\left[-\frac{1}{2}\sum_{i=1}^{n}(\mathbf{Y}_i - \boldsymbol{\mu}_i)^T \boldsymbol{\Sigma}^{-1}(\mathbf{Y}_i - \boldsymbol{\mu}_i)\right] \\
&\propto |\boldsymbol{\Sigma}|^{-n/2} \exp\left\{-\frac{1}{2}\sum_{i=1}^{n}\text{Trace}\left[(\mathbf{Y}_i - \boldsymbol{\mu}_i)^T \boldsymbol{\Sigma}^{-1}(\mathbf{Y}_i - \boldsymbol{\mu}_i)\right]\right\} \\
&\propto |\boldsymbol{\Sigma}|^{-n/2} \exp\left\{-\frac{1}{2}\sum_{i=1}^{n}\text{Trace}\left[\boldsymbol{\Sigma}^{-1}(\mathbf{Y}_i - \boldsymbol{\mu}_i)(\mathbf{Y}_i - \boldsymbol{\mu}_i)^T\right]\right\} \\
&\propto |\boldsymbol{\Sigma}|^{-n/2} \exp\left\{-\frac{1}{2}\text{Trace}\left[\sum_{i=1}^{n}\boldsymbol{\Sigma}^{-1}(\mathbf{Y}_i - \boldsymbol{\mu}_i)(\mathbf{Y}_i - \boldsymbol{\mu}_i)^T\right]\right\} \\
&\propto |\boldsymbol{\Sigma}|^{-n/2} \exp\left\{-\frac{1}{2}\text{Trace}\left[\boldsymbol{\Sigma}^{-1}\sum_{i=1}^{n}(\mathbf{Y}_i - \boldsymbol{\mu}_i)(\mathbf{Y}_i - \boldsymbol{\mu}_i)^T\right]\right\} \\
&\propto |\boldsymbol{\Sigma}|^{-n/2} \exp\left[-\frac{1}{2}\text{Trace}(\boldsymbol{\Sigma}^{-1}\mathbf{W})\right]
\end{aligned}$$

where $\mathbf{W} = \sum_{i=1}^{n}(\mathbf{Y}_i - \boldsymbol{\mu}_i)(\mathbf{Y}_i - \boldsymbol{\mu}_i)^T$. The inverse Wishart prior is

$$\pi(\boldsymbol{\Sigma}) \propto |\boldsymbol{\Sigma}|^{-(\nu+p+1)/2} \exp\left[-\frac{1}{2}\text{Trace}(\boldsymbol{\Sigma}^{-1}\mathbf{R})\right].$$

Combining the likelihood and prior, the posterior is

$$\begin{aligned}
p(\boldsymbol{\Sigma}|\mathbf{Y}) &\propto f(\mathbf{Y}|\boldsymbol{\Sigma})\pi(\boldsymbol{\Sigma}) \\
&\propto |\boldsymbol{\Sigma}|^{-(n+\nu+p+1)/2} \exp\left\{-\frac{1}{2}\text{Trace}[\boldsymbol{\Sigma}^{-1}(\mathbf{W}+\mathbf{R})]\right\}.
\end{aligned}$$

Therefore, $\boldsymbol{\Sigma}|\mathbf{Y} \sim \text{InvWishart}_p(n+\nu, \sum_{i=1}^{n}(\mathbf{Y}_i - \boldsymbol{\mu}_i)(\mathbf{Y}_i - \boldsymbol{\mu}_i)^T + \mathbf{R})$.

Jeffreys prior for a normal model

The Gaussian model is $Y_i \overset{iid}{\sim} \text{Normal}(\mu, \sigma^2)$. Denote $\tau = \sigma^2$. The log likelihood is

$$\log f(\mathbf{Y}|\mu, \tau) = -\frac{n}{2}\log(\tau) - \frac{1}{2\tau}\sum_{i=1}(Y_i - \mu)^2.$$

The information matrix depends on both second derivatives and the cross derivative. The second derivatives are

$$\frac{\partial^2 \log f(\mathbf{Y}|\mu, \tau)}{\partial \mu^2} = \frac{\partial}{\partial \mu}\frac{1}{\tau}\sum_{i=1}(Y_i - \mu) = -\frac{n}{\tau}$$

and

$$\begin{aligned}
\frac{\partial^2 \log f(\mathbf{Y}|\mu, \tau)}{\partial \tau^2} &= \frac{\partial}{\partial \tau}\frac{-n}{2\tau} + \frac{1}{2\tau^2}\sum_{i=1}(Y_i - \mu)^2 \\
&= \frac{n}{2\tau^2} - \frac{1}{\tau^3}\sum_{i=1}(Y_i - \mu)^2.
\end{aligned}$$

The cross derivative is

$$\frac{\partial^2 \log f(\mathbf{Y}|\mu,\tau)}{\partial\mu\partial\tau} = \frac{\partial}{\partial\tau}\frac{1}{\tau}\sum_{i=1}(Y_i-\mu) = -\frac{1}{\tau^2}\sum_{i=1}(Y_i-\mu)$$

Since $\text{E}(Y_i) = \mu$ and $\text{E}(Y_i - \mu)^2 = \tau$, the elements of the information matrix are

$$\begin{aligned}
-\text{E}\left(\frac{\partial^2 \log f(\mathbf{Y}|\mu,\tau)}{\partial\mu^2}\right) &= \frac{n}{\tau} \\
-\text{E}\left(\frac{\partial^2 \log f(\mathbf{Y}|\mu,\tau)}{\partial\tau^2}\right) &= -\frac{n}{2\tau^2} + \frac{n\tau}{\tau^3} = \frac{n}{2\tau^2} \\
-\text{E}\left(\frac{\partial^2 \log f(\mathbf{Y}|\mu,\tau)}{\partial\mu\partial\tau}\right) &= 0.
\end{aligned}$$

The determinant of the 2×2 information matrix is thus

$$|I(\mu,\tau)| = \left(\frac{n}{\tau}\right)\left(\frac{n}{2\tau^2}\right) - 0^2 = \frac{n^2}{2\tau^3},$$

and the JP is

$$\pi(\mu,\sigma^2) \propto \sqrt{\frac{n^2}{2\tau^3}} \propto \frac{1}{(\sigma^2)^{3/2}}.$$

Jeffreys prior for multiple linear regression

Assume $\mathbf{Y}|\boldsymbol{\beta},\sigma^2 \sim \text{Normal}(\mathbf{X}\boldsymbol{\beta},\sigma^2\mathbf{I}_n)$, and denote $\tau = \sigma^2$. The log likelihood is

$$\log f(\mathbf{Y}|\boldsymbol{\beta},\tau) = -\frac{n}{2}\log(\tau) - \frac{1}{2\tau}(\mathbf{Y}-\mathbf{X}\boldsymbol{\beta})^T(\mathbf{Y}-\mathbf{X}\boldsymbol{\beta}).$$

The second derivative with respect to τ is

$$\begin{aligned}
\frac{\partial^2 \log f(\mathbf{Y}|\boldsymbol{\beta},\tau)}{\partial\tau^2} &= \frac{\partial}{\partial\tau}\frac{-n}{2\tau} + \frac{1}{2\tau^2}(\mathbf{Y}-\mathbf{X}\boldsymbol{\beta})^T(\mathbf{Y}-\mathbf{X}\boldsymbol{\beta}) \\
&= \frac{n}{2\tau^2} - \frac{1}{\tau^3}(\mathbf{Y}-\mathbf{X}\boldsymbol{\beta})^T(\mathbf{Y}-\mathbf{X}\boldsymbol{\beta}).
\end{aligned}$$

Taking derivatives with respect to $\boldsymbol{\beta}$ requires using matrix calculus identities including the formula for the derivative of a quadratic form,

$$\frac{\partial^2 \log f(\mathbf{Y}|\boldsymbol{\beta},\tau)}{\partial\boldsymbol{\beta}^2} = \frac{\partial}{\partial\boldsymbol{\beta}}\frac{1}{\tau}\mathbf{X}^T(\mathbf{Y}-\mathbf{X}\boldsymbol{\beta}) = -\frac{1}{\tau}\mathbf{X}^T\mathbf{X}.$$

The cross derivative is

$$\frac{\partial^2 \log f(\mathbf{Y}|\boldsymbol{\beta},\tau)}{\partial\boldsymbol{\beta}\partial\tau} = \frac{\partial}{\partial\tau}\frac{1}{\tau}\mathbf{X}^T(\mathbf{Y}-\mathbf{X}\boldsymbol{\beta}) = -\frac{1}{\tau^2}\mathbf{X}^T(\mathbf{Y}-\mathbf{X}\boldsymbol{\beta}).$$

Since $\text{E}(\mathbf{Y}) = \mathbf{X}\boldsymbol{\beta}$ and $\text{E}(\mathbf{Y}-\mathbf{X}\boldsymbol{\beta})^T(\mathbf{Y}-\mathbf{X}\boldsymbol{\beta}) = n\tau$, the elements of the information matrix are

$$\begin{aligned}
-\text{E}\left(\frac{\partial^2 \log f(\mathbf{Y}|\mu,\tau)}{\partial\boldsymbol{\beta}^2}\right) &= \frac{1}{\tau}\mathbf{X}^T\mathbf{X} \\
-\text{E}\left(\frac{\partial^2 \log f(\mathbf{Y}|\mu,\tau)}{\partial\tau^2}\right) &= -\frac{n}{2\tau^2} + \frac{n\tau}{\tau^3} = \frac{n}{2\tau^2} \\
-\text{E}\left(\frac{\partial^2 \log f(\mathbf{Y}|\boldsymbol{\beta},\tau)}{\partial\mu\partial\tau}\right) &= 0.
\end{aligned}$$

The determinant of the $(p+1) \times (p+1)$ block-diagonal information matrix is thus

$$|I(\boldsymbol{\beta}, \tau)| = \left| \frac{1}{\tau} \mathbf{X}^T \mathbf{X} \right| \frac{n}{2\tau^2} = \frac{n}{2\tau^{p+2}} |\mathbf{X}^T \mathbf{X}|,$$

and the JP is

$$\pi(\boldsymbol{\beta}, \sigma^2) \propto \sqrt{\frac{n}{2\tau^{p+2}} \left| \mathbf{X}^T \mathbf{X} \right|} \propto \frac{1}{(\sigma^2)^{p/2+1}}.$$

Convergence of the Gibbs sampler

Here we provide: (1) a proof that the Gibbs sampler generates posterior samples after convergence and (2) a discussion of the theory of Markov processes that ensures that Gibbs sampler converges to the posterior distribution.

Part (1): The proof of (1) is equivalent to showing that the posterior distribution is the stationary distribution of this Markov chain. That is, if we make a draw from the posterior distribution and then iterate the Gibbs sampler forward one iteration from this starting point, the next iteration also follows the posterior distribution. To make the derivations tractable, we restrict the proof to the bivariate case with $p = 2$ and thus $\boldsymbol{\theta} = (\theta_1, \theta_2)$ and denote the posterior density as $p(\theta_1, \theta_2) = p(\boldsymbol{\theta}|\mathbf{Y})$, the full conditional density as $p(\theta_1|\theta_2) = p(\theta_1|\theta_2, \mathbf{Y})$, and the marginal density as $p(\theta_1) = \int p(\theta_1, \theta_2) d\theta_2 = \int p(\theta_1|\theta_2)p(\theta_2)d\theta_2$. Assume we have reached convergence and so one draw in the chain is a realization from the posterior distribution, say $\boldsymbol{\theta}^* = (\theta_1^*, \theta_2^*) \sim p(\theta_1, \theta_2)$. We would like to show that the subsequent sample also follows the posterior distribution. By recursion, this shows that once the algorithm has converged, all samples follow the posterior.

The next sample, (θ_1', θ_2'), drawn from Gibbs sampling has density

$$q(\theta_1', \theta_2'|\theta_1^*, \theta_2^*) = p(\theta_1'|\theta_2^*)p(\theta_2'|\theta_1'),$$

where the two elements of the product represent the updates of the two parameters from their full conditional distribuitons given the current value of the parameters in the chain. We want to show that the marginal distribution of (θ_1', θ_2') integrating over (θ_1^*, θ_2^*) follows the posterior. The marginal distribution is

$$g(\theta_1', \theta_2') = \int \int q(\theta_1', \theta_2'|\theta_1^*, \theta_2^*) f(\theta_1^*, \theta_2^*) d\theta_1^* d\theta_2^*.$$

The integral reduces to

$$\begin{aligned} g(\theta_1', \theta_2') &= \int \int p(\theta_1'|\theta_2^*)p(\theta_2'|\theta_1')p(\theta_1^*|\theta_2^*)p(\theta_2^*)d\theta_2^* d\theta_1^* \\ &= p(\theta_2'|\theta_1') \int p(\theta_1'|\theta_2^*)p(\theta_2^*) \left[\int p(\theta_1^*|\theta_2^*) d\theta_1^* \right] d\theta_2^* \\ &= p(\theta_2'|\theta_1') \int p(\theta_1'|\theta_2^*)p(\theta_2^*) d\theta_2^* \\ &= p(\theta_2'|\theta_1') \int p(\theta_1', \theta_2^*) d\theta_2^* \\ &= p(\theta_2'|\theta_1')p(\theta_1') = p(\theta_1', \theta_2'), \end{aligned}$$

as desired. The proof for $p > 2$ similar but involves higher-order integration.

Part (2): Part (1) shows (for a special case) that the stationary distribution of the Gibbs sampler is the posterior distribution. The proof that the Gibbs sampler converges to its stationary (posterior) distribution draws heavily from Markov chain theory. Given that the posterior distribution is the stationary distribution, [156] proves that a Gibbs sampler converges to the posterior distribution if the chain is aperiodic and p-irreducible. A chain is *aperiodic* if for any partition of the posterior domain of $\boldsymbol{\theta}$, say $\{\mathcal{A}_1, ..., \mathcal{A}_m\}$, so that each subset has positive posterior probability, then the probability of the chain transitioning from $\mathcal{A}_i$ to $\mathcal{A}_j$ is positive for any i and j. A chain is *p-irreducible* if for any initial value $\boldsymbol{\theta}^{(0)}$ in the support of the posterior distribution and set $\mathcal{A}$ with positive posterior probability, i.e., $\text{Prob}(\boldsymbol{\theta} \in \mathcal{A}) > 0$, there exists an s so that there is a positive probability that the chain will visit $\mathcal{A}$ at iteration s. Proving convergence then requires showing that the Gibbs sampler is aperiodic and p-irreducible, which is discussed, e.g., in [156] and [130]. A sufficient condition is that for any set $\mathcal{A}$ with positive posterior probability and any initial value $\boldsymbol{\theta}^{(0)}$ in the support of the posterior distribution, the probability under the Gibbs sampler that $\theta^{(1)} \in \mathcal{A}$ is positive. This condition is met in all but exotic cases where support of full conditional distributions depends on the values of other parameters, in which case convergence should be studied carefully.

Marginal distribution of a normal mean under Jeffreys prior

Assume $Y_i \stackrel{iid}{\sim} \text{Normal}(\mu, \sigma^2)$ and Jeffreys prior $\pi(\mu, \sigma^2) \propto (\sigma^2)^{-3/2}$. Denoting $\tau = \sigma^2$, the joint posterior is

$$\begin{aligned} p(\mu, \tau|\mathbf{Y}) &\propto \left\{\tau^{-n/2} \exp\left[-\frac{\sum_{i=1}^n (Y_i - \mu)^2}{2\tau}\right]\right\}\left\{\tau^{-3/2}\right\} \\ &\propto \tau^{-(n+1)/2-1} \exp\left[-\frac{\sum_{i=1}^n (Y_i - \mu)^2}{2\tau}\right] \\ &\propto \tau^{-A-1} \exp\left[-\frac{B}{\tau}\right], \end{aligned}$$

where $A = (n+1)/2$ and $B = \sum_{i=1}^n (Y_i - \mu)^2/2$. As a function of τ, the joint distribution resembles an InvGamma(A, B) PDF. Integrating over τ gives

$$\begin{aligned} p(\mu|\mathbf{Y}) &\propto \int p(\mu, \tau|bY)d\tau \\ &\propto \int \tau^{-A-1} \exp(-B/\tau)d\tau \\ &\propto \frac{\Gamma(A)}{B^A} \int \frac{B^A}{\Gamma(A)} \tau^{-A-1} \exp(-B/\tau)d\tau \\ &\propto \frac{\Gamma(A)}{B^A} \\ &\propto \left[\sum_{i=1}^n (Y_i - \mu)^2\right]^{-(n+1)/2}. \end{aligned}$$

The marginal PDF is a quadratic function of μ raised to the power $-(n+1)/2$, suggesting that the posterior is a t distribution with n degrees of freedom. Defining $\hat{\sigma}^2 = \sum_{i=1}^n (Y_i -$

$\bar{Y})^2/n = \sum_{i=1}^n Y_i^2/n - \bar{Y}^2$ and completing the square gives

$$\begin{aligned}
\sum_{i=1}^n (Y_i - \mu)^2 &= \sum_{i=1}^n Y_i^2 - 2\sum_{i=1}^n Y_i\mu + n\mu^2 \\
&= n\left[\sum_{i=1}^n Y_i^2/n - 2\bar{Y}\mu + \mu^2\right] \\
&= n\left[\sum_{i=1}^n Y_i^2/n - \bar{Y}^2 + \bar{Y}^2 - 2\bar{Y}\mu + \mu^2\right] \\
&= n\left[\sum_{i=1}^n Y_i^2/n - \bar{Y}^2 + (\mu - \bar{Y})^2\right] \\
&= n\left[\hat{\sigma}^2 + (\mu - \bar{Y})^2\right].
\end{aligned}$$

Inserting this expression back into the marginal posterior gives

$$\begin{aligned}
p(\mu|\mathbf{Y}) &\propto \left[\sum_{i=1}^n (Y_i - \mu)^2\right]^{-(n+1)/2} \\
&\propto \left[\hat{\sigma}^2 + (\mu - \bar{Y})^2\right]^{-(n+1)/2} \\
&\propto \left[1 + \frac{1}{n}\left(\frac{\mu - \bar{Y}}{\hat{\sigma}/\sqrt{n}}\right)^2\right]^{-(n+1)/2}.
\end{aligned}$$

This is Student's t-distribution with location parameter $\bar{Y}$, scale parameter $\hat{\sigma}/\sqrt{n}$, and n degrees of freedom.

Marginal posterior of the regression coefficients under Jeffreys prior

Assume $\mathbf{Y}|\boldsymbol{\beta}, \sigma^2 \sim \text{Normal}(\mathbf{X}\boldsymbol{\beta}, \sigma^2\mathbf{I}_n)$ and Jeffreys prior $\pi(\boldsymbol{\beta}, \sigma^2) \propto (\sigma^2)^{-p/2-1}$. Denoting $\tau = \sigma^2$, the joint posterior is

$$\begin{aligned}
p(\boldsymbol{\beta}, \tau|\mathbf{Y}) &\propto \left\{\tau^{-n/2}\exp\left[-\frac{1}{2\tau}(\mathbf{Y} - \mathbf{X}\boldsymbol{\beta})^T(\mathbf{Y} - \mathbf{X}\boldsymbol{\beta})\right]\right\}\tau^{-p/2-1} \\
&\propto \tau^{-A-1}\exp\left[-\frac{B}{\tau}\right],
\end{aligned}$$

where $A = (n+p)/2$ and $B = (\mathbf{Y} - \mathbf{X}\boldsymbol{\beta})^T(\mathbf{Y} - \mathbf{X}\boldsymbol{\beta})/2$. Marginalizing over σ^2 gives

$$\begin{aligned}
p(\boldsymbol{\beta}|\mathbf{Y}) &= \int p(\boldsymbol{\beta}, \tau|\mathbf{Y})d\tau \\
&\propto \frac{\Gamma(A)}{B^A}\int \frac{B^A}{\Gamma(A)}\tau^{-A-1}\exp\left[-\frac{B}{\tau}\right]d\tau \\
&\propto B^{-A} \\
&\propto \left[(\mathbf{Y} - \mathbf{X}\boldsymbol{\beta})^T(\mathbf{Y} - \mathbf{X}\boldsymbol{\beta})\right]^{-(n+p)/2}.
\end{aligned}$$

The quadratic form is factored as

$$\begin{aligned}(\mathbf{Y}-\mathbf{X}\boldsymbol{\beta})^T(\mathbf{Y}-\mathbf{X}\boldsymbol{\beta}) &= \mathbf{Y}^T\mathbf{Y} - 2\mathbf{Y}^T\mathbf{X}\boldsymbol{\beta} + \boldsymbol{\beta}^T\mathbf{W}\boldsymbol{\beta}\\ &= \mathbf{Y}^T\mathbf{Y} - 2\hat{\boldsymbol{\beta}}^T\mathbf{W}\boldsymbol{\beta} + \boldsymbol{\beta}^T\mathbf{W}\boldsymbol{\beta}\\ &= \mathbf{Y}^T\mathbf{Y} - \hat{\boldsymbol{\beta}}^T\mathbf{W}\hat{\boldsymbol{\beta}} + \hat{\boldsymbol{\beta}}^T\mathbf{W}\hat{\boldsymbol{\beta}} - 2\hat{\boldsymbol{\beta}}^T\mathbf{W}\boldsymbol{\beta} + \boldsymbol{\beta}^T\mathbf{W}\boldsymbol{\beta}\\ &= n\hat{\sigma}^2 + (\boldsymbol{\beta}-\hat{\boldsymbol{\beta}})^T\mathbf{W}(\boldsymbol{\beta}-\hat{\boldsymbol{\beta}})\end{aligned}$$

where $\mathbf{W} = \mathbf{X}^T\mathbf{X}$, $\hat{\boldsymbol{\beta}} = (\mathbf{W})^{-1}\mathbf{X}^T\mathbf{Y}$ is the usual least squares estimator, and $n\hat{\sigma}^2 = (\mathbf{Y}-\mathbf{X}\hat{\boldsymbol{\beta}})^T(\mathbf{Y}-\mathbf{X}\hat{\boldsymbol{\beta}}) = \mathbf{Y}^T\mathbf{Y} - \hat{\boldsymbol{\beta}}^T\mathbf{W}\hat{\boldsymbol{\beta}}$. Therefore,

$$\begin{aligned}p(\boldsymbol{\beta}|\mathbf{Y}) &\propto \left[(\mathbf{Y}-\mathbf{X}\boldsymbol{\beta})^T(\mathbf{Y}-\mathbf{X}\boldsymbol{\beta})\right]^{-(n+p)/2}\\ &\propto \left[n\hat{\sigma}^2 + (\boldsymbol{\beta}-\hat{\boldsymbol{\beta}})^T\mathbf{W}(\boldsymbol{\beta}-\hat{\boldsymbol{\beta}})\right]^{-(n+p)/2}\\ &\propto \left[1 + \frac{1}{n\hat{\sigma}^2}(\boldsymbol{\beta}-\hat{\boldsymbol{\beta}})^T\mathbf{W}(\boldsymbol{\beta}-\hat{\boldsymbol{\beta}})\right]^{-(n+p)/2}.\end{aligned}$$

The marginal posterior of $\boldsymbol{\beta}$ is thus the p-dimensional t-distribution with location vector $\hat{\boldsymbol{\beta}}$, scale matrix $\hat{\sigma}^2(\mathbf{X}^T\mathbf{X})^{-1}$, and n degrees of freedom.

The two-sample t-test in Section 4.1.2 is a special case. If we parameterize the means as μ for the first group and $\mu+\delta$ for the second group, then $\mathbf{X}$'s first column has all ones and its second column is n_1 zeros followed by n_2 ones, and $\boldsymbol{\beta} = (\mu,\delta)^T$. This gives the least squares estimator as $\hat{\boldsymbol{\beta}} = (\bar{Y}, \bar{Y}_2 - \bar{Y}_1)^T$ and $\hat{\sigma}^2 = (n_1\hat{\sigma}_1^2 + n_2\hat{\sigma}_2^2)/(n_1+n_2)$ and

$$(\mathbf{X}^T\mathbf{X})^{-1} = \begin{bmatrix} n_1+n_2 & n_2\\ n_2 & n_2\end{bmatrix}^{-1} = \frac{1}{n_1n_2}\begin{bmatrix} n_2 & -n_2\\ -n_2 & n_1+n_2\end{bmatrix} = \begin{bmatrix} \frac{1}{n_1} & -\frac{1}{n_1}\\ -\frac{1}{n_1} & \frac{1}{n_1}+\frac{1}{n_2}.\end{bmatrix}.$$

This gives the components of the joint posterior distribution of $\boldsymbol{\beta}$, and since the marginal distributions of multivariate t are univariate, we have $\delta|\mathbf{Y} \sim \text{t}_n\left[\bar{Y}_2 - \bar{Y}_1, \hat{\sigma}\left(\frac{1}{n_1}+\frac{1}{n_2}\right)\right]$

A Gamma scale-mixture is half-Cauchy

Say $\tau|b \sim \text{Gamma}(0.5, b)$ and $b \sim \text{Gamma}(0.5, 1)$. Then the marginal distribution of τ is

$$\begin{aligned}f(\tau) &= \int_0^\infty f(\tau|b)f(b)db\\ &\propto \int \left\{b^{1/2}\tau^{-1/2}\exp(-b\tau)\right\}\left\{b^{-1/2}\exp(-b)\right\}db\\ &\propto \tau^{-1/2}\int \exp\{-b(\tau+1)\}db\\ &\propto \tau^{-1/2}(\tau+1)^{-1}.\end{aligned}$$

Transforming to $\sigma = \tau^{-1/2}$ so $\tau = \sigma^{-2}$ gives

$$\begin{aligned}f(\sigma) &\propto (\sigma^{-2})^{-1/2}(\sigma^{-1}+1)^{-1}\left|\frac{d\tau}{d\sigma}\right|\\ &\propto (\sigma^{-2})^{-1/2}(\sigma^{-2}+1)^{-1}\sigma^{-3}\\ &\propto (1+\sigma^2)^{-1}\end{aligned}$$

which is the kernel of the HC(1) PDF.

Variance derivation for the R^2 prior

Assuming the set-up in (4.18) of Section 4.2,

$$\begin{aligned}
\text{Var}(\eta_i) &= \text{E}\left\{\text{Var}\left(\sum_{j=1}^{p} X_{ij}\beta_j|\mathbf{X}_i\right)\right\} + \text{Var}\left\{\text{E}\left(\sum_{j=1}^{p} X_{ij}\beta_j|\mathbf{X}_i\right)\right\} \\
&= \text{E}\left\{\sum_{j=1}^{p} X_{ij}^2\text{Var}\left(\beta_j|\mathbf{X}_i\right)\right\} + \text{Var}\left\{\sum_{j=1}^{p} X_{ij}\text{E}\left(\beta_j|\mathbf{X}_i\right)\right\} \\
&= \sum_{j=2}^{p}\text{E}\left(X_{ij}^2\right)\sigma^2\tau^2,
\end{aligned}$$

since β_1 is fixed, and for $j > 1$ the β_j are independent with $\text{E}(\beta_j|\mathbf{X}_i) = 0$ and $\text{Var}(\beta_j|\mathbf{X}_i) = \sigma^2\tau^2$. If we further assume that the covariates are standardized with X_{ij} having mean 0 and variance 1, then $\text{E}(X_{ij}^2) = 1$ and $\text{Var}(\eta_i) = (p-1)\sigma^2\tau^2$.

Proof of posterior consistency

Here we prove posterior consistency in the general but simple case with independent data and parameter with discrete support. Assume that:

(A1) $Y_i \overset{iid}{\sim} f(y|\boldsymbol{\theta})$ for $i = 1, ..., n$

(A2) The support is discrete, $\boldsymbol{\theta} \in \{\boldsymbol{\theta}_1, \boldsymbol{\theta}_2, ...\} = \mathcal{S}$

(A3) The prior is proper and the true value $\theta_0 \in \mathcal{S}$ has positive prior probability, $\pi(\boldsymbol{\theta}_0) > 0$

(A4) The Kullback-Leibler divergence

$$KL(\boldsymbol{\theta}) = \text{E}_{Y|\theta_0}\left[\log\left(\frac{f(Y|\boldsymbol{\theta}_0)}{f(Y|\boldsymbol{\theta})}\right)\right]$$

satisfies $KL(\boldsymbol{\theta}) > 0$ for all $\boldsymbol{\theta} \neq \boldsymbol{\theta}_0$.

Assumption (A4) ensures that the parameter is identifiable by asserting that on average the likelihood is higher for true value and any other value.

Theorem 1 *Assuming (A1)-(A4), Prob*$(\boldsymbol{\theta} = \boldsymbol{\theta}_0|Y_1, ..., Y_n) \to 1$ *as* $n \to \infty$.

Proof 1 *For any* $\boldsymbol{\theta} \in \mathcal{S}$,

$$\log\left[\frac{p(\boldsymbol{\theta}|\mathbf{Y})}{p(\boldsymbol{\theta}_0|\mathbf{Y})}\right] = \log\left[\frac{\pi(\boldsymbol{\theta})}{\pi(\boldsymbol{\theta}_0)}\right] + \sum_{i=1}^{n}\log\left[\frac{f(Y_i|\boldsymbol{\theta})}{f(Y_i|\boldsymbol{\theta}_0)}\right].$$

By the law of large numbers, $\frac{1}{n}\sum_{i=1}^{n}\log\left[\frac{f(Y_i|\boldsymbol{\theta})}{f(Y_i|\boldsymbol{\theta}_0)}\right] \to -KL(\boldsymbol{\theta})$, *and thus*

$$\log\left[\frac{p(\boldsymbol{\theta}|\mathbf{Y})}{p(\boldsymbol{\theta}_0|\mathbf{Y})}\right] \approx \log\left[\frac{\pi(\boldsymbol{\theta})}{\pi(\boldsymbol{\theta}_0)}\right] - nKL(\boldsymbol{\theta}).$$

Therefore, as $n \to \infty$, $p(\boldsymbol{\theta}|\mathbf{Y})/p(\boldsymbol{\theta}_0|\mathbf{Y}) \to 0$ *for any* $\boldsymbol{\theta} \neq \boldsymbol{\theta}_0$, *and Prob*$(\boldsymbol{\theta} = \boldsymbol{\theta}_0|\mathbf{Y})$ *converges to one.*

This proof can be generalized to continuous parameters by discretizng the support and making additional assumptions about the smoothness of the likelihood and prior density functions.

A.4: Computational algorithms

Integrated nested Laplace approximation (INLA)

INLA [135] is a deterministic approximation to the marginal posterior of each parameter that combines many of the ideas discussed in Section 3.1. The method is most fitting in the special but common case where the parameter vector $\boldsymbol{\theta} = (\boldsymbol{\alpha}, \boldsymbol{\beta})$ can be divided into a low-dimensional $\boldsymbol{\alpha}$ and a high-dimensional $\boldsymbol{\beta}$ whose posterior is approximately Gaussian. For example, in a random effects model (Section 4.4) $\boldsymbol{\alpha}$ might include the variance components, and $\boldsymbol{\beta}$ might include all of the Gaussian random effects.

Evoking the Bayesian CLT (i.e., Laplace approximation) in Section 3.1.3, assume that the conditional posterior of $\boldsymbol{\beta}$ conditioned on $\boldsymbol{\alpha}$ is approximately

$$\boldsymbol{\beta}|\boldsymbol{\alpha}, \mathbf{Y} \sim \text{Normal}\left(\mu(\boldsymbol{\alpha}), \Sigma(\boldsymbol{\alpha})\right),$$

and denote the corresponding density function as $\phi\left(\boldsymbol{\beta}; \mu(\boldsymbol{\alpha}), \Sigma(\boldsymbol{\alpha})\right)$. We first use this approximation for the marginal distribution of the low-dimensional parameter $\boldsymbol{\alpha}$. Since $p(\boldsymbol{\alpha}, \boldsymbol{\beta}|\mathbf{Y}) = p(\boldsymbol{\beta}|\boldsymbol{\alpha}, \mathbf{Y})p(\boldsymbol{\alpha}|\mathbf{Y})$, the marginal posterior of $\boldsymbol{\alpha}$ can be written

$$p(\boldsymbol{\alpha}|\mathbf{Y}) = \frac{p(\boldsymbol{\alpha}, \boldsymbol{\beta}|\mathbf{Y})}{p(\boldsymbol{\beta}|\boldsymbol{\alpha}, \mathbf{Y})}.$$

Expanding around the MAP estimate $\boldsymbol{\beta} = \mu(\boldsymbol{\alpha})$ and using the Laplace approximation for the denominator gives the approximation

$$p(\boldsymbol{\alpha}|\mathbf{Y}) \approx \left.\frac{f(\mathbf{Y}|\boldsymbol{\alpha}, \boldsymbol{\beta})\pi(\boldsymbol{\alpha}, \boldsymbol{\beta}|\mathbf{Y})}{\phi(\boldsymbol{\beta}; \mu(\boldsymbol{\alpha}), \Sigma(\boldsymbol{\alpha}))}\right|_{\boldsymbol{\beta}=\mu(\boldsymbol{\alpha})}.$$

This low-dimensional distribution and can be evaluated using the methods in Section 3.1, e.g., grid approximations or numerical integration.

The Laplace approximation can also be used to approximate the marginal distribution of each element of $\boldsymbol{\beta}$. Let $\boldsymbol{\beta}_{-i}$ be the elements of $\boldsymbol{\beta}$ excluding β_i. Following arguments similar to the approximation of the posterior of $\boldsymbol{\alpha}$,

$$p(\beta_i|\boldsymbol{\alpha}, \mathbf{Y}) \propto \frac{f(\mathbf{Y}|\boldsymbol{\alpha}, \boldsymbol{\beta})\pi(\boldsymbol{\alpha}, \boldsymbol{\beta})}{p(\boldsymbol{\beta}_{-i}|\beta_i, \boldsymbol{\alpha}, \mathbf{Y})}.$$

This can be approximated using a Laplace approximation for $p(\boldsymbol{\beta}_{-i}|\beta_i, \boldsymbol{\alpha}, \mathbf{Y})$ around its posterior mode ([135] also consider faster approximations). Finally, to obtain $p(\boldsymbol{\beta}_{-i}|\mathbf{Y})$ requires numerical integration over $\boldsymbol{\alpha}$, and therefore the Laplace approximation is nested within numerical integration.

Metropolis–adjusted Langevin algorithm

Metropolis–Hastings sampling (Section 3.2.1.2) is a flexible algorithm but depends on finding a reasonable candidate distribution. A Gaussian random walk distribution for the candidate $\boldsymbol{\theta}^* = (\theta_1^*, ..., \theta_p^*)$ given the current value at the onset of iteration s, $\boldsymbol{\theta}^{(s-1)}$, is

$$\boldsymbol{\theta}^*|\boldsymbol{\theta}^{(s-1)} \sim \text{Normal}(\boldsymbol{\theta}^{(s-1)}, c^2\mathbf{I}_p),$$

where $c > 0$ is a tuning parameter. This candidate is easy to code, very general and surprisingly effective. However, convergence can be improved by tailoring the candidate distribution to the problem at hand. We saw in Section 3.2.1 that if the candidate distribution is taken to be the full conditional distribution, Metropolis–Hastings sampling becomes Gibbs sampling. While Gibbs sampling is free from tuning parameters and is thus easier to implement, it requires derivation of full conditional distributions which can be tedious and is not always possible.

The Metropolis-adjusted Langevin (MALA) algorithm [129] balances the strengths of random-walk Metropolis and Gibbs sampling. Rather than simply centering the candidate distribution on the current value, MALA uses the gradient of the posterior to push the candidate distribution towards the center of the distribution. This requires computing the gradient of the posterior, and thus the algorithm is more complex than a random walk, but the gradient is typically easier to derive and more generally available than the full conditional distribution required for Gibbs sampling.

Define the gradient vector of the log posterior as $\nabla(\boldsymbol{\theta}) = [\nabla_1(\boldsymbol{\theta}), ..., \nabla_p(\boldsymbol{\theta})]^T$ where

$$\nabla_j(\boldsymbol{\theta}) = \frac{\partial}{\partial\theta_j}\{\log[f(\mathbf{Y}|\boldsymbol{\theta})] + \log[\pi(\boldsymbol{\theta})]\}$$

is the partial derivative with respect to the jth parameter. The candidate is

$$\boldsymbol{\theta}^*|\boldsymbol{\theta}^{(s-1)} \sim \text{Normal}\left(\boldsymbol{\theta}^{(s-1)} + \frac{c^2}{2}\nabla(\boldsymbol{\theta}^{(s-1)}), c^2\mathbf{I}_p\right).$$

Unlike the random-walk candidate distribution, the MALA candidate distribution is asymmetric and requires including the candidate distribution in the acceptance ratio,

$$R = \frac{f(\mathbf{Y}|\boldsymbol{\theta}^*)\pi(\boldsymbol{\theta}^*)}{f(\mathbf{Y}|\boldsymbol{\theta}^{(s-1)})\pi(\boldsymbol{\theta}^{(s-1)})}\frac{\prod_{j=1}^p \phi\{\theta_j^{(s-1)}; \theta_j^* + \frac{c^2}{2}\nabla_j(\boldsymbol{\theta}^*), c^2\}}{\prod_{j=1}^p \phi\{\theta_j^*; \theta_j^{(s-1)} + \frac{c^2}{2}\nabla_j(\boldsymbol{\theta}^{(s-1)}), c^2\}}$$

where $\phi(y; \mu, \sigma^2)$ is the Gaussian density function with mean μ and variance σ^2.

As with the standard Metropolis algorithm, the tuning parameter c should be adjusted to give reasonable acceptance probability. Roberts and Rosenthal [129] argue that 0.574 is the optimal acceptance probability, but they claim that acceptance probabilities between 0.4 and 0.8 work well. In this chapter we have assumed that the candidate standard deviation c is the same for all p parameters and that the candidates are independent across parameters. Convergence can often be improved by adapting the candidate covariance to resemble the posterior covariance. Finally, we note that since MALA is simply a special type of MH sampling, it can be used within a larger Gibbs sampling algorithm just like MH sampling steps.

Hamiltonian Monte Carlo (HMC)

MALA improves on random-walk Metropolis sampling by fitting the candidate distribution to the posterior by incorporating the gradient of the log posterior. However, for highly irregular posterior distributions (e.g., a U-shaped or donut-shaped posterior), one step along the gradient may not be sufficient to traverse the posterior. Hybrid Monte Carlo (HMC; also called Hamiltonian Monte Carlo) sampling [113] generalizes MALA to take multiple random steps guided by the gradient. The simple version in Algorithm 7 has two tuning parameters: the step size c and the number of steps L. If $L = 1$, then this algorithm reduces to MALA with c as the candidate standard deviation. Motivation, extensions and tuning of this algorithm are beyond the scope of this text but form the basis for the software STAN [28].

Algorithm 7 Hamiltonian MCMC

Initialize $\boldsymbol{\theta}^{(0)} = (\theta_1^{(0)}, ..., \theta_p^{(0)})$
for $s = 1, ..., S$ **do**
  sample $\mathbf{z} \sim \text{Normal}(\mathbf{0}, \mathbf{I}_p)$
  set $\boldsymbol{\theta}^* = \boldsymbol{\theta}^{(s-1)}$
  set $\mathbf{z}^* = \mathbf{z} + c\nabla(\boldsymbol{\theta}^*)/2$
  for $l = 1, ..., L$ **do**
    set $\boldsymbol{\theta}^* = \boldsymbol{\theta}^* + c\mathbf{z}^*$
    set $\mathbf{z}^* = \mathbf{z}^* + c\nabla(\boldsymbol{\theta}^*)$
  end for
  set $\mathbf{z}^* = \mathbf{z}^* - c\nabla(\boldsymbol{\theta}^*)/2$
  set $R = \frac{f(\mathbf{Y}|\boldsymbol{\theta}^*)\pi(\boldsymbol{\theta}^*)}{f\left(\mathbf{Y}|\boldsymbol{\theta}^{(s-1)}\right)\pi\left(\boldsymbol{\theta}^{(s-1)}\right)} \cdot \frac{\exp(-\mathbf{z}^{*T}\mathbf{z}^*/2)}{\exp(-\mathbf{z}^T\mathbf{z}/2)}$
  sample $U \sim \text{Uniform}(0, 1)$
  if $U < R$ **then**
    $\boldsymbol{\theta}^{(s)} = \boldsymbol{\theta}^*$
  else
    $\boldsymbol{\theta}^{(s)} = \boldsymbol{\theta}^{(s-1)}$
  end if
end for

Delayed rejection and adaptive Metropolis

Delayed Rejection and Adaptive Metropolis (DRAM, [69]) is a combination of two ideas: delayed rejection Metropolis [107] and adaptive Metropolis [70]. Adaptive Metropolis allows the covariance of the candidate distribution to evolve across iterations. The intuition is that if the posterior is irregularly shaped, then a different proposal distribution is needed depending on the current state of the chain. Assuming a Gaussian random-walk candidate distribution, $\boldsymbol{\theta}^*|\boldsymbol{\theta}^{(s-1)} \sim \text{Normal}\left(\boldsymbol{\theta}^{(s-1)}, \mathbf{V}^{(s-1)}\right)$, the user sets an initial $p \times p$ covariance matrix $\mathbf{V}^{(0)}$ that is then adapted as

$$\mathbf{V}^{(s)} = c\left(\hat{\mathbf{V}}^{(s)} + \delta\mathbf{I}\right)$$

where $\hat{\mathbf{V}}^{(s)}$ is the sample covariance of the previous samples $\boldsymbol{\theta}^{(1)}, ..., \boldsymbol{\theta}^{(s-1)}$, $\delta > 0$ is a small constant to avoid singularities and $c = 2.4^2/p$ [56].

Delayed rejection Metropolis replaces the standard single proposal in Metropolis–Hastings sampling with multiple proposals considered sequentially. The first stage is a usual Metropolis–Hasting step with candidate $\boldsymbol{\theta}^*|\boldsymbol{\theta}^{(s-1)} \sim q\left(\boldsymbol{\theta}^*|\boldsymbol{\theta}^{(s-1)}\right)$ and acceptance probability

$$R\left(\boldsymbol{\theta}^*, \boldsymbol{\theta}^{(s-1)}\right) = \min\left\{1, \frac{p(\boldsymbol{\theta}^*|\mathbf{Y})\, q\left(\boldsymbol{\theta}^{(s-1)}|\boldsymbol{\theta}^*\right)}{p\left(\boldsymbol{\theta}^{(s-1)}|\mathbf{Y}\right) q\left(\boldsymbol{\theta}^*|\boldsymbol{\theta}^{(s-1)}\right)}\right\}.$$

If the first candidate is rejected, a second candidate is proposed as $\boldsymbol{\theta}'|\boldsymbol{\theta}^*, \boldsymbol{\theta}^{(s-1)} \sim Q\left(\boldsymbol{\theta}'|\boldsymbol{\theta}^*, \boldsymbol{\theta}^{(s-1)}\right)$ and accepted with probability

$$\min\left\{1, \frac{p(\boldsymbol{\theta}'|\mathbf{Y})\, q(\boldsymbol{\theta}'|\boldsymbol{\theta}^*)\, Q\left(\boldsymbol{\theta}'|\boldsymbol{\theta}^*, \boldsymbol{\theta}^{(s-1)}\right)\left[1 - R(\boldsymbol{\theta}', \boldsymbol{\theta}^*)\right]}{p\left(\boldsymbol{\theta}^{(s-1)}|\mathbf{Y}\right) q\left(\boldsymbol{\theta}^{(s-1)}|\boldsymbol{\theta}^*\right) Q\left(\boldsymbol{\theta}^{(s-1)}|\boldsymbol{\theta}^*, \boldsymbol{\theta}'\right)\left[1 - R\left(\boldsymbol{\theta}^{(s-1)}, \boldsymbol{\theta}^*\right)\right]}\right\}.$$

The notation becomes cumbersome, but this can be iterated beyond two candidates. DRAM combines these two ideas by using adaptive Metropolis to tune the Gaussian candidate distributions used for q and Q.

Slice sampling

Slice sampling [112] is a clever way to apply Gibbs sampling when the full conditional distributions do not belong to known parametric families of distributions. Slice sampling introduces an auxiliary variable (i.e., a variable that is not an actual parameter) $U > 0$ and draws samples from the joint distribution

$$p^*(\boldsymbol{\theta}, U) = I[0 < U < p(\boldsymbol{\theta}|\mathbf{Y})].$$

By construction, under p^* the marginal distribution of $\boldsymbol{\theta}$ is

$$\int I[0 < U < p(\boldsymbol{\theta}|\mathbf{Y})]dU = p(\boldsymbol{\theta}|\mathbf{Y}),$$

and therefore if samples of $(\boldsymbol{\theta}, U)$ are drawn from p^*, then the samples of $\boldsymbol{\theta}$ follow the posterior distribution. Also, Gibbs sampling can be used to draw samples from p^* since the full conditional distributions are both uniform

1. $U|\boldsymbol{\theta}, \mathbf{Y} \sim$ Uniform on $[0, p(\boldsymbol{\theta}|\mathbf{Y})]$
2. $\boldsymbol{\theta}|U, \mathbf{Y} \sim$ Uniform on $P(U) = \{\boldsymbol{\theta}; p(\boldsymbol{\theta}|\mathbf{Y}) > U\}$.

Therefore, slice sampling works by drawing from the joint distribution of $(\boldsymbol{\theta}, U)$, discarding the samples of U and retaining the samples from $\boldsymbol{\theta}$. The most challenging step is to make a draw from $P(U)$ (see the figure below). For some posteriors $P(U)$ has a simple form, and samples can be drawn directly. In other cases, $\boldsymbol{\theta}$ can be drawn from a uniform distribution with a domain that includes $P(U)$ until a sample falls in $P(U)$.

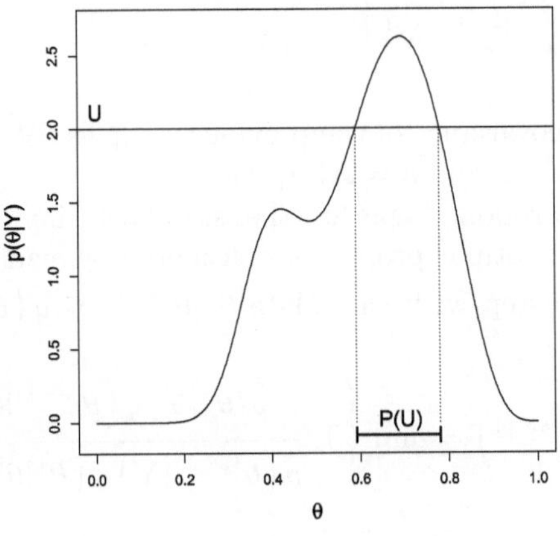

Illustration of slice sampling. The curve is the posterior density $p(\theta|\mathbf{Y})$, the horizontal line represents the auxiliary variable U (i.e., the "slice"), and the bold interval is $P(U) = \{\theta; p(\theta|\mathbf{Y}) > U\}$.

Bibliography

[1] Mohamed A. Abba, Jonathan P. Williams, and Brian J. Reich. A penalized complexity prior for deep Bayesian transfer learning with application to materials informatics. *The Annals of Applied Statistics*, 17(4):3241–3256, 2023.

[2] Mohamed A. Abba, Jonathan P. Williams, and Brian J. Reich. A Bayesian shrinkage estimator for transfer learning. *arXiv preprint arXiv:2403.17321*, 2024.

[3] Helen Abbey. An examination of the Reed-Frost theory of epidemics. *Human Biology*, 24(3):201, 1952.

[4] Hirotogu Akaike. Information theory and an extension of the maximum likelihood principle. In *Selected Papers of Hirotugu Akaike*, pages 199–213. Springer, 1998.

[5] James H. Albert and Siddhartha Chib. Bayesian analysis of binary and polychotomous response data. *Journal of the American Statistical Association*, 88(422):669–679, 1993.

[6] Stefano Alessandrini, Luca Delle Monache, Christopher M. Rozoff, and William E. Lewis. Probabilistic prediction of tropical cyclone intensity with an analog ensemble. *Monthly Weather Review*, 146(6):1723–1744, 2018.

[7] Dan Apley. *ALEPlot: Accumulated Local Effects (ALE) Plots and Partial Dependence (PD) Plots*, 2018. R package version 1.1.

[8] Daniel W. Apley and Jingyu Zhu. Visualizing the effects of predictor variables in black box supervised learning models. *Journal of the Royal Statistical Society: Series B*, 82:1059–1086, 2020.

[9] Lindsey R. Baden, Hana M. El Sahly, Brandon Essink, Karen Kotloff, Sharon Frey, Rick Novak, David Diemert, Stephen A. Spector, Nadine Rouphael, C. Buddy Creech, et al. Efficacy and safety of the mRNA-1273 SARS-CoV-2 vaccine. *New England Journal of Medicine*, 384(5):403–416, 2021.

[10] Ray Bai and Malay Ghosh. On the beta prime prior for scale parameters in high-dimensional Bayesian regression models. *Statistica Sinica*, 31(2):843–865, 2021.

[11] Sudipto Banerjee, Bradley P. Carlin, and Alan E. Gelfand. *Hierarchical Modeling and Analysis for Spatial Data.* Chapman and Hall/CRC, Boca Raton, 2014.

[12] Albert Barberán, Robert R. Dunn, Brian J. Reich, Krishna Pacifici, Eric B. Laber, Holly L. Menninger, James M. Morton, Jessica B. Henley, Jonathan W. Leff, Shelly L. Miller, and Noah Fierer. The ecology of microscopic life in household dust. In *Proceedings of the Royal Society B*, volume 282, page 20151139. The Royal Society, 2015.

[13] Maria Maddalena Barbieri and James O. Berger. Optimal predictive model selection. *The Annals of Statistics*, 32(3):870–897, 2004.

[14] Daryl J. Bem. Feeling the future: Experimental evidence for anomalous retroactive influences on cognition and affect. *Journal of Personality and Social Psychology*, 100(3):407, 2011.

[15] James Berger. The case for objective Bayesian analysis. *Bayesian Analysis*, 1(3):385–402, 2006.

[16] James O. Berger, Luis R. Pericchi, JK Ghosh, Tapas Samanta, Fulvio De Santis, JO Berger, and LR Pericchi. Objective Bayesian methods for model selection: Introduction and comparison. *Institute of Mathematical Statistics Lecture Notes - Monograph Series*, pages 135–207, 2001.

[17] Jose M. Bernardo. Reference posterior distributions for Bayesian inference. *Journal of the Royal Statistical Society: Series B (Methodological)*, pages 113–147, 1979.

[18] Anirban Bhattacharya, Debdeep Pati, Natesh S. Pillai, and David B. Dunson. Dirichlet–Laplace priors for optimal shrinkage. *Journal of the American Statistical Association*, 110(512):1479–1490, 2015.

[19] Howard D. Bondell and Brian J. Reich. Consistent high-dimensional Bayesian variable selection via penalized credible regions. *Journal of the American Statistical Association*, 107(500):1610–1624, 2012.

[20] Josh Bongard and Hod Lipson. Automated reverse engineering of nonlinear dynamical systems. *Proceedings of the National Academy of Sciences*, 104(24):9943–9948, 2007.

[21] Dennis D. Boos and Leonard A. Stefanski. *Essential Statistical Inference: Theory and Methods*, volume 120. Springer Science & Business Media, New York, 2013.

[22] Carlos A. Botero, Beth Gardner, Kathryn R. Kirby, Joseph Bulbulia, Michael C. Gavin, and Russell D Gray. The ecology of religious beliefs. *Proceedings of the National Academy of Sciences*, 111(47):16784–16789, 2014.

[23] Fred Brauer. Compartmental models in epidemiology. *Mathematical Epidemiology*, pages 19–79, 2008.

[24] Steven L. Brunton, Joshua L. Proctor, and J. Nathan Kutz. Discovering governing equations from data by sparse identification of nonlinear dynamical systems. *Proceedings of the National Academy of Sciences*, 113(15):3932–3937, 2016.

[25] Jonathan Buckley and Ian James. Linear regression with censored data. *Biometrika*, 66(3):429–436, 1979.

[26] Paul-Christian Bürkner. Bayesian item response modeling in R with brms and Stan. *Journal of Statistical Software*, 100(5):1–54, 2021.

[27] Niamh Cahill, Jacky Croke, Micheline Campbell, Kate Hughes, John Vitkovsky, Jack Eaton Kilgallen, and Andrew Parnell. A Bayesian hierarchical time series model for reconstructing hydroclimate from multiple proxies. *Environmetrics*, 34:1–19, 2023.

[28] Bob Carpenter, Andrew Gelman, Matt Hoffman, Daniel Lee, Ben Goodrich, Michael Betancourt, Michael A. Brubaker, Jiqiang Guo, Peter Li, and Allen Riddell. STAN: A probabilistic programming language. *Journal of Statistical Software*, 20(2):1–37, 2016.

[29] Raymond J Carroll, Jianqing Fan, Irene Gijbels, and Matt P. Wand. Generalized partially linear single-index models. *Journal of the American Statistical Association*, 92(438):477–489, 1997.

[30] Carlos M. Carvalho, Nicholas G. Polson, and James G. Scott. The horseshoe estimator for sparse signals. *Biometrika*, 97(2):465–480, 2010.

[31] Fang Chen. Bayesian modeling using the MCMC procedure. In *Proceedings of the SAS Global Forum 2008 Conference, Cary NC: SAS Institute Inc*, 2009.

[32] Zhe Chen. Bayesian filtering: From Kalman filters to particle filters, and beyond. *Statistics*, 182(1):1–69, 2003.

[33] Hugh A. Chipman, Edward I. George, and Robert E. McCulloch. BART: Bayesian additive regression trees. *The Annals of Applied Statistics*, 4:266–298, 2010.

[34] B Clarke and Ao Yuan. Closed form expressions for Bayesian sample size. *Annals of Statistics*, 34:1293–1330, 2006.

[35] Kirkwood A. Cloud, Brian J. Reich, Christopher M. Rozoff, Stefano Alessandrini, William E. Lewis, and Luca Delle Monache. A feed forward neural network based on model output statistics for short-term hurricane intensity prediction. *Weather and Forecasting*, 34(4):985–997, 2019.

[36] Ciprian M. Crainiceanu, David Ruppert, and Matthew P. Wand. Bayesian analysis for penalized spline regression using WinBUGS. *Journal of Statistical Software*, 14(1):1–24, 2005.

[37] Kyle Cranmer, Johann Brehmer, and Gilles Louppe. The frontier of simulation-based inference. *Proceedings of the National Academy of Sciences*, 117(48):30055–30062, 2020.

[38] Noel Cressie and Christopher Wikle. *Statistics for Spatio-Temporal Data.* John Wiley and Sons, New York, 2011.

[39] Jacky Croke, John Vítkovskỳ, Kate Hughes, Micheline Campbell, Sahar Amirnezhad-Mozhdehi, Andrew Parnell, Niamh Cahill, and Ramona Dalla Pozza. A palaeoclimate proxy database for water security planning in Queensland Australia. *Scientific Data*, 8(1):292, 2021.

[40] A. Philip Dawid. Present position and potential developments: Some personal views: Statistical theory: The prequential approach. *Journal of the Royal Statistical Society: Series A (General)*, pages 278–292, 1984.

[41] Gustavo de los Campos and Paulino Perez Rodriguez. *BLR: Bayesian Linear Regression*, 2014. R package version 1.4.

[42] Perry de Valpine, Daniel Turek, Christopher J. Paciorek, Clifford Anderson-Bergman, Duncan Temple Lang, and Rastislav Bodik. Programming with models: Writing statistical algorithms for general model structures with NIMBLE. *Journal of Computational and Graphical Statistics*, 26(2):403–413, 2017.

[43] Jia Deng, Wei Dong, Richard Socher, Li-Jia Li, Kai Li, and Li Fei-Fei. Imagenet: A large-scale hierarchical image database. In *2009 IEEE conference on computer vision and pattern recognition*, pages 248–255. IEEE, 2009.

[44] Peter Diggle, Rana Moyeed, Barry Rowlingson, and Madeleine Thomson. Childhood malaria in the Gambia: A case-study in model-based geostatistics. *Journal of the Royal Statistical Society: Series C (Applied Statistics)*, 51(4):493–506, 2002.

[45] Arnaud Doucet, Nando De Freitas, and Neil James Gordon. *Sequential Monte Carlo Methods in Practice*, volume 1. Springer, New York, 2001.

[46] David B. Dunson, Natesh Pillai, and Ju-Hyun Park. Bayesian density regression. *Journal of the Royal Statistical Society: Series B*, 69:163–183, 2007.

[47] Stewart M. Edie, Peter D. Smits, and David Jablonski. Probabilistic models of species discovery and biodiversity comparisons. *Proceedings of the National Academy of Sciences*, 114(14):3666–3671, 2017.

[48] Gregory M. Erickson, Peter J. Makovicky, Philip J. Currie, Mark A. Norell, Scott A. Yerby, and Christopher A. Brochu. Gigantism and comparative life-history parameters of tyrannosaurid dinosaurs. *Nature*, 430(7001):772–775, 2004.

[49] Mark A. Finney, Sara S. McAllister, Jason M. Forthofer, and Torben P. Grumstrup. *Wildland Fire Behaviour: Dynamics, Principles and Processes.* CSIRO publishing, Melbourne, 2021.

[50] Vincent Fortuin. Priors in Bayesian deep learning: A review. *International Statistical Review*, 90(3):563–591, 2022.

[51] Kevin R. Forward, David Haldane, Duncan Webster, Carolyn Mills, Cherly Brine, and Diane Aylward. A comparison between the Strep A Rapid Test Device and conventional culture for the diagnosis of streptococcal pharyngitis. *Canadian Journal of Infectious Diseases and Medical Microbiology*, 17:221–223, 2004.

[52] Seymour Geisser. Discussion on sampling and Bayes inference in scientific modeling and robustness (by GEP Box). *Journal of the Royal Statistical Society: Series A (General)*, 143:416–417, 1980.

[53] Alan E. Gelfand, Peter Diggle, Peter Guttorp, and Montserrat Fuentes. *Handbook of Spatial Statistics.* Chapman and Hall/CRC, Boca Raton, 2010.

[54] Andrew Gelman. Prior distributions for variance parameters in hierarchical models (comment on article by Browne and Draper). *Bayesian Analysis*, 1(3):515–534, 2006.

[55] Andrew Gelman, Jessica Hwang, and Aki Vehtari. Understanding predictive information criteria for Bayesian models. *Statistics and Computing*, 24(6):997–1016, 2014.

[56] Andrew Gelman, Gareth O. Roberts, and Walter R. Gilks. Efficient Metropolis jumping rules. *Bayesian Statistics*, 5(599-608):42, 1996.

[57] Andrew Gelman and Donald B. Rubin. Inference from iterative simulation using multiple sequences. *Statistical Science*, pages 457–472, 1992.

[58] Andrew Gelman and Cosma Rohilla Shalizi. Philosophy and the practice of Bayesian statistics. *British Journal of Mathematical and Statistical Psychology*, 66(1):8–38, 2013.

[59] Stuart Geman and Donald Geman. Stochastic relaxation, Gibbs distributions, and the Bayesian restoration of images. In *Readings in Computer Vision*, pages 564–584. Elsevier, 1987.

[60] Edward I. George and Robert E. McCulloch. Variable selection via Gibbs sampling. *Journal of the American Statistical Association*, 88(423):881–889, 1993.

[61] John Geweke. *Evaluating the accuracy of sampling-based approaches to the calculation of posterior moments*, volume 196. Federal Reserve Bank of Minneapolis, Research Department, Minneapolis, MN, USA, 1991.

[62] Subhashis Ghosal and Aad van der Vaart. *Fundamentals of Nonparametric Bayesian Inference*, volume 44. Cambridge University Press, Cambridge, 2017.

[63] Soumya Ghosh, Jiayu Yao, and Finale Doshi-Velez. Model selection in Bayesian neural networks via horseshoe priors. *Journal of Machine Learning Research*, 20(182):1–46, 2019.

[64] Tilmann Gneiting and Adrian E. Raftery. Strictly proper scoring rules, prediction, and estimation. *Journal of the American Statistical Association*, 102(477):359–378, 2007.

[65] Ian Goodfellow, Yoshua Bengio, and Aaron Courville. *Deep Learning.* MIT Press, Cambridge, 2016.

[66] Y Yi Grace, Aurore Delaigle, and Paul Gustafson. *Handbook of Measurement Error Models.* Chapman and Hall/CRC, Boca Raton, 2021.

[67] Robert B. Gramacy. *Surrogates: Gaussian Process Modeling, Design, and Optimization for the Applied Sciences.* Chapman and Hall/CRC, Boca Raton, 2020.

[68] Brandon M. Greenwell, Bradley C. Boehmke, and Andrew J. McCarthy. A simple and effective model-based variable importance measure. *arXiv:1805.04755*, 2018.

[69] Heikki Haario, Marko Laine, Antonietta Mira, and Eero Saksman. DRAM: Efficient adaptive MCMC. *Statistics and Computing*, 16(4):339–354, 2006.

[70] Heikki Haario, Eero Saksman, and Johanna Tamminen. An adaptive Metropolis algorithm. *Bernoulli*, 7(2):223–242, 2001.

[71] P. Richard Hahn and Carlos M. Carvalho. Decoupled shrinkage and selection in Baysian linear models. *Journal of the American Statistical Association*, 110:435–448, 2015.

[72] P. Richard Hahn, Jared S Murray, and Carlos M. Carvalho. Bayesian regression tree models for causal inference: Regularization, confounding, and heterogeneous effects (with discussion). *Bayesian Analysis*, 15(3):965–1056, 2020.

[73] Niels Richard Hansen. Computational Statistics with R. `https://cswr.nrhstat.org/`.

[74] Wilfred K. Hastings. Monte Carlo sampling methods using Markov chains and their applications. *Biometrika*, 57(1):97–109, 1970.

[75] Jennifer L. Hill. Bayesian nonparametric modeling for causal inference. *Journal of Computational and Graphical Statistics*, 20(1):217–240, 2011.

[76] James S. Hodges. *Richly Parameterized Linear Models: Additive, Time Series, and Spatial Models using Random Effects.* Chapman and Hall/CRC, Boca Raton, 2016.

[77] James S. Hodges and Murray K. Clayton. Random effects old and new. *Statistical Science*, 433, 2011.

[78] James S. Hodges and Brian J. Reich. Adding spatially-correlated errors can mess up the fixed effect you love. *The American Statistician*, 64(4):325–334, 2010.

[79] Arthur E. Hoerl and Robert W. Kennard. Ridge regression: Biased estimation for nonorthogonal problems. *Technometrics*, 12(1):55–67, 1970.

[80] Jennifer A. Hoeting, David Madigan, Adrian E. Raftery, and Chris T. Volinsky. Bayesian model averaging: A tutorial. *Statistical Science*, pages 382–401, 1999.

[81] Kurt Hornik, Maxwell Stinchcombe, and Halbert White. Multilayer feedforward networks are universal approximators. *Neural Networks*, 2(5):359–366, 1989.

[82] Joseph G. Ibrahim, Ming-Hui Chen, and Debajyoti Sinha. *Bayesian Survival Analysis*, volume 2. Springer, New York, 2001.

[83] A. Jasra, C. Holmes, and D. Stephens. Markov Chain Monte Carlo methods and the label switching problem in Bayesian mixture modeling. *Statistical Science*, 20(1), 2005.

[84] Ying Jin, Weilin Fu, Jian Kang, Jiadong Guo, and Jian Guo. Bayesian symbolic regression. *arXiv preprint arXiv:1910.08892*, 2019.

[85] Michael I. Jordan, Zoubin Ghahramani, Tommi S. Jaakkola, and Lawrence K. Saul. An introduction to variational methods for graphical models. *Machine Learning*, 37(2):183–233, 1999.

[86] Bindu Kalesan, Matthew E. Mobily, Olivia Keiser, Jeffrey A. Fagan, and Sandro Galea. Firearm legislation and firearm mortality in the USA: A cross-sectional, state-level study. *The Lancet*, 387(10030):1847–1855, 2016.

[87] Robert E. Kass and Adrian E. Raftery. Bayes factors. *Journal of the American Statistical Association*, 90(430):773–795, 1995.

[88] Robert E. Kass and Larry Wasserman. A reference Bayesian test for nested hypotheses and its relationship to the Schwarz criterion. *Journal of the American Statistical Association*, 90(431):928–934, 1995.

[89] Robert E. Kass and Larry Wasserman. The selection of prior distributions by formal rules. *Journal of the American Statistical Association*, 91:1343–1370, 1996.

[90] Marc C. Kennedy and Anthony O'Hagan. Bayesian calibration of computer models. *Journal of the Royal Statistical Society: Series B (Statistical Methodology)*, 63(3):425–464, 2001.

[91] Nadja Klein, Manuel Carlan, Thomas Kneib, Stefan Lang, and Helga Wagner. Bayesian effect selection in structured additive distributional regression models. *Bayesian Analysis*, 16(2):545–573, 2021.

[92] Roger Koenker, Victor Chernozhukov, Xuming He, and Limin Peng. *Handbook of Quantile Regression*. Chapman and Hall/CRC, Boca Raton, 2018.

[93] Michael R. Kosorok and Eric B. Laber. Precision medicine. *Annual Review of Statistics and Its Application*, 6:263–286, 2019.

[94] Hong Lan, Meng Chen, Jessica B. Flowers, Brian S. Yandell, Donnie S. Stapleton, Christine M. Mata, Eric Ton-Keen Mui, Matthew T. Flowers, Kathryn L. Schueler, Kenneth F. Manly, et al. Combined expression trait correlations and expression quantitative trait locus mapping. *PLoS Genetics*, 2(1):e6, 2006.

[95] Jaehoon Lee, Yasaman Bahri, Roman Novak, Samuel S. Schoenholz, Jeffrey Pennington, and Jascha Sohl-Dickstein. Deep neural networks as Gaussian processes, 2018. arXiv:1711.00165.

[96] Fan Li, Peng Ding, and Fabrizia Mealli. Bayesian causal inference: a critical review. *Philosophical Transactions of the Royal Society A*, 381(2247):20220153, 2023.

[97] Dennis V. Lindley. A statistical paradox. *Biometrika*, 44(1/2):187–192, 1957.

[98] Roderick Little. Calibrated Bayes, for statistics in general, and missing data in particular. *Statistical Science*, 26(2):162–174, 2011.

[99] Reetam Majumder, Brian J. Reich, and Benjamin A. Shaby. Modeling extremal streamflow using deep learning approximations and a flexible spatial process. *arXiv preprint arXiv:2208.03344*, 2022.

[100] Jean-Michel Marin, Pierre Pudlo, Christian P. Robert, and Robin J. Ryder. Approximate Bayesian computational methods. *Statistics and Computing*, 22(6):1167–1180, 2012.

[101] Simon Martina-Perez, Matthew J. Simpson, and Ruth E. Baker. Bayesian uncertainty quantification for data-driven equation learning. *Proceedings of the Royal Society A*, 477(2254):20210426, 2021.

[102] Yogesh Chandrakant Mathpati, Kalpesh Sanjay More, Tapas Tripura, Rajdip Nayek, and Souvik Chakraborty. MAntRA: A framework for model agnostic reliability analysis. *Reliability Engineering & System Safety*, 235:109233, 2023.

[103] Peter McCullagh and John Nelder. *Generalized Linear Models, Second Edition.* Chapman and Hall/CRC, Boca Raton, 1989.

[104] Nicholas Metropolis, Arianna W. Rosenbluth, Marshall N. Rosenbluth, Augusta H. Teller, and Edward Teller. Equation of state calculations by fast computing machines. *The Journal of Chemical Physics*, 21(6):1087–1092, 1953.

[105] Mary C. Meyer, Amber J. Hackstadt, and Jennifer A. Hoeting. Bayesian estimation and inference for generalised partial linear models using shape-restricted splines. *Journal of Nonparametric Statistics*, 23(4):867–884, 2011.

[106] Greg Miller. ESP paper rekindles discussion about statistics. *Science*, 331(6015):272–273, 2011.

[107] Antonietta Mira. On Metropolis-Hastings algorithms with delayed rejection. *Metron*, 59(3-4):231–241, 2001.

[108] Christoph Molnar. *Interpretable Machine Learning, Second Edition.* 2022. https://christophm.github.io/interpretable-ml-book.

[109] Kalpesh Sanjay More, Tapas Tripura, Rajdip Nayek, and Souvik Chakraborty. A Bayesian framework for learning governing partial differential equation from data. *Physica D: Nonlinear Phenomena*, 456:133927, 2023.

[110] Frederick Mosteller and David L. Wallace. Inference in an authorship problem: A comparative study of discrimination methods applied to the authorship of the disputed Federalist Papers. *Journal of the American Statistical Association*, 58(302):275–309, 1963.

[111] Rajdip Nayek, Ramon Fuentes, Keith Worden, and Elizabeth J. Cross. On spike-and-slab priors for Bayesian equation discovery of nonlinear dynamical systems via sparse linear regression. *Mechanical Systems and Signal Processing*, 161:107986, 2021.

[112] Radford M. Neal. Slice sampling. *Annals of Statistics*, pages 705–741, 2003.

[113] Radford M. Neal. MCMC using Hamiltonian dynamics. *Handbook of Markov Chain Monte Carlo*, 2(11):2, 2011.

[114] Radford M. Neal. *Bayesian Learning for Neural Networks*, volume 118. Springer Science & Business Media, New York, 2012.

[115] Christopher Nemeth and Paul Fearnhead. Stochastic gradient Markov Chain Monte Carlo. *Journal of the American Statistical Association*, 116(533):433–450, 2021.

[116] Andrew Y. Ng. *Shaping and Policy Search in Reinforcement Learning.* University of California, Berkeley, 2003.

[117] Jorge Nocedal and Stephen J. Wright. *Sequential Quadratic Programming.* Springer, New York, 2006.

[118] Anthony O'Hagan. Fractional Bayes factors for model comparison. *Journal of the Royal Statistical Society: Series B (Methodological)*, 57(1):99–118, 1995.

[119] Krishna Pacifici, Brian J. Reich, David A. W. Miller, Beth Gardner, Glenn Stauffer, Susheela Singh, Alexa McKerrow, and Jaime A. Collazo. Integrating multiple data sources in species distribution modeling: A framework for data fusion. *Ecology*, 98(3):840–850, 2017.

[120] Trevor Park and George Casella. The Bayesian Lasso. *Journal of the American Statistical Association*, 103:681–686, 2008.

[121] Anand Patil, David Huard, and Christopher J. Fonnesbeck. PyMC: Bayesian stochastic modelling in Python. *Journal of Statistical Software*, 35(4):1, 2010.

[122] LI Pettit. The conditional predictive ordinate for the normal distribution. *Journal of the Royal Statistical Society: Series B (Methodological)*, pages 175–184, 1990.

[123] Martyn Plummer. JAGS Version 4.0.0 user manual. *See https://sourceforge.net/projects/mcmc-jags/files/Manuals/4.x*, 2015.

[124] S. James Press and Judith M. Tanur. *The Subjectivity of Scientists and the Bayesian Approach.* J. Wiley, New York, 2001.

[125] Tom Rainforth, Adam Foster, Desi R. Ivanova, and Freddie Bickford Smith. Modern Bayesian experimental design. *Statistical Science*, 39(1):100–114, 2024.

[126] Carl Edward Rasmussen and Christopher K. I Williams. *Gaussian Processes for Machine Learning.* The MIT Press, Cambridge, 2005.

[127] Brian J. Reich, Montserrat Fuentes, and David B. Dunson. Bayesian spatial quantile regression. *Journal of the American Statistical Association*, 106:6–20, 2011.

[128] Brian J. Reich, Shu Yang, Yawen Guan, Andrew B. Giffin, Matthew J. Miller, and Ana Rappold. A review of spatial causal inference methods for environmental and epidemiological applications. *International Statistical Review*, 89(3):605–634, 2021.

[129] Gareth O. Roberts and Jeffrey S. Rosenthal. Optimal scaling of discrete approximations to Langevin diffusions. *Journal of the Royal Statistical Society: Series B (Statistical Methodology)*, 60(1):255–268, 1998.

[130] Gareth O. Roberts and Adrian F. M. Smith. Simple conditions for the convergence of the Gibbs sampler and Metropolis-Hastings algorithms. *Stochastic Processes and Their Applications*, 49(2):207–216, 1994.

[131] Veronika Ročková and Edward I. George. The spike-and-slab LASSO. *Journal of the American Statistical Association*, 113(521):431–444, 2018.

[132] Donald B. Rubin. Bayesian inference for causal effects: The role of randomization. *The Annals of Statistics*, pages 34–58, 1978.

[133] Donald B. Rubin. Bayesianly justifiable and relevant frequency calculations for the applied statistician. *The Annals of Statistics*, pages 1151–1172, 1984.

[134] Donald B. Rubin. Multiple imputation after 18+ years. *Journal of the American Statistical Association*, 91(434):473–489, 1996.

[135] Håvard Rue, Sara Martino, and Nicolas Chopin. Approximate Bayesian inference for latent Gaussian models by using integrated nested Laplace approximations. *Journal of the Royal Statistical Society: Series B (Statistical Methodology)*, 71(2):319–392, 2009.

[136] Daniel J. Russo, Benjamin Van Roy, Abbas Kazerouni, Ian Osband, Zheng Wen, et al. A tutorial on Thompson sampling. *Foundations and Trends in Machine Learning*, 11(1):1–96, 2018.

[137] SK Sahu and TMF Smith. A Bayesian method of sample size determination with practical applications. *Journal of the Royal Statistical Society Series A: Statistics in Society*, 169(2):235–253, 2006.

[138] John R. Sauer, James E. Hines, and Jane E. Fallon. *The North American Breeding Bird Survey, Results and Analysis 1966–2005*. Version 6.2.2006. USGS Patuxent Wildlife Research Center, Laurel, Maryland, USA, 2005.

[139] Michael Schmidt and Hod Lipson. Distilling free-form natural laws from experimental data. *Science*, 324(5923):81–85, 2009.

[140] Steven L. Scott, Alexander W. Blocker, Fernando V. Bonassi, Hugh A. Chipman, Edward I. George, and Robert E. McCulloch. Bayes and big data: The consensus Monte Carlo algorithm. *International Journal of Management Science and Engineering Management*, 11(2):78–88, 2016.

[141] Daniel Simpson, Håvard Rue, Andrea Riebler, Thiago G. Martins, and Sigrunn H. Sørbye. Penalising model component complexity: A principled, practical approach to constructing priors. *Statistical Science*, 32(1):1–28, 2017.

[142] Aleksandrs Slivkins. Introduction to multi-armed bandits. *Foundations and Trends in Machine Learning*, 12(1-2):1–286, 2019.

[143] Ralph C Smith. *Uncertainty Quantification: Theory, Implementation, and Applications*, volume 12. SIAM, Philadelphia, 2013.

[144] Rodney Sparapani, Charles Spanbauer, and Robert McCulloch. Nonparametric machine learning and efficient computation with Bayesian additive regression trees: The BART R package. *Journal of Statistical Software*, 97(1):1–66, 2021.

[145] David J. Spiegelhalter, Nicola G. Best, Bradley P. Carlin, and Angelika Van Der Linde. Bayesian measures of model complexity and fit. *Journal of the Royal Statistical Society: Series B (Statistical Methodology)*, 64(4):583–639, 2002.

[146] Mervyn Stone. Necessary and sufficient condition for convergence in probability to invariant posterior distributions. *The Annals of Mathematical Statistics*, pages 1349–1353, 1970.

[147] John D Storey. The positive false discovery rate: a Bayesian interpretation and the q-value. *The Annals of Statistics*, 31(6):2013–2035, 2003.

[148] Sibylle Sturtz, Uwe Ligges, and Andrew Gelman. R2OpenBUGS: A package for running OpenBUGS from R. *URL http://cran.rproject.org/web/packages/R2OpenBUGS/vignettes/R2OpenBUGS.pdf*, 2010.

[149] Piotr M. Suder, Jason Xu, and David B. Dunson. Bayesian transfer learning. *arXiv preprint arXiv:2312.13484*, 2023.

[150] Brian L. Sullivan, Christopher L. Wood, Marshall J. Iliff, Rick E. Bonney, Daniel Fink, and Steve Kelling. eBird: A citizen-based bird observation network in the biological sciences. *Biological Conservation*, 142(10):2282–2292, 2009.

[151] Wenguang Sun, Brian J. Reich, T. Tony Cai, Michele Guindani, and Armin Schwartzman. False discovery control in large-scale spatial multiple testing. *Journal of the Royal Statistical Society: Series B (Statistical Methodology)*, 77(1):59–83, 2015.

[152] Richard S. Sutton and A.G. Barto. *Reinforcement Learning: An Introduction.* MIT Press, Cambridge, 2018.

[153] Terry Therneau and Beth Atkinson. *rpart: Recursive Partitioning and Regression Trees*, 2022. R package version 4.1.19.

[154] William R. Thompson. On the likelihood that one unknown probability exceeds another in view of the evidence of two samples. *Biometrika*, 25(3-4):285–294, 1933.

[155] Robert Tibshirani. Regression shrinkage and selection via the LASSO. *Journal of the Royal Statistical Society: Series B (Methodological)*, pages 267–288, 1996.

[156] Luke Tierney. Markov chains for exploring posterior distributions. *Annals of Statistics*, pages 1701–1728, 1994.

[157] Tapas Tripura and Souvik Chakraborty. A sparse Bayesian framework for discovering interpretable nonlinear stochastic dynamical systems with Gaussian white noise. *Mechanical Systems and Signal Processing*, 187:109939, 2023.

[158] Tapas Tripura, Aarya Sheetal Desai, Sondipon Adhikari, and Souvik Chakraborty. Probabilistic machine learning based predictive and interpretable digital twin for dynamical systems. *Computers & Structures*, 281:107008, 2023.

[159] Erik van Zwet, Simon Schwab, and Stephen Senn. The statistical properties of RCTs and a proposal for shrinkage. *Statistics in Medicine*, 40(27):6107–6117, 2021.

[160] Fei Wang and Alan E. Gelfand. A simulation-based approach to Bayesian sample size determination for performance under a given model and for separating models. *Statistical Science*, pages 193–208, 2002.

[161] Hao Wang. Bayesian graphical LASSO models and efficient posterior computation. *Bayesian Analysis*, 7(4):867–886, 2012.

[162] Ran Wei, Brian J. Reich, Jane A. Hoppin, and Subhashis Ghosal. Sparse Bayesian additive nonparametric regression with application to healtheffect of pesticide mixtures. *Statistica Sinica*, 30:55–79, 2020.

[163] Mike West and Jeff Harrison. *Bayesian Forecasting and Dynamic Models.* Springer Science & Business Media, New York, 2006.

[164] Shaun Wilkinson. aphid: an R package for analysis with profile hidden Markov models. *Bioinformatics*, 35:3829–3830, 2019.

[165] Simon Wood. *mgcv: Mixed GAM Computation Vehicle with GCV/AIC/REML smoothness estimation and GAMMs by REML/PQL*, 2023. R package version 1.9-1.

[166] Simon N. Wood. *Generalized Additive Models: An Introduction with R, Second Edition.* Chapman and Hall/CRC, Boca Raton, 2017.

[167] Steven G. Xu and Brian J. Reich. Bayesian nonparametric quantile process regression and estimation of marginal quantile effects. *Biometrics*, 79:151–164, 2023.

[168] Eric Yanchenko, Howard D. Bondell, and Brian J. Reich. The R2D2 prior for generalized linear mixed models. *The American Statistician*, 79:40–49, 2025.

[169] Yun Yang and Surya T. Tokdar. Joint estimation of quantile planes over arbitrary predictor spaces. *Journal of the American Statistical Association*, 112:1107–1120, 2017.

[170] Keming Yu and Rana A. Moyeed. Bayesian quantile regression. *Statistics & Probability Letters*, 54:437–447, 2001.

[171] Arnold Zellner. On assessing prior distributions and Bayesian regression analysis with g-prior distributions. *Bayesian Inference and Decision Techniques*, 1986.

[172] Yan Dora Zhang, Brian P. Naughton, Howard D. Bondell, and Brian J. Reich. Bayesian regression using a prior on the model fit: The R2-D2 shrinkage prior. *Journal of the American Statistical Association*, 117(538):862–874, 2022.

[173] Corwin M. Zigler. The central role of Bayes' theorem for joint estimation of causal effects and propensity scores. *The American Statistician*, 70:47–54, 2016.

[174] Walter Zucchini, Iain L. MacDonald, and Roland Langrock. Hidden Markov Models for Time Series: An Introduction Using R, 2016.

Index